Heavy Metals Accumulation, Toxicity and Detoxification in Plants

Heavy Metals Accumulation, Toxicity and Detoxification in Plants

Editors

Luigi De Bellis
Alessio Aprile

MDPI • Basel • Beijing • Wuhan • Barcelona • Belgrade • Manchester • Tokyo • Cluj • Tianjin

Editors
Luigi De Bellis
University of Salento
Italy

Alessio Aprile
University of Salento
Italy

Editorial Office
MDPI
St. Alban-Anlage 66
4052 Basel, Switzerland

This is a reprint of articles from the Special Issue published online in the open access journal *International Journal of Molecular Sciences* (ISSN 1422-0067) (available at: https://www.mdpi.com/journal/ijms/special_issues/plant_heavy_metals).

For citation purposes, cite each article independently as indicated on the article page online and as indicated below:

LastName, A.A.; LastName, B.B.; LastName, C.C. Article Title. *Journal Name* **Year**, *Article Number*, Page Range.

ISBN 978-3-03936-633-0 (Hbk)
ISBN 978-3-03943-478-7 (PDF)

Cover image courtesy of Alessio Aprile.

© 2020 by the authors. Articles in this book are Open Access and distributed under the Creative Commons Attribution (CC BY) license, which allows users to download, copy and build upon published articles, as long as the author and publisher are properly credited, which ensures maximum dissemination and a wider impact of our publications.
The book as a whole is distributed by MDPI under the terms and conditions of the Creative Commons license CC BY-NC-ND.

Contents

About the Editors

Luigi De Bellis (Professor) completed his degree in agricultural science with distinction, at the University of Pisa on October 30th, 1981. He then went on to study for his PhD in crop and fruit trees science (curriculum propagation), also at the University of Pisa, 1983–1986. He was a researcher at the Dept. of Crop Biology, University of Pisa, from January 1st, 1988 to October 31st, 1998. He was employed as a postdoctoral fellow of the Japan Society for Promotion of Science (JSPS) in the lab directed by Prof. M. Nishimura, at the National Institute for Basic Biology (NIBB), Okazaki, Japan, for 18 months, from March 1990 to October 1991. He was then employed as a postdoctoral fellow in Dr. S.M. Smith's lab, Institute of Cell and Molecular Biology (ICMB), at the University of Edinburgh, UK, for 12 months, from October 1993 to October 1994; this fellowship was sponsored by the EU's Human Capital Mobility Program. He enrolled on a short-term training fellowship sponsored by the EU (Technical Priority Program) in Dr. S.M. Smith's lab, ICMB, at the University of Edinburgh, UK, for 2 months, from July to September 1995. He was then granted a professorship contract (Fellow of the Japanese Minister of Culture and Research) at Prof. M. Nishimura's lab, NIBB, Okazaki, Japan, for 12 months, from March 1997 to March 1998. De Bellis was an associate professor of plant physiology at the University of Lecce, Faculty of Science, starting on November 1st, 1998. On November 1st, 2002, he became Chair Professor of Plant Physiology at the University of Lecce, Faculty of Science. From March 2012 to February 2020, he was Head of the Dept. of Biological and Environmental Science and Technologies (DiSTeBA) at the University of Salento. Additionally, he has been a research unit leader for several regional, national and European/international research projects. He is also a referee of the main major plant physiology journals and other journals, and a member of the Editorial Board of Biology, Horticulturae, and Plants. He is the author of over 100 scientific publications. De Bellis' main scientific interests are the enzymes of the glyoxylate cycle, the intracellular localization of aconitase isoforms, carbohydrates and the control of gene expression, the study of wheat and barley flours for the preparation of fresh pasta, the valorisation of Salento olive oil and the development of oil wastewater purification systems, the valorisation of agro-food products and the evaluation of organoleptic qualities, the genetics of plant species such as olive trees, durum wheat and bread wheat, plant response to heavy metals, *Xylella fastidiosa* as an olive tree pest, and the selection of tolerant olive varieties.

Alessio Aprile is a researcher. He completed his PhD in 2008, after defending a thesis about the transcriptome changes after abiotic stresses in crops. Since 2015, he has been a plant physiology researcher at the Department of Biological and Environmental Sciences and Technologies at the University of Salento. He was the principal investigator of the project "Genetic and breeding evaluation of durum wheat cv Cappelli", funded by the Apulia region as part of the program "Future in Research". The research activity, documented by more than 30 publications, is focused on the study of the physiology of abiotic stresses in plants. Part of his works has investigated the effects of drought, heat stress and CO_2 in wheat. Recently, his research activity has been focused on the effects of heavy metals in durum wheat. He is also the co-author of papers describing olive secondary metabolites and the pathogen *Xylella fastidiosa*. He is an assistant professor in plant physiology at the University of Salento.

International Journal of
Molecular Sciences

Editorial

Editorial for Special Issue "Heavy Metals Accumulation, Toxicity, and Detoxification in Plants" †

Alessio Aprile and Luigi De Bellis *

Department of Biological and Environmental Sciences and Technologies, University of Salento, I-73100 Lecce, Italy; alessio.aprile@unisalento.it
* Correspondence: luigi.debellis@unisalento.it
† This article is dedicated to Antonio Michele Stanca, eminent plant geneticist, friend, and mentor.

Received: 1 June 2020; Accepted: 5 June 2020; Published: 9 June 2020

"Heavy metals" is a collective term widely applied for the group of metals and metalloids with an atomic density above 4 g/cm^3 [1]. Non-essential toxic plant heavy metals include arsenic (As), cadmium (Cd), chromium (Cr), cobalt (Co), lead (Pb), mercury (Hg), nickel (Ni), and vanadium (V); whereas others are essential, such as copper (Cu), iron (Fe), manganese (Mn), and zinc (Zn). Heavy metals cause harmful effects in plants, animals, and humans as a result of long-term or acute exposure. Toxicity from heavy metals is increasing due to the extensive release from industrial, agricultural, chemical, domestic, and technological sources, which in turn contaminate the water, soil, and air. Natural phenomena, such as volcanic eruptions and sea movements, also contribute to the natural cyclization of metals on the earth, and human activities often alter the rate of release and transport by increasing emissions by a few orders of magnitude.

Heavy metals penetrate the human body through water, food, and air. Inside an organism, they bind to cellular structures, thereby damaging the performance of essential biological functions. Metals, for example, easily bind to the sulfhydryl groups of several enzymes that control the speed of metabolic reactions: the "new" metal-enzyme complex leads to the loss of the catalytic activity of the enzyme. The level of toxicity from heavy metals depends on several factors, including time of exposure, dose, and the health status of the people exposed.

The European Environment Agency (EEA) reported that of the 1000 industrial plants that released heavy metals into the air in 2016, eighteen accounted for more than half of the total pollution, suggesting a great responsibility on the part of a few large companies (Figure 1) [2].

An additional issue is the biomagnification (or bioaccumulation) caused by the very slow rate of elimination of heavy metals from an organism. Bioaccumulation, in ecology and biology, is the process whereby the accumulation of toxic substances in living beings increases in concentration following a rise in the trophic level: the higher the trophic level, the stronger the concentration of heavy metals. Biomagnification is also expressed as the concentration increase of a pollutant in a biological organism over time.

To limit the risks for humans and the environment, many countries have legislated limits for each heavy metal. Specific limits have been defined in drinking, waste, and surface waters (lakes, rivers, seas). There are also limits in foods and animal feed, because heavy metals can easily enter the food chain through plants (or algae) and are subsequently bioaccumulated into the higher trophic levels. The risk for human health is due to directly eating edible plant tissues, or indirectly through eating animals that have in turn fed on herbivores or directly on edible plant tissues. Understanding the mechanisms for regulating the storage and distribution of heavy metals in plants is the basis for improving the safety of the food chain.

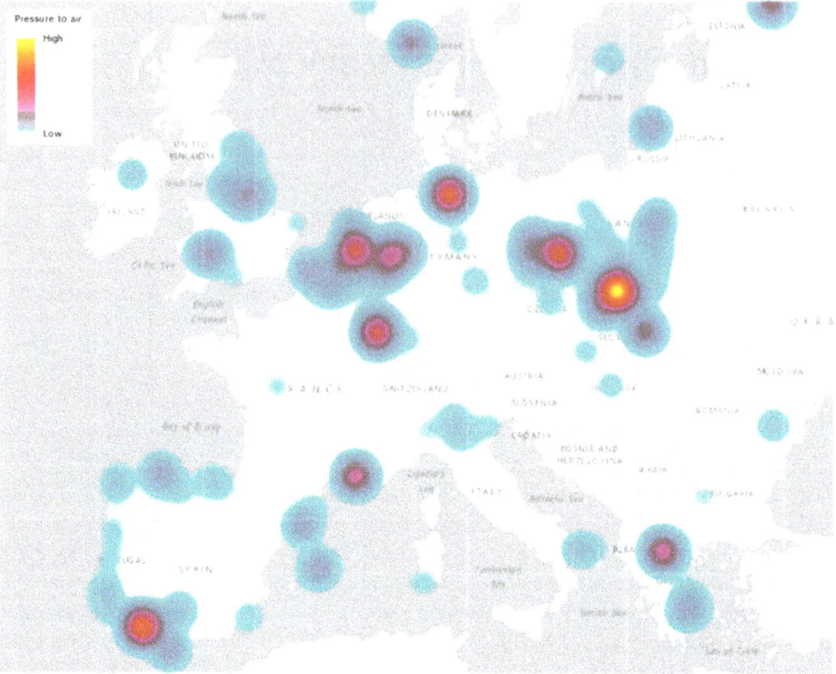

Figure 1. Environmental pressures of heavy metal releases to air, 2016 [2]. An eco-toxicity approach (USEtox model, https://usetox.org/model) was applied to illustrate spatially the combined environmental pressures on Europe's environment caused by releases of the selected pollutants. This gives information about the location of source of heavy metals and the low or high levels in air as indicated in the upper left corner of the figure.

This special issue, entitled "Heavy Metals Accumulation, Toxicity, and Detoxification in Plants", explores three main issues concerning heavy metals: (a) the accumulation and partitioning of heavy metals in crops and wild plants; (b) the toxicity and molecular behaviors of cells, tissues, and their effects on physiology and plant growth; and (c) detoxification strategies, plant tolerance, and phytoremediation.

The issue contains a total of 19 articles (Table 1). There are four reviews covering the following topics: phytoremediation [3], manganese phytotoxicity in plants [4], cadmium effect on plant development [5], the genetic characteristics of Cd accumulation and the research status of genes and quantitative trait loci (QTLs) in rice [6], and fifteen original research articles, mainly regarding the impact of cadmium on plants [7–21].

Table 1. Contributors to the special issue "Heavy Metals Accumulation, Toxicity, and Detoxification in Plants". ABC: ATP-binding cassette.

Authors	Title	Heavy Metals	Type
Małkowski et al. [7]	Hormesis in Plants: The Role of Oxidative Stress, Auxins and Photosynthesis in Corn Treated with Cd or Pb	Cadmium Lead	Original Research
Hu et al. [8]	Full-Length Transcriptome Assembly of Italian Ryegrass Root Integrated with RNA-Seq to Identify Genes in Response to Plant Cadmium Stress	Cadmium	Original Research
Sun et al. [9]	Comparative Transcriptome Analysis of the Molecular Mechanism of the Hairy Roots of *Brassica campestris* L. in Response to Cadmium Stress	Cadmium	Original Research
Zúñiga et al. [10]	Isolation and Characterization of Copper- and Zinc-Binding Metallothioneins from the Marine Alga *Ulva compressa* (Chlorophyta)	Copper, Zinc	Original Research
Cui et al. [11]	OsMSR3, a Small Heat Shock Protein, Confers Enhanced Tolerance to Copper Stress in *Arabidopsis thaliana*	Copper	Original Research
Aprile et al. [12]	Combined Effect of Cadmium and Lead on Durum Wheat	Cadmium, Lead	Original Research
Shafiq et al. [13]	Lead, Cadmium and Zinc Phytotoxicity Alter DNA Methylation Levels to Confer Heavy Metal Tolerance in Wheat	Cadmium, Lead, Zinc	Original Research
Celis-Plá et al. [14]	MAPK Pathway under Chronic Copper Excess in Green Macroalgae (Chlorophyta): Influence on Metal Exclusion/Extrusion Mechanisms and Photosynthesis	Copper	Original Research
Rodríguez-Rojas et al. [15]	MAPK Pathway under Chronic Copper Excess in Green Macroalgae (Chlorophyta): Involvement in the Regulation of Detoxification Mechanisms	Copper	Original Research
Małecka et al. [16]	Insight into the Phytoremediation Capability of *Brassica juncea* (v. Malopolska): Metal Accumulation and Antioxidant Enzyme Activity	Cadmium, Copper, Lead, Zinc	Original Research
Luo et al. [17]	Selenium Modulates the Level of Auxin to Alleviate the Toxicity of Cadmium in Tobacco	Cadmium	Original Research
Wang et al. [18]	Ectopic Expression of Poplar ABC Transporter PtoABCG36 Confers Cd Tolerance in *Arabidopsis thaliana*	Cadmium	Original Research
Shu et al. [19]	Comparative Transcriptomic Studies on a Cadmium Hyperaccumulator *Viola baoshanensis* and Its Non-Tolerant Counterpart *V. inconspicua*	Cadmium	Original Research
He et al. [20]	Exogenous Glycinebetaine Reduces Cadmium Uptake and Mitigates Cadmium Toxicity in Two Tobacco Genotypes Differing in Cadmium Tolerance	Cadmium	Original Research
Han et al. [21]	Transcriptome Analysis Reveals Cotton (*Gossypium hirsutum*) Genes That Are Differentially Expressed in Cadmium Stress Tolerance	Cadmium	Original Research
Li et al. [4]	Advances in the Mechanisms of Plant Tolerance to Manganese Toxicity	Manganese	Review
Huybrechts et al. [5]	Cadmium and Plant Development: An Agony from Seed to Seed	Cadmium	Review
Chen et al. [6]	Advances in the Uptake and Transport Mechanisms and QTLs Mapping of Cadmium in Rice	Cadmium	Review
Dal Corso et al. [3]	Heavy Metal Pollutions: State of the Art and Innovation in Phytoremediation	All	Review

Cadmium is therefore the predominant topic of this special issue, thus confirming the focus of the research community on the negative impacts determined by cadmium or cadmium associated with other heavy metals. Interestingly, we did not receive any manuscripts on other heavy metals such as arsenic, chromium and mercury despite their danger for human health.

The cadmium research articles come from China, Poland, Italy, Canada, Pakistan, and the United States. These studies investigate different molecular mechanisms or approaches, using model plants such as *Arabidopsis* and tobacco [17,18,20] or hyperaccumulator plant species [9,16,19,21] to unravel their molecular strategies in heavy metal accumulation. Other articles focus on how to prevent cadmium from entering the food chain by investigating edible plants such as *Zea mays* [7], durum and bread wheat [12,13], or animal feeding plants such as *Lolium multiflorum*.

The studies reveal some common strategies in terms of the molecular mechanisms involved. Some plants activate the production of small proteins such as glutathione S-transferase (GST) and small heat shock protein (sHSP) [9,11,21] or antioxidants [16]. In order to alleviate heavy metal toxicity, other plants respond by activating a complex metabolism-like auxin pathway [7,8,17]. Plants also produce specific metallothionines and phytosiderophores [10,12] to chelate heavy metals or to activate heavy metals transporters such as heavy metal ATPase (e.g., HMA2 and HMA4) and ATP-binding cassette (ABC) transporters [12,13,18,19,21].

The studies in this special issue highlight considerable genetic variability, suggesting different possibilities for accumulation, translocation, and reducing or controlling heavy metals toxicity in plants.

Heavy metal pollution is still one of the world's great challenges. In the future, the main research objective should be to identify and characterize the genes controlling the uptake and translocation of heavy metals in a plant's above-ground organs in order to produce (i) phytoremediation plants that efficiently move heavy metals in the stem and leaves or (ii) plants dedicated to human nutrition that transport heavy metals only in trace amounts to seeds or fruits.

Conflicts of Interest: The authors declare no conflict of interest.

Abbreviations

QTLs	Quantitative trait loci
sHSP	small heat shock protein
GST	glutathione s-transferase
HMA	heavy metal ATPase
ABC	ATP-binding cassette

References

1. Hawkes, J.S. Heavy metals. *J. Chem. Educ.* **1997**, *74*, 1369–1374. [CrossRef]
2. European Environment Agency (EEA). Environmental Pressures of Heavy Metal Releases from Europe's Industry. Available online: https://www.eea.europa.eu/themes/industry/industrial-pollution-in-europe (accessed on 27 May 2020).
3. DalCorso, G.; Fasani, E.; Manara, A.; Visioli, G.; Furini, A. Heavy Metal Pollutions: State of the Art and Innovation in Phytoremediation. *Int. J. Mol. Sci.* **2019**, *20*, 3412. [CrossRef] [PubMed]
4. Li, J.; Jia, Y.; Dong, R.; Huang, R.; Liu, P.; Li, X.; Wang, Z.; Liu, G.; Chen, Z. Advances in the Mechanisms of Plant Tolerance to Manganese Toxicity. *Int. J. Mol. Sci.* **2019**, *20*, 5096. [CrossRef] [PubMed]
5. Huybrechts, M.; Cuypers, A.; Deckers, J.; Iven, V.; Vandionant, S.; Jozefczak, M.; Hendrix, S. Cadmium and Plant Development: An Agony from Seed to Seed. *Int. J. Mol. Sci.* **2019**, *20*, 3971. [CrossRef] [PubMed]
6. Chen, J.; Zou, W.; Meng, L.; Fan, X.; Xu, G.; Ye, G. Advances in the Uptake and Transport Mechanisms and QTLs Mapping of Cadmium in Rice. *Int. J. Mol. Sci.* **2019**, *20*, 3417. [CrossRef] [PubMed]
7. Małkowski, E.; Sitko, K.; Szopiński, M.; Gieroń, Ż.; Pogrzeba, M.; Kalaji, H.M.; Zieleźnik-Rusinowska, P. Hormesis in Plants: The Role of Oxidative Stress, Auxins and Photosynthesis in Corn Treated with Cd or Pb. *Int. J. Mol. Sci.* **2020**, *21*, 2099. [CrossRef] [PubMed]
8. Hu, Z.; Zhang, Y.; He, Y.; Cao, Q.; Zhang, T.; Lou, L.; Cai, Q. Full-Length Transcriptome Assembly of Italian Ryegrass Root Integrated with RNA-Seq to Identify Genes in Response to Plant Cadmium Stress. *Int. J. Mol. Sci.* **2020**, *21*, 1067. [CrossRef] [PubMed]
9. Sun, Y.; Lu, Q.; Cao, Y.; Wang, M.; Cheng, X.; Yan, Q. Comparative Transcriptome Analysis of the Molecular Mechanism of the Hairy Roots of *Brassica campestris* L. in Response to Cadmium Stress. *Int. J. Mol. Sci.* **2020**, *21*, 180. [CrossRef] [PubMed]
10. Zúñiga, A.; Laporte, D.; González, A.; Gómez, M.; Sáez, C.A.; Moenne, A. Isolation and Characterization of Copper- and Zinc- Binding Metallothioneins from the Marine Alga *Ulva compressa* (Chlorophyta). *Int. J. Mol. Sci.* **2020**, *21*, 153. [CrossRef] [PubMed]
11. Cui, Y.; Wang, M.; Yin, X.; Xu, G.; Song, S.; Li, M.; Liu, K.; Xia, X. OsMSR3, a Small Heat Shock Protein, Confers Enhanced Tolerance to Copper Stress in *Arabidopsis thaliana*. *Int. J. Mol. Sci.* **2019**, *20*, 6096. [CrossRef] [PubMed]

12. Aprile, A.; Sabella, E.; Francia, E.; Milc, J.; Ronga, D.; Pecchioni, N.; Ferrari, E.; Luvisi, A.; Vergine, M.; De Bellis, L. Combined Effect of Cadmium and Lead on Durum Wheat. *Int. J. Mol. Sci.* **2019**, *20*, 5891. [CrossRef] [PubMed]
13. Shafiq, S.; Zeb, Q.; Ali, A.; Sajjad, Y.; Nazir, R.; Widemann, E.; Liu, L. Lead, Cadmium and Zinc Phytotoxicity Alter DNA Methylation Levels to Confer Heavy Metal Tolerance in Wheat. *Int. J. Mol. Sci.* **2019**, *20*, 4676. [CrossRef] [PubMed]
14. Celis-Plá, P.S.M.; Rodríguez-Rojas, F.; Méndez, L.; Moenne, F.; Muñoz, P.T.; Lobos, M.G.; Díaz, P.; Sánchez-Lizaso, J.L.; Brown, M.T.; Moenne, A.; et al. MAPK Pathway under Chronic Copper Excess in Green Macroalgae (Chlorophyta): Influence on Metal Exclusion/Extrusion Mechanisms and Photosynthesis. *Int. J. Mol. Sci.* **2019**, *20*, 4547. [CrossRef]
15. Rodríguez-Rojas, F.; Celis-Plá, P.S.M.; Méndez, L.; Moenne, F.; Muñoz, P.T.; Lobos, M.G.; Díaz, P.; Sánchez-Lizaso, J.L.; Brown, M.T.; Moenne, A.; et al. MAPK Pathway under Chronic Copper Excess in Green Macroalgae (Chlorophyta): Involvement in the Regulation of Detoxification Mechanisms. *Int. J. Mol. Sci.* **2019**, *20*, 4546. [CrossRef]
16. Małecka, A.; Konkolewska, A.; Hanć, A.; Barałkiewicz, D.; Ciszewska, L.; Ratajczak, E.; Staszak, A.M.; Kmita, H.; Jarmuszkiewicz, W. Insight into the Phytoremediation Capability of *Brassica juncea* (v. Malopolska): Metal Accumulation and Antioxidant Enzyme Activity. *Int. J. Mol. Sci.* **2019**, *20*, 4355. [CrossRef]
17. Luo, Y.; Wei, Y.; Sun, S.; Wang, J.; Wang, W.; Han, D.; Shao, H.; Jia, H.; Fu, Y. Selenium Modulates the Level of Auxin to Alleviate the Toxicity of Cadmium in Tobacco. *Int. J. Mol. Sci.* **2019**, *20*, 3772. [CrossRef] [PubMed]
18. Wang, H.; Liu, Y.; Peng, Z.; Li, J.; Huang, W.; Liu, Y.; Wang, X.; Xie, S.; Sun, L.; Han, E.; et al. Ectopic Expression of Poplar ABC Transporter PtoABCG36 Confers Cd Tolerance in *Arabidopsis thaliana*. *Int. J. Mol. Sci.* **2019**, *20*, 3293. [CrossRef] [PubMed]
19. Shu, H.; Zhang, J.; Liu, F.; Bian, C.; Liang, J.; Liang, J.; Liang, W.; Lin, Z.; Shu, W.; Li, J.; et al. Comparative Transcriptomic Studies on a Cadmium Hyperaccumulator *Viola baoshanensis* and Its Non-Tolerant Counterpart *V. inconspicua*. *Int. J. Mol. Sci.* **2019**, *20*, 1906. [CrossRef] [PubMed]
20. He, X.; Richmond, M.E.; Williams, D.V.; Zheng, W.; Wu, F. Exogenous Glycinebetaine Reduces Cadmium Uptake and Mitigates Cadmium Toxicity in Two Tobacco Genotypes Differing in Cadmium Tolerance. *Int. J. Mol. Sci.* **2019**, *20*, 1612. [CrossRef] [PubMed]
21. Han, M.; Lu, X.; Yu, J.; Chen, X.; Wang, X.; Malik, W.A.; Wang, J.; Wang, D.; Wang, S.; Guo, L.; et al. Transcriptome Analysis Reveals Cotton (*Gossypium hirsutum*) Genes That Are Differentially Expressed in Cadmium Stress Tolerance. *Int. J. Mol. Sci.* **2019**, *20*, 1479. [CrossRef] [PubMed]

© 2020 by the authors. Licensee MDPI, Basel, Switzerland. This article is an open access article distributed under the terms and conditions of the Creative Commons Attribution (CC BY) license (http://creativecommons.org/licenses/by/4.0/).

International Journal of
Molecular Sciences

Review

Heavy Metal Pollutions: State of the Art and Innovation in Phytoremediation

Giovanni DalCorso [1,*], Elisa Fasani [1], Anna Manara [1], Giovanna Visioli [2] and Antonella Furini [1,*]

1 Department of Biotechnology, University of Verona, Strada Le Grazie 15, 37134 Verona, Italy
2 Department of Chemistry, Life Sciences and Environmental Sustainability, University of Parma, Parco Area delle Scienze, 11/A, 43124 Parma, Italy
* Correspondence: giovanni.dalcorso@univr.it (G.D.); antonella.furini@univr.it (A.F.); Tel.: +0039-045-802-7950 (G.D. & A.F.)

Received: 20 June 2019; Accepted: 10 July 2019; Published: 11 July 2019

Abstract: Mineral nutrition of plants greatly depends on both environmental conditions, particularly of soils, and the genetic background of the plant itself. Being sessile, plants adopted a range of strategies for sensing and responding to nutrient availability to optimize development and growth, as well as to protect their metabolisms from heavy metal toxicity. Such mechanisms, together with the soil environment, meaning the soil microorganisms and their interaction with plant roots, have been extensively studied with the goal of exploiting them to reclaim polluted lands; this approach, defined phytoremediation, will be the subject of this review. The main aspects and innovations in this field are considered, in particular with respect to the selection of efficient plant genotypes, the application of improved cultural strategies, and the symbiotic interaction with soil microorganisms, to manage heavy metal polluted soils.

Keywords: phytoremediation; heavy metals; hyperaccumulation; plant genotype improvement; soil management

1. Introduction

Like all living organisms, plants require chemical elements that are used as cofactors in biochemical reactions, as components of structural proteins and macromolecules, and as regulators of the electrochemical balance of cellular compartments [1]. Soil availability of nutrient elements fluctuate due to temperature, precipitation, soil type and pH, oxygen content, and the presence or absence of other inorganic and organic compounds. Being sessile organisms, plants developed adaptive and flexible strategies for sensing and responding to fluctuations in element availability to optimize growth, development, and reproduction under a dynamic range of environmental conditions. In addition, once taken up, elements must be allocated to different organs, cell types, and tissues through tight homeostasis mechanisms to ensure metal requirement, storage, and re-mobilization under different environmental conditions [2].

Heavy metals are naturally occurring elements, which are widely distributed in the Earth's crust; they derive from rocks of volcanic, sedimentary or metamorphic origin, but in recent years, the prevalence of heavy metals in areas of agricultural and industrial activities has increased because of human activity [3]. A limited number of heavy metal ions are water soluble upon physiological conditions and thus bioavailable to plants and other living organisms, being either essential or potential risks for life [4]. Indeed, many heavy metals (mainly Fe, Zn, Cu, Ni, Co, and Mo), which are toxic when present in excess, are essential for plant and cellular biochemistry being involved in cell protection, gene regulation, and signal transduction and their absence (or deficiency) inhibits plant growth, reproduction, and tolerance to environmental stresses [5]. Other heavy metals such as Cd, Hg, Ag, Pb, and Cr are biologically non-essential and show toxicity even at low concentrations. The similarity of certain non-essential metals to essential ones allows the latter to enter plants replacing their essential homolog and interfering with biological functions. To minimize the unfavorable effects of non-essential heavy metals, while maintaining the uptake of essential elements, plants have evolved a homeostatic network that controls metal uptake, trafficking, storage, and detoxification. Although a basal metal tolerance is usual, to guarantee the correct concentration of essential metal nutrients in different cell types at different stages of plant development, plants have acquired complex mechanisms to avoid or overcome the harmfulness of heavy metal excess. In metal-rich soils, plants have evolved mechanisms to tolerate, within a certain limit, metal toxicity. Plants encountering heavy metals employ two main approaches: the most common strategy is metal exclusion, in which metal accumulation is limited to the belowground organs. Uptake and root-to-shoot transport are regulated to maintain low shoot content over a wide range of external concentration. On the opposite, plants can accumulate metals, and an extreme evolution of this capacity is well represented in metal (hyper) accumulators, which are able to accumulate heavy metals in their shoots keeping low concentrations in roots. This trait is associated with the enhanced ability to detoxify high metal levels in the aboveground tissues [6]. Both strategies are regulated by finely tuned homeostatic mechanisms to guarantee sufficient metal uptake, transport, accumulation, and detoxification.

2. Plant Mechanisms for Heavy Metal Tolerance and Detoxification

Relevant components of homeostatic networks underneath metal tolerance and detoxification include ion transporters, metallo-chaperons, and ligands that act in concert to ensure metal uptake, transport to different cell types and delivery inside cells. Membrane proteins are able to transport different metals across cellular membranes, playing a pivotal role in each influx-efflux step of the translocation from roots to shoots. The function of several transporters involved in import, trafficking, sequestration, and export of essential metals across the plasma membrane, tonoplast, or chloroplast envelope has been clarified [2,7,8]. Metal transporters have been classified into families according to sequence homology. For example, the ZIP family (ZRT-IRT-like proteins) is involved in several homeostatic processes including uptake and translocation from root to shoot [9,10]. The NRAMPs (naturally resistant associated macrophage proteins) comprises members such as: NRAMP1, which when in *A. thaliana* is localized in the plasma membrane, is involved in Fe transport, and also shows high-affinity Mn uptake from soil [11]; NRAMP3 and NRAMP4, which are localized in the tonoplast and are essential for exporting stored Fe from the vacuole during seed germination [12]. The HMA proteins (heavy metals P1B-type ATPases) contribute to pump cations out of the cytoplasm by ATP hydrolysis. HMA1 localizes in the chloroplast envelope and is possibly involved in plastid Zn detoxification under Zn excess [13]. Similarly, HMA3 is involved in the detoxification of Zn, Cd, Co, and Pb by regulating their sequestration into the vacuole, and HMA4, a plasma membrane transporter, plays a role in Zn efflux from the cytoplasm and xylem loading/unloading [14,15]. Another group of transporters that tightly regulate metal homeostasis ensuring the appropriate metal supply to tissues is represented by the CDF (cation diffusion facilitator) family whose members are involved in the translocation of metals towards internal compartments and extracellular space [7]. Among them, several MTPs (metal tolerance proteins) have been described in a variety of plant species. The best characterized is MTP1, which is a vacuolar Zn^{2+}/H^{+} antiporter involved in Zn tolerance, which in case of Zn excess accumulates Zn into the vacuole [16].

In addition to metal trafficking, plant responses to heavy metal stress include a variety of mechanisms, ranging from changes in gene expression and methylation to metabolic and biochemical adjustments, with the final goal of scavenging toxic metal ions, and ameliorating stress symptoms and damages. The production of hormones such as ethylene, jasmonic acid, and abscisic acid is also induced, as well as molecules involved in chelation of metal ions, such as organic acid, specific amino acids, phytochelatins, and metallothioneins [17,18]. Proline and histidine induce tolerance by chelating ions within cells and xylem sap [19]. The induction of phytochelatins occurs because of high levels of different heavy metals although Cd seems to be the most effective stimulator [20].

As opposed to phytochelatins, which are produced enzymatically, metallothioneins are gene-encoded polypeptides that play a role in the homeostasis and sequestration of intracellular metal ions [21,22]. Chelating compounds contribute to heavy metal tolerance by removing toxic ions from sensitive sites through sequestration and subsequent vacuolar compartmentalization by tonoplast-localized transporters. When the above-mentioned strategies are insufficient to contain the damage, cells trigger the production of reactive oxygen species (ROS), which might potentially result in massive oxidative stress with cell homeostasis disruption, inhibition of most cellular processes, DNA damage and protein oxidation [23]. As a result, cells activate the ROS-scavenging machinery with the production of antioxidant compounds such as glutathione, flavonoids, and carotenoids as well as antioxidant enzymes including superoxide dismutases, catalases, and peroxidases.

3. Phytoremediation

As mentioned before, despite natural occurrence in soils, large quantities of heavy metals and metalloids have been dispersed into the environment by a variety of human activities including fertilizer use in agriculture, metal mining, and manufacturing by metallurgy, fossil fuel use, and military operations. Land contamination poses a serious risk to both human health and animal and plant biodiversity [24]. There are a variety of conventional approaches to reclaim contaminated sites that are usually based on physicochemical techniques, including soil washing, electric field application (electrokinetics), excavation and reburial of contaminated matrices, pumping and treating systems in case of polluted water. These approaches suffer from two main disadvantages, being expensive and frequently inefficient if pollutants are present at low concentrations. Moreover, harsh approaches cause significant changes to the physicochemical and biological characteristics of soils and landscapes. Ecological rehabilitation of contaminated sites may also be achieved by *phytoremediation*: an alternative in situ technology, which exploits plants and their rhizosphere to remove the contaminants or lower their bioavailability in soil and water with concurrent land revegetation [25].

3.1. Strategies for Phytoremediation

Once placed in loco, plants deepen their root system into the contaminated soil matrix, establishing ecosystems with soil bacteria and fungi. Into this context, plants and the rhizosphere, i.e., soil and microorganisms associated to roots, employ mechanisms that altogether are responsible for the soil reclamation: phytodegradation, phytoextraction, phytovolatilization, phyto(rhizo)stabilization and phyto(rhizo)filtration (Figure 1). Such mechanisms are usually considered separately just for sake of clarity, even if they act in concert on the metal decontamination. In the following lines, each of these aspects will be highlighted individually, with the exclusion of phytodegradation, which is applicable to organic contaminants, rather than heavy metals, which are not degradable [26]. Plants acquire mineral elements from the soil primarily in the form of inorganic ions. The extended root system and its ability to absorb ionic compounds even at low concentrations make mineral absorption highly efficient. Obviously, heavy metals and metalloids can be absorbed by the plant root system, but since some of them, such as Cd and Pb, have no known biological function, it is likely that specific transporters do not exist. Indeed, toxic metals enter into the cells through cation transporters with a wide range of substrate specificity [18]. The ability of plants to take metals up and to accumulate them into the aboveground harvestable tissues is the rationale behind the second mechanism in phytoremediation, the so-called phytoextraction. Effective phytoextraction of metal-contaminated matrixes requires plants, which are characterized by a) efficient metal uptake and translocation to shoots; b) the ability to accumulate and tolerate high levels of metals; c) rapidly-growing and abundant shoots and deep root system. Some particular plants, commonly described as hyperaccumulator, show the ability to accumulate metals in aboveground tissues at very high concentration, without phytotoxic effects [27]. Unfortunately, most plant species displaying the hyperaccumulation trait are biennial or short-lived perennial herbs, shrubs, or small trees, characterized by a low biomass and a slow growth rate, which are major limitations for phytoextraction purposes [26]. Plants, which are suitable for effective phytoremediation, are therefore selected considering their tolerance to metal stress and biomass of aerial organs. For instance, it has been shown that *Populus* spp. and *Salix* spp. are able to accumulate relatively high foliar concentrations of metals such as Cd and Zn [28] and are often associated with metal-contaminated lands in northern Europe [29]. Interestingly, the evaluation on a time span of ca. 30 years of natural colonization on a contaminated site by different plant species, i.e., *Populus* 'Robusta', *Quercus robur*, *Fraxinus excelsior*, and *Acer pseudoplatanus*, has shown that the tree species determined a redistribution of metals in the soil profile, which was dependent on two main processes: the accumulation of metals in the leaves (an enhanced metal deposit into leaves contributes to an increased metal amount in the upper soil layer, upon seasonal leaf fall) and species-specific soil acidification (higher soil acidification by the root metabolism resulted in higher metal leaching from the upper soil layer with subsequent lower metal concentrations in such soil layer) [28]. Other than such biological aspects, successful phytoextraction is also guaranteed by lowering the time constraint of the process itself, especially evaluating the rate of metal pollutant removal and any eventual pollutant inputs [30].

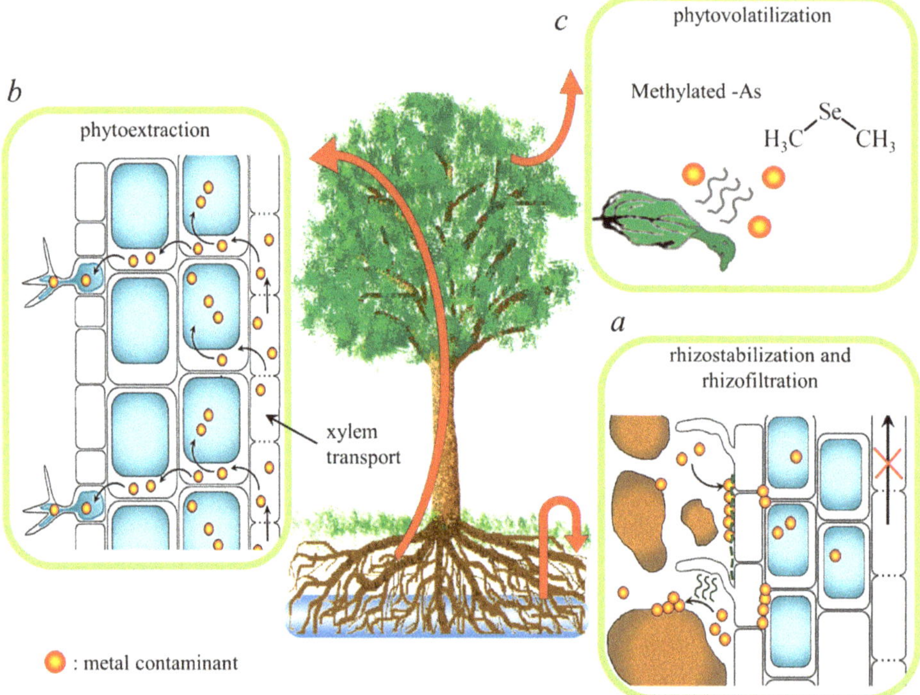

Figure 1. The main aspects of phytoremediation: the main steps involved in phytoremediation of heavy metals, which include (**a**) metal (yellow dots) adsorption on soil particles or cell walls (induced by rhizosphere metabolism) and compartmentalization of metals into root cell vacuoles (blue circles inside cells), preventing transport to the shoot; (**b**) metal accumulation in aerial organs (e.g., in vacuoles or trichomes) upon root-to-shoot xylem transport; (**c**) for particular metalloids (e.g., Se and As), leaf metabolism allows volatilization of the toxic compound.

When land contamination includes particular contaminants, such as Hg, As, and Se, plant metabolism is applicable to root absorption, translocation, and conversion of toxins into volatile compounds, which are released into the environment. Such a phenomenon, considered as a phytoremediation strategy, is called phytovolatilization. For instance, being chemically analog to sulfur, inorganic Se is converted to dimethyl selenide by plant enzymes involved in sulfur metabolism pathways, assimilation, and volatilization. Dimethyl selenide is dispersed into the air as a gas, which is significantly less toxic than inorganic Se [31]. Arsenic is another carcinogen and its contamination in soils is mainly due to natural sources and anthropogenic activities. Arsenite, formed in soils by the microbial activity, is readily taken up by plants, and some crops, e.g., rice, showed particular attitude to mobilize arsenite through the silicon uptake pathway, resulting in serious As poisoning to consumers [32]. The third metal that can be converted into volatile compounds is Hg, which is present in soils, waters and in the atmosphere. Leaf Hg content, mainly in the form of methyl-Hg, seems to derive almost entirely from leaf absorption by the atmosphere, since Hg transport through vascular tissues is very limited even considered that in particular paddy soils, chemical forms of water-soluble Hg can be promptly adsorbed and transferred to shoots, as observed in rice [33]. Experiments upon controlled conditions with wild-type *Brassica juncea* plants hydroponically treated with $HgCl_2$ confirmed that upon root uptake, phytovolatilization of Hg is indeed happening in roots, rather than shoots due to the low root-to-shoot transport of the metal, and seems to occur via the metabolic activity of the root-associated algal and microbial community [34].

In soil, plant metabolism may contribute to the chemical stabilization of metal ions within the vadose zone, limiting leaching, mobility, bioavailability, and ultimately hazard. This process is known as phytostabilization and is accomplished by both metal ions absorption and accumulation in and onto roots, and by their precipitation in the rhizosphere zone due to binding by organic compounds and changes of metal oxidative state. Positively charged metal ions effectively bind to pectins in plant cell walls and to the negatively charged plasma membranes [35]. Plant species that accumulate heavy metals in their belowground parts are recognized as the most effective for phytostabilization, also known as rhizostabilization. Enhancement of phytostabilization processes is commonly obtained by coupling biological activity with soil amendment, in particular when dealing with heavily polluted soils. The utilization of inorganic soil additives, which include phosphate fertilizers, manganese, and iron oxides, clay and other minerals, and organic compounds, such as coal, compost and manure, aids plants by metal sorption and/or chemical alteration, as well as by beneficial effects on plant growth [36,37]. By reducing contaminants mobility and eventually the associated risks without necessarily removing them from the site, phytostabilization does not produce contaminated waste, such as harvested materials, which would need further treatments.

The root metabolism of both terrestrial and aquatic plants can be also exploited to remediate polluted waters. This approach, named rhizofiltration, is used to absorb, concentrate, and precipitate metals from polluted water into plant biomass and its efficiency compares with currently employed water treatment technologies [38,39]. Early works demonstrated that a variety of aquatic plants, microorganisms, and seaweeds were able to biosorb metals and radionuclides dispersed in water (for a detailed description, refer to Section 3.2.2 in this review), but the lack of low-cost culturing, harvesting, and handling methods prevented full-scale testing [40]. It was in the early 1990s that the use of terrestrial plants grown hydroponically, able to achieve high above-water biomass and extensive root system to adsorb and absorb metals from contaminated liquids, got a foothold [41]. Mechanisms involved in rhizofiltration, also known as phytofiltration, mainly fall into three types characterized by different kinetics: a) sorption on the root surface, a quick component of metal removal, due to physical and chemical processes as chelation, ion exchange and specific adsorption (which do not include biological activity); b) processes that depend on plant metabolism, responsible for a slower metal removal from solutions and which rely on intracellular uptake, vacuolar deposition and eventually translocation to the shoots [42]; lastly, c) the slowest component of metal removal involves the release of root exudates which mediate metal precipitation from the solution in the form of insoluble compounds, as in case of phytostabilization.

An interesting corollary of phytofiltration is the use of microalgae to treat municipal, industrial, agro-industrial, and livestock wastewaters. Microalgal bioremediation has been effective in the removal of toxic minerals such as, Br, Cd, Hg, and Pb, from effluents of food-processing plants and different agricultural wastes [43]. Moreover, algal biomass has found application for the passive biosorption of heavy metals in wastewater [43]. Recently, microalgae have drawn researchers' attention due to their abilities in CO_2 mitigation with environmentally beneficial outcomes, considering that CO_2 is the largest contributor to the greenhouse effect [44]. Other than the above-mentioned features of microalgae, the main product of algal culture, i.e., the biomass, can be valorized for the production of biofuels, including biodiesel, biomethane, and biohydrogen [45], which is noteworthy in the context of the circular economy.

3.2. Advancement in the Field of Phytoremediation

3.2.1. Choosing the Best Plant Genotype

Enhancement of phytoremediation efficiency by increasing plant biomass, metal uptake or tolerance to metal toxicity is an important step in the development of new phytoremediation programs. The efficiency of phytoremediation can be improved through traditional approaches (such as plant breeding or hybridization and selection) or biotechnological techniques (i.e., the creation of engineered plants) that contribute to the development of plants with suitable phenotypes (Figure 2a).

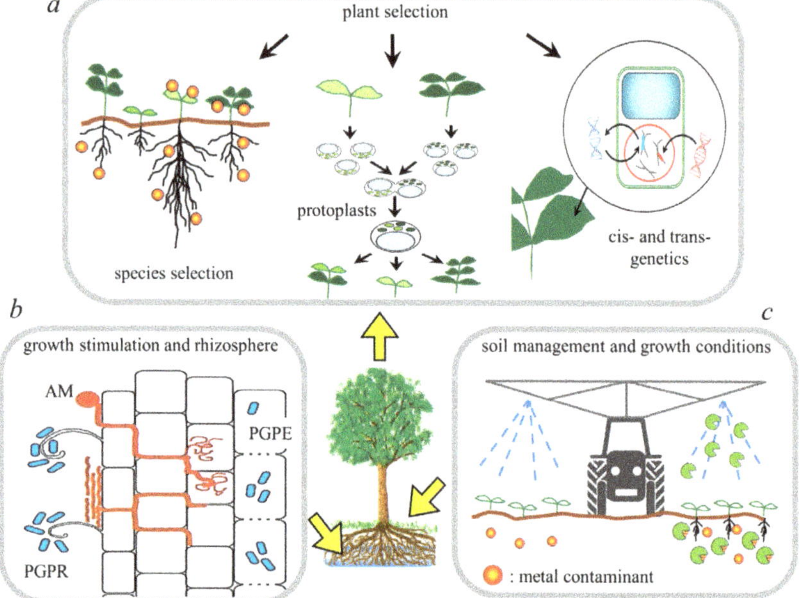

Figure 2. Main aspects of human intervention to enhance phytoremediation. Steps in which human activity can operate are, (**a**) the selection, through varietal choice, classical breeding (e.g., somatic hybridization), or transgenic approach, of the most useful plant species to be applied; (**b**) the enhancement of rhizosphere interconnections between plant growth promoting rhizo- and endophytic bacteria (PGPR and PGPE, respectively, blue dots), and mycorrhizal fungi (drawn in red -arbuscular mychorrizae, AM - and orange), exploiting ex-novo inoculum or the native microflora; (**c**) the management of the polluted site, in terms of both soil conditions and growth techniques, which can change metal availability, plant growth, and remediation effectiveness.

Considering traditional approaches, improvement of plant phytoremediation efficiency was realized by the selection of wild non-edible ecotypes naturally growing in contaminated sites. Native plant species and populations, growing in metalliferous or contaminated sites, are able to cope with the high metal levels present in these soils; for this reason, they are much more resistant to these conditions than other plants and can be used for reclamation purposes [46]. Singh et al. [47] analyzed native plants growing on a site near the Uranium mine tailing ponds in Jaduguda and Turamdih, in the Jharkand State (eastern India), contaminated with heavy metals (Al, V, Ni, Cu, Zn, Fe, Co, Se, Mn) and radionuclides. Among the plants able to accumulate toxic metals and remediate the contaminated site, the As hyperaccumulator *Pteris vittata* was identified as the most versatile as it could accumulate Al, V, Ni, Co, Se, and U. Barrutia et al. [48] identified and characterized native plants spontaneously growing on soils from an abandoned Pb-Zn mine containing toxic levels of Cd, Pb, and Zn in the Basque Country (northern Spain). Among these, 31 species were able to accumulate and tolerate metals, including *Festuca rubra*, *Noccaea caerulescens*, *Jasione montana*, *Rumex acetosa*, and *Plantago lanceolata*. Shoots of *N. caerulescens* accumulated the highest Zn concentrations. Moreover, in vitro and greenhouse selections are suitable for the obtainment of heavy metal-tolerant plants, useful for soil remediation. *Daphne jasminea* and *Daphne tangutica* shoots were cultivated in vitro in the presence of different concentrations of $Pb(NO_3)_2$. In these conditions, *D. tangutica* accumulated high Pb concentrations, and chlorophyll and carotenoid biosynthesis were higher in comparison to *D. jasminea* [49].

To enhance the phytoremediation efficiency, genetic determinants of heavy metal accumulation and tolerance associated with wild hyperaccumulator species can be introduced by introgression into the genome of plants with significantly higher biomass [50]. For instance, *Brassica juncea* protoplasts were fused with *N. caerulescens* protoplasts to transfer the metal-resistant ability of *N. caerulescens* into *B. juncea* by somatic hybridizations. Hybrid plants showed the high Zn and Ni accumulation potential and tolerance derived from *N. caerulescens*, and the high biomass production specific of *B. juncea* [51].

Despite positive results obtained using classical breeding and genetic approaches, molecular engineering may be helpful to enhance plant phytoremediation potential and efficiency for the reclamation of polluted sites. Recombinant DNA technologies currently used for both nuclear and cytoplasmic genome transformation and the availability of genome sequences for different plant species allow the transfer of desirable determinants from hyperaccumulator species to sexually incompatible and high-biomass crops, suitable for in field phytoremediation. For instance, genetic engineering can be exploited to enhance metal tolerance and accumulation by the introduction of genes responsible for metal uptake, transport, accumulation, and detoxification, and for the response to oxidative stress, or to increase biomass production of hyperaccumulator plants [26]. Considering the reclamation of metal contaminated sites, good prospects come from the genetic engineering of high biomass species and trees, such as poplar. Eastern cottonwood trees (*Populus deltoides* Bartr. ex Marsh.) were genetically engineered by the introduction of the bacterial *merA* (mercuric ion reductase that reduces Hg^{2+} to the less toxic Hg^0, which is volatized by plants) and *merB* (organomercury lyase that converts organic Hg to Hg^{2+}) genes isolated and modified from *Escherichia coli* for the reclamation of mercury-contaminated sites. In vitro, *merA*/*merB* plants were more resistant to phenylmercuric acetate than wild-type controls and could detoxify organic Hg more efficiently [52]. Furthermore, genes that are currently widely used to improve plant phytoremediation potential are those that encode transporters of metal ions [26]. For example, Shim et al. [53] produced genetically engineered Bonghwa poplar (*Populus alba* x *P. tremula* var. *glandulosa*) lines expressing the yeast *ScYCF1* gene (*Saccharomyces cerevisiae* - yeast cadmium factor 1), which encodes a vacuolar transporter involved in toxic metal sequestration into the vacuole. When grown on a heavy metal contaminated soil from a mining site in South Korea, *ScYCF1*-expressing plants showed reduced Cd toxicity symptoms and accumulated much more Cd in comparison to wild plants. When plants were tested in the field on contaminated soil, dry weight and accumulation of Cd, Zn, and Pb in transgenic roots were significantly higher than in wild-types, demonstrating a potential utilization of these lines in long-term phytoextraction and phytostabilization of highly contaminated lands [53].

Among the genetic engineering tools, the clustered regularly interspaced short palindromic repeats (CRISPR)-Cas9 system is highly attractive to introduce a wide range of genes in different candidate organisms. CRISPR-Cas9 technology is a revolutionary and versatile gene-editing tool that can be used to enhance selected traits in plants targeting highly specific sequences of DNA [54]; for this reason, the technique could be used to transfer or modulate a desired set of genes in the plant genome to enhance the phytoremediation potential toward polluted soils and waters. The availability of genome sequences from model hypertolerant/hyperaccumulator species that may be considered for phytoremediation (such as the Cd, Ni, and Zn hyperaccumulator *N. caerulescens*, the Cd and Zn hyperaccumulator *Arabidopsis halleri* or the As hyperaccumulator *P. vittata*) and the improvement of bioinformatic tools have opened new opportunities for the use of CRISPR-Cas9 genome editing in the improvement of plant phytoremediation potential [55]. CRISPR-Cas9 technology could be used to introduce or modulate the expression of genes coding for metal transport proteins or involved in the synthesis of metal ligands [55]. Nowadays, despite the potential of this technique, its application in genome editing is still at an early stage in this field. However, there are few reports to date in which the CRISPR/Cas9 system has been successfully used for the reduction of metal content in plants. New rice lines knockout for the metal transporter gene *OsNRAMP5* were generated using the CRISPR/Cas9 system. These plants showed low Cd accumulation in shoots, roots, and in grains, upon hydroponic culture and in Cd-contaminated paddy field trials, maintaining biomass similar to wild-type [56]. This work provides a good example of the potential of the CRISPR/Cas9 system for the development of plants with a modified heavy metal content that could be used to achieve sustainable environmental cleanup via phytoremediation. In addition, phytoremediation could also benefit from the plant association with plant growth-promoting rhizobacteria [PGPR] (see following Section 3.2.3), where the CRISPR technology could be used to create more competent bacterial strains [55]. The application of the CRISPR-Cas9 technology to plants and PGPR, therefore, could be used to increase biomass yield and/or heavy metal tolerance, accumulation and detoxification, and thus to enhance their potential for the application in phytoremediation programs.

3.2.2. Changing the Growth Conditions

In addition to the plant genotype used, cultural strategies have an enormous impact on phytoremediation efficiency (Figure 2c). Several phytotypologies, i.e., planting strategies, can be suitable to treat different pollutants taking into account the characteristics of the polluted matrixes [57]. Among the systems for the decontamination of liquid polluted matrixes, such as sewages, landfill leachates or storm water, constructed wetlands have been extensively applied. This strategy relies on floating and/or rooted hydrophytes and the associated microbiota to detain and remove pollutants mainly by rhizofiltration, although other remediation processes such as phytostabilization and extraction also occur. Free-floating macrophytes (e.g., *Lemna* spp., *Eichhornia crassipes*, and others) have been demonstrated to be excellent metal accumulators [58]; however, the rapid growth and invasiveness of many of these species are a double-edged sword and raise controversies regarding their application for phytoremediation [59]. For this reason, rooted plant species have found a wider application. Rooted hydrophytes (e.g., *Phragmites* spp., *Thypha* spp., *Cyperus alternifolius*, and others) and flood-tolerant species (e.g., *Chrysopogon zizanioides*) have been employed with excellent results for the construction of surface- and subsurface-flow wetlands (reviewed by [60,61]), as well as floating bed systems [62,63]. In constructed wetlands, the composition of growth beds plays a pivotal role. Almost complete removal of metal pollutants has been achieved by using porous materials such as crushed sea shell grits [64], stratified pomice and loamy soil [65], and zeolite [66], whereas composted green waste and gravel limit the remediation performance [64,66]. However, it is important to notice that this fundamental role of growth bed materials is imputable to their different absorption capacities. Indeed, evidence show that, in constructed wetlands, metal pollutants are partially retained by the medium depending on its properties, opening the question of growth bed disposal after the treatment [60].

Another strategy proposed for liquid waste treatment and disposal is that of land treatment, i.e., the application of polluted wastewater as irrigation for plant cultures. Although this system has been tested on different scales with some success [67,68], serious issues remain regarding its influence on soil characteristics, leaching in groundwater and overall impact on the environment [69,70]. As for polluted soils, the application of plant covers has been widely used for inorganic contaminants and relies on a variety of phytoremediation mechanisms, including phytoextraction and phytostabilization (refer to Section 3.1). Indeed, the plant covers effectively prevent pollutant dispersion and leaching in the groundwater, in addition to playing a more active role by removing it through uptake and detoxification [71]. Since metal-polluted soils often offer prohibitive growth conditions due to low nutrient content in addition to high metal levels, phytoremediation can be aided by good soil management practices to enhance global soil quality. In particular, the application of organic amendments, as for example manure or waste compost, has a positive effect on plant growth in phytoremediation [71,72]. Moreover, amendments can alter metal speciation, solubility, and bioavailability by altering water holding capacity, pH, and redox status of the soil [71], influencing the predominant phytoremediation strategy and the efficiency of the system. For example, cow manure, sewage sludge, and forest litter have been reported to enhance As extractability in As-polluted soils, favoring phytoextraction by ryegrass [73]. On the contrary, the application of sewage sludge and municipal waste compost reduced Cu, Zn, and Pb mobility in acidic metal-contaminated soil, leading to a more phytostabilization-oriented strategy [74].

Another widely considered practice to alter metal availability in polluted soils is the chelate-induced phytoextraction. The synthetic chelating agent ethylenediaminetetraacetic acid (EDTA) has proven highly efficient in metal mobilization [75], but its poor biodegradability causes concern due to its persistence in the environment, possible metal leaching and negative effects on soil properties and microbial communities [76]. As an alternative, biodegradable chelating agents as ethylenediaminedisuccinic acid (EDDS) and nitrilotriacetic acid (NTA), as well as organic acids, have been employed in several studies, enhancing phytoextraction efficiency [77–79].

Another supplementation that has been considered is that of exogenous phytohormones, with the aim of enhancing plant biomass, fitness and stress tolerance, thus improving phytoremediation efficiency. Cytokinins were demonstrated to produce generally positive effects in terms of plant growth and phytoextraction capacity [80], whereas conflicting results were achieved using auxins [81,82]. The application of gibberellins or some commercial growth regulator mixtures reduced metal accumulation in plant tissues, but the increased plant biomass brought still to a general enhancement of metal extraction per plant [83]. Moreover, the combined application of phytohormones with other treatments, such as nitrogen fertilization and chelating agents, produced a synergistic action resulting in a significant increase in metal phytoextraction [83].

In recent years, a novel strategy has been proposed to overcome some of the flaws of traditional phytoremediation, namely the long time and limited treatment depths. This technique, named electro kinetic-enhanced phytoremediation, relies on the combined application of plants and a physicochemical treatment, i.e., low-intensity electric fields, to the metal-polluted soil, favoring metal mobilization and bioavailability [84]. The application of different electrode materials and distribution, and different types of electric field have been considered with variable results [84]; however, a significant increase in Pb, As, and Cs phytoextraction has been achieved by DC electric field applied with inert low-cost graphite electrodes, due to the alteration of soil pH and metal solubility resulting from the electric field [85]. Interestingly, the application of a solar cell-powered electric field has been tested in a real-scale field trial for the phytoremediation of a metal polluted electronic waste recycling center by *Eucalyptus globulus*: the application of the electric field resulted in increased plant growth and metal accumulation. Moreover, although traditional power supply systems are more efficient in metal mobilization and containment, solar cells make this strategy significantly more sustainable on the economical level [86].

Finally, *nanoparticles* have also been considered for their possible use in assisting phytoremediation. Different types of nanomaterials have been applied for the decontamination of metal-polluted substrates thanks to their absorption capacity or redox catalytic activity [87]. In combination with plants, nanoparticles can be employed to improve the effectiveness of phytostabilization by absorbing metal ions [88], or phytoextraction by improving plant fitness and stress tolerance and increasing metal bioavailability [89,90].

3.2.3. Enhancing the Plant-Microorganism Interactions

In recent years, researchers have focused their attention on the interactions between plant and metal resistant soil microorganisms, in particular, those colonizing roots (i.e., the rhizobiome) [91]. The synergism between plant roots and microorganisms can implement the remediation process by enhancing phytostabilization, as in the case of arbuscular mycorrhizae (AM) fungi, and phytoextraction employing plant growth promoting rhizo- and endo-bacteria [91] (Figure 2b). AM fungi establish mutual symbiosis with higher plants, improving mineral nutrition. Thus, AM contributes to plant growth in heavy metal contaminated sites by increasing plant access to nutrients such as P, by improving soil texture through the stable aggregation of soil particles and by binding heavy metals into roots restricting their translocation to shoot tissues. In that respect, AM fungi have been reported to reduce metal uptake and distribution in sunflower plants [92,93]. Therefore, AM fungi promote phytostabilization of heavy metals, accelerating the revegetation of severely degraded lands, such as coalmines or waste sites [94].

On the other hand, plant growth promoting rhizobacteria and endophytes (PGPR and PGPE, respectively) are able to increase the phytoremediation competence of plants by promoting their growth and health even under hazardous levels of heavy metals, by means of traits, such as organic acid production, secretion of siderophores, indole-3-acetic acid (IAA) production and 1-aminocyclopropane-1-carboxylate (ACC) deaminase activity. In most metalliferous soils, metals are strongly bound to soil particles, being not promptly available for plant uptake. Various PGPR and PGPE (e.g., *Arthrobacter*, *Microbacterium*, *Bacillus*, *Kocuria* and *Pseudomonas* spp.) can solubilize water-insoluble Zn, Ni, and Cu by local soil acidification through the secretion of protons and/or organic anions (e.g., acetate, lactate, oxalate, tartrate, succinate, citrate, gluconate, ketogluconate, and glycolate) [95]. Moreover, metal bioavailability in soils can be further increased by inoculating PGPR able to secrete biosurfactants, which can aid in metal ion release from soil particles [96]. Under iron-limiting conditions, PGPR secrete low molecular weight siderophores, which are iron chelators with an exceptionally strong affinity for ferric iron (Fe^{3+}), enhancing its availability to both microorganisms and, indirectly, plants [97]. Siderophores are able to chelate several other metal species, such as Mg, Mn, Cr(III), Cd, Zn, Cu, Ni, As, and Pb with variable affinity. For instance, *B. juncea* plants inoculated with the mutant SD1 of the phosphate-solubilizing *Enterobacter* sp. NBRI K28, characterized by an enhanced siderophore production, showed increased biomass and phytoextraction of Ni, Zn, and Cr [98]. In addition to altering metal availability, a great majority of root-associated PGPR also produces the main bacterial auxin IAA, which promotes plant growth, stimulating root cell proliferation, lateral root initiation and overproduction of root hairs. Generally, bacterial IAA facilitates the adaptation of host plants in metal-contaminated sites by triggering physiological changes in plant cell metabolism under metal stress and helping plants to withstand high concentrations of heavy metals [99]. Several PGPR and PGPE are also able to synthesize the enzyme ACC deaminase, which degrades ACC (an immediate precursor of plant ethylene) into 2-oxobutanoate and ammonia, hence inhibiting ethylene production in plants, which is usually induced by heavy metal stress. It has been demonstrated that inoculation with ACC deaminase-producing PGPR resulted in extensive root proliferation in hyperaccumulator plants and efficient phytoremediation in metal-polluted soils [100]. From a technological point of view, microorganisms, which can be exploited for soil remediation or phytoextraction technologies, are usually members of complex metal-tolerant populations associated with tolerant and/or hyperaccumulator plant species growing in metalliferous soils. In some cases, PGPR and PGPE originally isolated from hyperaccumulator plants have been shown to promote growth and phytoextraction of diverse plant species grown in single and multiple metal-contaminated soils [101,102]. However, the impact of PGPR and AM fungi on different plants varies depending on the plant and microbial species and soil types. Several authors have tested PGPR and PGPE as bio-inoculum to remove different heavy metals from soils [103–105]. Zn-mobilizing bacteria, isolated from serpentine soils, promoted Zn, Cu, and Ni accumulation in *Ricinus communis* [106], while *Rahnella* sp. JN6, isolated from *Polygonum pubescens*, can promote growth and Cd, Pb, and Zn uptake in

B. napus [107]. A bacterial consortium, isolated from the rhizosphere of the pseudometallophyte *Betula celtiberica* growing in an As-polluted site, enhanced As accumulation in leaves and roots, whereas the rhizobacterium *Ensifer adhaerens* strain 91R mainly promoted plant growth upon laboratory conditions [108]. Moreover, field experimentation showed that additional factors, such as soil As content and pH, influenced As uptake in the plant, attesting the relevance of field conditions in the success of phytoextraction strategies [108]. As for a phytostabilization-oriented strategy, AM fungi associated to the metallophyte non-accumulator *Viola calaminaria* inhabiting Zn- and Pb-rich soils were shown to improve maize growth in a polluted soil reducing heavy metal concentrations in plant tissues [109,110].

Nevertheless, it must be considered that the details of the interaction between plant roots and root-associated microorganisms are still rather unknown. Moreover, the rhizosphere is an extremely complex and still poorly characterized community: roughly, 99% of soil microbial taxa are yet to be cultured and can only be investigated using culture-independent methods [111]. At this purpose, approaches with *Omics* technologies, based on DNA, RNA, and protein sequencing have advanced our understanding of plant and microbial responses to pollutants and of plant–microbe interactions. For instance, high-throughput sequencing of bacterial 16S rDNA allows defining the composition of the microbial community and how heavy metals drive the selection toward microorganisms, which are more suitable for phytoremediation purposes [92]. In addition, transcriptomic and proteomic studies on rhizosphere communities in contaminated soils are instrumental to predict valuable microbial functions directly [112,113]. The integration of these strategies allows creating a complete picture of how cohabiting and symbiotic biological communities interact to adapt to metal stress and could enhance phytoremediation [114]. Eventually, these data need to be combined with high-throughput isolation and screening for key microbial characteristics such as growth rate, to target microbes that are perhaps not naturally dominant but have valuable traits for their application in phytoremediation [114].

Despite all the research in the field of plant-microorganism interaction, applications of PGPR and mycorrhizal consortia in assisted phytoremediation in contaminated soils are still scarce and the performance of these microorganisms under natural conditions needs to be more deeply investigated. A particular consideration is the biosafety linked to the release of non-autochthonous bacterial strains. In addition, even though such strains might be superior in terms of metal resistance and mobilization effectiveness, the competition with the native microbial population can reduce the efficacy of the inoculated strains. Despite these concerns, co-inoculation with PGPR and mycorrhizal consortia might partially mimic the natural conditions of contaminated soils, in which multiple microorganism interactions occur, helping plants to cope with the toxic effects of heavy metals. Co-inoculation can also improve the phytostabilization or phytoextraction efficiency for various metals at the same time, which indicates the possibility of exporting the technology to multi-metal contaminated sites [67,105].

4. Conclusions

Summarizing, plants and associated microorganisms surely are of great interest for their potential application in polluted soil reclamation. A variety of options are available when considering a phytoremediation approach, including the utilization of wild plant-microorganisms associations, or the implementation by applying particular planting and culturing techniques. Researchers can develop the best suitable plant lines or microorganisms to be exploited, as well as the best fertilizers or soil conditioners. Interestingly, attention moved also on the fate of contaminated biomass, particularly when dealing with approaches of phytoextraction. Indeed, recent research is aimed to valorize metal-rich biomass rather than simply dispose of it, coupling land reclamation with non-food products (e.g., timber) and energy production, a concept known as integrated phytoremediation. In this view, harvested plant biomass has a substantial calorific value in terms of renewable energy production. Therefore, long-term operations of planting, maintaining the phytoremediation site (otherwise unsuitable for remunerative and productive uses) and fruitfully converting harvested biomass, are grouped into the new idea of phyto-management, in which the major goal is mitigating environmental risk and making contaminated lands economically valuable [26,30].

Conflicts of Interest: The authors declare no conflict of interest.

References

1. Baxter, I. Ionomics: Studying the social network of mineral nutrients. *Curr. Opin. Plant Biol.* **2009**, *12*, 381–386. [CrossRef] [PubMed]
2. Krämer, U.; Talke, I.N.; Hanikenne, M. Transition metal transport. *FEBS Lett.* **2007**, *581*, 2263–2272. [CrossRef] [PubMed]
3. DalCorso, G.; Manara, A.; Furini, A. An overview of heavy metal challenge in plants: From roots to shoots. *Metallomics* **2013**, *5*, 1117–1132. [CrossRef] [PubMed]
4. Boughriet, A.; Proix, N.; Billon, G.; Recourt, P.; Ouddane, B. Environmental impacts of heavy metal discharges from a smelter in Deûle-canal sediments (Northern France): Concentration levels and chemical fractionation. *Water Air Soil Pollut.* **2007**, *180*, 83–95. [CrossRef]
5. Broadley, M.; Brown, P.; Cakmak, I.; Rengel, Z.; Zhao, F. Function of nutrients: Micronutrients. In *Mineral Nutrition of Higher Plants*; Marschner, P., Ed.; Academic Press Inc.: San Diego, CA, USA, 2012; pp. 191–248.
6. Krämer, U. Metal hyperaccumulation in plants. *Annu. Rev. Plant Biol.* **2010**, *61*, 517–534. [CrossRef] [PubMed]
7. Gustin, J.L.; Zanis, M.J.; Salt, D.E. Structure and evolution of the plant cation diffusion facilitator family of ion transporters. *BMC Evol. Biol.* **2011**, *11*, 76–89. [CrossRef] [PubMed]
8. White, P.J.; Broadley, M.R. Physiological limits to zinc biofortification of edible crops. *Front. Plant Sci.* **2011**, *2*, 80. [CrossRef] [PubMed]
9. Guerinot, M.L. The ZIP family of metal transporters. *Biochim. Biophys. Acta* **2000**, *1465*, 190–198. [CrossRef]
10. Milner, M.J.; Seamon, J.; Craft, E.; Kochian, L.V. Transport properties of members of the ZIP family in plants and their role in Zn and Mn homeostasis. *J. Exp. Bot.* **2013**, *64*, 369–381. [CrossRef]
11. Cailliatte, R.; Schikora, A.; Briat, J.F.; Mari, S.; Curie, C. High-affinity manganese uptake by the metal transporter NRAMP1 is essential for Arabidopsis growth in low manganese conditions. *Plant Cell* **2010**, *22*, 904–917. [CrossRef]
12. Bastow, E.L.; de la Torre, V.S.G.; Maclean, A.E.; Green, R.T.; Merlot, S.; Thomine, S.; Balk, J. Vacuolar iron stores gated by NRAMP3 and NRAMP4 are the primary source of iron in germinating seeds. *Plant Physiol.* **2018**, *177*, 1267–1276. [CrossRef] [PubMed]
13. Kim, Y.Y.; Choi, H.; Segami, S.; Cho, H.T.; Martinoia, E.; Maeshima, M.; Lee, Y. AtHMA1contributes to the detoxification of excess Zn(II) in Arabidopsis. *Plant J.* **2009**, *58*, 737–753. [CrossRef] [PubMed]
14. Williams, L.E.; Mills, R.F. P(1B)-ATPases—An ancient family of transition metal pumps with diverse functions in plants. *Trends Plant. Sci.* **2005**, *10*, 491–502. [CrossRef] [PubMed]
15. Hanikenne, M.; Baurain, D. Origin and evolution of metal P-type ATPases in Plantae (Archaeplastida). *Front. Plant Sci.* **2014**, *4*, 544. [CrossRef] [PubMed]
16. Desbrosses-Fonrouge, A.G.; Voigt, K.; Schröder, A.; Arrivault, S.; Thomine, S.; Krämer, U. *Arabidopsis thaliana* MTP1 is a Zn transporter in the vacuolar membrane which mediates Zn detoxification and drives leaf Zn accumulation. *FEBS Lett.* **2005**, *579*, 4165–4174. [CrossRef] [PubMed]
17. Dalvi, A.A.; Bhalerao, S.A. Response of plants towards heavy metal toxicity: An overview of avoidance, tolerance and uptake mechanism. *Ann. Plant Sci.* **2013**, *2*, 362–368.
18. Manara, A. Plant Responses to Heavy Metal Toxicity. In *Plants and Heavy Metals*; Furini, A., Ed.; Springer: Cham, Switzerland, 2012; pp. 27–53.
19. Rai, V.K. Role of amino acids in plant responses to stress. *Biol. Plant.* **2002**, *45*, 481–487. [CrossRef]
20. Howden, R.; Goldsbrough, P.B.; Andersen, C.R.; Cobbett, C.S. Cadmium-sensitive, *cad1* mutants of *Arabidopsis thaliana* are phytochelatin deficient. *Plant Physiol.* **1995**, *107*, 1059–1066. [CrossRef]
21. Kohler, A.; Blaudez, D.; Chalot, M.; Martin, F. Cloning and expression of multiple metallothioneins from hybrid poplar. *New Phytol.* **2004**, *164*, 83–93. [CrossRef]
22. Guo, J.; Xu, L.; Su, Y.; Wang, H.; Gao, S.; Xu, J.; Que, Y. ScMT2-1-3, A metallothionein gene of sugarcane, plays an important role in the regulation of heavy metal tolerance/accumulation. *BioMed Res. Int.* **2013**, *2013*, 904769. [CrossRef]

23. Huang, H.; Gupta, D.K.; Tian, S.; Yang, X.; Li, T. Lead tolerance and physiological adaptation mechanism in roots of accumulating and non-accumulating ecotypes of *Sedum alfredii*. *Environ. Sci. Pollut. Res.* **2012**, *19*, 1640–1651. [CrossRef] [PubMed]
24. Wijnhoven, S.; Leuven, R.S.E.W.; van der Velde, G.; Jungheim, G.; Koelemij, E.I.; de Vries, F.T.; Eijsackers, H.J.P.; Smits, A.J.M. Heavy-metal concentrations in small mammals from a diffusely polluted floodplain: Importance of species- and location-specific characteristics. *Arch. Environ. Contam. Toxicol.* **2007**, *52*, 603–613. [CrossRef] [PubMed]
25. Ensley, B.D. Rationale for use of phytoremediation. In *Phytoremediation of Toxic Metals: Using Plants to Clean up the Environment*; Raskin, I., Ensley, B.D., Eds.; Wiley: New York, NY, USA, 2000; pp. 3–11.
26. Fasani, E.; Manara, A.; Martini, F.; Furini, A.; DalCorso, G. The potential of genetic engineering of plants for the remediation of soils contaminated with heavy metals. *Plant Cell Environ.* **2018**, *41*, 1201–1232. [CrossRef] [PubMed]
27. Fasani, E. Plants that hyperaccumulate heavy metals. In *Plants and Heavy Metals*; Furini, A., Ed.; Springer: Cham, Switzerland, 2012; pp. 55–74.
28. Mertens, J.; Van Nevel, L.; De Schrijver, A.; Piesschaert, F.; Oosterbaan, A.; Tack, F.M.; Verheyen, K. Tree species effect on the redistribution of soil metals. *Environ. Pollut.* **2007**, *149*, 173–181. [CrossRef]
29. Dickinson, N.M.; Baker, A.J.; Doronila, A.; Laidlaw, S.; Reeves, R.D. Phytoremediation of inorganics: Realism and synergies. *Int. J. Phytoremed.* **2009**, *11*, 97–114. [CrossRef] [PubMed]
30. Robinson, B.H.; Anderson, C.W.N.; Dickinson, N.M. Phytoextraction: Where's the action? *J. Geochem. Explor.* **2015**, *151*, 34–40. [CrossRef]
31. De Souza, M.P.; Lytle, C.M.; Mulholland, M.M.; Otte, M.L.; Terry, N. Selenium assimilation and volatilization from dimethylselenoniopropionate by Indian mustard. *Plant Physiol.* **2000**, *122*, 1281–1288. [CrossRef] [PubMed]
32. Zhu, Y.G.; Williams, P.N.; Meharg, A.A. Exposure to inorganic arsenic from rice: A global health issue? *Environ. Pollut.* **2008**, *154*, 169–171. [CrossRef] [PubMed]
33. Meng, M.; Li, B.; Shao, J.J.; Wang, T.; He, B.; Shi, J.B.; Ye, Z.H.; Jiang, G.B. Accumulation of total mercury and methylmercury in rice plants collected from different mining areas in China. *Environ. Pollut.* **2014**, *184*, 179–186. [CrossRef]
34. Moreno, F.N.; Anderson, C.W.N.; Stewart, R.B.; Robinson, B.H. Phytofiltration of mercury-contaminated water: Volatilisation and plant-accumulation aspects. *Environ. Exp. Bot.* **2008**, *62*, 78–85. [CrossRef]
35. Brunner, I.; Luster, J.; Günthardt-Goerg, M.S.; Frey, B. Heavy metal accumulation and phytostabilisation potential of tree fine roots in a contaminated soil. *Environ. Pollut.* **2008**, *152*, 559–568. [CrossRef] [PubMed]
36. Berti, W.R.; Cunningham, S.D. Phytostabilization of metals. In *Phytoremediation of Toxic Metals: Using Plants to Clean up the Environment*; Raskin, I., Ensley, B.D., Eds.; Wiley: New York, NY, USA, 2000; pp. 71–88.
37. Mench, M.; Vangronsveld, J.; Bleeker, P.; Ruttens, A.; Geebelen, W.; Lepp, N. Phytostabilisation of metal-contaminated sites. In *Phytoremediation of Metal-Contaminated Soils*; Morel, J.L., Echevarria, G., Goncharova, N., Eds.; Springer: Dordrecht, The Netherlands, 2006; pp. 109–190.
38. Yadav, B.K.; Siebel, M.A.; Van Bruggen, J.J.A. Rhizofiltration of a Heavy Metal (Lead) Containing Wastewater Using the Wetland Plant *Carex pendula*. *CLEAN Soil Air Water* **2011**, *39*, 467–474. [CrossRef]
39. Dushenkov, V.; Kumar, P.B.; Motto, H.; Raskin, I. Rhizofiltration: The use of plants to remove heavy metals from aqueous streams. *Environ. Sci. Technol.* **1995**, *29*, 1239–1245. [CrossRef] [PubMed]
40. Dushenkov, S.; Kapulnik, Y. Phytofiltration of metals. In *Phytoremediation of Toxic Metals—Using Plants to Clean up the Environment*; Raskin, I., Ensley, B.D., Eds.; Wiley: New York, NY, USA, 2000; pp. 89–106.
41. Raskin, I.; Kumar, P.B.A.N.; Dushenkov, S.; Salt, D.E. Bioconcentration of heavy metals by plants. *Curr. Opin. Biotechnol.* **1994**, *5*, 285–290. [CrossRef]
42. Salt, D.E.; Blaylock, M.; Kumar, N.P.; Dushenkov, V.; Ensley, B.D.; Chet, I.; Raskin, I. Phytoremediation: A novel strategy for the removal of toxic metals from the environment using plants. *Biotechnology* **1995**, *13*, 468–474. [CrossRef] [PubMed]
43. Abdel-Raoufa, N.; Al-Homaidan, A.A.; Ibraheem, I.B.M. Microalgae and wastewater treatment. *Saudi J. Biol. Sci.* **2012**, *19*, 257–275. [CrossRef] [PubMed]
44. Molazadeh, M.; Ahmadzadeh, H.; Pourianfar, H.R.; Lyon, S.; Rampelotto, P.H. The use of microalgae for coupling wastewater treatment with CO_2 biofixation. *Front. Bioeng. Biotechnol.* **2019**, *7*, 42. [CrossRef] [PubMed]

45. Judd, S.; van den Broeke, L.J.; Shurair, M.; Kuti, Y.; Znad, H. Algal remediation of CO_2 and nutrient discharges: A review. *Water Res.* **2015**, *87*, 356–366. [CrossRef] [PubMed]
46. Roccotiello, E.; Marescotti, P.; Di Piazza, S.; Cecchi, G.; Mariotti, M.G.; Zotti, M. Biodiversity in Metal-Contaminated Sites-Problem and Perspective—A Case Study. In *Biodiversity in Ecosystems—Linking Structure and Function*; Blanco, J.A., Ed.; IntechOpen: London, UK, 2015; pp. 563–582.
47. Laxman Singh, K.; Sudhakar, G.; Swaminathan, S.K.; Muralidhar Rao, C. Identification of elite native plants species for phytoaccumulation and remediation of major contaminants in uranium tailing ponds and its affected area. *Environ. Dev. Sustain.* **2015**, *17*, 57–81. [CrossRef]
48. Barrutia, O.; Artetxe, U.; Hernández, A.; Olano, J.M.; García-Plazaola, J.I.; Garbisu, C.; Becerril, J.M. Native plant communities in an abandoned Pb-Zn mining area of northern Spain: Implications for phytoremediation and germplasm preservation. *Int. J. Phytoremed.* **2011**, *13*, 256–270. [CrossRef]
49. Wiszniewska, A.; Hanus-Fajerska, E.; Smoleń, S.; Muszyńska, E. In vitro selection for lead tolerance in shoot culture of Daphne species. *Acta Sci. Pol. Hortorum Cultus* **2015**, *14*, 129–142.
50. Dushenkov, S.; Skarzhinskaya, M.; Glimelius, K.; Gleba, D.; Raskin, I. Bioengineering of a phytoremediation plant by means of somatic hybridization. *Int. J. Phytoremed.* **2002**, *4*, 117–126. [CrossRef] [PubMed]
51. Brewer, E.P.; Saunders, J.A.; Scott, A.J.; Chaney, R.L.; McIntosh, M.S. Somatic hybridization between the zinc accumulator *Thlaspi caerulescens* and *Brassica napus*. *Theor. Appl. Genet.* **1999**, *99*, 761–771. [CrossRef]
52. Lyyra, S.; Meagher, R.B.; Kim, T.; Heaton, A.; Montello, P.; Balish, R.S.; Merkle, S.A. Coupling two mercury resistance genes in Eastern cottonwood enhances the processing of organomercury. *Plant Biotechnol. J.* **2007**, *5*, 254–262. [CrossRef] [PubMed]
53. Shim, D.; Kim, S.; Choi, Y.I.; Song, W.Y.; Park, J.; Youk, E.S.; Jeong, S.C.; Martinoia, E.; Noh, E.W.; Lee, Y. Transgenic poplar trees expressing yeast cadmium factor 1 exhibit the characteristics necessary for the phytoremediation of mine tailing soil. *Chemosphere* **2013**, *90*, 1478–1486. [CrossRef] [PubMed]
54. Didovyk, A.; Borek, B.; Tsimring, L.; Hasty, J. Transcriptional regulation with CRISPR-Cas9: Principles, advances, and applications. *Curr. Opin. Biotechnol.* **2016**, *40*, 177–184. [CrossRef] [PubMed]
55. Basharat, Z.; Novo, L.A.B.; Yasmin, A. Genome editing weds CRISPR: What is in it for phytoremediation? *Plants* **2018**, *7*, 51. [CrossRef] [PubMed]
56. Tang, L.; Mao, B.; Li, Y.; Lv, Q.; Zhang, L.P.; Chen, C.; He, H.; Wang, W.; Zeng, X.; Shao, Y.; et al. Knockout of OsNramp5 using the CRISPR/Cas9 system produces low Cd-accumulating indica rice without compromising yield. *Sci. Rep.* **2017**, *7*, 14438. [CrossRef] [PubMed]
57. Sleegers, F.; Hisle, M. Redesigning Abandoned Gas Stations Through Phytotechnologies. In *Phytoremediation*; Ansari, A.A., Gill, S.S., Gill, R., Lanza, G.R., Newman, L., Eds.; Springer: Cham, Switzerland, 2018; Volume 6, pp. 3–20.
58. Rezania, S.; Taib, S.M.; Din, M.F.M.; Dahalan, F.A.; Kamyab, H. Comprehensive review on phytotechnology: Heavy metals removal by diverse aquatic plants species from wastewater. *J. Hazard. Mater.* **2016**, *318*, 587–599. [CrossRef] [PubMed]
59. Newete, S.W.; Byrne, M.J. The capacity of aquatic macrophytes for phytoremediation and their disposal with specific reference to water hyacinth. *Environ. Sci. Pollut. Res.* **2016**, *23*, 10630–10643. [CrossRef] [PubMed]
60. Marchand, L.; Mench, M.; Jacob, D.L.; Otte, M.L. Metal and metalloid removal in constructed wetlands, with emphasis on the importance of plants and standardized measurements: A review. *Environ. Pollut.* **2010**, *158*, 3447–3461. [CrossRef] [PubMed]
61. Odinga, C.A.; Swalaha, F.M.; Otieno, F.A.O.; Ranjith, K.R.; Bux, F. Investigating the efficiency of constructed wetlands in the removal of heavy metals and enteric pathogens from wastewater. *Environ. Technol. Rev.* **2013**, *2*, 1–16. [CrossRef]
62. Ladislas, S.; Gerente, C.; Chazarenc, F.; Brisson, J.; Andres, Y. Floating treatment wetlands for heavy metal removal in highway stormwater ponds. *Ecol. Eng.* **2015**, *80*, 85–91. [CrossRef]
63. Ning, D.; Huang, Y.; Pan, R.; Wang, F.; Wang, H. Effect of eco-remediation using planted floating bed system on nutrients and heavy metals in urban river water and sediment: A field study in China. *Sci. Total Environ.* **2014**, *485*, 596–603. [CrossRef] [PubMed]
64. Bavandpour, F.; Zou, Y.; He, Y.; Saeed, T.; Sun, Y.; Sun, G. Removal of dissolved metals in wetland columns filled with shell grits and plant biomass. *Chem. Eng. J.* **2018**, *331*, 234–241. [CrossRef]
65. Dan, A.; Oka, M.; Fujii, Y.; Soda, S.; Ishigaki, T.; Machimura, T.; Ike, M. Removal of heavy metals from synthetic landfill leachate in lab-scale vertical flow constructed wetlands. *Sci. Total Environ.* **2017**, *584*, 742–750.

66. Lizama Allende, K.; Fletcher, T.D.; Sun, G. Enhancing the removal of arsenic, boron and heavy metals in subsurface flow constructed wetlands using different supporting media. *Water Sci. Technol.* **2011**, *63*, 2612–2618. [CrossRef] [PubMed]
67. Fasani, E.; DalCorso, G.; Zerminiani, A.; Ferrarese, A.; Campostrini, P.; Furini, A. Phytoremediatory efficiency of *Chrysopogon zizanioides* in the treatment of landfill leachate: A case study. *Environ. Sci. Pollut. Res.* **2019**, *26*, 10057–10069. [CrossRef] [PubMed]
68. Zalesny, R.S., Jr.; Bauer, E.O. Evaluation of Populus and Salix continuously irrigated with landfill leachate I. Genotype-specific elemental phytoremediation. *Int. J. Phytoremed.* **2007**, *9*, 281–306. [CrossRef]
69. Aronsson, P.; Dahlin, T.; Dimitriou, I. Treatment of landfill leachate by irrigation of willow coppice—Plant response and treatment efficiency. *Environ. Pollut.* **2010**, *158*, 795–804. [CrossRef]
70. Coyle, D.R.; Zalesny, J.A.; Zalesny, R.S., Jr.; Wiese, A.H. Irrigating poplar energy crops with landfill leachate negatively affects soil micro-and meso-fauna. *Int. J. Phytoremed.* **2011**, *13*, 845–858. [CrossRef] [PubMed]
71. Burges, A.; Alkorta, I.; Epelde, L.; Garbisu, C. From phytoremediation of soil contaminants to phytomanagement of ecosystem services in metal contaminated sites. *Int. J. Phytoremed.* **2018**, *20*, 384–397. [CrossRef] [PubMed]
72. Burges, A.; Epelde, L.; Benito, G.; Artetxe, U.; Becerril, J.M.; Garbisu, C. Enhancement of ecosystem services during endophyte-assisted aided phytostabilization of metal contaminated mine soil. *Sci. Total Environ.* **2016**, *562*, 480–492. [CrossRef] [PubMed]
73. Karczewska, A.; Gałka, B.; Dradrach, A.; Lewińska, K.; Mołczan, M.; Cuske, M.; Gersztyn, L.; Litak, K. Solubility of arsenic and its uptake by ryegrass from polluted soils amended with organic matter. *J. Geochem. Explor.* **2017**, *182*, 193–200. [CrossRef]
74. Alvarenga, P.; Gonçalves, A.P.; Fernandes, R.M.; De Varennes, A.; Vallini, G.; Duarte, E.; Cunha-Queda, A.C. Organic residues as immobilizing agents in aided phytostabilization: (I) Effects on soil chemical characteristics. *Chemosphere* **2009**, *74*, 1292–1300. [CrossRef] [PubMed]
75. Duo, L.A.; Lian, F.; Zhao, S.L. Enhanced uptake of heavy metals in municipal solid waste compost by turfgrass following the application of EDTA. *Environ. Monit. Assess.* **2010**, *165*, 377–387. [CrossRef] [PubMed]
76. Lee, J.; Sung, K. Effects of chelates on soil microbial properties, plant growth and heavy metal accumulation in plants. *Ecol. Eng.* **2014**, *73*, 386–394. [CrossRef]
77. Zhao, S.; Jia, L.; Duo, L. The use of a biodegradable chelator for enhanced phytoextraction of heavy metals by Festuca arundinacea from municipal solid waste compost and associated heavy metal leaching. *Bioresour. Technol.* **2013**, *129*, 249–255. [CrossRef] [PubMed]
78. Attinti, R.; Barrett, K.R.; Datta, R.; Sarkar, D. Ethylenediaminedisuccinic acid (EDDS) enhances phytoextraction of lead by vetiver grass from contaminated residential soils in a panel study in the field. *Environ. Pollut.* **2017**, *225*, 524–533. [CrossRef]
79. Ehsan, S.; Ali, S.; Noureen, S.; Mahmood, K.; Farid, M.; Ishaque, W.; Shakoor, M.B.; Rizwan, M. Citric acid assisted phytoremediation of cadmium by *Brassica napus* L. *Ecotoxicol. Environ. Saf.* **2014**, *106*, 164–172. [CrossRef]
80. Aderholt, M.; Vogelien, D.L.; Koether, M.; Greipsson, S. Phytoextraction of contaminated urban soils by *Panicum virgatum* L. enhanced with application of a plant growth regulator (BAP) and citric acid. *Chemosphere* **2017**, *175*, 85–96. [CrossRef] [PubMed]
81. Fässler, E.; Evangelou, M.W.; Robinson, B.H.; Schulin, R. Effects of indole-3-acetic acid (IAA) on sunflower growth and heavy metal uptake in combination with ethylene diamine disuccinic acid (EDDS). *Chemosphere* **2010**, *80*, 901–907. [CrossRef] [PubMed]
82. Vamerali, T.; Bandiera, M.; Hartley, W.; Carletti, P.; Mosca, G. Assisted phytoremediation of mixed metal (loid)-polluted pyrite waste: Effects of foliar and substrate IBA application on fodder radish. *Chemosphere* **2011**, *84*, 213–219. [CrossRef] [PubMed]
83. Hadi, F.; Bano, A.; Fuller, M.P. The improved phytoextraction of lead (Pb) and the growth of maize (*Zea mays* L.): The role of plant growth regulators (GA3 and IAA) and EDTA alone and in combinations. *Chemosphere* **2010**, *80*, 457–462. [CrossRef] [PubMed]
84. Cameselle, C.; Gouveia, S. Phytoremediation of mixed contaminated soil enhanced with electric current. *J. Hazard. Mater.* **2019**, *361*, 95–102. [CrossRef] [PubMed]
85. Mao, X.; Han, F.X.; Shao, X.; Guo, K.; McComb, J.; Arslan, Z.; Zhang, Z. Electro-kinetic remediation coupled with phytoremediation to remove lead, arsenic and cesium from contaminated paddy soil. *Ecotoxicol. Environ. Saf.* **2016**, *125*, 16–24. [CrossRef]

86. Luo, J.; Yang, D.; Qi, S.; Wu, J.; Gu, X.S. Using solar cell to phytoremediate field-scale metal polluted soil assisted by electric field. *Ecotox. Environ. Saf.* **2018**, *165*, 404–410. [CrossRef]
87. Mueller, N.C.; Nowack, B. Nanoparticles for remediation: Solving big problems with little particles. *Elements* **2010**, *6*, 395–400. [CrossRef]
88. Liang, S.X.; Jin, Y.; Liu, W.; Li, X.; Shen, S.G.; Ding, L. Feasibility of Pb phytoextraction using nano-materials assisted ryegrass: Results of a one-year field-scale experiment. *J. Environ. Manag.* **2017**, *190*, 170–175. [CrossRef]
89. Moameri, M.; Khalaki, M.A. Capability of *Secale montanum* trusted for phytoremediation of lead and cadmium in soils amended with nano-silica and municipal solid waste compost. *Environ. Sci. Pollut. R.* **2017**, 1–8. [CrossRef]
90. Singh, J.; Lee, B.K. Influence of nano-TiO_2 particles on the bioaccumulation of Cd in soybean plants (Glycine max): A possible mechanism for the removal of Cd from the contaminated soil. *J. Environ. Manag.* **2016**, *170*, 88–96. [CrossRef] [PubMed]
91. Sessitsch, A.; Kuffner, M.; Kidd, P.; Vangronsveld, J.; Wenzel, W.W.; Fallman, K.; Puschenreiter, M. The role of plant-associated bacteria in the mobilization and phytoextraction of trace elements in contaminated soils. *Soil Biol. Biochem.* **2013**, *60*, 182–194. [CrossRef] [PubMed]
92. Hassan, S.E.D.; Boon, E.; St-Arnaud, M.; Hijri, M. Molecular biodiversity of arbuscular mycorrhizal fungi in trace metal-polluted soils. *Mol. Ecol.* **2011**, *20*, 3469–3483. [CrossRef] [PubMed]
93. Ker, K.; Charest, C. Nickel remediation by AM-colonized sunflower. *Mycorrhiza* **2010**, *20*, 399–406. [CrossRef] [PubMed]
94. Enkhtuya, B.; Rydlová, J.; Vosátka, M. Effectiveness of indigenous and non-indigenous isolates of arbuscular mycorrhizal fungi in soils from degraded ecosystems and man-made habitats. *Appl. Soil Ecol.* **2002**, *14*, 201–211. [CrossRef]
95. Becerra-Castro, C.; Prieto-Fernández, Á.; Álvarez-López, V.; Cabello-Conejo, M.I.; Acea, M.J.; Kidd, P.S. Nickel solubilizing capacity and characterization of rhizobacteria isolated from hyperaccumulating and non-hyperaccumulating subspecies of *Alyssum serpyllifolium*. *Int. J. Phytoremed.* **2011**, *13*, 229–244. [CrossRef] [PubMed]
96. Gamalero, E.; Glick, B.R. Plant growth-promoting bacteria and metals phytoremediation. In *Phytotechnologies: Remediation of Environmental Contaminants*; Anjum, N.A., Pereira, M.E., Ahmad, I., Duarte, A.C., Umar, S., Khan, N.A., Eds.; CRC Press: Boca Raton, FL, USA, 2012; pp. 361–376.
97. Schalk, I.J.; Hannauer, M.; Braud, A. New roles for bacterial siderophores in metal transport and tolerance. *Environ. Microbiol.* **2011**, *13*, 2844–2854. [CrossRef] [PubMed]
98. Kumar, K.V.; Singh, N.; Behl, H.M.; Srivastava, S. Influence of plant growth promoting bacteria and its mutant on heavy metal toxicity in *Brassica juncea* grown in fly ash amended soil. *Chemosphere* **2008**, *72*, 678–683. [CrossRef]
99. Glick, B.R. Using soil bacteria to facilitate phytoremediation. *Biotechnol. Adv.* **2010**, *28*, 367–374. [CrossRef]
100. Arshad, M.; Saleem, M.; Hussain, S. Perspectives of bacterial ACC deaminase in phytoremediation. *Trends Biotechnol.* **2007**, *25*, 356–362. [CrossRef]
101. Ma, Y.; Rajkumar, M.; Luo, Y.; Freitas, H. Inoculation of endophytic bacteria on host and non-host plants—Effects on plant growth and Ni uptake. *J. Hazard. Mater.* **2011**, *195*, 230–237. [CrossRef] [PubMed]
102. Ma, Y.; Rajkumar, M.; Freitas, H. Isolation and characterization of Ni mobilizing PGPB from serpentine soils and their potential in promoting plant growth and Ni accumulation by *Brassica* spp. *Chemosphere* **2009**, *75*, 719–725. [CrossRef] [PubMed]
103. Sheng, X.F.; Xia, J.J.; Jiang, C.Y.; He, L.Y.; Qian, M. Characterization of heavy metal-resistant endophytic bacteria from rape (*Brassica napus*) roots and their potential in promoting the growth and lead accumulation of rape. *Environ. Pollut.* **2008**, *156*, 1164–1170. [CrossRef] [PubMed]
104. Jiang, C.Y.; Sheng, X.F.; Qian, M.; Wang, Q.Y. Isolation and characterization of a heavy metal resistant *Burkholderia* sp. from heavy metal-contaminated paddy field soil and its potential in promoting plant growth and heavy metal accumulation in metal polluted soil. *Chemosphere* **2008**, *72*, 157–164. [CrossRef] [PubMed]
105. Visioli, G.; Vamerali, T.; Mattarozzi, M.; Dramis, L.; Sanangelantoni, A.M. Combined endophytic inoculants enhance nickel phytoextraction from serpentine soil in the hyperaccumulator *Noccaea caerulescens*. *Front. Plant Sci.* **2015**, *6*, 638. [CrossRef] [PubMed]
106. Rajkumar, M.; Freitas, H. Influence of metal resistant-plant growth promoting bacteria on the growth of *Ricinus communis* in soil contaminated with heavy metals. *Chemosphere* **2008**, *71*, 834–842. [CrossRef]

107. He, H.; Ye, Z.; Yang, D.; Yan, J.; Xiao, L.; Zhong, T.; Yuan, M.; Cai, X.; Fang, Z.; Jing, Y. Characterization of endophytic *Rahnella* sp. JN6 from *Polygonum pubescens* and its potential in promoting growth and Cd, Pb, Zn uptake by *Brassica napus*. *Chemosphere* **2013**, *90*, 1960–1965. [CrossRef] [PubMed]
108. Mesa, V.; Navazas, A.; González-Gil, R.; González, A.; Weyens, N.; Lauga, B.; Gallego, J.L.R.; Sánchez, J.; Peláez, A. Use of endophytic and rhizosphere bacteria to improve phytoremediation of arsenic-contaminated industrial soils by autochthonous *Betula celtiberica*. *Appl. Environ. Microbiol.* **2017**, *83*, e03411-16. [CrossRef] [PubMed]
109. Kaldorf, M.; Kuhn, A.J.; Schroder, W.H.; Hildebrandt, U.; Bothe, H. Selective element deposits in maize colonized by a heavy metal tolerance conferring arbuscular mycorrhizal fungus. *J. Plant Physiol.* **1999**, *154*, 718–728. [CrossRef]
110. Tonin, C.; Vandenkoornhuyse, P.; Joner, E.J.; Straczek, J.; Leyval, C. Assessment of arbuscular mycorrhizal fungi diversity in the rhizosphere of *Viola calaminaria* and effect of these fungi on heavy metal uptake by clover. *Mycorrhiza* **2001**, *10*, 161–168. [CrossRef]
111. Pham, V.J.T.; Kim, J. Cultivation of unculturable soil bacteria. *Trends Biotechnol.* **2012**, *30*, 475–484. [CrossRef] [PubMed]
112. Mattarozzi, M.; Manfredi, M.; Montanini, B.; Gosetti, F.; Sanangelantoni, A.M.; Marengo, E.; Careri, M.; Visioli, G. A metaproteomic approach dissecting major bacterial functions in the rhizosphere of plants living in serpentine soil. *Anal. Bioanal. Chem.* **2017**, *409*, 2327–2339. [CrossRef] [PubMed]
113. Yergeau, E.; Sanschagrin, S.; Maynard, C.; St-Arnaud, M.; Greer, C.W. Microbial expression profiles in the rhizopshere of willows depend on soil contamination. *ISME J.* **2014**, *8*, 344–358. [CrossRef] [PubMed]
114. Bell, T.H.; Joly, S.; Pitre, F.E.; Yergeau, E. Increasing phytoremediation efficiency and reliability using novel omics approaches. *Trends Biotechnol.* **2014**, *32*, 271–280. [CrossRef] [PubMed]

© 2019 by the authors. Licensee MDPI, Basel, Switzerland. This article is an open access article distributed under the terms and conditions of the Creative Commons Attribution (CC BY) license (http://creativecommons.org/licenses/by/4.0/).

International Journal of
Molecular Sciences

Article

OsMSR3, a Small Heat Shock Protein, Confers Enhanced Tolerance to Copper Stress in *Arabidopsis thaliana*

Yanchun Cui [1,*], Manling Wang [1], Xuming Yin [1], Guoyun Xu [2], Shufeng Song [3], Mingjuan Li [1], Kai Liu [1] and Xinjie Xia [1]

1 Key Laboratory of Agro-ecological Processes in Subtropical Region, Institute of Subtropical Agriculture, Chinese Academy of Sciences, Changsha 410125, China; mlwang@isa.ac.cn (M.W.); yxm@isa.ac.cn (X.Y.); limingjuan5002@163.com (M.L.); lk15111238792@163.com (K.L.); jxxia@isa.ac.cn (X.X.)

2 Zhengzhou Tobacco Research Institute of China National Tobacco Corporation, Zhengzhou 450001, China; gyxu@isa.ac.cn

3 State Key Laboratory of Hybrid Rice, Hunan Hybrid Rice Research Centre, Changsha 410125, China; shufengsong@126.com

* Correspondence: cuiyanchun@isa.ac.cn

Received: 21 October 2019; Accepted: 30 November 2019; Published: 3 December 2019

Abstract: Copper is a mineral element essential for the normal growth and development of plants; however, excessive levels can severely affect plant growth and development. *Oryza sativa* L. multiple stress-responsive gene 3 *(OsMSR3)* is a small, low-molecular-weight heat shock protein (HSP) gene. A previous study has shown that *OsMSR3* expression improves the tolerance of *Arabidopsis* to cadmium stress. However, the role of *OsMSR3* in the Cu stress response of plants remains unclear, and, thus, this study aimed to elucidate this phenomenon in *Arabidopsis thaliana*, to further understand the role of small HSPs (sHSPs) in heavy metal resistance in plants. Under Cu stress, transgenic *A. thaliana* expressing *OsMSR3* showed higher tolerance to Cu, longer roots, higher survival rates, biomass, and relative water content, and accumulated more Cu, abscisic acid (ABA), hydrogen peroxide, chlorophyll, carotenoid, superoxide dismutase, and peroxidase than wild-type plants did. Moreover, *OsMSR3* expression in *A. thaliana* increased the expression of antioxidant-related and ABA-responsive genes. Collectively, our findings suggest that *OsMSR3* played an important role in regulating Cu tolerance in plants and improved their tolerance to Cu stress through enhanced activation of antioxidative defense mechanisms and positive regulation of ABA-responsive gene expression.

Keywords: *Arabidopsis*; small heat shock protein; *OsMSR3*; copper stress; reactive oxygen species

1. Introduction

Copper is an essential mineral element for the normal growth and development of plants. In plants, Cu functions as an important cofactor for metalloproteins and participates in numerous biological processes, including photosynthesis, respiration, oxygen superoxide scavenging, cell wall metabolism and lignification, and ethylene perception [1–6]. Cu-deficient soils not only affect the quality and quantity of plant food crops but also reduce their nutritional value as the main source of essential minerals for humans [7]. In addition, exposure of plants to excess Cu interferes with normal growth, proliferation, and differentiation of most plant cells [8–14]. One of the earliest and most obvious symptoms of Cu stress is inhibition of primary root elongation [15–17], while its prominent manifestations are decreased proliferation of root meristem cells [18], impaired cell integrity [19], and cell death [20]. Excessive accumulation of Cu in plants leads to the production of reactive oxygen species (ROS), which are toxic owing to their high redox activity [21].

Plants have developed specific mechanisms to prevent Cu toxicity by tightly regulating Cu homeostasis, including Cu uptake, translocation, efflux, and sequestration [22]. Plants also activate antioxidant defense responses to mitigate oxidative damage caused by free Cu ions in the cytosol [23]. The defense system includes ROS-removing enzymes, such as superoxide dismutase (SOD), peroxidase (POD), and catalase (CAT), as well as low-molecular-weight antioxidants, such as ascorbic acid (ASC) and glutathione (GSH) [24]. These antioxidant compounds and enzymes can be used as physiological indicators for evaluating plant antioxidant defense ability [15–26].

The stress hormone abscisic acid (ABA) plays an important role in plant stress tolerance. Heavy metals such as Cd, Hg, and Cu can induce the expression of ABA synthesis genes, which in turn increase the endogenous level of ABA [27,28]. A previous study showed that cadmium treatment increases endogenous ABA levels in cattail and reed roots [29], potato tubers [30], and rice plants [31]. Exposure to high Cu concentrations also increased ABA levels in rice [31]. During the germination of wheat seeds, ABA levels increased in the presence of Hg, Cd, and Cu stress [32]. Plants exposed to heavy metal stress showed an increased concentration of ABA, which indicates that the hormone is involved in the protective mechanism against heavy metal toxicity [27,33,34].

In plants, small heat shock proteins (sHSPs) with monomer sizes ranging from 12 to 42 kDa are more diverse and abundant than those in other organisms. There are 23 sHSPs in rice, and these are proposed to be categorized into fourteen classes [35]. Classes CI-CXI (nine subfamilies) are localized in the nucleus or cytoplasm, whereas the other five are positioned in the endoplasmic reticulum, mitochondria, plastid, and peroxisome [25–36]. These sHSPs are stimulated in response to a wide range of abiotic stresses. For instance, *Oshsp26*, which encodes a chloroplast-localized sHSP, has been shown to enhance tolerance against oxidative and heat stress in tall fescue [37]. Overexpression of sHSP17.7 enhances drought tolerance in transgenic rice [38]. Overexpression of *OsHsp18.0-CI*, an sHSP-CI family gene, enhances resistance to bacterial leaf streak in rice [39]. However, there are only a few studies on the sHSPs involved in heavy metal resistance.

Our previous studies have shown that OsMSR3 belongs to the class I sHSP family [40]. The expression of the *OsMSR3* gene significantly enhances tolerance of *Arabidopsis thaliana* (L.) Heynh (*A. thaliana*) to cadmium stress [41]. However, the molecular mechanism of *OsMSR3*-induced Cu tolerance is poorly understood. In this study, we determined that the expression of *OsMSR3* was upregulated by Cu stress. Therefore, we speculated that *OsMSR3* plays an important role in plant tolerance to Cu. Expression of *OsMSR3* enhanced the Cu stress tolerance of *A. thaliana*. Our research enhances the understanding of the role of sHSPs in heavy metal resistance in plants.

2. Results

2.1. Expression of OsMSR3 is Induced by Cu Stress

The quantitative reverse transcription-polymerase chain reaction (qRT-PCR) showed that the expression of *OsMSR3* increased rapidly after 6 h of Cu stress and peaked at 12 h in Pei'ai 64S rice seedlings (Figure 1). Subsequently, *OsMSR3* expression decreased by nearly 3.5-fold compared to the control levels at 48 h (Figure 1).

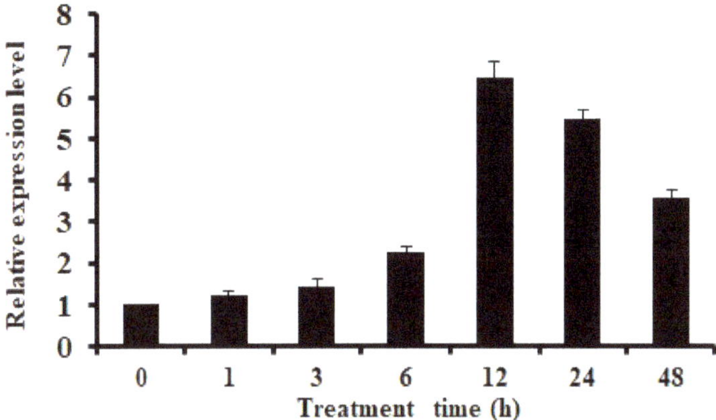

Figure 1. Expression analysis of the *Oryza sativa* L. multiple stress-responsive gene 3 (*OsMSR3*) gene. Expression analysis of *OsMSR3* in Pei'ai 64S rice seedlings at different time points (0, 1, 3, 6, 12, 24, and 48 h) under Cu stress using quantitative reverse transcription-polymerase reaction (qRT-PCR). The *ACTIN1* gene was used as an internal control. Error bars indicate standard deviations (SD) of three independent experiments.

2.2. Expression of OsMSR3 Enhances Cu Tolerance of Transgenic A. thaliana

To assess whether upregulation of *OsMSR3* enhances tolerance to Cu, transgenic *Arabidopsis* plants expressing *OsMSR3* were generated and analyzed. Based on a previous study [41], two independent transgenic lines, L-5 and L-7, were chosen for further experiments. We examined survival rates and root lengths of both *OsMSR3*-expressing and wild-type plants treated with Cu (50 μM). As shown in Figure 2A, there was no significant difference in survival rate and root length between wild-type and transgenic seedlings in the absence of stress. However, in half-strength Murashige and Skoog ($\frac{1}{2}$ MS) medium supplemented with Cu, transgenic plants showed higher Cu tolerance than wild-type plants did, with a higher survival rate and longer root length (Figures 2B and 3B). Furthermore, the fresh and dry weight of wild-type and transgenic plants measured under normal and Cu stress did not differ significantly in the absence of stress, but fresh and dry weight were higher in both transgenic lines than they were in wild-type plants under Cu stress (Figure 3C,D). After the application of Cu stress, the relative water content (RWC) of transgenic plants was >35%. As shown in Figure 3E, under control conditions, the RWC was almost the same for all tested lines. In the presence of 50 μM copper chloride ($CuCl_2$), all plants showed a reduction in RWC. However, water loss was higher in the wild-type than it was in the transgenic lines.

Figure 2. Performance of transgenic plants and wild-types under normal and Cu stress conditions. (**A**) Left panel, seedlings reared under normal conditions (0 μM $CuCl_2$) for 30 days; right panel, seedlings exposed to 50 μM $CuCl_2$ for 30 days. WT, wild-type *Arabidopsis thaliana*; L-5 and L-7, transgenic lines 5 and 7. (**B**) Survival rates of plants after growth under Cu stress for 30 days. Each column represents an average of three replicates, and bars indicate ± standard deviation (SD); and ** $p < 0.01$ indicate significant differences compared to wild-type T plants under the same conditions determined using Student's *t*-tests.

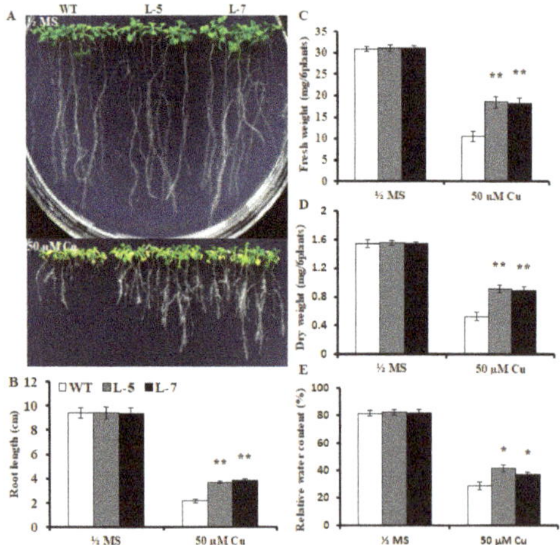

Figure 3. Improved tolerance to Cu stress induced by *Oryza sativa* L. multiple stress-responsive gene 3 (*OsMSR3*) expression. (**A**) Images of representative plants grown on half-strength Murashige and Skoog ($\frac{1}{2}$ MS) medium with or without 50 μM copper chloride ($CuCl_2$) for 21 days. (**B**) Effect of Cu treatment on root length of plants presented in panel A. (**C**) Fresh weight (FW) and (**D**) dry weight (DW) of wild-type and transgenic line plants treated with or without 50 μM $CuCl_2$ for 21 days. (**E**) Relative water content (RWC) of wild-type and transgenic plants treated with or without 50 μM $CuCl_2$ for 21 days. Values are means ± standard deviation (SD) of three independent biological replicates; * $p < 0.05$ and ** $p < 0.01$ indicate significant differences from the wild-type determined using Student's *t*-test.

2.3. Expression of OsMSR3 in Arabidopsis Causes Higher Accumulation of Cu

To determine whether the enhanced Cu tolerance of transgenic plants was associated with their lower Cu accumulation, Cu content was determined in the different lines at the end of Cu treatment. Cu accumulation was higher in the roots and shoots of transgenic plants than that of wild-type plants. As shown in Figure 4A,B, the Cu content of transgenic lines L-5 and L-7 was approximately 1.2 and 1.1 times higher in the roots, and 1.66 and 1.59 times higher in the shoots, respectively, than it was in the wild-type plants.

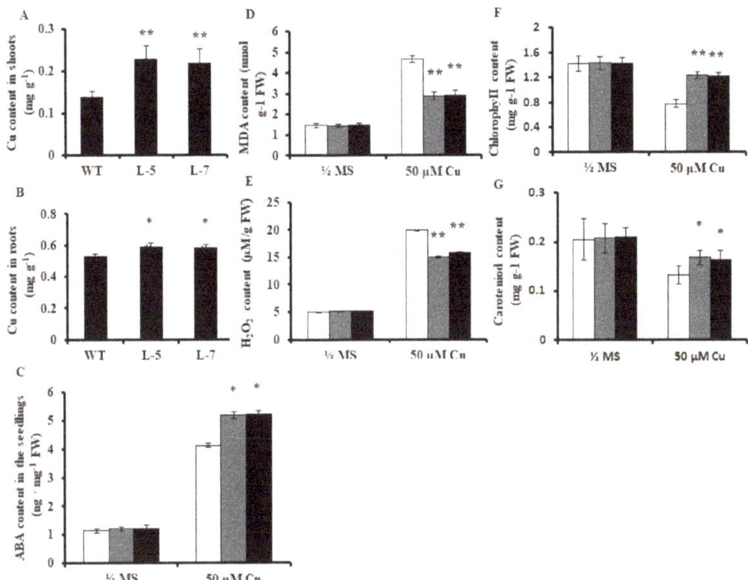

Figure 4. Quantitative analysis of various physiological indexes in wild-type and transgenic plants. (**A**,**B**) Cu content in wild-type and transgenic plant shoots and roots treated with 50 μM copper chloride ($CuCl_2$) for 21 days, respectively. (**C**) abscisic acid (ABA), (**D**) malondialdehyde (MDA), and (**E**) hydrogen peroxide (H_2O_2) content in wild-type and transgenic plants treated with or without 50 μM $CuCl_2$ for 24 h. (**F**,**G**) Chlorophyll and carotenoid content in wild-type and transgenic plants treated with or without 50 μM $CuCl_2$ for 21 days. Values are means ± standard deviation (SD) of three independent biological replicates; * $p < 0.05$ and ** $p < 0.01$ indicate significant differences from wild-type plants under the same conditions determined using Student's *t*-test.

2.4. Effects of OsMSR3 Expression on ABA, Malondialdehyde (MDA), and Hydrogen Peroxide (H_2O_2) Content in A. thaliana

To determine whether OsMSR3 affects ABA content in *A. thaliana* under Cu stress, endogenous ABA content in transgenic and wild-type plants was measured. Under normal conditions, there was almost no difference in ABA content between the wild-type and transgenic lines, whereas Cu treatment increased the levels in both plant types (Figure 4C). Specifically, the mean ABA content, which was 4.12 ng g^{-1} fresh weight (FW) in wild-type plants, increased to 5.18 and 5.21 ng g^{-1} FW in the L-5 and L-7 lines, respectively (Figure 4C).

To examine the oxidative damage induced by excess Cu, we monitored the accumulation of malondialdehyde (MDA) and hydrogen peroxide (H_2O_2) in wild-type and transgenic plants. Under normal conditions, differences in MDA levels were not apparent between wild-type and transgenic plants, but levels were significantly increased by Cu stress (Figure 4D). Wild-type plants had a higher MDA content than transgenic plants did (Figure 4D). As shown in Figure 4D, MDA levels were approximately 1.61 and 1.59 times higher in wild-type plants than they were in transgenic lines L-5 and L-7 plants, respectively.

Cu stress can lead to H_2O_2 generation, which can be used to examine the oxidative damage induced by excess Cu [42]. In our study, H_2O_2 levels were not significantly different between transgenic and wild-type plants under controlled conditions. However, under Cu stress, H_2O_2 levels were lower in both transgenic lines than in wild-type plants, but no significant difference was observed between the L-5 and L-7 lines (Figure 4E).

2.5. Effects of OsMSR3 Expression in A. thaliana on Chlorophyll and Carotenoid Content

To determine whether the chlorophyll content is altered in transgenic plants under salt stress, we detected chlorophyll and carotenoid content in the leaves of wild-type and transgenic seedlings under normal conditions and Cu stress. As shown in Figure 4F,G, there was no significant difference in chlorophyll and carotenoid content between the transgenic and wild-type lines under normal growth conditions. In contrast, following Cu treatment, the chlorophyll content in the leaves of the *OsMSR3* transgenic lines (L-5 and L-7) was 1.59 and 1.57 times higher than that in the wild-type plants, although the content decreased in both transgenic and wild-type plants. The carotenoid content in L-5 and L-7 plants was 1.26 and 1.23 times higher than that in the wild-type.

2.6. Antioxidant Enzyme Activities are Altered in Transgenic A. thaliana

To determine whether increased Cu tolerance in transgenic plants is related to changes in oxidase activity in vivo, SOD, POD, and CAT activities were measured in wild-type and transgenic plants grown in medium without (CK) or with 50 μM $CuCl_2$. The data showed that under normal conditions, SOD and POD activities in the transgenic lines were slightly higher than those in wild-type plants (Figure 5A,B), whereas CAT activity was slightly lower in the transgenic lines than in the wild-type plants (Figure 5C). Under Cu stress, SOD and POD activities increased in both wild-type and transgenic lines. However, the SOD activity of L-5 and L-7 was 1.07 times higher (Figure 5A), and the POD activity was 1.12 and 1.14 times higher (Figure 5B) than that of wild-type plants, respectively. Cu stress decreases CAT activity in both wild-type and transgenic lines with a greater decrease in the transgenic lines than in the wild-type plants (Figure 5C). The POD activity of the two transgenic lines was only 0.87 and 0.88 times higher than that of the wild-type plants (Figure 5C).

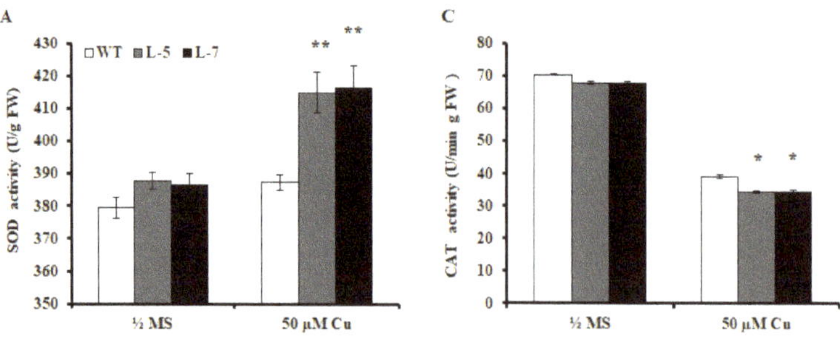

Figure 5. *Cont.*

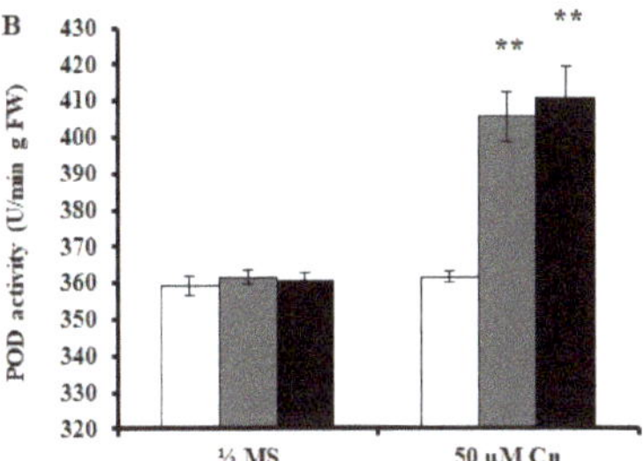

Figure 5. Antioxidant enzyme activity determination. Quantitative analysis of (**A**) superoxide dismutase (SOD), (**B**) peroxidase (POD), (**C**) and catalase (CAT) activity in wild-type and transgenic plants treated with or without 50 μM copper chloride ($CuCl_2$) for 24 h. Values are means ± standard deviation (SD) of three independent experiments; * $p < 0.05$ and ** $p < 0.01$ indicate significant differences from wild-type plants under the same conditions determined using the Student's *t*-test.

2.7. Expression of OsMSR3 Increases Expression of Antioxidant-Related and ABA-Responsive Genes

To determine the performance of transgenic plants under Cu stress and elucidate the molecular mechanism underlying the resistance of transgenic plants to Cu stress, the transcript levels of antioxidant-related (*AtCSD1*, *AtCSD2*, and *AtPOD*) and ABA-responsive (*AtRD29A*, *AtABA1*, and *AtABI5)* genes were assayed in wild-type and transgenic plants under normal and stress conditions. The expression levels of *AtCSD1*, *AtCSD2*, and *AtPOD* were higher in wild-type plants than they were in transgenic plants under control conditions. However, higher gene expression was observed in transgenic plants than in wild-type plants under Cu stress conditions (Figure 6A–C). Compared to the expression under normal conditions, Cu stress inhibited the expression of these genes in wild-type plants but activated their expression in transgenic lines. The ABA-responsive genes, *AtRD29A*, *AtABA1*, and *AtABI5*, showed no significantly different expression levels between transgenic and wild-type plants under normal conditions (Figure 6D–F). Under Cu stress conditions, the expression levels of the three genes were higher in the transgenic lines than in the wild-type plants (Figure 6D–F).

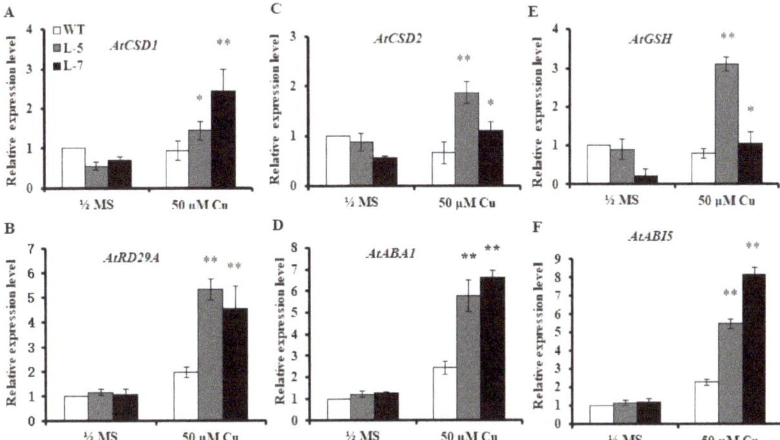

Figure 6. Quantitative reverse transcription-polymerase reaction (qRT-PCR) gene analysis. qRT-PCR analysis of relative expression of (**A**) *AtCSD1*, (**B**) *AtRD29A*, (**C**) *AtCSD2*, (**D**) *AtABA1*, (**E**) *AtGSH*, (**F**) and *AtABI5*, in two-week-old transgenic and wild-type (WT) plants treated with (Cu) or without (Control) 50 μM copper chloride ($CuCl_2$) for 24 h. Values are means ± standard deviation (SD) of three independent biological replicates; * $p < 0.05$ and ** $p < 0.01$ indicate significant differences from wild-type plants under the same conditions, determined using the Student's *t*-test.

3. Discussion

Heavy metal pollution in soils is an emerging worldwide threat owing to its adverse effects on environmental safety [43]. Currently, cadmium and lead pollution in soils and their harmful effects on humans are attracting the attention of researchers globally. Cu pollution has become an important problem in the soil environment; however, studies of this phenomenon are still in infancy [44]. With the development of modern molecular biology, transgenic technology has emerged as an effective method to discover new Cu-tolerant genes in plants and cultivate plants that are highly efficient at repairing damage due to Cu contamination. In the previous study, we found that the expression of *OsMSR3* in *Arabidopsis* significantly enhanced tolerance to cadmium stress [41]. As Cu and cadmium belong to the group of heavy metal elements, we evaluated if the transgenic lines expressing *OsMSR3* in *Arabidopsis* could enhance the ability of copper tolerance.

The expression of *OsMSR3* was enhanced by Cu stress (Figure 1), which indicated that *OsMSR3* is involved in the response to Cu. Then, two transgenic lines L-5 and L-7 were used to perform the Cu tolerance experiment. The results showed that the expression of *OsMSR3* enhanced the tolerance of *A. thaliana* to Cu stress than that of wild-type plants, manifested as higher survival rate (Figure 2), higher biomass (Figure 3), and longer root length (Figure 3B). Moreover, transgenic plants accumulated less MDA than wild-type plants under Cu stress (Figure 4D). It is well known that Cu damages cell membranes by inducing lipid peroxidation [45]. The MDA level is used to detect membrane lipid peroxidation and permeability [46]. These results suggest that the expression of *OsMSR3* alleviated Cu-induced damage to the cell membrane of *A. thaliana*. Cell membrane stability is a major factor contributing to the maintenance of water status in plants during water deficit [47,48]. Therefore, the RWC of transgenic lines under Cu stress was higher than that of wild-type plants (Figure 3E).

Although the transgenic plants accumulated more Cu in the root and shoot (Figure 4A,B), their growth was significantly better than that of the wild-type. The accumulation of Cu in plant roots may inhibit the development of fine roots and reduce the absorption of iron and other trace elements [49]. To some extent, OsMSR3 protein may reduce the inhibition of Cu transport from the root to the shoot. We suggest that OsMSR3 may be helpful in maintaining the homeostasis of Cu metal ions at the cell and plant levels. In addition, we found that chlorophyll and carotenoid content in the leaves of *OsMSR3* transgenic lines was higher than that in those of the wild-type plants under Cu stress (Figure 4F,G). Chlorophyll is an important part of the light-harvesting complex (LHCII). As an antenna for capturing light energy and transferring it to the reaction center, the chlorophyll content reflects the intensity of plant photosynthesis [50]. Carotenoids play an important role in plant growth and development. For example, they can act as a haptokine by transferring captured light to chlorophyll and can also act as a scavenger of free radicals in plant cells [51,52]. We speculated that the expression of *OsMSR3* reduced the damage to chlorophyll and carotenoids under Cu stress.

SOD activity and SOD-related gene expression in *OsMSR3* transgenic lines were significantly higher than those in wild-type plants (Figure 5A, Figure 6A,C). The presence of excess Cu causes the generation of ROS, such as superoxide radical (O_2^-), H_2O_2, singlet oxygen ($^1O_2$), and hydroxyl radicals (OH) [53]. To scavenge ROS and alleviate their deleterious effects, plants stimulate ROS-scavenging systems such as CAT, SOD, and POD [24], to combat the oxidative injury induced by heavy metal exposure [54]. SOD is the first line of defense against ROS and catalyzes O_2^- to produce H_2O_2 and O_2 [55]. An increase in SOD activity in stressed plants is an important indicator of superoxide ion production and enhancement of oxidative tolerance [55]. Therefore, the enhanced Cu tolerance of transgenic lines is related to the expression of SOD-related genes and SOD activity in *A. thaliana* under Cu stress induced by *OsMSR3* expression.

Cu stress can lead to the generation of ROS, such as H_2O_2 [56]. In this study, we found that H_2O_2 accumulated in wild-type and transgenic seedlings significantly during Cu stress, albeit to a lower extent in the two *OsMSR3* transgenic lines than in the wild-type (Figure 4E). The primary H_2O_2-scavenging enzymes in plant cells are CAT and POD; the former degrades H_2O_2 into water and oxygen. No studies, to date, have confirmed that a change in CAT activity is necessary to eliminate

H_2O_2 in rice plants under Cu stress [55]. However, the current study revealed that Cu significantly increased POD activity in the transgenic lines but had little effect on CAT activity (Figure 5B,C). Moreover, *POD* gene expression was upregulated under Cu stress (Figure 6E). This is consistent with the increase in POD activity. Studies have shown that CAT has a high capacity but low affinity, whereas POD has a high affinity for H_2O_2 [57]. Thus, POD is the most effective H_2O_2-scavenging enzyme to reduce H_2O_2 content in plant cells under Cu stress.

Heavy metal exposure induces the expression of ABA synthesis-related genes in plants, which eventually leads to an increase in endogenous ABA levels [28]. In this study, under Cu stress conditions, ABA content and ABA-related gene expression levels in transgenic plants were significantly higher than those in wild-type plants (Figures 4C and 6B,D,F). Therefore, we also propose that the expression of *OsMSR3* leads to the upregulation of ABA-related genes and an increase in endogenous ABA level under Cu stress, which may partly explain the increased tolerance of transgenic plants to Cu stress.

4. Materials and Methods

4.1. Plant Material and Growth Conditions

The seeds of rice (*Oryza sativa* ssp. *indica*) cultivar Pei'ai 64S were surface-sterilized with 75% ethanol for 2 min, treated with 50% sodium hypochlorite for 20 min, and then washed with distilled water at least thrice. The sterilized seeds were germinated on half-strength Murashige and Skoog ($\frac{1}{2}$ MS) medium and grown in a greenhouse under conditions of a light intensity of 600 μmol/m2/s 70% relative humidity, and 28 °C temperature with a 12-h light/dark photoperiod. For the Cu stress experiment, two-week-old seedlings were exposed to a nutrient solution containing 50 μM $CuCl_2$ for 48 h. The leaves were harvested as a pool for each sample at 0, 1, 3, 6, 12, 24, 36, and 48 h after Cu treatment.

4.2. Cu Tolerance Assay

We used 50 μM and 100 μM $CuCl_2$ to do the pre-experiment and then selected the concentration of 50 μM $CuCl_2$ as the most suitable. The seeds of T3 transgenic and wild-type *A. thaliana* (ecotype Columbia-0) were surface-sterilized and sown in Petri dishes containing $\frac{1}{2}$ MS media with or without 50 μM $CuCl_2$. The seeds were incubated in the dark at 4 °C for two days to break the dormancy and then transferred to a growth chamber. After incubation for 30 days, the survival rate of *A. thaliana* was determined. For measurement of root growth under Cu treatment, three-day-old *A. thaliana* seedlings were transferred onto $\frac{1}{2}$ MS medium with or without 50 μM $CuCl_2$, in vertically placed dishes. After incubation for 21 days, the root length (from the base of the root to the tip) and FW of six plants were measured.

Next, whole plants were rehydrated with distilled water at 4 °C for 12 h, blotted dry, and then the turgid weight (TW) was recorded. Rehydrated whole plants were oven-dried at 80 °C for 24 h, and the dry weight (DW) was recorded. RWC was calculated as follows: RWC (%) = (FW − DW)/(TW − DW) × 100. For the qRT-PCR analysis of selected genes, three-day-old *A. thaliana* seedlings were transferred onto $\frac{1}{2}$ MS medium with or without 50 μM $CuCl_2$. After 21 days of treatment, plant materials were harvested, and qRT-PCR was performed. The detailed procedure is provided in the next section.

4.3. RNA Extraction and qRT-PCR Analysis

Total RNAs were extracted with TRIzol reagent (Invitrogen, Burlington, ON, Canada), as described previously [58]. qPCR analysis was conducted using AceQ qPCR SYBR Green Master Mix (Vazyme Biotech, Nanjing, China), and the reactions were performed using an ABI7900HT (Applied Biosystems, Foster City, CA, USA) and run on the following schedule: 95 °C for 10 min, followed by 40 cycles at 95 °C for 15 s and 58 °C for 30 s. The internal controls were *ACTIN1* and *β-TUBULIN* for rice and *A. thaliana*, respectively. The data for relative expression were analyzed using the comparative Ct method [59]. The primer pairs used in the qPCR analysis are listed in Table S1.

4.4. Measurement of Cu Content

Cu content was determined according to the method described by Li et al. [24]. Briefly, three-day-old *A. thaliana* seedlings were transferred to $\frac{1}{2}$ MS medium with or without 50 μM $CuCl_2$. After 21 days of treatment, the roots and shoots were harvested and dried at 80 °C for two days. Dried plant tissues (50–100 mg roots; 100–200 mg shoots) were digested with 11 N HNO_3 at 200 °C for 10 h. The digested samples were then diluted with 0.1 N HNO_3 and analyzed using an atomic absorption spectrometer (Solaar M6; Thermo Fisher, Boston, MA, USA). The experiments were performed in triplicate.

4.5. Measurement of MDA Content

Two-week-old transgenic and wild-type plants were cultivated on $\frac{1}{2}$ MS medium with or without 50 μM $CuCl_2$ for 24 h. Then, 0.3 g of the seedlings was harvested and ground into a powder for the determination of MDA content, which was measured according to a previously standardized method [24].

4.6. Measurement of ABA Content

Two-week-old transgenic and wild-type plants were cultivated on $\frac{1}{2}$ MS media with or without 50 μM $CuCl_2$ for 24 h. Approximately 0.2 g of the leaf tissue was harvested, ground into a powder, and then suspended in 1.8 mL 100 mM sodium phosphate buffer (PBS, pH = 7.4) for ABA leaf content detection, using a previously published method [60].

4.7. Measurement of Chlorophylls and Carotenoids

The chlorophyll and carotenoid content were determined according to a previous method [60]. Briefly, three-day-old *A. thaliana* seedlings were transferred onto $\frac{1}{2}$ MS medium with or without 50 μM $CuCl_2$ in vertically placed dishes. After incubation for 21 days, chlorophyll and carotenoids were extracted from the rosette leaves of the wild-type and transgenic plants with 100% alcohol. An ultraviolet-visible (UV-vis) spectrometer (UV-2600; Shimadzu Co., Kyoto, Japan) was used to measure the absorption of the extracts. The total chlorophyll and carotenoid content were calculated according to a previously published method [61].

4.8. Quantitative Analysis of H_2O_2

The H_2O_2 concentration was determined using a commercially available kit (Nanjing Jiancheng Bioengineering Institute, Nanjing, China). Briefly, two-week-old plants (wild-type and transgenic lines) cultivated in $\frac{1}{2}$ MS medium were treated for 24 h with or without 50 μM $CuCl_2$. Then, 0.5 g of the seedlings was harvested, weighed, immediately ground, and then suspended in 5 mL 0.9% sodium chloride solution. The supernatant was collected after centrifugation for 10 min at 4 °C and 3000× *g*, and the H_2O_2 content was measured according to the protocol provided by the manufacturer of the kit.

4.9. Assay of Antioxidant Enzyme Activities

To measure antioxidant enzyme activities, two-week-old plants (wild-type and transgenic lines) cultivated in $\frac{1}{2}$ MS medium were treated for 24 h with or without 50 μM $CuCl_2$. Seedling samples (0.5 g) were frozen in liquid nitrogen, rapidly ground into powder, and then homogenized in 100 mM sodium phosphate buffer (pH 7.4) on ice. After centrifugation at 3000× *g* for 15 min at 4 °C, the supernatant samples were immediately used for the detection of antioxidant enzymes. The activities of SOD, POD, and CAT were measured using specific assay kits (A001-1, A084-3, and A007-1, respectively) from Nanjing Jiancheng Bioengineering Institute (Nanjing, China) according to the manufacturer's instructions.

4.10. Statistical Analysis

The experimental data were analyzed using the statistical package for the social sciences (SPSS) 17.0 statistical software (*SPSS* Inc., Chicago, IL, USA). At least three independent experiments were performed, and the average results are presented. Error bars represent standard deviation (SD, $n > 3$). Furthermore, * $p < 0.05$ or ** $p < 0.01$ indicate statistically significant means.

5. Conclusions

In conclusion, we showed the involvement of *OsMSR3* in Cu tolerance in *A. thaliana*. *OsMSR3*-expressing lines exhibited enhanced Cu stress tolerance, possibly through enhanced activation of antioxidative defense mechanisms and positive regulation of ABA-responsive gene expression. In view of the good performance of the transgenic lines, *OsMSR3* can be used to modify plants for remediation of Cu pollution in the soil. Therefore, this study provides an important insight into plant biology and mechanisms to overcome increasing heavy metal pollution in soils.

Supplementary Materials: Supplementary materials can be found at http://www.mdpi.com/1422-0067/20/23/6096/s1. Table S1. Primer sequences used for quantitative reverse transcription polymerase reaction (qRT-PCR).

Author Contributions: Y.C. designed and performed the experiments; Y.C. and X.X. analyzed the data and wrote the paper; M.W., X.Y., G.X., S.S., M.L., and K.L. contributed reagents/materials/analysis tools; X.X. supervised the work and revised the manuscript. All the authors agreed on the contents of the paper and declared no conflicting interests.

Funding: This work was financially supported by the National Natural Science Foundation of China (Grant Nos. 31301253 and 31671671) and Youth Innovation Team Project of ISA, CAS (2017QNCXTD_GTD).

Conflicts of Interest: The authors declare no conflict of interest.

Abbreviations

ABA	abscisic acid
SOD	superoxide dismutase
POD	peroxidase
CAT	catalase
ROS	reactive oxygen species
ASC	ascorbic acid
GSH	glutathione
sHSPs	small heat shock protei
SD	standard deviation

References

1. Yruela, I.; Pueyo, J.J.; Alonso, P.J.; Picorel, R. Photoinhibition of photosystem II from higher plants effect of copper inhibition. *J. Biol. Chem.* **1996**, *271*, 27408–27415. [CrossRef]
2. Rodriguez, F.I.; Esch, J.J.; Hall, A.E.; Binder, B.M.; Schaller, G.E.; Bleecker, A.B. A copper cofactor for the ethylene receptor ETR1 from Arabidopsis. *Science* **1999**, *283*, 996–998. [CrossRef]
3. Himelblau, E.; Amasino, R.M. Delivering copper within plant cells. *Curr. Opin. Plant Biol.* **2000**, *3*, 205–210. [CrossRef]
4. Pilon, M.; Abdel-Ghany, S.E.; Cohu, C.M.; Gogolin, K.A.; Ye, H. Copper cofactor delivery in plant cells. *Curr. Opin. Plant Biol.* **2006**, *9*, 256–263. [CrossRef]
5. Burkhead, J.L.; Reynolds, K.A.; Abdel-Ghany, S.E.; Cohu, C.M.; Pilon, M. Copper homeostasis. *New Phytol.* **2009**, *182*, 799–816. [CrossRef]
6. Bernal, M.; Casero, D.; Singh, V.; Wilson, G.T.; Grande, A.; Yang, H.; Dodani, S.C.; Pellegrini, M.; Huijser, P.; Connolly, E.L.; et al. Transcriptome sequencing identifies *SPL7*-regulated copper acquisition genes *FRO4/FRO5* and the copper dependence of iron homeostasis in Arabidopsis. *Plant Cell* **2012**, *24*, 738–761. [CrossRef]
7. Song, Y.; Zhou, L.; Yang, S.; Wang, C.; Zhang, T.; Wang, J. Dose-dependent sensitivity of *Arabidopsis* thaliana seedling root to copper is regulated by auxin homeostasis. *Environ. Exp. Bot.* **2017**, *139*, 23–30. [CrossRef]

8. Luna, C.M.; Gonzalez, C.A.; Trippi, V.S. Oxidative damage caused by an excess of copper in oat leaves. *Plant Cell Physiol.* **1994**, *35*, 11–15. [CrossRef]
9. Shen, Z.G.; Zhang, F.Q.; Zhang, F.S. Toxicity of copper and zinc in seedlings of mung bean and inducing accumulation of polyamine. *J. Plant Nutr.* **1998**, *21*, 1153–1162. [CrossRef]
10. Patsikka, E.; Kairavuo, M.; Sersen, F.; Aro, E.M.; Tyystjarvi, E. Excess copper predisposes photosystem II to photoinhibition in vivo by outcompeting iron and causing decrease in leaf chlorophyll. *Plant Physiol.* **2002**, *129*, 1359–1367. [CrossRef]
11. Nielsen, H.D.; Brownlee, C.; Coelho, S.M.; Brown, M.T. Inter-population differences in inherited copper tolerance involve photosynthetic adaptation and exclusion mechanisms in *Fucus serratus*. *New Phytol.* **2003**, *160*, 157–165. [CrossRef]
12. Demirevska-Kepovaa, K.; Simova-Stoilovaa, L.; Stoyanovaa, Z.; Hölzerb, R.; Feller, U. Biochemical changes in barley plants after excessive supply of copper and manganese. *Environ. Exp. Bot.* **2004**, *52*, 253–266. [CrossRef]
13. Drazkiewicz, M.; Skorzynska-Polit, E.; Krupa, Z. Copper-induced oxidative stress and antioxidant defence in *Arabidopsis thaliana*. *Biometals* **2004**, *17*, 379–387. [CrossRef]
14. Wang, P.; De Schamphelaere, K.A.; Kopittke, P.M.; Zhou, D.M.; Peijnenburg, W.J.; Lock, K. Development of an electrostatic model predicting copper toxicity to plants. *J. Exp. Bot.* **2012**, *63*, 659–668. [CrossRef]
15. Navari-Izzo, F.; Cestone, B.; Cavallini, A.; Natali, L.; Giordani, T.; Quartacci, M.F. Copper excess triggers phospholipase D activity in wheat roots. *Phytochemistry* **2006**, *67*, 1232–1242. [CrossRef]
16. Peto, A.; Lehotai, N.; Lozano-Juste, J.; Leon, J.; Tari, I.; Erdei, L.; Kolbert, Z. Involvement of nitric oxide and auxin in signal transduction of copper-induced morphological responses in *Arabidopsis* seedlings. *Ann. Bot.* **2011**, *108*, 449–457. [CrossRef]
17. Marchand, L.; Nsanganwimana, F.; Lamy, J.B.; Quintela-Sabaris, C.; Gonnelli, C.; Colzi, I.; Fletcher, T.; Oustriere, N.; Kolbas, A.; Kidd, P.; et al. Root biomass production in populations of six rooted macrophytes in response to Cu exposure: Intra-specific variability versus constitutive-like tolerance. *Environ. Pollut.* **2014**, *193*, 205–215. [CrossRef]
18. Liu, D.; Jiang, W.; Meng, Q.; Zou, J.; Gu, J.; Zeng, M. Cytogenetical and ultrastructural effects of copper on root meristem cells of *Allium sativum* L. *Biocell* **2009**, *33*, 25–32.
19. Madejón, P.; Ramírez-Benítez, J.E.; Corrales, I.; Barceló, J.; Poschenrieder, C. Copper-induced oxidative damage and enhanced antioxidant defenses in the root apex of maize cultivars differing in Cu tolerance. *Environ. Exp. Bot.* **2009**, *67*, 415–420. [CrossRef]
20. Yeh, C.M.; Hung, W.C.; Huang, H.J. Copper treatment activates mitogen-activated protein kinase signalling in rice. *Physiol. Plant.* **2003**, *119*, 392–399. [CrossRef]
21. Zhan, E.; Zhou, H.; Li, S.; Liu, L.; Tan, T.; Lin, H. OTS1-dependent deSUMOylation increases tolerance to high copper levels in Arabidopsis. *J. Integr. Plant Biol.* **2018**, *60*, 310–322. [CrossRef] [PubMed]
22. Clemens, S. Molecular mechanisms of plant metal tolerance and homeostasis. *Planta* **2001**, *212*, 475–486. [CrossRef] [PubMed]
23. Sandalio, L.M.; Dalurzo, H.C.; Gomez, M.; Romero-Puertas, M.C.; Del Rio, L.A. Cadmium-induced changes in the growth and oxidative metabolism of pea plants. *J. Exp. Bot.* **2001**, *52*, 2115–2126. [CrossRef] [PubMed]
24. Li, M.J.; Xu, G.Y.; Xia, X.J.; Wang, M.L.; Yin, X.M.; Zhang, B.; Zhang, X.; Cui, Y.C. Deciphering the physiological and molecular mechanisms for copper tolerance in autotetraploid Arabidopsis. *Plant Cell Rep.* **2017**, *36*, 1585–1597. [CrossRef] [PubMed]
25. Jaleel, C.A.; Riadh, K.; Gopi, R.; Manivannan, P.; Ine's, J.; Al-Juburi, H.J.; Zhao, C.X.; Shao, H.B.; Rajaram, P. Antioxidant defense responses: Physiological plasticity in higher plants under abiotic constraints. *Acta Physiol. Plant.* **2009**, *31*, 427–436. [CrossRef]
26. Maksymiec, W.; Krupa, Z. The effects of short-term exposition to Cd, excess Cu ions and jasmonate on oxidative stress appearing in *Arabidopsis thaliana*. *Environ. Exp. Bot.* **2006**, *57*, 187–194. [CrossRef]
27. Hollenbach, B.; Schreiber, L.; Hartung, W.; Dietz, K.J. Cadmium leads to stimulated expression of the lipid transfer protein genes in barley: Implications for the involvement of lipid transfer proteins in wax assembly. *Planta* **1997**, *203*, 9–19. [CrossRef]
28. Bücker-Neto, L.; Paiva, A.L.S.; Machado, R.D.; Arenhart, R.; Margis-Pinheiro, M. Interactions between plant hormones and HMs responses. *Genet. Mol. Biol.* **2017**, *40*, 373–386. [CrossRef]

29. Fediuc, E.; Lips, S.H.; Erdei, L. O-acetylserine (thiol) lyase activity in Phragmites and Typha plants under cadmium and NaCl stress conditions and the involvement of ABA in the stress response. *J. Plant Physiol.* **2005**, *162*, 865–872. [CrossRef]
30. Stroinski, A.; Chadzinikolau, T.; Gizewska, K.; Zielezinska, M. ABA or cadmium induced phytochelatin synthesis in potato tubers. *Biol. Plant.* **2010**, *54*, 117–120. [CrossRef]
31. Kim, Y.H.; Khan, A.L.; Kim, D.H.; Lee, S.Y.; Kim, K.M.; Waqas, M.; Jung, H.Y.; Shin, J.H.; Kim, J.G.; Lee, I.J. Silicon mitigates heavy metal stress by regulating P-type heavy metal ATPases, *Oryza sativa* low silicon genes, and endogenous phytohormones. *BMC Plant Biol.* **2014**, *14*, 13. [CrossRef] [PubMed]
32. Munzuro, Ö.; Fikriye, K.Z.; Yahyagil, Z. The abscisic acid levels of wheat (*Triticum aestivum* L. cv. Çakmak 79) seeds that were germinated under heavy metal (Hg++, Cd++, Cu++) stress. *Gazi Univ. J. Sci.* **2008**, *21*, 1–7.
33. Rauser, W.E.; Dumbroff, E.B. Effects of excess cobalt, nickel and zinc on the water relations of *Phaseolus vulgaris*. *Environ. Exp. Bot.* **1981**, *21*, 249–255. [CrossRef]
34. Poschenrieder, C.; Gunsé, B.; Barceló, J. Influence of cadmium on water relations, stomatal resistance, and abscisic acid content in expanding bean leaves. *Plant Physiol.* **1989**, *90*, 1365–1371. [CrossRef]
35. Sarkar, N.K.; Kim, Y.K.; Grover, A. Rice sHsp genes: Genomic organization and expression profiling under stress and development. *BMC Genom.* **2009**, *10*, 393. [CrossRef]
36. Siddique, M.; Gernhard, S.; von Koskull-Doring, P.; Vierling, E.; Scharf, K.D. The plant sHSP superfamily: Five new members in *Arabidopsis* thaliana with unexpected properties. *Cell Stress Chaperon.* **2008**, *13*, 183–197. [CrossRef]
37. Kim, K.H.; Alam, I.; Kim, Y.G.; Sharmin, S.A.; Lee, K.W.; Lee, S.H.; Lee, B.H. Overexpression of a chloroplast-localized small heat shock protein OsHSP26 confers enhanced tolerance against oxidative and heat stresses in tall fescue. *Biotechnol. Lett.* **2012**, *34*, 371–377. [CrossRef]
38. Sato, Y.; Yokoya, S. Enhanced tolerance to drought stress in transgenic rice plants overexpressing a small heat-shock protein, sHSP17.7. *Plant Cell Rep.* **2008**, *27*, 329–334. [CrossRef]
39. Ju, Y.; Tian, H.; Zhang, R.; Zuo, L.; Chu, Z. Overexpression of oshsp18.0-CI enhances resistance to bacterial leaf streak in rice. *Rice* **2017**, *10*, 12. [CrossRef]
40. Cui, Y.C.; Xu, M.L.; Li, L.Y.; Wang, M.L.; Xu, G.Y.; Xia, X.J. Expression and cloning of a multiple stress responsive gene (*OsMSR3*) in rice. *J. Wuhan. Bot. Res.* **2009**, *6*, 574–581.
41. Cui, Y.C.; Xu, G.Y.; Wang, M.L.; Yu, Y.; Li, M.J.; da Rocha, P.S.C.F.; Xia, X.J. Expression of *OsMSR3* in *Arabidopsis* enhances tolerance to cadmium stress. *Plant Cell Tiss. Org. Cult.* **2013**, *113*, 331–340. [CrossRef]
42. Kumar, G.; Kushwaha, H.R.; Panjabi-Sabharwal, V.; Kumari, S.; Joshi, R.; Karan, R.; Mittal, S.; Singla Pareek, S.L.; Pareek, A. Clustered metallothionein genes are co-regulated in rice and ectopic expression of OsMT1e-P confers multiple abiotic stress tolerance in tobacco via ROS scavenging. *BMC Plant Biol.* **2012**, *12*, 107. [CrossRef] [PubMed]
43. Wang, R.; Shafi, M.; Ma, J.; Zhong, B.; Guo, J.; Hu, X.; Xu, W.; Yang, Y.; Ruan, Z.; Wang, Y.; et al. Effect of amendments on contaminated soil of multiple heavy metals and accumulation of heavy metals in plants. *Environ. Sci. Pollut. Res.* **2018**, *25*, 28695–28704. [CrossRef] [PubMed]
44. Wang, M.; Li, S.S.; Li, X.Y.; Zhao, Z.Q.; Chen, S.B. An overview of current status of copper pollution in soil and remediation efforts in China. *Earth Sci. Front.* **2018**, *25*, 305–313.
45. De Vos, C.H.R.; Ten Bookum, W.M.; Vooijs, R.; Schat, H.; Dekok, L.J. Effect of copper on fatty acid composition and peroxidation of lipids in the roots of copper tolerant and sensitive *Silenecucubalus*. *Plant Physiol. Biochem.* **1993**, *31*, 151–158.
46. Heath, R.L.; Packer, L. Photoperoxidation in isolated chloroplasts: I. Kinetics and stoichiometry of fatty acid peroxidation. *Arch. Biochem. Biophys.* **1968**, *125*, 189–198. [CrossRef]
47. Nishiyama, R.; Watanabe, Y.; Leyva-Gonzalez, M.A.; Van Ha, C.; Fujita, Y.; Tanaka, M.; Seki, M.; Yamaguchi-Shinozaki, K.; Shinozaki, K.; Herrera-Estrella, L.; et al. Arabidopsis AHP2, AHP3, and AHP5 histidine phosphotransfer proteins function as redundant negative regulators of drought stress response. *Proc. Natl. Acad. Sci. USA* **2013**, *110*, 4840–4845. [CrossRef]
48. Demidchik, V.; Straltsova, D.; Medvedev, S.S.; Pozhvanov, G.A.; Sokolik, A.; Yurin, V. Stress-induced electrolyte leakage: The role of K+-permeable channels and involvement in programmed cell death and metabolic adjustment. *J. Exp. Bot.* **2014**, *65*, 1259–1270. [CrossRef]

49. Rehman, M.; Liu, L.; Wang, Q.; Saleem, M.H.; Bashir, S.; Ullah, S.; Peng, D. Copper environmental toxicology, recent advances, and future outlook: A review. *Environ. Sci. Pollut. Res. Int.* **2019**, *26*, 18003–18016. [CrossRef]
50. Sui, N.; Tian, S.; Wang, W.; Wang, M.; Fan, H. Overexpression of glycerol-3-phosphate acyltransferase from *Suaeda salsa* improves salt tolerance in Arabidopsis. *Front. Plant Sci.* **2017**, *8*, 1337. [CrossRef]
51. Polívka, T.; Frank, H.A. Molecular factors controlling photosynthetic light harvesting by carotenoids. *Acc. Chem. Res.* **2010**, *43*, 1125–1134. [CrossRef] [PubMed]
52. Polívka, T.; Pullerits, T.; Frank, H.A.; Cogdell, R.J.; Sundström, V. Ultrafast formation of a carotenoid radical in LH2 antenna complexes of purple bacteria. *J. Phys. Chem. B* **2004**, *108*, 15398–15407. [CrossRef]
53. Saglam, A.; Yetiddin, F.; Demiralay, M.; Terzi, R. Copper stress and responses in plants. In *Plant Metal Interaction: Emerging Remediation Techniques*; Ahmad, P., Ed.; Elsevier: Amsterdam, The Netherlands, 2016; pp. 21–40. [CrossRef]
54. Zabalza, A.; Galvez, L.; Marino, D.; Royuela, M.; Arrese-Igor, C.; Gonzalez, E.M. The application of ascorbate or its immediate precursor, galactono-1, 4-lactone, does not affect the response of nitrogen-fixing pea nodules to water stress. *J. Plant Physiol.* **2008**, *165*, 805–812. [CrossRef]
55. Thounaojam, T.C.; Panda, P.; Mazumdar, P.; Kumar, D.; Sharma, G.; Sahoo, L.; Panda, S. Excess copper induced oxidative stress and response of antioxidants in rice. *Plant Physiol. Biochem.* **2012**, *53*, 33–39. [CrossRef] [PubMed]
56. Gu, C.S.; Liu, L.Q.; Deng, Y.M.; Zhu, X.D.; Huang, S.Z.; Lu, X.Q. The heterologous expression of the Iris lactea var. chinensis type 2 metallothionein IlMT2b gene enhances copper tolerance in *Arabidopsis thaliana*. *Bull. Environ. Contam. Toxicol.* **2015**, *94*, 247–253. [CrossRef] [PubMed]
57. Mittler, R. Oxidative stress, antioxidants and stress tolerance. *Trends Plant Sci.* **2002**, *7*, 405–410. [CrossRef]
58. Xu, G.Y.; Rocha, P.S.C.F.; Wang, M.L.; Xu, M.L.; Cui, Y.C.; Li, L.Y.; Zhu, Y.X.; Xia, X.J. A novel rice calmodulin-like gene, *OsMSR2*, enhances drought and salt tolerance and increases ABA sensitivity in *Arabidopsis*. *Planta* **2011**, *234*, 47–59. [CrossRef]
59. Xu, M.L.; Chen, R.J.; Rocha, P.S.C.F.; Wang, M.L.; Xu, G.Y.; Xia, X.J. Expression and cloning of a novel stress responsive gene (*OsMsr1*) in rice. *Acta Agron. Sin.* **2008**, *10*, 1712–1718. [CrossRef]
60. Cui, Y.C.; Li, M.J.; Yin, X.M.; Song, S.F.; Xu, G.Y.; Wang, M.L.; Li, C.Y.; Peng, C.; Xia, X.J. OsDSSR1, a novel small peptide, enhances drought tolerance in transgenic rice. *Plant Sci.* **2018**, *270*, 85–96. [CrossRef]
61. Lichtenthaler, H.K.; Wellburn, A.R. Determinations of total carotenoids and chlorophylls a and b of leaf extracts in different solvents. *Biochem. Soc. Trans.* **1983**, *11*, 591e592. [CrossRef]

© 2019 by the authors. Licensee MDPI, Basel, Switzerland. This article is an open access article distributed under the terms and conditions of the Creative Commons Attribution (CC BY) license (http://creativecommons.org/licenses/by/4.0/).

International Journal of
Molecular Sciences

Review

Cadmium and Plant Development: An Agony from Seed to Seed

Michiel Huybrechts, Ann Cuypers, Jana Deckers, Verena Iven, Stéphanie Vandionant, Marijke Jozefczak and Sophie Hendrix *

Environmental Biology, Centre for Environmental Sciences, Hasselt University, B-3590 Diepenbeek, Belgium
* Correspondence: sophie.hendrix@uhasselt.be

Received: 5 July 2019; Accepted: 9 August 2019; Published: 15 August 2019

Abstract: Anthropogenic pollution of agricultural soils with cadmium (Cd) should receive adequate attention as Cd accumulation in crops endangers human health. When Cd is present in the soil, plants are exposed to it throughout their entire life cycle. As it is a non-essential element, no specific Cd uptake mechanisms are present. Therefore, Cd enters the plant through transporters for essential elements and consequently disturbs plant growth and development. In this review, we will focus on the effects of Cd on the most important events of a plant's life cycle covering seed germination, the vegetative phase and the reproduction phase. Within the vegetative phase, the disturbance of the cell cycle by Cd is highlighted with special emphasis on endoreduplication, DNA damage and its relation to cell death. Furthermore, we will discuss the cell wall as an important structure in retaining Cd and the ability of plants to actively modify the cell wall to increase Cd tolerance. As Cd is known to affect concentrations of reactive oxygen species (ROS) and phytohormones, special emphasis is put on the involvement of these compounds in plant developmental processes. Lastly, possible future research areas are put forward and a general conclusion is drawn, revealing that Cd is agonizing for all stages of plant development.

Keywords: cadmium; oxidative stress; cell cycle; cell wall; germination; reproduction; plant growth and development

1. Introduction

Cadmium (Cd) pollution, as a consequence of both geological and anthropogenic activities, affects many regions worldwide [1]. Although Cd is non-essential and no specific Cd uptake mechanisms have been identified in plants, it is taken up in root cells through transporters for essential bivalent cations such as calcium (Ca), iron (Fe), manganese (Mn) and zinc (Zn) [2]. Depending on the plant species and growth conditions, plants can endure low Cd concentrations, but in general Cd disturbs photosynthesis, respiration and the uptake of water and nutrients. As a consequence, Cd pollution negatively affects plant growth and development, thereby significantly reducing crop yield [3].

This Cd-induced phytotoxicity is related to its ability to bind to thiol, histidyl and carboxyl groups of structural proteins and enzymes, thereby interfering with their function. Furthermore, Cd can also disturb protein function by replacing essential ions in their active sites due to its strong chemical similarity with other divalent cations. Despite its non-redox-active nature, Cd indirectly induces the production of reactive oxygen species (ROS), resulting in an oxidative challenge, which is defined as an imbalance between cellular pro- and antioxidants in favor of the former [4,5]. This Cd-induced ROS production is achieved through multiple mechanisms. Firstly, Cd is able to replace Fe in various proteins, thereby increasing free cellular Fe levels. As a redox-active metal, Fe can directly induce ROS production through Fenton and Haber-Weiss reactions [6]. Furthermore, Cd indirectly induces ROS production by depleting cellular levels of the non-enzymatic antioxidant glutathione (GSH), as a consequence of increased phytochelatin (PC) synthesis. The latter contributes to Cd chelation but reduces the amount of GSH available for antioxidative defense [7,8]. In addition, indirect Cd-induced ROS production can result from its ability to inhibit enzymes involved in antioxidative defense mechanisms [8]. Cadmium can also contribute to ROS production through its effects on ROS-producing enzymes such as nicotinamide adenine dinucleotide phosphate (NADPH) oxidases and increases ROS production in plant organelles by interfering with metabolic processes such as photosynthesis, respiration and photorespiration [4].

When present in elevated concentrations, ROS can evoke damage to a multitude of cellular macromolecules including lipids, proteins and DNA. However, ROS are not only damaging agents, but are also key players in signal transduction during physiological processes as well as responses to biotic and abiotic stresses [9–11]. In order to enable ROS-induced signaling and prevent damage to cellular macromolecules, ROS levels should be tightly controlled. To this end, plants have developed an extensive antioxidative defense system consisting of both enzymatic antioxidants such as superoxide dismutase (SOD), catalase (CAT) and several peroxidases and non-enzymatic antioxidants such as the water-soluble GSH and ascorbate (AsA) and the lipid-soluble carotenes and α-tocopherol [9].

Besides interfering with ROS homeostasis, Cd exposure also affects phytohormone signaling [12,13]. It was shown, for example, that Cd exposure induces a fast and transient increase in ethylene levels in *Arabidopsis thaliana* through increasing the expression of 1-aminocyclopropane-1-carboxylate (ACC) synthase 2 (*ACS2*) and *ACS6*, which are involved in the biosynthesis of the ethylene precursor ACC [14]. Furthermore, Cd was shown to enhance ethylene and jasmonic acid (JA) concentrations in *Pisum sativum* [15] and abscisic acid (ABA), salicylic acid (SA) and JA levels in *Oryza sativa* [16]. It also interfered with auxin (AUX) homeostasis in *A. thaliana* seedlings, where it significantly altered AUX concentrations and distribution in primary root tips and cotyledons. It decreased indole-3-acetic acid (IAA) content, enhanced IAA oxidase activity and affected transcript levels of putative AUX biosynthetic and catabolic genes [17]. Similarly, a negative effect of Cd exposure on AUX biosynthesis, transport and distribution was also reported in *O. sativa* [18]. Interestingly, extensive cross-talk exists between ROS and phytohormones in plant development and stress tolerance, as was recently reviewed by Xia and colleagues (2015) [19].

In order to improve plant growth on Cd-polluted soils, for example for phytoremediation purposes [20], it is of crucial importance to increase our knowledge on the mechanisms underlying the negative impact of Cd exposure on plant development. Therefore, the main aim of this work is to provide an overview of Cd-induced effects on plant growth and development, specifically addressing (1) seed dormancy and germination, (2) vegetative plant growth and (3) reproductive plant growth. Important underlying mechanisms are discussed, highlighting the involvement of ROS and phytohormones. Furthermore, perspectives for future research are proposed, focusing on the investigation of transgenerational effects of Cd exposure and the involvement of the plant microbiome in response to Cd stress.

2. Seed Germination

Seed germination, accompanied with a release from seed dormancy, is one of the most important events in a plant's life cycle. It determines whether a plant is willing to take the risk of environmental exposure in order to reach reproductive maturity and produce seeds of its own. In the past, this trait has been exploited in cereals to obtain seeds with a low dormancy level, so that when the seeds were sown onto the field, they would germinate quickly and evenly. This led to the problem known as pre-harvest sprouting, during which seeds already germinate when they are still attached to the mother plant, resulting in a significant reduction of yield and quality of the seeds [21–23].

Abscisic acid and gibberellic acid (GA) are the major phytohormones regulating germination. They act antagonistically, with high levels of ABA causing the preservation of a dormant seed and high levels of GA initiating germination [24,25]. The ABA/GA ratio thereby acts as a central hub integrating environmental signals [26]. However, recent studies also indicate a role for other plant hormones.

Auxin, previously thought not to have an important role in seed germination, acts alongside ABA in keeping dormancy high [27]. Furthermore, AUX and ABA are dependent on each other in this process, as AUX operates by keeping the expression of a major ABA signaling downstream regulator, *ABSCISIC ACID INSENSITIVE 3* (*ABI3*), high [28]. In addition, ABA represses the elongation of the embryonic axis through AUX-regulated signaling [29]. Another hormone, ethylene, works in the opposite way. Ethylene is able to stop the inhibitory effect of ABA on endosperm weakening, thereby facilitating seed germination [30,31]. Several other phytohormones including brassinosteroids, cytokinins (CK), JA, strigolactones and SA, have been shown to either stimulate or inhibit germination. However, their mutual interactions and importance in this process are still unclear [27].

Next to the hormonal regulation of seed germination, the event is also characterized by an increase in ROS levels, especially hydrogen peroxide (H_2O_2) [32]. Reactive oxygen species are assumed to function upstream of the hormonal interactions by stimulating GA biosynthesis and metabolism and inducing ABA catabolism [33,34]. However, Bahin and colleagues (2011) stated that ROS mainly function through GA signaling since H_2O_2 did not influence ABA metabolism and signaling in *Hordeum vulgare* seeds [35]. On the other hand, ABA was shown to prevent the accumulation of ROS within the embryonic axis of the seeds [33]. Currently, the interactions between ROS and ABA/GA are still under debate [36]. The role of ROS in seed germination is dual and is defined as the oxidative window of germination [37]. Too low concentrations of ROS within the seed will fail to induce seed germination, whereas excessive ROS levels will lead to irreversible seed damage. In dry seeds, ROS levels tend to be rather low [37]. This is accomplished by maintaining a high level of antioxidant capacity scavenging ROS molecules. Depletion of antioxidants leads to a ROS increase, which initiates germination under favorable conditions [38]. Once H_2O_2 is elevated, it has the capacity to selectively oxidize mRNAs and proteins [39]. In *A. thaliana* seeds, ROS-induced carboxylation of 12S cruciferins, the major storage proteins, occurs. Upon seed imbibition, these oxidized molecules disappear rapidly, indicating their role in early seedling establishment [40]. A similar mechanism of proteome oxidation was found in sunflower seeds [41]. Furthermore, upon exposure to methylviologen, a ROS-inducing agent, genes involved in Ca and redox signaling were differentially expressed [33]. Hou and colleagues (2019) found that three major events were important in releasing *Leymus chinensis* seeds from dormancy [42]. A decrease in proteins related to AsA and aldarate metabolism accompanied with an increase of thioredoxins (Trx) changed the antioxidant system and led to an increase in ROS. This was followed by the oxidation of stored mRNAs and proteins. Furthermore, an increase in β-tubulin led to cytoskeleton changes and resulted in physical dormancy release by transporting substances related to germination and cell wall loosening. Thirdly, these cytoskeleton changes in turn affected chromatin remodeling and proteins [42].

Surrounded by a rigid seed coat, the plant embryo is well protected from environmental stresses [43,44]. Germination is initiated with the uptake of water followed by embryonic expansion [45]. Metabolic reactivation accompanied with metabolic respiration is characterized by a steep increase in oxygen consumption just after imbibition of the seed. Furthermore, protein synthesis, DNA repair and remobilization of stored reserves are essential processes in the successful germination of the seeds [46]. Physical constraints are imposed by outer seed tissue, and seed germination involves rupture of the testa and the endosperm [47]. Upon imbibition of the seed, the testa becomes more permeable over time and Cd content begins to increase in inner seed tissues [48]. Thereby, genotypic variations of seed coat permeability might be an important factor contributing to the effects of metals on seed germination [49,50]. Subsequently, Cd-inhibited seed germination occurs in a dose-dependent manner [51–53]. Seeds of *Trigonella foenum-graecum* exposed to solutions of chromium (Cr), lead (Pb) and Cd showed that Cd had the strongest germination inhibition effect at 10 mg L^{-1}, which was the highest concentration tested [54]. Likewise, in *Triticum aestivum*, less Cd than Pb was needed to inhibit the seed germination process [55]. Nevertheless, large differences can occur between plant species. Ahsan and colleagues (2007) reported that adding 1 mM of $CdCl_2$ to the solution completely inhibited seed germination of the *O. sativa* cultivar Hwayeong [56]. Seeds of *H. vulgare* were apparently more tolerant to Cd and their germination rate was only fully inhibited around 9.5 mM $CdCl_2$ [57]. In addition, Cd tolerance of germination might differ largely within one plant species, as was shown within both *T. aestivum* and *O. sativa* cultivars [49,51]. Some cultivars still germinate vigorously with high Cd concentrations, while others will fail. Interestingly, under the threshold of 0.5 mM $CdCl_2$, most tested *O. sativa* cultivars showed an increased germination compared to the control seeds [49]. A similar observation was made by Lefèvre and colleagues (2009) with *Dorycnium pentaphyllum* Scop. seeds [58]. Here, adding 10 µM $CdCl_2$ also significantly increased germination compared to control seeds, while after exposure to 1 mM $CdCl_2$ for 17 days, no more than 40% germination was obtained. These last findings could be indicative for seed germination stimulation at low Cd concentrations, i.e., hormesis, which are undoubtedly dependent on the plant species examined.

Cadmium is known to inhibit seed germination through different mechanisms (Figure 1). In *Vigna unguiculata* seeds, the inhibitory effect of Cd was proposed to be due to an impairment of water uptake, thereby limiting the water availability for the developing embryo [59]. A limited water supply is not the only problem for proper germination. An inhibition of starch mobilization from the endosperm accompanied with an impaired translocation of soluble sugars to the embryonic axis can lead to further starvation of this embryonic axis [50]. A reduction of hydrolyzing enzymes, such as α-amylase, proteases and acid phosphatases, in *Sorghum bicolor* seeds was suggested to be responsible for this reduced storage mobilization. A decrease in α-amylase activity has been reported multiple times in relation to a decrease in starch release from cotyledons [50,60,61]. He and colleagues (2008) pointed out that Ca is a vital element for amylase activity and the replacement of the chemically similar Cd ion could disrupt normal enzyme functioning [62]. Furthermore, in radish seeds, a direct competition for Ca-calmodulin binding sites occurred between Ca and Cd ions [63]. The interaction between Ca and calmodulin is suggested to serve a role in metabolic activation during the early phases of seed germination [64]. An alteration in the remobilization process is also observed in *Vicia faba* seeds by the leakage of soluble sugars and amino acids into the imbibition medium, which is probably related to the loss of membrane integrity [65]. An increase of malondialdehyde content is observed in Cd-exposed *P. sativum* embryos, which might indicate membrane lipid peroxidation [66].

Cadmium is a known inducer of oxidative stress resulting in elevated ROS levels [9,15,67]. Cadmium-induced oxidative stress was able to oxidize Trx isoforms in *P. sativum* seeds [68]. These proteins are potentially involved in monitoring the redox state of storage proteins in both cereals and dicotyledons [69]. Furthermore, the GSH levels were twofold lower, accompanied with a decrease in glutathione reductase (GR) activity, which suggests that intracellular oxidative stress occurred in seeds under Cd exposure [70]. This depletion of the reduced GSH pool might be partially compensated by a higher level of glutaredoxin (Grx) level, which is able to bind Cd at its active site. Peroxiredoxin (Prx) expression was elevated under Cd exposure in both cotyledons and embryonic axes of *P. sativum* seeds. Cadmium could bind to the cysteine residues of Prx, which could serve as a Cd sink [71]. This mechanism might protect the seed from Cd-induced oxidative stress.

Studies that link the interaction between Cd and phytohormones during seed germination are still scarce. Treatment with ethylene is shown to have an alleviating role on Cd-inhibited germination in *Cajanus cajan* [72]. In *O. sativa*, the α-amylase activity is enhanced when Cd-exposed seeds are pre-treated with SA and seedlings show a reduced Cd uptake [73]. In conclusion, seed germination comes forward as a tightly regulated orchestra between hormones and the ROS balance. Within the oxidative window of seed germination hypothesis, one might assume that a small amount of Cd could stimulate the initiation of germination as long as the adverse effects of Cd are not impossible to overcome by the plant's antioxidant defense mechanisms.

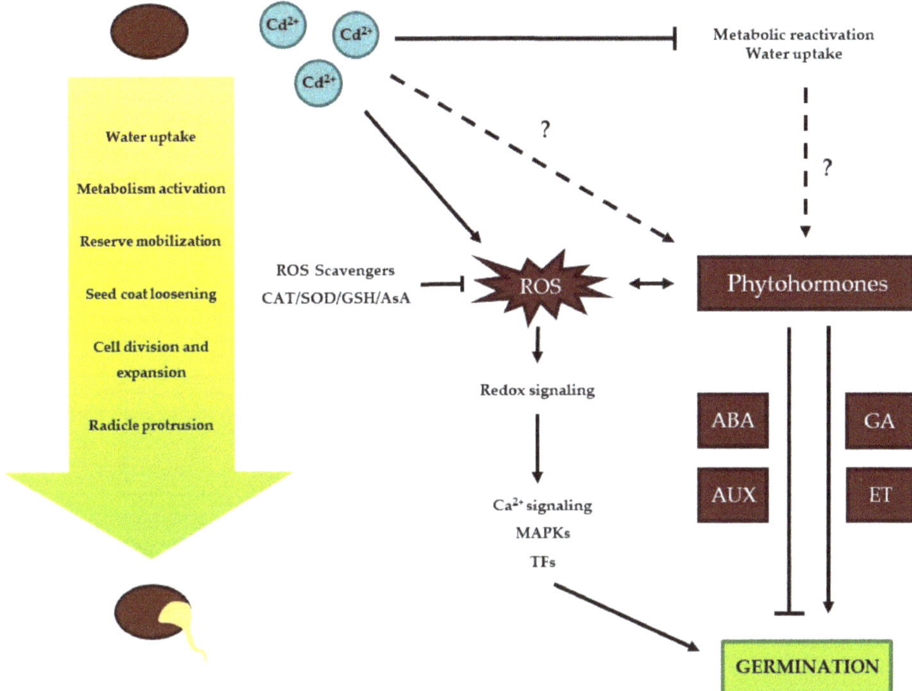

Figure 1. Possible interference mechanisms of cadmium on the process of seed germination. Cadmium (Cd) negatively affects metabolic reactivation by reducing levels of hydrolyzing enzymes, starch mobilization and seed imbibition. Furthermore, it can alter redox signaling via calcium (Ca), mitogen-activated protein kinases (MAPKs) and transcription factors (TFs) and the level of phytohormones such as abscisic acid (ABA), auxin (AUX), giberrellic acid (GA) and ethylene (ET). Both are of major importance in the seed germination process. One-way arrows: indicate a stimulating effect, whereas T-shaped arrows represent an inhibitory effect. Two-way arrows signify an interaction and dashed lines indicate effects which are still uncertain.

3. Vegetative Plant Growth

Once germination has occurred, two major processes drive plant growth, namely cell division and cell expansion, which is limited by the cell wall during vegetative growth. In the following sections, the effect of Cd on these processes is uncovered in detail (Figure 2).

3.1. The DNA Damage Response

Stress-induced effects on cell cycle progression often result from the activation of the DNA damage response (DDR). Upon perceiving DNA damage, cells trigger this response, which includes the activation of DNA repair pathways. When the extent of DNA damage is low, cell cycle progression is transiently inhibited in order to repair the DNA before DNA replication or cell division take place. When the damaged DNA cannot be repaired, cells undergo terminal differentiation or programmed cell death (PCD) [74].

The induction of cell cycle arrest upon the perception of DNA damage requires the activation of one of two phosphatidylinositol-3-OH-kinase-like kinases: ataxia telangiectasia mutated (ATM) and ATM- and RAD3-related (ATR). Whereas ATM is mainly activated by the presence of DNA double-strand breaks (DSBs), ATR is involved in responses to stalled replication forks. However, both types of DNA damage often occur simultaneously, causing the activation of both kinases [74].

Whereas cell cycle regulation in response to DNA damage depends on p53 in animals, plants lack a p53 orthologue. Instead, SUPPRESSOR OF GAMMA RESPONSE 1 (SOG1) is considered as the plant counterpart of p53. This transcription factor belongs to the NO APICAL MERISTEM/*ARABIDOPSIS* TRANSCRIPTION ACTIVATION FACTOR/CUP-SHAPED COTYLEDON (NAC) domain family and is activated through phosphorylation by ATM. Once active, SOG1 induces the expression of a multitude of genes involved in DNA repair, cell cycle progression and cell death [74,75]. In addition to SOG1, also WEE1 plays a key role in the plant DDR. This kinase can be activated by both ATM and ATR and mainly controls S phase progression [74].

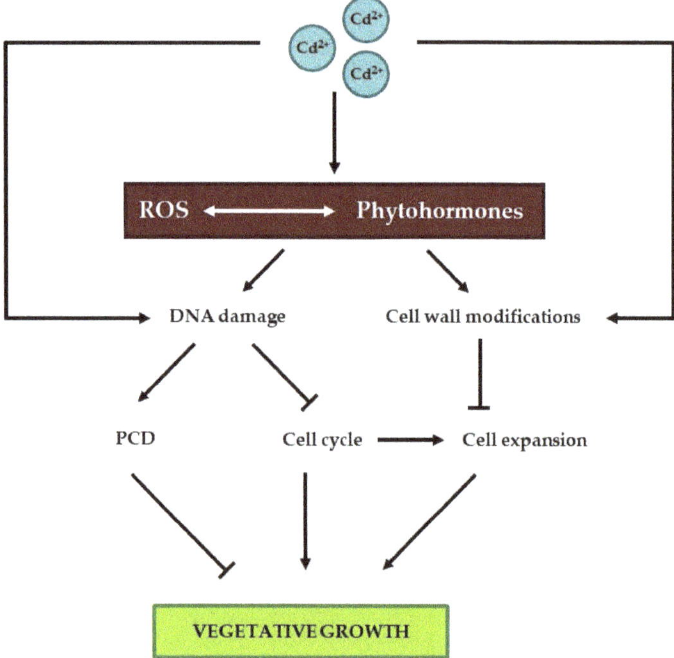

Figure 2. Schematic overview of important players affecting vegetative plant growth upon Cd exposure. Cadmium exposure is well known to affect concentrations of reactive oxygen species (ROS) and phytohormones, which are closely intertwined. Cadmium induces DNA damage, thereby activating the DNA damage response, which can either induce programmed cell death (PCD) or affect cell cycle progression, depending on the extent of DNA damage. In addition, Cd exposure induces cell wall modifications as a strategy to reduce Cd entry into cells. This in turn limits cell expansion, which is intertwined with the cell cycle and specifically endoreduplication. Cadmium-induced DNA damage and cell wall modifications could either result from its effects on ROS and phytohormone levels or arise through an alternative pathway. Together, Cd-induced PCD, cell cycle alterations and inhibition of cell expansion contribute to its negative effect on vegetative growth. One-way arrows indicate stimulating effects, whereas T-shaped arrows represent an inhibitory effect. Two-way arrows signify an interaction.

3.1.1. DNA Damage

Cadmium exposure is well known to induce DNA damage in mammalian cells, as reviewed by Bertin and Averbeck (2006) and Filipic (2012) [76,77]. The mechanisms responsible for this Cd-induced DNA damage include ROS-induced formation of 8-hydroxyguanosine and the inhibition of DNA repair systems. Furthermore, Cd is able to interfere with proteins containing a zinc finger motif, implicated in the maintenance of genome stability, DNA repair and DNA damage signaling [76,77].

At present, in-depth knowledge on Cd-induced DNA damage and its underlying mechanisms in plants is lacking. However, many studies have demonstrated that Cd exposure induces DNA damage in multiple plant species. Table 1 provides an overview of recently published research (since 2014) demonstrating different types of Cd-induced DNA damage in a broad range of plant species. Several authors reported that Cd exposure increases the percentage tail length, tail intensity and tail moment determined through single cell gel electrophoresis, also known as the comet assay (Table 1). The detection of DNA damage using this method relies on the fact that DNA strand breaks facilitate DNA migration from the nucleoids towards the anode during gel electrophoresis, thereby forming a "comet tail". Whereas the alkaline comet assay (performed at pH > 13) detects both DNA single-stranded breaks (SSBs) and DSBs, only DSBs are detected through the neutral comet assay [78,79]. Taken together, the results obtained through this method suggest that Cd exposure induces DNA strand breaks in different plant species including *V. faba* [80–82], *Nicotiana tabacum* [83,84], *Allium sativum* [82], *Solanum tuberosum* [85], *Allium cepa* [80,86,87], *Lemna minor* [88,89], *Lactuca sativa* [86,90], *H. vulgare* [91], *Brassica oleracea* and *Trifolium repens* [92]. In *S. tuberosum*, the Cd-induced increase in percentage tail DNA was more pronounced in roots as compared to leaves and appeared later in the latter organ [85]. Similarly, Cd exposure for 24 and 72 h caused significant increases in the percentage tail DNA and the tail moment in *N. tabacum* roots, whereas no effects were observed in leaves [83]. This is likely due to the fact that roots form the entry route for Cd into the plant and are therefore exposed earlier than leaves. Furthermore, roots of both plant species were shown to accumulate higher Cd levels than leaves [83,85].

In addition to inducing DNA strand breaks, Cd is also shown to induce micronucleus formation and chromosomal aberrations, which are both detected through microscopic analysis (Table 1). Micronuclei arise when chromosome fragments or entire chromosomes fail to be included in the nuclei of the daughter cells at the end of mitosis, as they do not manage to properly attach to the mitotic spindle during anaphase. They eventually become enclosed by a nuclear membrane and appear as small nuclei after conventional nuclear staining. Micronuclei are generally considered as biomarkers for genotoxicity [93,94]. Chromosomal aberrations frequently detected during mitosis in Cd-exposed plants include anaphase and telophase bridges, sticky chromosomes, chromosome breaks, non-oriented chromosomes and laggards chromosomes [80,86,95,96]].

Another approach frequently used to determine Cd-induced genotoxicity in plants is the investigation of random amplified polymorphic DNA (RAPD) profiles. This method consists of a PCR amplification of genomic DNA using multiple short (approximately 10 nucleotides) random primers, followed by gel electrophoresis and visualization of the amplified PCR products. Mutations and other types of DNA damage induced by exposure to stress factors such as Cd can result in the disappearance as well as *de novo* creation of primer annealing sites, thereby yielding an altered RAPD profile [97]. This subsequently allows for calculation of the genomic template stability (GTS) using the following formula: GTS (%) = $(1 - a/n) \times 100$, where "a" represents the number of polymorphic bands in the treated samples and "n" identifies the total number of bands in the control samples [98,99]. As indicated in Table 1, Cd exposure was shown to alter the RAPD profile in both roots and leaves of a broad range of plant species, resulting in a reduced GTS.

Table 1. Overview of recent research articles (published since 2014) demonstrating Cd-induced DNA damage, arranged by plant species. Cadmium is shown to induce different types of DNA damage, including DNA strand breaks, chromosomal aberrations and micronuclei in different plant species. Furthermore, it alters the expression of DNA repair genes and changes amplified fragment length polymorphism (AFLP), inter-simple sequence repeat (ISSR), random amplified polymorphic DNA (RAPD), sequence-related amplified polymorphism (SRAP) and simple sequence repeat (SSR) profiles, thereby reducing the genomic template stability (GTS). ↑ and ↓ symbols indicate increases and decreases, respectively.

Species	Organ	Cd Concentration	Exposure Duration	Effect	Detection Method	Reference
Allium cepa	Root tip	50–200 µM	2 h + 24 h recovery	Micronucleus formation Chromosomal aberrations	Microscopic analysis	Arya and Mukherjee, 2014 [80]
				% tail DNA ↑	Comet assay (alkaline)	
	Root tip	25 µM	48 h	Chromosomal aberrations Micronucleus formation	Microscopic analysis	Silveira et al., 2017 [86]
				% DNA damage ↑	Comet assay (alkaline)	
	Root tip	25 µM	48 h	Number of nucleoli ↑	Microscopic analysis	Lima et al., 2019 [100]
Arabidopsis thaliana	Root tip	0.125–2.5 mg L^{-1}	5 d	Altered expression DNA repair genes	qRT-PCR	Cui et al., 2017 [101]
	Root	1.25–4 mg L^{-1}	5 d	Altered RAPD profile	RAPD	Cao et al., 2018 [102]
				Altered expression DNA repair genes	qRT-PCR	
	Leaf	0.5–5 mg L^{-1}	16 d	Altered AFLP profile	AFLP	Li et al., 2015 [103]
	Leaf	0.25–8 mg L^{-1}	15 d	Microsatellite instability	SSR	Wang et al., 2016 [104]
				Altered RAPD profile	RAPD	
	Leaf	5 µM	72 h	Altered expression DNA repair genes	qRT-PCR	Hendrix et al., 2018 [105]
Brassica chinensis	Leaf	15–120 mg kg^{-1} soil	30 d	Altered RAPD profile	RAPD	Sudmoon et al., 2015 [106]
Brassica oleracea	Root	2.5–20 mg kg^{-1} soil	3–56 d	Altered % tail intensity	Comet assay (alkaline)	Lanier et al., 2019 [92]
Capsicum annuum	Root tip	20–100 ppm	24 h	Chromosomal aberrations	Microscopic analysis	Aslam et al., 2014 [96]
	Leaf	20–100 ppm	24 h	Altered RAPD profile	RAPD	
Hordeum vulgare	Root tip	75–225 µM	7 d	Altered RAPD profile (GTS ↓)	RAPD	Cenkci and Dogan, 2015 [98]
	Leaf	5 µM	15 d	DNA damage ↑	Comet assay (alkaline)	Cao et al., 2014 [91]
Ipomoea aquatica	Entire seedling	15–120 mg kg^{-1} soil	21 d	Altered RAPD profile (GTS ↓)	RAPD	Tanee et al., 2016 [99]
Lactuca sativa	Root tip	25 µM	48 h	Chromosomal aberrations Micronucleus formation	Microscopic analysis	Silveira et al., 2017 [86]
				% DNA damage ↑	Comet assay (alkaline)	
Lathyrus sativus	Root tip	5–50 µM	3–7 d	Chromosomal aberrations Micronucleus formation	Microscopic analysis	Talukdar, 2014 [95]
Leucaena leucocephala	Leaf	50 mg L^{-1}	15 d	Altered RAPD profile	RAPD	Venkatachalam et al., 2017 [107]
Nicotiana tabacum	Root and leaf	10–15 µM	7 d	% tail DNA ↑	Comet assay (alkaline)	Tkalec et al., 2014 [84]
Oryza sativa	Root tip	50–200 µM	48–96 h	Altered SRAP profil (GTS ↓)	SRAP	Zhang et al., 2015 [108]
Sphagnum palustre	Shoot	0.1–10 µM	24–48 h	Altered ISSR profile (GTS ↓)	ISSR	Sorrentino et al., 2017 [109]
Trifolium repens	Root	2.5–20 mg kg^{-1} soil	3–56 d	Altered % tail intensity	Comet assay (alkaline)	Lanier et al., 2019 [92]
	Root and leaf	20–60 mg kg^{-1} soil	2 weeks	Altered RAPD profile	RAPD	Ghiani et al., 2014 [110]
	Leaf	2.5–20 mg kg^{-1} soil	3–56 d	Tail moment ↑	Comet assay (alkaline)	Lanier et al., 2016 [111]
Urtica pilulifera	Root tip	100–200 µM	2 months	Altered RAPD profile	RAPD	Dogan et al., 2016 [112]
Vicia faba	Root tip	50–200 µM	2 h + 24 h recovery	Micronucleus formation Chromosomal aberrations	Microscopic analysis	Arya and Mukherjee, 2014 [80]
				% tail DNA ↑	Comet assay (alkaline)	

Other markers used to assess Cd-induced DNA damage are the full peak coefficient of variation determined via flow cytometric analysis [113] and DNA polymorphisms determined by PCR-based methods (besides RAPD) such as amplified fragment length polymorphism (AFLP), simple sequence repeat (SSR), inter-simple sequence repeat (ISSR) and sequence-related amplified polymorphism (SRAP) [103,104,108,109]]. However, in comparison to the other DNA damage indicators described, the use of these markers in studies evaluating Cd genotoxicity in plants is currently limited.

Despite the large number of studies indicating Cd-induced genotoxicity in plants, its underlying mechanisms are still largely unknown. However, as Cd exposure is well known to induce ROS production, it is likely that DNA oxidation contributes to the observed damage. Therefore, it would be of interest to study the extent of oxidative DNA damage in Cd-exposed plants. Furthermore, it would be interesting to assess the extent of DNA repair in Cd-exposed plants to gain an insight into plant responses to the observed Cd-induced damage. Although methods for the assessment of oxidative DNA damage and DNA repair are widely adopted in animal samples, these methods have so far received little attention in plant research. However, their optimization might significantly enhance our knowledge on DNA damage and repair induced by Cd and other stress factors in plants.

3.1.2. The Cell Cycle

As a consequence of inducing DNA damage, Cd exposure can affect cell cycle progression. The formation of new cells through cell division is the primary driving force for organ growth in plants. Similar to that in other eukaryotes, the classical plant cell cycle consists of four phases: gap 1 (G_1) phase, DNA synthesis (S) phase, gap 2 (G_2) phase and mitotic (M) phase [114]. The gap phases enable the control of accurate and full completion of previous phases. As a consequence, many important regulatory mechanisms controlling cell cycle progression operate at the G_1/S and G_2/M transitions. During the S phase, nuclear DNA is replicated, whereas the replicated sister chromatids are divided over the two daughter cells arising through cytokinesis during the M phase. The cell cycle is regulated by the activity of cyclin-dependent kinases (CDKs), which are serine/threonine protein kinases that phosphorylate target proteins crucial for cell cycle progression. As their name implies, CDKs form heterodimers with regulatory cyclins in order to become activated [115]. During progression throughout the cell cycle, CDK activity shows a typical pattern, reaching two thresholds: one for DNA replication (S phase) and one for cell division (M phase) [116].

In addition to dividing through the classical cell cycle, plant cells can also undergo endoreduplication. During this alternative cell cycle mode, plant cells replicate their nuclear DNA (S phase) without intermittent cell division (M phase), resulting in endopolyploidy (*i.e.*, the existence of different ploidy levels in adjacent cells of a species) [117].

Similar to the classical cell cycle, endoreduplication is also regulated by the action of CDK-cyclin complexes. During an endocycle, CDK activity only reaches the threshold for DNA replication but not for cell division. Endoreduplication is important for normal plant growth and development, as it is tightly related to cell differentiation and expansion [116]. The importance of endoreduplication in plant development and its connection to cell size become apparent during trichome development. Trichomes are large, single epidermal cells that develop on most aerial parts of *A. thaliana* plants and typically contain three to four branches. These highly specialized structures—involved in plant protection against external stress factors such as herbivory, frost and ultraviolet radiation—require endoreduplication for their development and reach a final DNA content of 32C, with C representing the haploid DNA content [118,119]. Interestingly, trichomes were previously shown to accumulate Cd, possibly to prevent Cd-induced damage at more sensitive sites within the plant [120,121]. In addition to its involvement in plant growth and development, endoreduplication could also serve as a strategy in plant defense against biotic and abiotic stress factors, as reviewed by Scholes and Paige (2015) [122]. Under stress conditions, an increased ploidy level and therefore a higher number of DNA templates could help to sustain genome integrity and stimulate genetic pathways responsible for plant defense [122].

Cadmium-induced disturbances of root and leaf growth were shown to coincide with effects on cell division in a wide range of plant species in a multitude of experimental set-ups. An overview of recently published research (since 2014) demonstrating Cd-induced effects on multiple cell cycle-related parameters in several plant species is provided in Table 2. In general, Cd negatively affects cell cycle progression, as becomes apparent from a decreased mitotic index (i.e., the ratio between the number of cells undergoing mitosis and the total cell number), determined via microscopic analysis. As suggested by Monteiro et al. (2012), Cd might bind to the sulfhydryl groups of cysteine residues present in tubulins, thereby affecting microtubule formation and disturbing cell division [90]. Although many studies have addressed the influence of Cd exposure on the cell cycle in roots, knowledge regarding its effects on this process in leaves is scarce. However, Baryla et al. (2001) demonstrated that leaves of Cd-exposed *Brassica napus* plants contained a smaller number of stomatal guard cells and were characterized by a larger mesophyll cell size as compared to their control counterparts, suggesting that Cd also affects the cell cycle in leaves [123]. Similarly, Cd exposure inhibited cell division and increased cell size in both the upper and lower cell layers of young *Elodea canadensis* leaves. Interestingly, this response coincided with a strong Cd-induced disturbance of the typical cell wall structure, which was likely the consequence of the accumulation of large amounts of Cd in the apoplast and the binding of Cd ions to cell wall components [124] (*cfr. infra*). Furthermore, Cd exposure was shown to reduce both adaxial pavement cell number and surface area in different leaves of *A. thaliana* in a time-dependent manner. The decreased cell surface area might be related to a lower extent of endoreduplication in leaves of Cd-exposed plants, as indicated by a significantly decreased endoreduplication factor (i.e., the average number of endocycles that has taken place per cell) [105]. In contrast, other studies reported a decreased percentage of cells with a 2C nuclear DNA content and an increased percentage of cells with a higher nuclear DNA content in roots of *A. thaliana* [101,102] and *P. sativum* [125–127]. Although an increased proportion of 4C cells might indicate a cell cycle arrest in G_2 phase, an increased level of 8C cells points towards an elevated extent of endoreduplication. These data suggest that Cd exposure stimulates this alternative cell cycle variant in roots, whereas inhibiting it in leaves. However, in young *A. thaliana* leaves, an increased extent of endoreduplication was observed shortly after the start of Cd exposure [Hendrix et al., personal communication]. It could be hypothesized that increased ploidy levels confer tolerance to Cd stress, since Talukdar (2014) demonstrated that roots of tetraploid and triploid *Lathyrus sativus* plants were less sensitive to Cd as compared to their diploid counterparts [95]. Furthermore, it is tempting to speculate that the increased extent of endoreduplication is related to a stimulation of trichome development, as these structures are characterized by a high nuclear DNA content [118,119] and provide sites for Cd sequestration [120,121]. However, this hypothesis requires further investigation.

Cadmium-induced effects on the cell cycle in roots and leaves often coincide with alterations in the expression of cell-cycle related genes, such as those encoding cyclins and CDKs (Table 2). Indeed, *CYCB1* expression decreased upon Cd exposure in *Glycine max* cell suspension cultures, possibly affecting G_2/M progression [128,129]. Exposure to 100 μM Cd for 15 days affected the expression of a large number of cell cycle-related genes in roots of *O. sativa*. Interestingly, Cd-induced effects on a number of these genes were altered by simultaneous treatment with 2,3,5-triiodobenzoic acid (an inhibitor of polar AUX transport), indole-3-butytric acid (an AUX hormone), Tiron (a superoxide ($O_2^{\bullet-}$) scavenger) or sodium diethyldithiocarbamate (an SOD inhibitor) [130]. In another study, the authors demonstrated that treatment with ABA or tungstate (an ABA inhibitor) also influenced Cd-induced effects on transcript levels of certain genes involved in cell cycle regulation [131]. These data emphasize the involvement of plant hormones and ROS in Cd-induced effects on the cell cycle. The significance of various phytohormones and their complex interactions in regulating cell division and endoreduplication was extensively reviewed by Tank et al. (2014) [132]. Although the precise role of ROS in cell cycle regulation is still unclear, their importance in this process is illustrated by the fact that oxidative and reductive signals are required for certain cell cycle transitions. In addition, transcript levels and activities of cyclins and CDKs were demonstrated to be altered upon redox perturbations. The antioxidative metabolite GSH might constitute an important redox-related cell cycle regulator, as it was reported to be translocated into the nucleus during cell division [133] and the severely GSH-deficient *root meristemless 1* (*rml1*) *A. thaliana* mutant is unable to form a root apical meristem due to a lack of cell division [134]. Furthermore, ROS and redox homeostasis are also crucial mediators of cytokinesis, as indicated by a disturbed cell division in mutants characterized by impaired ROS production or signal transduction [135,136]. The involvement of ROS and redox regulators in the cell cycle is discussed in more detail in a recent review by Mhamdi and Van Breusegem (2018) [137].

Additional evidence for Cd-induced effects on transcript levels of cell cycle-related genes comes from *A. thaliana*. In roots of this species, Cd exposure caused a downregulation of several G_1/S marker genes such as *HISTONE H4* and *E2Fa* and G_2/M marker genes, including *CYCB1;1* and *CYCB1;2*. This response coincided with an altered expression of DNA repair genes and was less pronounced in mutants with an impaired DNA mismatch repair pathway, suggesting that DNA damage contributes to the observed cell cycle arrest [101,102]. In agreement with this hypothesis, Hendrix et al. (2018) demonstrated significant increases in the expression of several DNA repair genes and genes encoding CDK inhibitors of the SIAMESE-related (SMR) family in leaves of *A. thaliana* exposed to 5 μM Cd for 72 h [105]. Interestingly, these *SMR* genes were previously reported to be transcriptionally upregulated in response to ROS-induced DNA damage in plants exposed to hydroxyurea [138], further pointing towards the involvement of DNA damage in the Cd-induced cell cycle inhibition.

Table 2. Overview of recent research articles (published since 2014) demonstrating Cd-induced effects on cell cycle-related parameters, arranged by plant species. Cadmium exposure is shown to reduce the mitotic index (i.e., the ratio between the number of cells undergoing mitosis and the total cell number), alter nuclear ploidy levels and affect the expression of cell cycle-related genes in different plant species. ↑ and ↓ symbols indicate increases and decreases, respectively. EdU: 5-ethynyl-2′-deoxyuridine; FCM: flow cytometry; qRT-PCR: quantitative reverse transcription polymerase chain reaction; RT-PCR: reverse transcription polymerase chain reaction.

Species	Organ	Cd Concentration	Exposure Duration	Effect	Detection Method	Reference
Allium cepa	Root tip	50–200 µM	2 h + 24 h recovery	Mitotic index ↓	Microscopic analysis	Arya and Mukherjee, 2014 [80]
	Root tip	25 µM	48 h	Mitotic index ↓	Microscopic analysis	Silveira et al., 2017 [86]
Arabidopsis thaliana	Root tip	0.125–2.5 mg L^{-1}	5 d	2C ↓, 4C ↑, 8C ↑ Altered cell cycle phase distribution	FCM	Cui et al., 2017 [101]
				Altered expression cell cycle-related genes	qRT-PCR	
	Root	1.25–4 mg L^{-1}	5 d	2C ↓, 4C ↑	FCM	Cao et al., 2018 [102]
				Altered expression of cell cycle-related genes	qRT-PCR	
	Leaf	5 µM	3–12 d	Endoreduplication factor ↓	FCM	Hendrix et al., 2018 [105]
				Epidermal cell number and cell surface area ↓	Microscopic analysis	
				Altered expression of cell-cycle related genes	qRT-PCR	
Capsicum annuum	Root tip	20–100 ppm	24 h	Mitotic index ↓	Microscopic analysis	Aslam et al., 2014 [96]
Lactuca sativa	Root tip	25 µM	48 h	Mitotic index ↓	Microscopic analysis	Silveira et al., 2017 [86]
Lathyrus sativus	Root tip	5–50 µM	3–7 d	Mitotic index ↓	Microscopic analysis	Talukdar, 2014 [95]
Oryza sativa	Root	200 µM	7 d	Cortex cell length in elongation zone ↓ Cortex cell number in elongation zone ↓	Microscopic analysis	Zhao et al., 2014 [131]
			7–11 d	Altered expression of cell cycle-related genes	RT-PCR	
Sorghum bicolor	Root tip	50–200 µM	5 d	Inhibition of S phase progression	EdU assay	Zhan et al., 2017 [139]

3.1.3. Cell Death

In case DNA damage is severe and cannot be repaired, the DDR is responsible for activating PCD [74]. Plant PCD is defined as a genetically encoded and actively controlled form of cellular suicide and is generally subdivided into developmental PCD and environmentally induced PCD. Developmental PCD plays a crucial role in normal plant growth and development, being involved in seed development and germination, as well as vegetative development. During the latter, PCD is for example involved in the differentiation of xylem tracheary elements, the emergence of lateral and adventitious roots and in senescence [140]. It is well-known that ROS and NO are important regulators of both developmental and environmentally induced plant PCD, as was recently reviewed by Locato et al. (2016) [140].

Interestingly, Cd-induced DNA damage and cell cycle effects observed in different plant species often coincide with a decrease in cell viability. For example, Kuthanova et al. (2008) demonstrated that exposure to 50 μM Cd affected cell cycle progression in synchronized tobacco BY-2 cell cultures and significantly reduced cell viability [141]. The type of cell death induced upon Cd exposure strongly depends on the cell cycle phase at which Cd exposure started. Whereas Cd application during S and G_2 phase resulted in an apoptosis-like PCD type, characterized by DNA fragmentation, no such effect was observed in cells exposed to Cd during the other cell cycle phases. Instead, cells exposed during M phase displayed a rapid cell death, coinciding with fragmented late telophasic nuclei, while the decrease in viability of cells treated with Cd at G_1 phase took place at a slower rate [141]. Similarly, the decreased mitotic index and chromosomal aberrations observed in *A. cepa* root tips upon exposure to a range of Cd concentrations was accompanied by a strong increase in Evans blue staining, indicative of cell death [80]. However, no effect on cell viability was observed in root tips of *T. aestivum* exposed to a range of Cd concentrations for 48 h, although these conditions induced oxidative modifications of cell cycle-related proteins and cell cycle arrest [142]. These data emphasize that the occurrence of Cd-induced cell death depends on many factors including the plant species, Cd concentration and exposure duration.

The Cd-induced PCD response in tobacco BY-2 was shown to depend on three ROS waves: (1) an initial transient NADPH oxidase-dependent increase in H_2O_2 levels, (2) the accumulation of mitochondrial $O_2^{\bullet-}$ and (3) fatty acid hydroperoxide accumulation, which coincides with the occurrence of cell death [143]. The importance of ROS in Cd-induced PCD is also supported by the work from Tamás et al. (2017) [144], who showed that cell death in roots of Cd-exposed *H. vulgare* is mostly pronounced at locations where $O_2^{\bullet-}$ generation was observed (i.e., the transition and distal elongation zones). However, Cd-induced cell death was only observed in roots of plants exposed to 60 μM Cd, whereas no effect was observed upon exposure to lower concentrations, suggesting that cell death only occurs when ROS concentrations exceed a certain threshold [144]. In an *Arabidopsis* cell suspension culture, the H_2O_2 production required for PCD was shown to depend on NO production, as inhibition of NO synthesis partially prevented H_2O_2 accumulation and cell death [145]. As summarized by Locato et al. (2016) [140], different underlying mechanisms have been proposed for the observed NO-dependency of Cd-induced PCD. First, NO-dependent nitrosylation of PCs could reduce their ability to chelate Cd, thereby increasing free Cd levels and enhancing toxicity symptoms [145,146]. Furthermore, NO could increase Cd accumulation by affecting transcript levels of genes involved in Cd uptake and detoxification [147,148]. A final mechanism proposed is the Cd-induced activation of MPK6, which subsequently activates caspase-3-like, a PCD executor [149].

It is likely that phytohormones are also involved in regulating Cd-induced PCD, possibly through interactions with ROS. Simultaneous exposure of a tomato cell suspension culture to Cd and ethylene caused a stronger Cd-induced PCD response as compared to treatment with Cd alone. Furthermore, Cd-induced PCD was mitigated by application of the ethylene biosynthesis inhibitor 2-aminoethoxyvinyl glycine and the ethylene receptor blocker silver thiosulfate [150], clearly indicating the key role of ethylene signaling in Cd-induced PCD. Furthermore, a Cd-induced increase in cellular SA concentrations of tobacco cells was shown to induce a MAPK signaling pathway involved in mediating

PCD [151]. Similarly, exogenously applied SA was able to alleviate Cd-induced ROS production, photosynthetic damage and cell death in Cd-exposed *A. thaliana* [152]. The cell death observed in this study was likely a consequence of a strong Cd-induced activation of autophagy. This process involves the vacuolar degradation and recycling of cellular macromolecules or entire organelles and was previously reported to be induced upon Cd exposure in various plant species including *A. thaliana* [152], *G. max* [153], *T. aestivum* [154] and *Theobroma cacao* [155]. Although their interplay has not yet been studied in Cd-exposed plants, crosstalk between autophagy and several phytohormones exists in plants exposed to other environmental stimuli or during normal plant development [156]. Furthermore, an interplay between autophagy and ROS was also shown, with ROS contributing to the establishment of autophagy and autophagy contributing to ROS scavenging. A comprehensive overview of the involvement of ROS and phytohormones and their interplay in autophagy in stress-exposed plants was recently published by Signorelli et al. (2019) [157]. However, the involvement and interplay of ROS and phytohormones in Cd-induced autophagy remains largely unknown. Interestingly, autophagy contributes to nutrient recycling and remobilization during senescence [157], a process which was shown to be prematurely induced upon Cd exposure in different plant species, as indicated by increases in senescence-related parameters such as protease activity, lipid peroxidation and the expression of senescence-associated genes (SAGs) [9].

Taken together, these data suggest that Cd-induced DNA damage is an important trigger for effects on plant growth either through the induction of cell death or through influences on the cell cycle. In order to prevent Cd entry into cells and subsequent DNA damage, cell wall structure can be altered to increase the number of Cd binding sites. This often coincides with an increased cell wall rigidity, which can in turn limit cell expansion and hence plant growth.

3.2. The Cell Wall

The primary cell wall consists of structural proteins that are embedded in a matrix of polysaccharides which includes cellulose, hemicellulose and pectin [158–160]. These components ensure that inner structures have sufficient support, but at the same time remain adjustable for the expanding cell. Furthermore, these polysaccharides contain functional groups such as hydroxyl, thiol and carboxyl groups that enable the cell wall to bind large amounts of divalent and trivalent metals including Cd [161]. The homogalacturonan (HG) domain of pectin is created by the golgi apparatus and secreted into the cell wall in a highly methylesterified form. Using a Ca ion, two HG domains can be linked together by their carboxylic groups [162]. Calcium can be replaced within this structure by other metals such as Pb, Cu, Cd and Zn that show greater affinity [161].

3.2.1. The Cell Wall as Major Storage Compartment for Cadmium

The cell wall is the primary defense structure of a plant's cell against pathogen attacks or unfavorable environmental conditions such as drought and metals [163–165]. By keeping excess Cd out of the cytoplasm, it prevents damage to macromolecules, proteins and DNA caused directly or upon Cd-induced oxidative stress. Since roots are the organs in direct contact with Cd from the soil, their cell walls play an important role in this process. Many studies show that most Cd is stored within the cell walls of roots by various plant species [166,167]. For example, in the plant *Coptis chinensis*, the roots and rhizomes retained between 62 and 77% of all Cd [168]. When the capacity of the cell wall is exceeded, Cd might additionally form complexes with PCs and is subsequently sequestered within the vacuole [167,169,170]. However, Dong and colleagues (2016) reported that upon exposure to 200 μM $CdCl_2$, most Cd in the leaves was present in the soluble fraction in *Arachis hypogaea* and compartmentalization in vacuoles might be of greater importance here [171].

3.2.2. Cadmium-Induced Cell Wall Modifications

In addition to the formation of a rigid barrier, the cell wall might also be actively modified under Cd exposure [172–174]. By increasing the pectin methylesterase (PME) activity, the amount of de-esterified pectin increases, thereby creating more negative charges to bind Cd [175,176]. Moreover, an increase in pectin and hemicellulose content can also contribute to the latter [177,178]. Recently, in *O. sativa* seedlings it was shown that root aeration increases pectin content and PME activity resulting in Cd toxicity alleviation [179]. Since pectin is especially synthesized in young, expanding cells, a delay in root maturation due to aeration might enhance pectin deposit. Next to cell wall modifications related to primary components, another effect that occurs under Cd exposure, is the activation of lignin biosynthesis [180,181]. Lignin is incorporated in the secondary cell walls of specialized cells such as tracheids and vessel elements of the xylem and is created by the oxidative polymerization of monolignol subunits by class III peroxidases (PODs) and laccases [182]. Adding lignin to the cells' walls decreases the permeability to Cd, but at the same time restricts cell elongation, thereby inhibiting growth. An increase of POD activity in response to Cd has been observed in various plant species [180,181,183]. It is of particular interest that POD uses H_2O_2 as a substrate, but at the same time H_2O_2 comes forward as a major signaling molecule within the redox network, which is strongly disturbed by Cd exposure. These close interactions between lignification and redox regulation have been extensively reviewed by Loix and colleagues (2017) [173]. Cell wall expansion is largely regulated by ROS homeostasis in the apoplast [184]. Whereas H_2O_2 contributes to cell wall stiffening, hydroxyl radicals ($^{\bullet}OH$) lead to cell wall loosening due to pectin and xyloglucan cleavage [185,186].

The role of cellulose under Cd exposure is still a matter of debate. The cell wall cellulose content was shown to decrease in *O. sativa* and *Zea mays* under Cd exposure, but an increase was observed in *Linum usitatissimum*. Therefore, it is proposed that different defense strategies are present between monocots and dicots [175,177,187]. Recently, it was shown that Cd accumulates preferably within cellulose in the Cd-sensitive *G. max* BX10 variety, but contrarily, the Cd-tolerant *G. max* HX3 accumulated more Cd in cell wall-related pectin [188]. This might even suggest intraspecies differences in sequestration of Cd in different cell wall components. Cellulose microfibrils are cable-like structures composed of many β-1,4-linked glucose molecules, which are added together by cellulose synthase using uridine diphosphate glucose as a donor molecule [189]. Cadmium is known to reduce the amount of cytosolic sucrose, which could have detrimental effects on cellulose biosynthesis, as UDP-glucose is thought to be mainly produced by sucrose synthase (SUSY) from sucrose [174,190]. In a recent proteomic study by Gutsch and colleagues (2018), SUSY was shown to be upregulated in response to long-term Cd exposure in *Medicago sativa* [191]. The photosynthetic capacity of these plants was inhibited, as indicated by a lower abundance of photosynthetic proteins, which cuts off glucose supply for cell wall biosynthesis. However, a higher SUSY activity might suggest that the plants still invested in cellulose synthesis. In *Miscanthus sacchariflorus*, several genes involved in cellulose biosynthesis were upregulated under Cd exposure as well. These included cellulose synthase A, cellulose synthase-like protein D4 and cellulose synthase-like protein H1 [192].

In addition to these polysaccharides and lignin, two protein families, i.e., xyloglucan endotransglucosylases/hydrolases (XTH) and expansins (EXP), were demonstrated to affect cell wall expansion as well [193]. Expansins have a function in breaking the non-covalent binding of polysaccharides resulting in cell wall loosening [158,194]. In *N. tabacum*, *NtEXP1*, *NtEXP4* and *NtEXP5* transcripts were abundant in the shoot apices and young leaves, but not in roots and mature leaves, supporting their function in cell wall extension [195]. Furthermore, these genes were induced by growth-stimulating hormones such as AUX, GA and CK, but Cd exposure inhibited their transcription. This was supported by a study in *Brassica juncea* by Sun and colleagues (2011) who found that overexpression of *BjEXPA1* led to a higher Cd sensitivity [196]. However, the opposite was found for the *TaEXP2* gene in *T. aestivum* [197]. This gene was upregulated under Cd exposure and its overexpression led to an increase in biomass and root elongation. This improved plant performance was stated to be due to an enhanced translocation of Cd to the vacuoles, a higher antioxidant capacity and a higher water retention. Secondly, XTHs have the ability to cut and rejoin xyloglucan, which locks cellulose microfibrils and thereby contributes to cell wall loosening [198]. In *A. thaliana*, *XTH33* was required for Cd accumulation within the roots [199]. In addition, *XTH33* was shown to be a direct target of EIN3 which acts as a master transcription factor in ethylene-mediated Cd-induced root growth inhibition. When *PeXTH* from *Populus euphratica* was introduced in *N. tabacum*, a decrease of xyloglucan was observed in the cell wall of the roots, leading to a reduced number of Cd binding sites, thereby reducing Cd influx into the roots and limiting Cd toxicity [200]. Lastly, in a proteomic study with two cultivars of *A. hypogaea*, it was hypothesized that XTHs and α-expansins might be important in keeping cell wall extensibility under Cd exposure in the Cd-sensitive cultivar, which showed a higher capacity of cell wall modification [201]. Both XTH and EXP proteins are activated by AUX, inducing acidic growth of the cell wall [202].

3.2.3. The Role of The Cell Wall in Cadmium Hyperaccumulators

Plants that are able to withstand very high metal concentrations in the soil and show an enhanced metal accumulation within the aboveground organs are defined as hyperaccumulators [169]. These remarkable plant species account for less than 0.2% of all angiosperms, although additional heavy metal-accumulating plants are likely to be identified [203]. The trait is expected to have evolved multiple times but is of very high occurrence within the Brassicaceae family [169]. It is a nice coincidence that frequently studied hyperaccumulators such as *Arabidopsis halleri* and *Noccea caerulescens*, previously known as *Thlaspi caerulescens*, are phylogenetic closely related to the model plant *A. thaliana*, which does not accumulate metals. As shown by Benzarti et al. (2008), the Cd hyperacummulator *N. caerulescens* has a more than tenfold higher EC50 value for Cd-induced root growth inhibition in comparison to several non-accumulator plants [204]. However, large differences in the ability to accumulate Cd also exist between populations of hyperaccumulating species [203]. In *A. halleri*, some individuals of metallicolous populations were able to survive in the presence of 450 μM $CdSO_4$, whereas concentrations as low as 100 μM $CdSO_4$ caused mortality within non-metallicolous populations [205]. Tolerance mechanisms to cope with high soil metal concentrations include enhanced metal uptake and xylem loading, followed by detoxification in the shoot, as reviewed by Verbruggen and colleagues (2009) [206].

The role of Cd storage in the cell wall of hyperaccumulators might not be that straightforward. In a comparative study between a non-hyperaccumulating and a hyperaccumulating ecotype of *Sedum alfredii*, it was shown that for the latter more cell wall-bound Cd was available for xylem loading [207]. Furthermore, no increase of pectin or hemicellulose 2 was detected in the hyperaccumulating ecotype, which was accompanied with a lower PME activity. This led the authors to the conclusion that Cd translocation could at least partly be due to a difference in cell wall modification regulation [207]. In addition, cell wall modifications in the shoot cell walls of *A. halleri* under Cd exposure were more pronounced in less tolerant populations [205]. A Cd tolerance mechanism might already be present under controlled conditions in *A. halleri* in contrast to *A. thaliana* [208]. Gene expression related to redox balance, Ca signaling, and cell wall remodeling was more affected in *A. thaliana*, yet *PME* and *CESA* transcripts (encoding for cellulose synthases) were more upregulated in *A. halleri*. This could result in a larger extent of Cd binding to the cell wall and cell wall stiffening, which in turn leads to a greater barrier for cytosolic entering [208]. Moreover, Peng and colleagues (2017) showed that the cell wall of a newly discovered hyperaccumulator *Sedum plumbizincicola* plays a crucial role in Cd tolerance [209]. In contrast to a non-hyperaccumulating ecotype of *S. alfredii*, Fourier transform infrared (FT-IR) spectroscopy displayed a higher absorbance ratio of $–COO^-$ against –COOR, indicating a lower esterification of pectin and a more efficient binding of metal ions for *S. plumbizincicola*.

4. Reproductive Growth

In order for a plant to complete its life cycle, it must commence the reproductive phase, which in higher plants involves the setting of flowers, pollination and fertilization, followed by the production of seeds. In *Brassica campestris*, the GSH and AsA content dropped the most when plants were exposed to Cd at their flowering stage, indicating that the reproductive phase is highly susceptible [210].

In *L. usitatissimum* plants grown in Cd-contaminated soils, a decrease in fitness was noted due to 31.8% fewer seeds and 25.6% fewer fruits [211].

Although *A. thaliana* plants showed a reduction in silique counts under long-term Cd exposure, they were still able to complete their life cycle, thereby producing seeds with equal germination capacity as control plants [212]. The transition from the vegetative to the reproductive phase under 10 μM Cd was unaffected, which was in agreement with the results from Maistri and colleagues (2011), however an accelerated emergence of inflorescence was observed under 5 μM Cd [213]. Limited research has been done on the later stages of plant development in relation to Cd exposure. Nevertheless, accumulation of large amounts of Cd in the seeds, even without visible toxic effects to the plant, can be a great threat to human health [214]. Cadmium uptake by roots and translocation to aboveground parts is a process that continues throughout the plant's life cycle. In *O. sativa*, most of the Cd accumulated in the grains during the early phases of grain development either directly via the xylem or through remobilization through the phloem [215]. However, in *Solanum lycopersicum* the majority of Cd was taken up in the final stage of fruit development [216]. This co-occurred with a disturbance in nutritional status, as potassium (K), Fe and Zn contents in fruits decreased, while Ca and magnesium (Mg) increased. Exposure to 100 μM Cd resulted in an absence of fruit setting. Furthermore, an interesting finding was reported in the monoecious plant *Crocus sativus*, where a shift to more male flowers was apparent upon Cd exposure [217]. This was also observed in a study with *Cannabis sativa*, where treatment with Pb resulted likewise in male flowers and was demonstrated to be due to a hormonal shift with increasing GA levels in these plants, while zeatine (a specific form of CK) decreased [218].

Cadmium has been shown to negatively affect pollen germination accompanied with a disruption of pollen tube morphology in multiple plant species [219,220]. Even very low Cd concentrations of 0.01 μg mL^{-1} were able to inhibit either pollen germination or tube growth in *Vicia angustifolia* and *Vicia tetrasperma*, indicating that this process is very sensitive to Cd [221]. The impaired cell elongation of the pollen tube by Cd is a consequence of its interference with the anionic content of secretory vesicles and its interaction with the cell wall, which contains large quantities of pectin and callose [222]. Cell wall thickening, an increase in cell diameter and abnormal pollen tube growth were observed in all *Prunus avium* cultivars tested in vitro under Cd exposure [223]. Pollen tubes of *Picea wilsonii* showed swelling of the tips accompanied by cytoplasmic vacuolization [224]. Furthermore, the importance of ROS/Ca signaling in pollen tube formation has been well documented [137,225]. Exogenously applied $^{\bullet}OH$ to pollen of *N. tabacum* caused loosening of the intine (i.e., the inner layer of the pollen tube cell wall), leading to a disrupted polar growth, while H_2O_2 stiffened the cell wall [226]. Both treatments resulted in a reduced pollen germination, since only the region of the germinating pore should be weakened through $^{\bullet}OH$-dependent reactions. In conclusion, Cd negatively affects plant fitness by interfering with various processes including pollen tube formation and pollen germination resulting in smaller numbers of seeds that show a reduced germination as was described in the section on seed germination. In conclusion, knowledge on the effects of Cd on the reproductive phase of plant development is still relatively scarce in contrast to information on its effects on vegetative growth and deserves more attention in the future.

5. Conclusion and Future Perspectives

Cadmium exposure interferes with all stages of plant development, inhibiting seed germination, vegetative growth and reproductive growth. Important players in the Cd-induced disturbance of root and leaf growth are DNA damage, which subsequently affects cell cycle progression or even causes cell death, and structural alterations of the cell wall. Furthermore, Cd alters pollen tube morphology, inhibits pollen germination and can be transported into seeds and fruits. The type and extent of Cd-induced effects strongly depends on many factors, including the plant species, organ, cell type, Cd concentration and exposure duration.

Additional research on the mechanisms underlying these Cd-induced disturbances of plant growth and development is required to develop and further improve strategies to enhance plant growth on Cd-polluted soils. In this context, it would be interesting to also focus on transgenerational effects involved in plant adaptation to Cd exposure. Although research in this field is currently limited, a recent study showed that Cd-induced alterations of RAPD profiles in *Urtica pilulifera* parent plants are transmitted to the next generation [112]. Similarly, Carvalho et al. (2018) demonstrated that exposure of *S. lycopersicum* to Cd resulted in Cd accumulation in seeds, which altered their nutrient profile and caused a decreased mitotic index in root tips of the offspring [227]. Although the mechanisms underlying these transgenerational effects are currently unknown, epigenetic changes might be involved, as Cd exposure was previously shown to alter DNA methylation patterns in *A. thaliana* [104]. In addition, the possible involvement of the microbiome in plant adaptation to Cd exposure should also be taken into account, as transgenerational exposure of *A. thaliana* to Cd significantly altered the seed endophytic community [228,229]. Furthermore, inoculation with specific plant-associated bacteria improved root growth of Cd-exposed *A. thaliana* plants [230], emphasizing the importance of the plant microbiome in coping with environmental stress factors and optimizing plant growth and development under suboptimal conditions. Finally, the use of soil amendments such as biochar and silicon-based fertilizers is promising for future applications to alleviate Cd toxicity, thereby improving crop growth and quality [231,232].

Author Contributions: All authors participated in the conception of the topic. M.H., A.C. and S.H. wrote the manuscript. M.H. and S.H. made the figures and tables. All authors read and approved the final manuscript after critically revising it for important intellectual content.

Funding: This work was supported by the Research Foundation Flanders (FWO) by project funding for M.H. [G0B6716], S.H. [G0C7518CUY], M.J. [SBOS000119N], J.D. [FWO PhD grant], and bijzonder onderzoeksfonds (BOF) from Hasselt University to V.I. [BOF-18D02CUYA] and S.V. [PhD grant].

Conflicts of Interest: The authors declare no conflict of interest.

Abbreviations

ABA	Abscisic acid
ABI3	Abscisic acid insensitive 3
ACC	1-aminocyclopropane-1-carboxylate
ACS	ACC synthase
AFLP	Amplified fragment length polymorphism
AsA	Ascorbate
AUX	Auxin
ATM	Ataxia telangiectasia mutated
ATR	RAD3-related
Ca	Calcium
CAT	Catalase
Cd	Cadmium
CDKs	Cyclin-dependent kinases
Cr	Chromium
CK	Cytokinin
DDR	DNA damage response
DSBs	Double-stranded breaks
EdU	5-ethynyl-2′-deoxyuridine
ET	Ethylene
EXP	Expansin
FCM	Flow cytometry
Fe	Iron
FT-IR	Fourier transform infrared
GA	Gibberellic acid
GR	Glutathione reductase
Grx	Glutaredoxin
GSH	Glutathione

GTS	Genomic template stability
H_2O_2	Hydrogen peroxide
HG	Homogalacturonan
IAA	Indole-3-acetic acid
ISSR	Inter-simple sequence repeat
JA	Jasmonic acid
Mn	Manganese
NAC	NO APICAL MERISTEM/ARABIDOPSIS TRANSCRIPTION ACTIVATION FACTOR/CUP-SHAPED COTYLEDON
NADPH	Nicotinamide adenine dinucleotide phosphate
•OH	Hydroxyl radical
$O_2^{\bullet}$	Superoxide
Pb	Lead
PC	Phytochelatin
PCD	Programmed cell death
PME	Pectin methylesterase
POD	Class III peroxidase
Prx	Peroxiredoxin
qRT-PCR	Quantitative reverse transcription polymerase chain reaction
RAPD	Random amplified polymorphic DNA
ROS	Reactive oxygen species
RT-PCR	Reverse transcription polymerase chain reaction
SA	Salicylic acid
SAGs	Senescence-associated genes
SMR	SIAMESE-related
SOD	Superoxide dismutase
SOG1	SUPPRESSOR OF GAMMA RESPONSE 1
SRAP	Sequence-related amplified polymorphism
SSBs	Single-stranded breaks
SSR	Simple sequence repeat
SUSY	Sucrose synthase
TFs	Transcription factors
Trx	Thioredoxin
XTH	Endotransglucosylases/hydrolases
Zn	Zinc

References

1. Khan, M.A.; Khan, S.; Khan, A.; Alam, M. Soil contamination with cadmium, Consequences and remediation using organic amendments. *Sci. Total Environ.* **2017**, *601*, 1591–1605. [CrossRef] [PubMed]
2. Verbruggen, N.; Hermans, C.; Schat, H. Mechanisms to cope with arsenic or cadmium excess in plants. *Curr. Opin. Plant Biol.* **2009**, *12*, 364–372. [CrossRef] [PubMed]
3. DalCorso, G.; Farinati, S.; Maistri, S.; Furini, A. How plants cope with cadmium: Staking all on metabolism and gene expression. *J. Integr. Plant Biol.* **2008**, *50*, 1268–1280. [CrossRef] [PubMed]
4. Cuypers, A.; Keunen, E.; Bohler, S.; Jozefczak, M.; Opdenakker, K.; Gielen, H.; Vercampt, H.; Bielen, A.; Schellingen, K.; Vangronsveld, J.; et al. Cadmium and copper stress induce a cellular oxidative challenge leading to damage versus signalling. In *Metal Toxicity in Plants: Perception, Signaling and Remediation*; Gupta, D.K., Sandalio, L.M., Eds.; Springer: Berlin/Heidelberg, Germany, 2012; pp. 65–90. [CrossRef]
5. Sharma, S.S.; Dietz, K.J. The relationship between metal toxicity and cellular redox imbalance. *Trends Plant Sci.* **2009**, *14*, 43–50. [CrossRef] [PubMed]
6. Cuypers, A.; Plusquin, M.; Remans, T.; Jozefczak, M.; Keunen, E.; Gielen, H.; Opdenakker, K.; Nair, A.R.; Munters, E.; Artois, T.J.; et al. Cadmium stress: An oxidative challenge. *Biometals* **2010**, *23*, 927–940. [CrossRef] [PubMed]

7. Jozefczak, M.; Keunen, E.; Schat, H.; Bliek, M.; Hernandez, L.E.; Carleer, R.; Remans, T.; Bohler, S.; Vangronsveld, J.; Cuypers, A. Differential response of *Arabidopsis* leaves and roots to cadmium: Glutathione-related chelating capacity vs antioxidant capacity. *Plant Physiol. Biochem.* **2014**, *83*, 1–9. [CrossRef] [PubMed]
8. Schutzendubel, A.; Polle, A. Plant responses to abiotic stresses: Heavy metal-induced oxidative stress and protection by mycorrhization. *J. Exp. Bot.* **2002**, *53*, 1351–1365. [CrossRef] [PubMed]
9. Cuypers, A.; Hendrix, S.; dos Reis, R.A.; De Smet, S.; Deckers, J.; Gielen, H.; Jozefczak, M.; Loix, C.; Vercampt, H.; Vangronsveld, J.; et al. Hydrogen peroxide, Signaling in disguise during metal phytotoxicity. *Front. Plant Sci.* **2016**, *7*, 25. [CrossRef] [PubMed]
10. Moller, I.M.; Jensen, P.E.; Hansson, A. Oxidative modifications to cellular components in plants. *Annu. Rev. Plant Biol.* **2007**, *58*, 459–481. [CrossRef] [PubMed]
11. Petrov, V.D.; Van Breusegem, F. Hydrogen peroxide-A central hub for information flow in plant cells. *AoB Plants* **2012**. [CrossRef] [PubMed]
12. Bucker-Neto, L.; Paiva, A.L.S.; Machado, R.D.; Arenhart, R.A.; Margis-Pinheiro, M. Interactions between plant hormones and heavy metals responses. *Genet. Mol. Biol.* **2017**, *40*, 373–386. [CrossRef] [PubMed]
13. De Smet, S.; Cuypers, A.; Vangronsveld, J.; Remans, T. Gene networks involved in hormonal control of root development in *Arabidopsis thaliana*: A framework for studying its disturbance by metal stress. *Int. J. Mol. Sci.* **2015**, *16*, 19195–19224. [CrossRef] [PubMed]
14. Schellingen, K.; Van Der Straeten, D.; Vandenbussche, F.; Prinsen, E.; Remans, T.; Vangronsveld, J.; Cuypers, A. Cadmium-induced ethylene production and responses in *Arabidopsis thaliana* rely on *ACS2* and *ACS6* gene expression. *BMC Plant Biol.* **2014**, *14*, 14. [CrossRef] [PubMed]
15. Rodriguez-Serrano, M.; Romero-Puertas, M.C.; Pazmino, D.M.; Testillano, P.S.; Risueno, M.C.; del Rio, L.A.; Sandalio, L.M. Cellular response of pea Plants to cadmium toxicity: Cross talk between reactive oxygen species, nitric oxide, and calcium. *Plant Physiol.* **2009**, *150*, 229–243. [CrossRef] [PubMed]
16. Kim, Y.H.; Khan, A.L.; Kim, D.H.; Lee, S.Y.; Kim, K.M.; Waqas, M.; Jung, H.Y.; Shin, J.H.; Kim, J.G.; Lee, I.J. Silicon mitigates heavy metal stress by regulating P-type heavy metal ATPases, *Oryza sativa* low silicon genes, and endogenous phytohormones. *BMC Plant Biol.* **2014**, *14*, 13. [CrossRef] [PubMed]
17. Hu, Y.F.; Zhou, G.Y.; Na, X.F.; Yang, L.J.; Nan, W.B.; Liu, X.; Zhang, Y.Q.; Li, J.L.; Bi, Y.R. Cadmium interferes with maintenance of auxin homeostasis in *Arabidopsis* seedlings. *J. Plant Physiol.* **2013**, *170*, 965–975. [CrossRef]
18. Ronzan, M.; Piacentini, D.; Fattorini, L.; Della Rovere, F.; Eiche, E.; Rieman, M.; Altamura, M.M.; Falasca, G. Cadmium and arsenic affect root development in *Oryza sativa* L. negatively interacting with auxin. *Environ. Exp. Bot.* **2018**, *151*, 64–75. [CrossRef]
19. Xia, X.J.; Zhou, Y.H.; Shi, K.; Zhou, J.; Foyer, C.H.; Yu, J.Q. Interplay between reactive oxygen species and hormones in the control of plant development and stress tolerance. *J. Exp. Bot.* **2015**, *66*, 2839–2856. [CrossRef]
20. Vangronsveld, J.; Herzig, R.; Weyens, N.; Boulet, J.; Adriaensen, K.; Ruttens, A.; Thewys, T.; Vassilev, A.; Meers, E.; Nehnevajova, E.; et al. Phytoremediation of contaminated soils and groundwater: Lessons from the field. *Environ. Sci. Pollut. Res.* **2009**, *16*, 765–794. [CrossRef]
21. Gubler, F.; Millar, A.A.; Jacobsen, J.V. Dormancy release, ABA and pre-harvest sprouting. *Curr. Opin. Plant Biol.* **2005**, *8*, 183–187. [CrossRef]
22. Nee, G.; Xiang, Y.; Soppe, W.J.J. The release of dormancy, A wake-up call for seeds to germinate. *Curr. Opin. Plant Biol.* **2017**, *35*, 8–14. [CrossRef] [PubMed]
23. Rodriguez, M.V.; Barrero, J.M.; Corbineau, F.; Gubler, F.; Benech-Arnold, R.L. Dormancy in cereals (not too much, not so little): About the mechanisms behind this trait. *Seed Sci. Res.* **2015**, *25*, 99–119. [CrossRef]
24. Finkelstein, R.; Reeves, W.; Ariizumi, T.; Steber, C. Molecular aspects of seed dormancy. *Annu. Rev. Plant Biol.* **2008**, *59*, 387–415. [CrossRef] [PubMed]
25. Graeber, K.; Nakabayashi, K.; Miatton, E.; Leubner-Metzger, G.; Soppe, W.J.J. Molecular mechanisms of seed dormancy. *Plant Cell Environ.* **2012**, *35*, 1769–1786. [CrossRef] [PubMed]
26. Bassel, G.W. To grow or not to grow? *Trends Plant Sci.* **2016**, *21*, 498–505. [CrossRef] [PubMed]
27. Shu, K.; Liu, X.D.; Xie, Q.; He, Z.H. Two faces of one seed: Hormonal regulation of dormancy and germination. *Mol. Plant.* **2016**, *9*, 34–45. [CrossRef] [PubMed]

28. Liu, X.D.; Zhang, H.; Zhao, Y.; Feng, Z.Y.; Li, Q.; Yang, H.Q.; Luan, S.; Li, J.M.; He, Z.H. Auxin controls seed dormancy through stimulation of abscisic acid signaling by inducing ARF-mediated ABI3 activation in *Arabidopsis*. *Proc. Natl. Acad. Sci. USA* **2013**, *110*, 15485–15490. [CrossRef] [PubMed]
29. Belin, C.; Megies, C.; Hauserova, E.; Lopez-Molina, L. Abscisic acid represses growth of the *Arabidopsis* embryonic axis after germination by enhancing auxin signaling. *Plant Cell* **2009**, *21*, 2253–2268. [CrossRef]
30. Arc, E.; Sechet, J.; Corbineau, F.; Rajjou, L.; Marion-Poll, A. ABA crosstalk with ethylene and nitric oxide in seed dormancy and germination. *Front. Plant Sci.* **2013**, *4*, 19. [CrossRef]
31. Linkies, A.; Muller, K.; Morris, K.; Tureckova, V.; Wenk, M.; Cadman, C.S.C.; Corbineau, F.; Strnad, M.; Lynn, J.R.; Finch-Savage, W.E.; et al. Ethylene interacts with abscisic acid to regulate endosperm rupture during germination: A comparative approach using *Lepidium sativum* and *Arabidopsis thaliana*. *Plant Cell* **2009**, *21*, 3803–3822. [CrossRef]
32. Wojtyla, L.; Lechowska, K.; Kubala, S.; Garnczarska, M. Different modes of hydrogen peroxide action during seed germination. *Front. Plant Sci.* **2016**, *7*, 16. [CrossRef] [PubMed]
33. El-Maarouf-Bouteau, H.; Sajjad, Y.; Bazin, J.; Langlade, N.; Cristescu, S.M.; Balzergue, S.; Baudouin, E.; Bailly, C. Reactive oxygen species, abscisic acid and ethylene interact to regulate sunflower seed germination. *Plant Cell Environ.* **2015**, *38*, 364–374. [CrossRef] [PubMed]
34. Huang, Y.T.; Lin, C.; He, F.; Li, Z.; Guan, Y.J.; Hu, Q.J.; Hu, J. Exogenous spermidine improves seed germination of sweet corn via involvement in phytohormone interactions, H_2O_2 and relevant gene expression. *BMC Plant Biol.* **2017**, *17*, 1. [CrossRef]
35. Bahin, E.; Bailly, C.; Sotta, B.; Kranner, I.; Corbineau, F.; Leymarie, J. Crosstalk between reactive oxygen species and hormonal signalling pathways regulates grain dormancy in barley. *Plant Cell Environ.* **2011**, *34*, 980–993. [CrossRef] [PubMed]
36. Li, Z.; Gao, Y.; Zhang, Y.C.; Lin, C.; Gong, D.T.; Guan, Y.J.; Hu, J. Reactive oxygen species and gibberellin acid mutual induction to regulate tobacco seed germination. *Front. Plant Sci.* **2018**, *9*, 1279. [CrossRef] [PubMed]
37. Bailly, C.; El-Maarouf-Bouteau, H.; Corbineau, F. From intracellular signaling networks to cell death: The dual role of reactive oxygen species in seed physiology. *C. R. Biol.* **2008**, *331*, 806–814. [CrossRef] [PubMed]
38. Bykova, N.V.; Hoehn, B.; Rampitsch, C.; Banks, T.; Stebbing, J.A.; Fan, T.; Knox, R. Redox-sensitive proteome and antioxidant strategies in wheat seed dormancy control. *Proteomics* **2011**, *11*, 865–882. [CrossRef] [PubMed]
39. El-Maarouf-Bouteau, H.; Meimoun, P.; Job, C.; Job, D.; Bailly, C. Role of protein and mRNA oxidation in seed dormancy and germination. *Front. Plant Sci.* **2013**, *4*, 77. [CrossRef]
40. Job, C.; Rajjou, L.; Lovigny, Y.; Belghazi, M.; Job, D. Patterns of protein oxidation in *Arabidopsis* seeds and during germination. *Plant Physiol.* **2005**, *138*, 790–802. [CrossRef]
41. Oracz, K.; Bouteau, H.E.M.; Farrant, J.M.; Cooper, K.; Belghazi, M.; Job, C.; Job, D.; Corbineau, F.; Bailly, C. ROS production and protein oxidation as a novel mechanism for seed dormancy alleviation. *Plant J.* **2007**, *50*, 452–465. [CrossRef]
42. Hou, L.Y.; Wang, M.Y.; Wang, H.; Zhang, W.H.; Mao, P.S. Physiological and proteomic analyses for seed dormancy and release in the perennial grass of *Leymus chinensis*. *Environ. Exp. Bot.* **2019**, *162*, 95–102. [CrossRef]
43. Li, W.Q.; Khan, M.A.; Yamaguchi, S.; Kamiya, Y. Effects of heavy metals on seed germination and early seedling growth of *Arabidopsis thaliana*. *Plant Growth Regul.* **2005**, *46*, 45–50. [CrossRef]
44. Raviv, B.; Aghajanyan, L.; Granot, G.; Makover, V.; Frenkel, O.; Gutterman, Y.; Grafi, G. The dead seed coat functions as a long-term storage for active hydrolytic enzymes. *PLoS ONE* **2017**, *12*, e0181102. [CrossRef] [PubMed]
45. Kucera, B.; Cohn, M.A.; Leubner-Metzger, G. Plant hormone interactions during seed dormancy release and germination. *Seed Sci. Res.* **2005**, *15*, 281–307. [CrossRef]
46. Bewley, J.D. Seed germination and dormancy. *Plant Cell* **1997**, *9*, 1055–1066. [CrossRef] [PubMed]
47. Bentsink, L.; Koornneef, M. Seed dormancy and germination. *Arabidopsis Book* **2008**, *6*. [CrossRef] [PubMed]
48. Sfaxi-Bousbih, A.; Chaoui, A.; El Ferjani, E. Cadmium impairs mineral and carbohydrate mobilization during the germination of bean seeds. *Ecotoxicol. Environ. Safe.* **2010**, *73*, 1123–1129. [CrossRef] [PubMed]
49. Cheng, W.D.; Zhang, G.P.; Yao, H.G.; Zhang, H.M. Genotypic difference of germination and early seedling growth in response to Cd stress and its relation to Cd accumulation. *J. Plant Nutr.* **2008**, *31*, 702–715. [CrossRef]

50. Kuriakose, S.V.; Prasad, M.N.V. Cadmium stress affects seed germination and seedling growth in *Sorghum bicolor* (L.) Moench by changing the activities of hydrolyzing enzymes. *Plant Growth Regul.* **2008**, *54*, 143–156. [CrossRef]
51. Ahmad, I.; Akhtar, M.J.; Zahir, Z.A.; Jamil, A. Effects of cadmium on seed germination and seedling growth of four wheat (*Triticum aestivum* L. *) cultivars Pak. J. Bot.* **2012**, *44*, 1569–1574.
52. Farooqi, Z.R.; Iqbal, M.Z.; Kabir, M.; Shafiq, M. Toxic effects of lead and cadmium on germination and seedling growth *of Albizia Lebbeck* (L.) benth. *Pak. J. Bot.* **2009**, *41*, 27–33.
53. Liu, S.J.; Yang, C.Y.; Xie, W.J.; Xia, C.H.; Fan, P. The effects of cadmium on germination and seedling growth of *Suaeda salsa*. In Proceedings of the Seventh International Conference on Waste Management and Technology (ICWMT 7), Beijing, China, 5–7 September 2012; Volume 16, pp. 293–298. [CrossRef]
54. Alaraidh, I.A.; Alsahli, A.A.; Razik, E.S.A. Alteration of antioxidant gene expression in response to heavy metal stress in *Trigonella foenum-graecum* L. *S. Afr. J. Bot.* **2018**, *115*, 90–93. [CrossRef]
55. Titov, A.F.; Talanova, V.V.; Boeva, N.P. Growth responses of barley and wheat seedlings to lead and cadmium. *Biol. Plant.* **1996**, *38*, 431–436. [CrossRef]
56. Ahsan, N.; Lee, S.H.; Lee, D.G.; Lee, H.; Lee, S.W.; Bahk, J.D.; Lee, B.H. Physiological and protein profiles alternation of germinating rice seedlings exposed to acute cadmium toxicity. *C. R. Biol.* **2007**, *330*, 735–746. [CrossRef] [PubMed]
57. Munzuroglu, O.; Zengin, F.K. Effect of cadmium on germination, coleoptile and root growth of barley seeds in the presence of gibberellic acid and kinetin. *J. Environ. Biol.* **2006**, *27*, 671–677. [PubMed]
58. Lefevre, I.; Marchal, G.; Correal, E.; Zanuzzi, A.; Lutts, S. Variation in response to heavy metals during vegetative growth *in Dorycnium pentaphyllum* Scop. *Plant Growth Regul.* **2009**, *59*, 1–11. [CrossRef]
59. Vijayaragavan, M.; Prabhahar, C.; Sureshkumar, J.; Natarajan, A.; Vijayarengan, P.; Sharavanan, S. Toxic effect of cadmium on seed germination, growth and biochemical contents of cowpea (*Vigna unguiculata* L.) plants. *Int. Multidiscip. Res. J.* **2011**, *1*, 1–6.
60. De Lespinay, A.; Lequeux, H.; Lambillotte, B.; Lutts, S. Protein synthesis is differentially required for germination in *Poa pratensis* and *Trifolium repens* in the absence or in the presence of cadmium. *Plant Growth Regul.* **2010**, *61*, 205–214. [CrossRef]
61. Kalai, T.; Bouthour, D.; Manai, J.; Ben Kaab, L.B.; Gouia, H. Salicylic acid alleviates the toxicity of cadmium on seedling growth, amylases and phosphatases activity in germinating barley seeds. *Arch. Agron. Soil Sci.* **2016**, *62*, 892–904. [CrossRef]
62. He, J.Y.; Ren, Y.F.; Zhu, C.; Jiang, D. Effects of cadmium stress on seed germination, seedling growth and seed amylase activities in rice (*Oryza sativa*). *Rice Sci.* **2008**, *15*, 319–325. [CrossRef]
63. Rivetta, A.; Negrini, N.; Cocucci, M. Involvement of Ca^{2+}-calmodulin in Cd^{2+} toxicity during the early phases of radish (*Raphanus sativus* L) seed germination. *Plant Cell Environ.* **1997**, *20*, 600–608. [CrossRef]
64. Cocucci, M.; Negrini, N. Change in the levels of calmodulin and of a calmodulin inhibitor in the early phases of radish (*Raphanus sativus* L) seed germination—Effects of ABA and fusicoccin. *Plant Physiol.* **1988**, *88*, 910–914. [CrossRef] [PubMed]
65. Rahoui, S.; Chaoui, A.; El Ferjani, E. Reserve mobilization disorder in germinating seeds of *Vicia faba* L. exposed to cadmium. *J. Plant Nutr.* **2010**, *33*, 809–817. [CrossRef]
66. Rahoui, S.; Chaoui, A.; El Ferjani, E. Membrane damage and solute leakage from germinating pea seed under cadmium stress. *J. Hazard. Mater.* **2010**, *178*, 1128–1131. [CrossRef] [PubMed]
67. Jalmi, S.K.; Bhagat, P.K.; Verma, D.; Noryang, S.; Tayyeba, S.; Singh, K.; Sharma, D.; Sinha, A.K. Traversing the links between heavy metal stress and plant signaling. *Front. Plant Sci.* **2018**, *9*, 12. [CrossRef] [PubMed]
68. Smiri, M.; Jelali, N.; El Ghoul, J. Cadmium affects the NADP-thioredoxin reductase/thioredoxin system in germinating pea seeds. *J. Plant Interact.* **2013**, *8*, 125–133. [CrossRef]
69. Alkhalfioui, F.; Renard, M.; Vensel, W.H.; Wong, J.; Tanaka, C.K.; Hurkman, W.J.; Buchanan, B.B.; Montrichard, F. Thioredoxin-linked proteins are reduced during germination of *Medicago truncatula* seeds. *Plant Physiol.* **2007**, *144*, 1559–1579. [CrossRef]
70. Smiri, M.; Chaoui, A.; Rouhier, N.; Gelhaye, E.; Jacquot, J.P.; El Ferjani, E. Cadmium affects the glutathione/glutaredoxin system in germinating pea seeds. *Biol. Trace Elem. Res.* **2011**, *142*, 93–105. [CrossRef]
71. Smiri, M.; Jelali, N.; El Ghoul, J. Role for plant peroxiredoxin in cadmium chelation. *J. Plant Interact.* **2013**, *8*, 255–262. [CrossRef]

72. Sneideris, L.C.; Gavassi, M.A.; Campos, M.L.; D'Amico-Damiao, V.; Carvalho, R.F. Effects of hormonal priming on seed germination of pigeon pea under cadmium stress. *An. Acad. Bras. Cienc.* **2015**, *87*, 1847–1852. [CrossRef]

73. He, J.Y.; Ren, Y.F.; Pan, X.B.; Yan, Y.P.; Zhu, C.; Jiang, D. Salicylic acid alleviates the toxicity effect of cadmium on germination, seedling growth, and amylase activity of rice. *J. Plant Nutr. Soil Sci.* **2010**, *173*, 300–305. [CrossRef]

74. Hu, Z.B.; Cools, T.; De Veylder, L. Mechanisms used by plants to cope with DNA Damage. *Annu. Rev. Plant Biol.* **2016**, *67*, 439–462. [CrossRef] [PubMed]

75. Yoshiyama, K.O. SOG1: A master regulator of the DNA damage response in plants. *Genes Genet. Syst.* **2015**, *90*, 209–216. [CrossRef] [PubMed]

76. Bertin, G.; Averbeck, D. Cadmium: Cellular effects, modifications of biomolecules, modulation of DNA repair and genotoxic consequences (A review). *Biochimie* **2006**, *88*, 1549–1559. [CrossRef] [PubMed]

77. Filipic, M. Mechanisms of cadmium induced genomic instability. *Mutat. Res.-Fundam. Mol. Mech. Mutagen.* **2012**, *733*, 69–77. [CrossRef] [PubMed]

78. Afanasieva, K.; Sivolob, A. Physical principles and new applications of comet assay. *Biophys. Chem.* **2018**, *238*, 1–7. [CrossRef]

79. Ventura, L.; Giovannini, A.; Savio, M.; Dona, M.; Macovei, A.; Buttafava, A.; Carbonera, D.; Balestrazzi, A. Single Cell Gel Electrophoresis (Comet) assay with plants: Research on DNA repair and ecogenotoxicity testing. *Chemosphere* **2013**, *92*, 1–9. [CrossRef]

80. Arya, S.K.; Mukherjee, A. Sensitivity of *Allium cepa* and *Vicia faba* towards cadmium toxicity. *J. Soil Sci. Plant Nutr.* **2014**, *14*, 447–458. [CrossRef]

81. Koppen, G.; Verschaeve, L. The alkaline comet test on plant cells: A new genotoxicity test for DNA strand breaks in *Vicia faba* root cells. *Mutat. Res.-Environ. Mutagen. Rel. Subj.* **1996**, *360*, 193–200. [CrossRef]

82. Lin, A.J.; Zhang, X.H.; Chen, M.M.; Cao, Q. Oxidative stress and DNA damages induced by cadmium accumulation. *J. Environ. Sci.* **2007**, *19*, 596–602. [CrossRef]

83. Gichner, T.; Patkova, Z.; Szakova, J.; Demnerova, K. Cadmium induces DNA damage in tobacco roots, but no DNA damage, somatic mutations or homologous recombination in tobacco leaves. *Mutat. Res. Genet. Toxicol. Environ. Mutagen.* **2004**, *559*, 49–57. [CrossRef] [PubMed]

84. Tkalec, M.; Stefanic, P.P.; Cvjetko, P.; Sikic, S.; Pavlica, M.; Balen, B. The effects of cadmium-zinc interactions on biochemical responses in tobacco seedlings and adult plants. *PLoS ONE* **2014**, *9*, 13. [CrossRef] [PubMed]

85. Gichner, T.; Patkova, Z.; Szakova, J.; Znidar, I.; Mukherjee, A. DNA damage in potato plants induced by cadmium, ethyl methanesulphonate and gamma-rays. *Environ. Exp. Bot.* **2008**, *62*, 113–119. [CrossRef]

86. Silveira, G.L.; Lima, M.G.F.; dos Reis, G.B.; Palmieri, M.J.; Andrade-Vieria, L.F. Toxic effects of environmental pollutants: Comparative investigation using *Allium cepa* L. and *Lactuca sativa* L. *Chemosphere* **2017**, *178*, 359–367. [CrossRef] [PubMed]

87. Seth, C.S.; Misra, V.; Chauhan, L.K.S.; Singh, R.R. Genotoxicity of cadmium on root meristem cells of *Allium cepa*: Cytogenetic and Comet assay approach. *Ecotox. Environ. Safe.* **2008**, *71*, 711–716. [CrossRef] [PubMed]

88. Balen, B.; Tkalec, M.; Sikic, S.; Tolic, S.; Cvjetko, P.; Pavlica, M.; Vidakovic-Cifrek, Z. Biochemical responses of *Lemna minor* experimentally exposed to cadmium and zinc. *Ecotoxicology* **2011**, *20*, 815–826. [CrossRef] [PubMed]

89. Cvjetko, P.; Tolic, S.; Sikic, S.; Balen, B.; Tkalec, M.; Vidakovic-Cifrek, Z.; Pavlica, M. Effect of copper on the toxicity and genotoxicity of cadmium in duckweed (*Lemna minor* L.). *Arh. Hig. Rada. Toksikol.* **2010**, *61*, 287–296. [CrossRef] [PubMed]

90. Monteiro, C.; Santos, C.; Pinho, S.; Oliveira, H.; Pedrosa, T.; Dias, M.C. Cadmium-induced cyto- and genotoxicity are organ-dependent in lettuce. *Chem. Res. Toxicol.* **2012**, *25*, 1423–1434. [CrossRef] [PubMed]

91. Cao, F.B.; Chen, F.; Sun, H.Y.; Zhang, G.P.; Chen, Z.H.; Wu, F.B. Genome-wide transcriptome and functional analysis of two contrasting genotypes reveals key genes for cadmium tolerance in barley. *BMC Genomics* **2014**, *15*, 14. [CrossRef]

92. Lanier, C.; Bernard, F.; Dumez, S.; Leclercq-Dransart, J.; Lemiere, S.; Vandenbulcke, F.; Nesslany, F.; Platel, A.; Devred, I.; Hayet, A.; et al. Combined toxic effects and DNA damage to two plant species exposed to binary metal mixtures (Cd/Pb). *Ecotox. Environ. Safe.* **2019**, *167*, 278–287. [CrossRef]

93. Fenech, M.; Kirsch-Volders, M.; Natarajan, A.T.; Surralles, J.; Crott, J.W.; Parry, J.; Norppa, H.; Eastmond, D.A.; Tucker, J.D.; Thomas, P. Molecular mechanisms of micronucleus, nucleoplasmic bridge and nuclear bud formation in mammalian and human cells. *Mutagenesis* **2011**, *26*, 125–132. [CrossRef] [PubMed]
94. Vral, A.; Fenech, M.; Thierens, H. The micronucleus assay as a biological dosimeter of in vivo ionising radiation exposure. *Mutagenesis* **2011**, *26*, 11–17. [CrossRef] [PubMed]
95. Talukdar, D. Increasing nuclear ploidy enhances the capability of antioxidant defense and reduces chromotoxicity in *Lathyrus sativus* roots under cadmium stress. *Turk. J. Bot.* **2014**, *38*, 696–712. [CrossRef]
96. Aslam, R.; Ansari, M.Y.K.; Choudhary, S.; Bhat, T.M.; Jahan, N. Genotoxic effects of heavy metal cadmium on growth, biochemical, cyto-physiological parameters and detection of DNA polymorphism by RAPD in *Capsicum annuum* L. - An important spice crop of India. *Saudi J. Biol. Sci.* **2014**, *21*, 465–472. [CrossRef] [PubMed]
97. Enan, M.R. Application of random amplified polymorphic DNA (RAPD) to detect the genotoxic effect of heavy metals. *Biotechnol. Appl. Biochem.* **2006**, *43*, 147–154. [CrossRef] [PubMed]
98. Cenkci, S.; Dogan, N. Random amplified polymorphic DNA as a method to screen metal-tolerant barley (*Hordeum vulgare* L.) genotypes. *Turk. J. Bot.* **2015**, *39*, 747–756. [CrossRef]
99. Tanee, T.; Sudmoon, R.; Thamsenanupap, P.; Chaveerach, A. Effect of cadmium on DNA changes in *Ipomoea aquatica* Forssk. *Pol. J. Environ. Stud.* **2016**, *25*, 311–315. [CrossRef]
100. Lima, M.G.F.; Rocha, L.C.; Silveira, G.L.; Alvarenga, I.F.S.; Andrade-Vieria, L.F. Nucleolar alterations are reliable parameters to determine the cytogenotoxicity of environmental pollutants. *Ecotox. Environ. Safe.* **2019**, *174*, 630–636. [CrossRef]
101. Cui, W.N.; Wang, H.T.; Song, J.; Cao, X.; Rogers, H.J.; Francis, D.; Jia, C.Y.; Sun, L.Z.; Hou, M.F.; Yang, Y.S.; et al. Cell cycle arrest mediated by Cd-induced DNA damage in *Arabidopsis* root tips. *Ecotox. Environ. Safe.* **2017**, *145*, 569–574. [CrossRef]
102. Cao, X.; Wang, H.T.; Zhuang, D.F.; Zhu, H.; Du, Y.L.; Cheng, Z.B.; Cui, W.N.; Rogers, H.J.; Zhang, Q.R.; Jia, C.J.; et al. Roles of MSH2 and MSH6 in cadmium-induced G2/M checkpoint arrest in *Arabidopsis* roots. *Chemosphere* **2018**, *201*, 586–594. [CrossRef] [PubMed]
103. Li, Z.L.; Liu, Z.H.; Chen, R.J.; Li, X.J.; Tai, P.D.; Gong, Z.Q.; Jia, C.Y.; Liu, W. DNA damage and genetic methylation changes caused by Cd in *Arabidopsis thaliana* seedlings. *Environ. Toxicol. Chem.* **2015**, *34*, 2095–2103. [CrossRef] [PubMed]
104. Wang, H.T.; He, L.; Song, J.; Cui, W.N.; Zhang, Y.Z.; Jia, C.Y.; Francis, D.; Rogers, H.J.; Sun, L.Z.; Tai, P.D.; et al. Cadmium-induced genomic instability in *Arabidopsis*: Molecular toxicological biomarkers for early diagnosis of cadmium stress. *Chemosphere* **2016**, *150*, 258–265. [CrossRef] [PubMed]
105. Hendrix, S.; Keunen, E.; Mertens, A.I.G.; Beemster, G.T.S.; Vangronsveld, J.; Cuypers, A. Cell cycle regulation in different leaves of *Arabidopsis thaliana* plants grown under control and cadmium-exposed conditions. *Environ. Exp. Bot.* **2018**, *155*, 441–452. [CrossRef]
106. Sudmoon, R.; Neeratanaphan, L.; Thamsenanupap, P.; Tanee, T. Hyperaccumulation of cadmium and DNA changes in popular vegetable, *Brassica chinensis* L. *Int. J. Environ. Res.* **2015**, *9*, 433–438.
107. Venkatachalam, P.; Jayaraj, M.; Manikandan, R.; Geetha, N.; Rene, E.R.; Sharma, N.C.; Sahi, S.V. Zinc oxide nanoparticles (ZnONPs) alleviate heavy metal-induced toxicity in *Leucaena leucocephala* seedlings: A physiochemical analysis. *Plant Physiol. Biochem.* **2017**, *110*, 59–69. [CrossRef] [PubMed]
108. Zhang, X.Q.; Chen, H.N.; Jiang, H.; Lu, W.Y.; Pan, J.J.; Qian, Q.; Xue, D.W. Measuring the damage of heavy metal cadmium in rice seedlings by SRAP analysis combined with physiological and biochemical parameters. *J. Sci. Food Agric.* **2015**, *95*, 2292–2298. [CrossRef] [PubMed]
109. Sorrentino, M.C.; Capozzi, F.; Giordano, S.; Spagnuolo, V. Genotoxic effect of Pb and Cd on in vitro cultures of *Sphagnum palustre*: An evaluation by ISSR markers. *Chemosphere* **2017**, *181*, 208–215. [CrossRef] [PubMed]
110. Ghiani, A.; Fumagalli, P.; Van, T.N.; Gentili, R.; Citterio, S. The Combined toxic and genotoxic effects of Cd and As to plant bioindicator *Trifolium repens* L. *PLoS ONE* **2014**, *9*, 11. [CrossRef] [PubMed]
111. Lanier, C.; Bernard, F.; Dumez, S.; Leclercq, J.; Lemiere, S.; Vandenbulcke, F.; Nesslany, F.; Platel, A.; Devred, I.; Cuny, D.; et al. Combined effect of Cd and Pb spiked field soils on bioaccumulation, DNA damage, and peroxidase activities in *Trifolium repens*. *Environ. Sci. Pollut. Res.* **2016**, *23*, 1755–1767. [CrossRef] [PubMed]
112. Dogan, I.; Ozyigit, I.I.; Tombuloglu, G.; Sakcali, M.S.; Tombuloglu, H. Assessment of Cd-induced genotoxic damage in *Urtica pilulifera* L. using RAPD-PCR analysis. *Biotechnol. Biotechnol. Equip.* **2016**, *30*, 284–291. [CrossRef]

113. Monteiro, M.S.; Rodriguez, E.; Loureiro, J.; Mann, R.M.; Soares, A.; Santos, C. Flow cytometric assessment of Cd genotoxicity in three plants with different metal accumulation and detoxification capacities. *Ecotox. Environ. Safe.* **2010**, *73*, 1231–1237. [CrossRef] [PubMed]
114. Komaki, S.; Sugimoto, K. Control of the plant cell cycle by developmental and environmental cues. *Plant Cell Physiol.* **2012**, *53*, 953–964. [CrossRef]
115. Dewitte, W.; Murray, J.A.H. The plant cell cycle. *Annu. Rev. Plant Biol.* **2003**, *54*, 235–264. [CrossRef] [PubMed]
116. De Veylder, L.; Larkin, J.C.; Schnittger, A. Molecular control and function of endoreplication in development and physiology. *Trends Plant Sci.* **2011**, *16*, 624–634. [CrossRef] [PubMed]
117. Barow, M.; Meister, A. Endopolyploidy in seed plants is differently correlated to systematics, organ, life strategy and genome size. *Plant Cell Environ.* **2003**, *26*, 571–584. [CrossRef]
118. Bramsiepe, J.; Wester, K.; Weinl, C.; Roodbarkelari, F.; Kasili, R.; Larkin, J.C.; Hulskamp, M.; Schnittger, A. Endoreplication controls cell fate maintenance. *PLoS Genet.* **2010**, *6*, 14. [CrossRef] [PubMed]
119. Lee, H.O.; Davidson, J.M.; Duronio, R.J. Endoreplication: Polyploidy with purpose. *Genes Dev.* **2009**, *23*, 2461–2477. [CrossRef] [PubMed]
120. Ager, F.J.; Ynsa, M.D.; Dominguez-Solis, J.R.; Gotor, C.; Respaldiza, M.A.; Romero, L.C. Cadmium localization and quantification in the plant *Arabidopsis thaliana* using micro-PIXE. *Nucl. Instrum. Methods Phys. Res. Sect. B-Beam Interact. Mater. Atoms* **2002**, *189*, 494–498. [CrossRef]
121. Isaure, M.P.; Fayard, B.; Saffet, G.; Pairis, S.; Bourguignon, J. Localization and chemical forms of cadmium in plant samples by combining analytical electron microscopy and X-ray spectromicroscopy. *Spectroc. Acta Pt. B-Atom. Spectr.* **2006**, *61*, 1242–1252. [CrossRef]
122. Scholes, D.R.; Paige, K.N. Plasticity in ploidy: A generalized response to stress. *Trends Plant Sci.* **2015**, *20*, 165–175. [CrossRef]
123. Baryla, A.; Carrier, P.; Franck, F.; Coulomb, C.; Sahut, C.; Havaux, M. Leaf chlorosis in oilseed rape plants (*Brassica napus*) grown on cadmium-polluted soil: Causes and consequences for photosynthesis and growth. *Planta* **2001**, *212*, 696–709. [CrossRef] [PubMed]
124. Dalla Vecchia, F.; La Rocca, N.; Moro, I.; De Faveri, S.; Andreoli, C.; Rascio, N. Morphogenetic, ultrastructural and physiological damages suffered by submerged leaves of *Elodea canadensis* exposed to cadmium. *Plant Sci.* **2005**, *168*, 329–338. [CrossRef]
125. Fusconi, A.; Repetto, O.; Bona, E.; Massa, N.; Gallo, C.; Dumas-Gaudot, E.; Berta, G. Effects of cadmium on meristem activity and nucleus ploidy in roots of *Pisum sativum* L. cv. Frisson seedlings. *Environ. Exp. Bot.* **2006**, *58*, 253–260. [CrossRef]
126. Repetto, O.; Massa, N.; Gianinazzi-Pearson, V.; Dumas-Gaudot, E.; Berta, G. Cadmium effects on populations of root nuclei in two pea genotypes inoculated or not with the arbuscular mycorrhizal fungus *Glomus mosseae*. *Mycorrhiza* **2007**, *17*, 111–120. [CrossRef] [PubMed]
127. Wojtylakuchta, B.; Gabara, B. Changes in the content of DNA and NYS-stained nuclear, nucleolar and cytoplasmic proteins in cortex cells of pea (*Pisum sativum* cv. De Grace) roots treated with cadmium. *Biochem. Physiol. Pflanz.* **1991**, *187*, 67–76. [CrossRef]
128. Sobkowiak, R.; Deckert, J. Cadmium-induced changes in growth and cell cycle gene expression in suspension-culture cells of soybean. *Plant Physiol. Biochem.* **2003**, *41*, 767–772. [CrossRef]
129. Sobkowiak, R.; Deckert, J. The effect of cadmium on cell cycle control in suspension culture cells of soybean. *Acta Physiol. Plant.* **2004**, *26*, 335–344. [CrossRef]
130. Zhao, F.Y.; Hu, F.; Han, M.M.; Zhang, S.Y.; Liu, W. Superoxide radical and auxin are implicated in redistribution of root growth and the expression of auxin and cell-cycle genes in cadmium-stressed rice. *Russ. J. Plant Physiol.* **2011**, *58*, 851–863. [CrossRef]
131. Zhao, F.Y.; Wang, K.; Zhang, S.Y.; Ren, J.; Liu, T.; Wang, X. Crosstalk between ABA, auxin, MAPK signaling, and the cell cycle in cadmium-stressed rice seedlings. *Acta Physiol. Plant.* **2014**, *36*, 1879–1892. [CrossRef]
132. Tank, J.G.; Pandya, R.V.; Thaker, V.S. Phytohormones in regulation of the cell division and endoreduplication process in the plant cell cycle. *RSC Adv.* **2014**, *4*, 12605–12613. [CrossRef]
133. Vivancos, P.D.; Dong, Y.P.; Ziegler, K.; Markovic, J.; Pallardo, F.V.; Pellny, T.K.; Verrier, P.J.; Foyer, C.H. Recruitment of glutathione into the nucleus during cell proliferation adjusts whole-cell redox homeostasis in *Arabidopsis thaliana* and lowers the oxidative defence shield. *Plant J.* **2010**, *64*, 825–838. [CrossRef] [PubMed]

134. Vernoux, T.; Wilson, R.C.; Seeley, K.A.; Reichheld, J.P.; Muroy, S.; Brown, S.; Maughan, S.C.; Cobbett, C.S.; Van Montagu, M.; Inze, D.; et al. The *ROOT MERISTEMLESS1/CADMIUM SENSITIVE2* gene defines a glutathione-dependent pathway involved in initiation and maintenance of cell division during postembryonic root development. *Plant Cell* **2000**, *12*, 97–109. [CrossRef] [PubMed]
135. Kosetsu, K.; Matsunaga, S.; Nakagami, H.; Colcombet, J.; Sasabe, M.; Soyano, T.; Takahashi, Y.; Hirt, H.; Machida, Y. The MAP Kinase MPK4 Is required for cytokinesis in *Arabidopsis thaliana*. *Plant Cell* **2010**, *22*, 3778–3790. [CrossRef] [PubMed]
136. Yao, L.L.; Zhou, Q.; Pei, B.L.; Li, Y.Z. Hydrogen peroxide modulates the dynamic microtubule cytoskeleton during the defence responses to *Verticillium dahliae* toxins in *Arabidopsis*. *Plant Cell Environ.* **2011**, *34*, 1586–1598. [CrossRef] [PubMed]
137. Mhamdi, A.; Van Breusegem, F. Reactive oxygen species in plant development. *Development* **2018**, *145*. [CrossRef] [PubMed]
138. Yi, D.; Kamei, C.L.A.; Cools, T.; Vanderauwera, S.; Takahashi, N.; Okushima, Y.; Eekhout, T.; Yoshiyama, K.O.; Larkin, J.; Van den Daele, H.; et al. The *Arabidopsis* SIAMESE-RELATED cyclin-dependent Kinase inhibitors SMR5 and SMR7 regulate the DNA damage checkpoint in response to reactive oxygen species. *Plant Cell* **2014**, *26*, 296–309. [CrossRef]
139. Zhan, Y.H.; Zhang, C.H.; Zheng, Q.X.; Huang, Z.A.; Yu, C.L. Cadmium stress inhibits the growth of primary roots by interfering auxin homeostasis in *Sorghum bicolor* seedlings. *J. Plant Biol.* **2017**, *60*, 593–603. [CrossRef]
140. Locato, V.; Paradiso, A.; Sabetta, W.; De Gara, L.; de Pinto, M.C. Chapter nine - Nitric oxide and reactive oxygen species in PCD signaling. *Adv. Bot. Res.* **2016**, *77*, 165–192.
141. Kuthanova, A.; Fischer, L.; Nick, P.; Opatrny, Z. Cell cycle phase-specific death response of tobacco BY-2 cell line to cadmium treatment. *Plant Cell Environ.* **2008**, *31*, 1634–1643. [CrossRef]
142. Pena, L.B.; Barcia, R.A.; Azpilicueta, C.E.; Mendez, A.A.E.; Gallego, S.M. Oxidative post translational modifications of proteins related to cell cycle are involved in cadmium toxicity in wheat seedlings. *Plant Sci.* **2012**, *196*, 1–7. [CrossRef]
143. Garnier, L.; Simon-Plas, F.; Thuleau, P.; Agnel, J.P.; Blein, J.P.; Ranjeva, R.; Montillet, J.L. Cadmium affects tobacco cells by a series of three waves of reactive oxygen species that contribute to cytotoxicity. *Plant Cell Environ.* **2006**, *29*, 1956–1969. [CrossRef] [PubMed]
144. Tamas, L.; Mistrik, I.; Zelinova, V. Heavy metal-induced reactive oxygen species and cell death in barley root tip. *Environ. Exp. Bot.* **2017**, *140*, 34–40. [CrossRef]
145. De Michele, R.; Vurro, E.; Rigo, C.; Costa, A.; Elviri, L.; Di Valentin, M.; Careri, M.; Zottini, M.; Sanita di Toppi, L.; Lo Schiavo, F. Nitric oxide is involved in cadmium-induced programmed cell death in *Arabidopsis* suspension cultures. *Plant Physiol.* **2009**, *150*, 217–228. [CrossRef] [PubMed]
146. Elviri, L.; Speroni, F.; Careri, M.; Mangia, A.; Sanita di Toppi, L.; Zottini, M. Identification of in vivo nitrosylated phytochelatins in *Arabidopsis thaliana* cells by liquid chromatography-direct electrospray-linear ion trap-mass spectrometry. *J. Chromatogr. A* **2010**, *1217*, 4120–4126. [CrossRef] [PubMed]
147. Arasimowicz-Jelonek, M.; Floryszak-Wieczorek, J.; Deckert, J.; Rucinska-Sobkowiak, R.; Gzyl, J.; Pawlak-Sprada, S.; Abramowski, D.; Jelonek, T.; Gwozdz, E.A. Nitric oxide implication in cadmium-induced programmed cell death in roots and signaling response of yellow lupine plants. *Plant Physiol. Biochem.* **2012**, *58*, 124–134. [CrossRef] [PubMed]
148. Ma, W.W.; Xu, W.Z.; Xu, H.; Chen, Y.S.; He, Z.Y.; Ma, M. Nitric oxide modulates cadmium influx during cadmium-induced programmed cell death in tobacco BY-2 cells. *Planta* **2010**, *232*, 325–335. [CrossRef] [PubMed]
149. Ye, Y.; Li, Z.; Xing, D. Nitric oxide promotes MPK6-mediated caspase-3-like activation in cadmium-induced *Arabidopsis thaliana* programmed cell death. *Plant Cell Environ.* **2013**, *36*, 1–15. [CrossRef] [PubMed]
150. Iakimova, E.T.; Woltering, E.J.; Kapchina-Toteva, V.M.; Harren, F.J.M.; Cristescu, S.M. Cadmium toxicity in cultured tomato cells - Role of ethylene, proteases and oxidative stress in cell death signaling. *Cell Biol. Int.* **2008**, *32*, 1521–1529. [CrossRef]
151. Pormehr, M.; Ghanati, F.; Sharifi, M.; McCabe, P.F.; Hosseinkhani, S.; Zare-Maivan, H. The role of SIPK signaling pathway in antioxidant activity and programmed cell death of tobacco cells after exposure to cadmium. *Plant Sci.* **2019**, *280*, 416–423. [CrossRef]

152. Zhang, W.N.; Chen, W.L. Role of salicylic acid in alleviating photochemical damage and autophagic cell death induction of cadmium stress in *Arabidopsis thaliana*. *Photochem. Photobiol. Sci.* **2011**, *10*, 947–955. [CrossRef]
153. Gzyl, J.; Chmielowska-Bak, J.; Przymusinski, R. Gamma-tubulin distribution and ultrastructural changes in root cells of soybean (*Glycine max* L.) seedlings under cadmium stress. *Environ. Exp. Bot.* **2017**, *143*, 82–90. [CrossRef]
154. Yue, J.Y.; Wei, X.J.; Wang, H.Z. Cadmium tolerant and sensitive wheat lines: Their differences in pollutant accumulation, cell damage, and autophagy. *Biol. Plant.* **2018**, *62*, 379–387. [CrossRef]
155. De Araujo, R.P.; de Almeida, A.A.F.; Pereira, L.S.; Mangabeira, P.A.O.; Souza, J.O.; Pirovani, C.P.; Ahnert, D.; Baligar, V.C. Photosynthetic, antioxidative, molecular and ultrastructural responses of young cacao plants to Cd toxicity in the soil. *Ecotox. Environ. Safe.* **2017**, *144*, 148–157. [CrossRef] [PubMed]
156. Gou, W.; Li, X.; Guo, S.; Liu, Y.; Li, F.; Xie, Q. Autophagy in plant: A new orchestrator in the regulation of the phytohormones homeostasis. *Int. J. Mol. Sci.* **2019**, *20*, 2900. [CrossRef] [PubMed]
157. Signorelli, S.; Tarkowski, L.P.; Van den Ende, W.; Bassham, D.C. Linking autophagy to abiotic and biotic stress responses. *Trends Plant Sci.* **2019**, *24*, 413–430. [CrossRef] [PubMed]
158. Cosgrove, D.J. Growth of the plant cell wall. *Nat. Rev. Mol. Cell Biol.* **2005**, *6*, 850–861. [CrossRef] [PubMed]
159. Mohnen, D. Pectin structure and biosynthesis. *Curr. Opin. Plant Biol.* **2008**, *11*, 266–277. [CrossRef] [PubMed]
160. Scheller, H.V.; Ulvskov, P. Hemicelluloses. *Annu. Rev. Plant Biol.* **2010**, *61*, 263–289. [CrossRef]
161. Krzeslowska, M. The cell wall in plant cell response to trace metals: Polysaccharide remodeling and its role in defense strategy. *Acta Physiol. Plant.* **2011**, *33*, 35–51. [CrossRef]
162. Grant, G.T.; Morris, E.R.; Rees, D.A.; Smith, P.J.C.; Thom, D. Biological interactions between polysaccharides and divalent cations - Egg-box model. *FEBS Lett.* **1973**, *32*, 195–198. [CrossRef]
163. Hall, J.L. Cellular mechanisms for heavy metal detoxification and tolerance. *J. Exp. Bot.* **2002**, *53*, 1–11. [CrossRef] [PubMed]
164. Tenhaken, R. Cell wall remodeling under abiotic stress. *Front. Plant Sci.* **2015**, *5*, 9. [CrossRef] [PubMed]
165. Verslues, P.E.; Agarwal, M.; Katiyar-Agarwal, S.; Zhu, J.H.; Zhu, J.K. Methods and concepts in quantifying resistance to drought, salt and freezing, abiotic stresses that affect plant water status. *Plant J.* **2006**, *45*, 523–539. [CrossRef] [PubMed]
166. Van Belleghem, F.; Cuypers, A.; Semane, B.; Smeets, K.; Vangronsveld, J.; d'Haen, J.; Valcke, R. Subcellular localization of cadmium in roots and leaves of *Arabidopsis thaliana*. *New Phytol.* **2007**, *173*, 495–508. [CrossRef] [PubMed]
167. Vazquez, S.; Goldsbrough, P.; Carpena, R.O. Assessing the relative contributions of phytochelatins and the cell wall to cadmium resistance in white lupin. *Physiol. Plant.* **2006**, *128*, 487–495. [CrossRef]
168. Huang, W.L.; Bai, Z.Q.; Jiao, J.; Yuan, H.L.; Bao, Z.A.; Chen, S.N.; Ding, M.H.; Liang, Z.S. Distribution and chemical forms of cadmium in *Coptis chinensis* Franch. determined by laser ablation ICP-MS, cell fractionation, and sequential extraction. *Ecotox. Environ. Safe.* **2019**, *171*, 894–903. [CrossRef] [PubMed]
169. Kramer, U. Metal Hyperaccumulation in Plants. *Annu. Rev. Plant Biol.* **2010**, *61*, 517–534. [CrossRef]
170. Zhao, H.; Jin, Q.J.; Wang, Y.J.; Chu, L.L.; Li, X.; Xu, Y.C. Effects of nitric oxide on alleviating cadmium stress in *Typha angustifolia*. *Plant Growth Regul.* **2016**, *78*, 243–251. [CrossRef]
171. Dong, Y.J.; Chen, W.F.; Xu, L.L.; Kong, J.; Liu, S.; He, Z.L. Nitric oxide can induce tolerance to oxidative stress of peanut seedlings under cadmium toxicity. *Plant Growth Regul.* **2016**, *79*, 19–28. [CrossRef]
172. Berni, R.; Luyckx, M.; Xu, X.; Legay, S.; Sergeant, K.; Hausman, J.F.; Lutts, S.; Cai, G.; Guerriero, G. Reactive oxygen species and heavy metal stress in plants: Impact on the cell wall and secondary metabolism. *Environ. Exp. Bot.* **2019**, *161*, 98–106. [CrossRef]
173. Loix, C.; Huybrechts, M.; Vangronsveld, J.; Gielen, M.; Keunen, E.; Cuypers, A. Reciprocal interactions between cadmium-induced cell wall responses and oxidative stress in plants. *Front. Plant Sci.* **2017**, *8*, 19. [CrossRef] [PubMed]
174. Parrotta, L.; Guerriero, G.; Sergeant, K.; Cal, G.; Hausman, J.F. Target or barrier? The cell wall of early- and later-diverging plants vs cadmium toxicity: Differences in the response mechanisms. *Front. Plant Sci.* **2015**, *6*, 16. [CrossRef] [PubMed]
175. Douchiche, O.; Rihouey, C.; Schaumann, A.; Driouich, A.; Morvan, C. Cadmium-induced alterations of the structural features of pectins in flax hypocotyl. *Planta* **2007**, *225*, 1301–1312. [CrossRef] [PubMed]

176. Paynel, F.; Schaumann, A.; Arkoun, M.; Douchiche, O.; Morvan, C. Temporal regulation of cell-wall pectin methylesterase and peroxidase isoforms in cadmium-treated flax hypocotyl. *Ann. Bot.* **2009**, *104*, 1363–1372. [CrossRef] [PubMed]
177. Xiong, J.; An, L.Y.; Lu, H.; Zhu, C. Exogenous nitric oxide enhances cadmium tolerance of rice by increasing pectin and hemicellulose contents in root cell wall. *Planta* **2009**, *230*, 755–765. [CrossRef] [PubMed]
178. Zhao, Y.Y.; He, C.X.; Wu, Z.C.; Liu, X.W.; Cai, M.M.; Jia, W.; Zhao, X.H. Selenium reduces cadmium accumulation in seed by increasing cadmium retention in root of oilseed rape (*Brassica napus* L.). *Environ. Exp. Bot.* **2019**, *158*, 161–170. [CrossRef]
179. Li, H.B.; Zheng, X.W.; Tao, L.X.; Yang, Y.J.; Gao, L.; Xiong, J. Aeration increases cadmium (Cd) retention by enhancing iron plaque formation and regulating pectin synthesis in the roots of rice (*Oryza sativa*) seedlings. *Rice* **2019**, *12*, 14. [CrossRef]
180. Finger-Teixeira, A.; Ferrarese, M.D.L.; Soares, A.R.; da Silva, D.; Ferrarese, O. Cadmium-induced lignification restricts soybean root growth. *Ecotox. Environ. Safe.* **2010**, *73*, 1959–1964. [CrossRef]
181. Schutzendubel, A.; Schwanz, P.; Teichmann, T.; Gross, K.; Langenfeld-Heyser, R.; Godbold, D.L.; Polle, A. Cadmium-induced changes in antioxidative systems, hydrogen peroxide content, and differentiation in Scots pine roots. *Plant Physiol.* **2001**, *127*, 887–898. [CrossRef]
182. Liu, Q.Q.; Luo, L.; Zheng, L.Q. Lignins: Biosynthesis and biological functions in plants. *Int. J. Mol. Sci.* **2018**, *19*, 335. [CrossRef]
183. Rui, H.Y.; Chen, C.; Zhang, X.X.; Shen, Z.G.; Zhang, F.Q. Cd-induced oxidative stress and lignification in the roots of two *Vicia sativa* L. varieties with different Cd tolerances. *J. Hazard. Mater.* **2016**, *301*, 304–313. [CrossRef] [PubMed]
184. Schmidt, R.; Kunkowska, A.B.; Schippers, J.H.M. Role of reactive oxygen species during cell expansion in leaves. *Plant Physiol.* **2016**, *172*, 2098–2106. [CrossRef] [PubMed]
185. Muller, K.; Linkies, A.; Vreeburg, R.A.M.; Fry, S.C.; Krieger-Liszkay, A.; Leubner-Metzger, G. In vivo cell wall loosening by hydroxyl radicals during cress seed germination and elongation growth. *Plant Physiol.* **2009**, *150*, 1855–1865. [CrossRef] [PubMed]
186. Schopfer, P. Hydrogen peroxide-mediated cell-wall stiffening in vitro in maize coleoptiles. *Planta* **1996**, *199*, 43–49. [CrossRef]
187. Vatehova, Z.; Malovikova, A.; Kollarova, K.; Kucerova, D.; Liskova, D. Impact of cadmium stress on two maize hybrids. *Plant Physiol. Biochem.* **2016**, *108*, 90–98. [CrossRef] [PubMed]
188. Wang, P.; Yang, B.; Wan, H.B.; Fang, X.L.; Yang, C.Y. The differences of cell wall in roots between two contrasting soybean cultivars exposed to cadmium at young seedlings. *Environ. Sci. Pollut. Res.* **2018**, *25*, 29705–29714. [CrossRef]
189. Somerville, C. Cellulose synthesis in higher plants. *Annu. Rev. Cell Dev. Biol.* **2006**, *22*, 53–78. [CrossRef] [PubMed]
190. Haigler, C.H.; Ivanova-Datcheva, M.; Hogan, P.S.; Salnikov, V.V.; Hwang, S.; Martin, K.; Delmer, D.P. Carbon partitioning to cellulose synthesis. *Plant Mol. Biol.* **2001**, *47*, 29–51. [CrossRef] [PubMed]
191. Gutsch, A.; Keunen, E.; Guerriero, G.; Renaut, J.; Cuypers, A.; Hausman, J.F.; Sergeant, K. Long-term cadmium exposure influences the abundance of proteins that impact the cell wall structure in *Medicago sativa* stems. *Plant Biol.* **2018**, *20*, 1023–1035. [CrossRef]
192. Guo, H.P.; Hong, C.T.; Xiao, M.Z.; Chen, X.M.; Chen, H.M.; Zheng, B.S.; Jiang, D.A. Real-time kinetics of cadmium transport and transcriptomic analysis in low cadmium accumulator *Miscanthus sacchariflorus*. *Planta* **2016**, *244*, 1289–1302. [CrossRef]
193. Liu, Y.; Liu, D.C.; Zhang, H.Y.; Gao, H.B.; Guo, X.L.; Wang, D.M.; Zhang, X.Q.; Zhang, A.M. The alpha- and beta-expansin and xyloglucan endotransglucosylase/hydrolase gene families of wheat: Molecular cloning, gene expression, and EST data mining. *Genomics* **2007**, *90*, 516–529. [CrossRef] [PubMed]
194. Cosgrove, D.J. Plant expansins: Diversity and interactions with plant cell walls. *Curr. Opin. Plant Biol.* **2015**, *25*, 162–172. [CrossRef] [PubMed]
195. Kuluev, B.; Avalbaev, A.; Mikhaylova, E.; Nikonorov, Y.; Berezhneva, Z.; Chemeris, A. Expression profiles and hormonal regulation of tobacco expansin genes and their involvement in abiotic stress response. *J. Plant Physiol.* **2016**, *206*, 1–12. [CrossRef] [PubMed]
196. Sun, T.; Zhang, Y.X.; Chai, T.Y. Cloning, characterization, and expression of the BjEXPA1 gene and its promoter region from *Brassica juncea* L. *Plant Growth Regul.* **2011**, *64*, 39–51. [CrossRef]

197. Ren, Y.Q.; Chen, Y.H.; An, J.; Zhao, Z.X.; Zhang, G.Q.; Wang, Y.; Wang, W. Wheat expansin gene *TaEXPA2* is involved in conferring plant tolerance to Cd toxicity. *Plant Sci.* **2018**, *270*, 245–256. [CrossRef] [PubMed]
198. Van Sandt, V.S.T.; Suslov, D.; Verbelen, J.P.; Vissenberg, K. Xyloglucan endotransglucosylase activity loosens a plant cell wall. *Ann. Bot.* **2007**, *100*, 1467–1473. [CrossRef] [PubMed]
199. Kong, X.P.; Li, C.L.; Zhang, F.; Yu, Q.Q.; Gao, S.; Zhang, M.L.; Tian, H.Y.; Zhang, J.; Yuan, X.Z.; Ding, Z.J. Ethylene promotes cadmium-induced root growth inhibition through EIN3 controlled *XTH33* and *LSU1* expression in *Arabidopsis*. *Plant Cell Environ.* **2018**, *41*, 2449–2462. [CrossRef] [PubMed]
200. Han, Y.S.; Sa, G.; Sun, J.; Shen, Z.D.; Zhao, R.; Ding, M.Q.; Deng, S.R.; Lu, Y.J.; Zhang, Y.H.; Shen, X.; et al. Overexpression of *Populus euphratica* xyloglucan endotransglucosylase/hydrolase gene confers enhanced cadmium tolerance by the restriction of root cadmium uptake in transgenic tobacco. *Environ. Exp. Bot.* **2014**, *100*, 74–83. [CrossRef]
201. Yu, R.G.; Jiang, Q.; Xv, C.; Li, L.; Bu, S.J.; Shi, G.R. Comparative proteomics analysis of peanut roots reveals differential mechanisms of cadmium detoxification and translocation between two cultivars differing in cadmium accumulation. *BMC Plant Biol.* **2019**, *19*, 15. [CrossRef]
202. Majda, M.; Robert, S. The role of auxin in cell wall expansion. *Int. J. Mol. Sci.* **2018**, *19*, 951. [CrossRef]
203. Rascio, N.; Navari-Izzo, F. Heavy metal hyperaccumulating plants: How and why do they do it? And what makes them so interesting? *Plant Sci.* **2011**, *180*, 169–181. [CrossRef] [PubMed]
204. Benzarti, S.; Mohri, S.; Ono, Y. Plant response to heavy metal toxicity: Comparative study between the hyperaccumulator *Thlaspi caerulescens* (ecotype Ganges) and nonaccumulator plants: Lettuce, radish, and alfalfa. *Environ. Toxicol.* **2008**, *23*, 607–617. [CrossRef] [PubMed]
205. Meyer, C.L.; Juraniec, M.; Huguet, S.; Chaves-Rodriguez, E.; Salis, P.; Isaure, M.P.; Goormaghtigh, E.; Verbruggen, N. Intraspecific variability of cadmium tolerance and accumulation, and cadmium-induced cell wall modifications in the metal hyperaccumulator *Arabidopsis halleri*. *J. Exp. Bot.* **2015**, *66*, 3215–3227. [CrossRef] [PubMed]
206. Verbruggen, N.; Hermans, C.; Schat, H. Molecular mechanisms of metal hyperaccumulation in plants. *New Phytol.* **2009**, *181*, 759–776. [CrossRef] [PubMed]
207. Li, T.Q.; Tao, Q.; Shohag, M.J.I.; Yang, X.E.; Sparks, D.L.; Liang, Y.C. Root cell wall polysaccharides are involved in cadmium hyperaccumulation in *Sedum alfredii*. *Plant Soil* **2015**, *389*, 387–399. [CrossRef]
208. Baliardini, C.; Corso, M.; Verbruggen, N. Transcriptomic analysis supports the role of CATION EXCHANGER 1 in cellular homeostasis and oxidative stress limitation during cadmium stress. *Plant Signal. Behav.* **2016**, *11*, 5. [CrossRef]
209. Peng, J.S.; Wang, Y.J.; Ding, G.; Ma, H.L.; Zhang, Y.J.; Gong, J.M. A pivotal role of cell wall in cadmium accumulation in the Crassulaceae hyperaccumulator *Sedum plumbizincicola*. *Mol. Plant.* **2017**, *10*, 771–774. [CrossRef]
210. Anjum, N.A.; Umar, S.; Ahmad, A.; Iqbal, M.; Khan, N.A. Ontogenic variation in response of *Brassica campestris* L. to cadmium toxicity. *J. Plant Interact.* **2008**, *3*, 189–198. [CrossRef]
211. Hancock, L.M.S.; Ernst, C.L.; Charneskie, R.; Ruane, L.G. Effects of cadmium and mycorrhizal fungi on growth, fitness and cadmium accumulation in flax (*Linum unitatissimum*; Linaceae). *Am. J. Bot.* **2012**, *99*, 1445–1452. [CrossRef]
212. Keunen, E.; Truyens, S.; Bruckers, L.; Remans, T.; Vangronsveld, J.; Cuypers, A. Survival of Cd-exposed *Arabidopsis thaliana*: Are these plants reproductively challenged? *Plant Physiol. Biochem.* **2011**, *49*, 1084–1091. [CrossRef]
213. Maistri, S.; DalCorso, G.; Vicentini, V.; Furini, A. Cadmium affects the expression of *ELF4*, a circadian clock gene in *Arabidopsis*. *Environ. Exp. Bot.* **2011**, *72*, 115–122. [CrossRef]
214. Gusmão Lima, A.I.; Pereira, S.I.A.; de Almeida Figueira, E.M.; Caldeira, G.C.N.; Caldeira, H.D.Q. Cadmium uptake in pea plants under environmentally-relevant exposures: The risk of food-chain transfer. *J. Plant Nutr.* **2006**, *29*, 2165–2177. [CrossRef]
215. Rodda, M.S.; Li, G.; Reid, R.J. The timing of grain Cd accumulation in rice plants: The relative importance of remobilisation within the plant and root Cd uptake post-flowering. *Plant Soil* **2011**, *347*, 105–114. [CrossRef]
216. Hediji, H.; Djebali, W.; Belkadhi, A.; Cabasson, C.; Moing, A.; Rolin, D.; Brouquisse, R.; Gallusci, P.; Chaibi, W. Impact of long-term cadmium exposure on mineral content of *Solanum lycopersicum* plants: Consequences on fruit production. *S. Afr. J. Bot.* **2015**, *97*, 176–181. [CrossRef]

217. Shekari, L.; Aroiee, H.; Mirshekari, A.; Nemati, H. Protective role of selenium on cucumber (*Cucumis sativus* L.) exposed to cadmium and lead stress during reproductive stage role of selenium on heavy metals stress. *J. Plant Nutr.* **2019**, *42*, 529–542. [CrossRef]
218. Soldatova, N.A.; Khryanin, V.N. The effects of heavy metal salts on the phytohormonal status and sex expression in marijuana. *Russ. J. Plant Physiol.* **2010**, *57*, 96–100. [CrossRef]
219. Pan, J.T.; Li, D.Q.; Shu, Z.F.; Jiang, X.; Ma, W.W.; Wang, W.D.; Zhu, J.J.; Wang, Y.H. CsPDC-E1 alpha, a novel pyruvate dehydrogenase complex E1 alpha subunit gene from *Camellia sinensis*, is induced during cadmium inhibiting pollen tube growth. *Can. J. Plant Sci.* **2018**, *98*, 62–70. [CrossRef]
220. Sabrine, H.; Afif, H.; Mohamed, B.; Hamadi, B.; Maria, H. Effects of cadmium and copper on pollen germination and fruit set in pea (*Pisum sativum* L.). *Sci. Hortic.* **2010**, *125*, 551–555. [CrossRef]
221. Xiong, Z.T.; Peng, Y.H. Response of pollen germination and tube growth to cadmium with special reference to low concentration exposure. *Ecotox. Environ. Safe.* **2001**, *48*, 51–55. [CrossRef]
222. Sawidis, T. Effect of cadmium on pollen germination and tube growth in *Lilium longiflorum* and *Nicotiana tabacum*. *Protoplasma* **2008**, *233*, 95–106. [CrossRef]
223. Sharafi, Y.; Talebi, S.F.; Talei, D. Effects of heavy metals on male gametes of sweet cherry. *Caryologia* **2017**, *70*, 166–173. [CrossRef]
224. Wang, X.X.; Zhang, S.S.; Gao, Y.; Lu, W.G.; Sheng, X.Y. Different heavy metals have various effects on *Picea wilsonii* pollen germination and tube growth. *Plant Signal. Behav.* **2015**, *10*, 4. [CrossRef] [PubMed]
225. Wudick, M.M.; Feijo, J.A. At the Intersection: Merging Ca^{2+} and ROS signaling pathways in pollen. *Mol. Plant* **2014**, *7*, 1595–1597. [CrossRef] [PubMed]
226. Smirnova, A.V.; Matveyeva, N.P.; Yermakov, I.P. Reactive oxygen species are involved in regulation of pollen wall cytomechanics. *Plant Biol.* **2014**, *16*, 252–257. [CrossRef] [PubMed]
227. Carvalho, M.E.A.; Piotto, F.A.; Nogueira, M.L.; Gomes, F.G.; Chamma, H.; Pizzaia, D.; Azevedo, R.A. Cadmium exposure triggers genotype-dependent changes in seed vigor and germination of tomato offspring. *Protoplasma* **2018**, *255*, 989–999. [CrossRef] [PubMed]
228. Truyens, S.; Beckers, B.; Thijs, S.; Weyens, N.; Cuypers, A.; Vangronsveld, J. Cadmium-induced and trans-generational changes in the cultivable and total seed endophytic community of *Arabidopsis thaliana*. *Plant Biol.* **2016**, *18*, 376–381. [CrossRef] [PubMed]
229. Truyens, S.; Weyens, N.; Cuypers, A.; Vangronsveld, J. Changes in the population of seed bacteria of transgenerationally Cd-exposed *Arabidopsis thaliana*. *Plant Biol.* **2013**, *15*, 971–981. [CrossRef]
230. Remans, T.; Thijs, S.; Truyens, S.; Weyens, N.; Schellingen, K.; Keunen, E.; Gielen, H.; Cuypers, A.; Vangronsveld, J. Understanding the development of roots exposed to contaminants and the potential of plant-associated bacteria for optimization of growth. *Ann. Bot.* **2012**, *110*, 239–252. [CrossRef]
231. Bhat, J.A.; Shivaraj, S.M.; Singh, P.; Navadagi, D.B.; Tripathi, D.K.; Dash, P.K.; Solanke, A.U.; Sonah, H.; Deshmukh, R. Role of silicon in mitigation of heavy metal stresses in crop plants. *Plants* **2019**, *8*, 71. [CrossRef]
232. Gutsch, A.; Vandionant, S.; Sergeant, K.; Jozefczak, M.; Vangronsveld, J.; Hausman, J.-F.; Cuypers, A. Systems Biology of Metal Tolerance in Plants: A Case Study on the Effects of Cd Exposure on Two Model Plants. In *Plant Metallomics and Functional Omics: A System-Wide Perspective*; Sablok, G., Ed.; Springer International Publishing: Cham, Switzerland, 2019; pp. 23–37.

© 2019 by the authors. Licensee MDPI, Basel, Switzerland. This article is an open access article distributed under the terms and conditions of the Creative Commons Attribution (CC BY) license (http://creativecommons.org/licenses/by/4.0/).

International Journal of
Molecular Sciences

Review

Advances in the Uptake and Transport Mechanisms and QTLs Mapping of Cadmium in Rice

Jingguang Chen [1,2], Wenli Zou [1], Lijun Meng [1,*], Xiaorong Fan [2,*], Guohua Xu [2] and Guoyou Ye [1,3]

1. CAAS-IRRI Joint Laboratory for Genomics-Assisted Germplasm Enhancement, Agricultural Genomics Institute in Shenzhen, Chinese Academy of Agricultural Sciences, Shenzhen 518120, China
2. College of Resources and Environmental Sciences, Nanjing Agricultural University, Nanjing 210095, China
3. Strategic Innovation Platform, International Rice Research Institute, DAPO Box 7777, Metro Manila 1226, Philippines

* Correspondence: menglijun@caas.cn (L.M.); xiaorongfan@njau.edu.cn (X.F.)

Received: 28 June 2019; Accepted: 11 July 2019; Published: 11 July 2019

Abstract: Cadmium (Cd), as a heavy metal, presents substantial biological toxicity and has harmful effects on human health. To lower the ingress levels of human Cd, it is necessary for Cd content in food crops to be reduced, which is of considerable significance for ensuring food safety. This review will summarize the genetic traits of Cd accumulation in rice and examine the mechanism of Cd uptake and translocation in rice. The status of genes related to Cd stress and Cd accumulation in rice in recent years will be summarized, and the genes related to Cd accumulation in rice will be classified according to their functions. In addition, an overview of quantitative trait loci (QTLs) mapping populations in rice will be introduced, aiming to provide a theoretical reference for the breeding of rice varieties with low Cd accumulation. Finally, existing problems and prospects will be put forward.

Keywords: cadmium accumulation; absorption and transport; QTL location; mapping population; rice (*Oryza sativa* L.)

---

1. Introduction

Cadmium (Cd) is a soil contaminant and with a high mobility in living organisms, and is characterized as a toxic heavy metal [1,2]. In China, about 2.786×10^9 m^2 of agricultural land is contaminated by Cd [3]. Frequent applications of nitrogen fertilizer in the agricultural land of many areas of China have resulted in more acidic soil, and acidic soil means that cadmium is more easily absorbed by plants [4]. Rice (*Oryza sativa* L.) is the main food for more than half of the world's population. Cd is easily transferred from soil to rice and accumulates in rice plants and grains [2,3], and is then enriched in the human body through the food chain, thereby threatening human health [5–7], and causing effects such as anemia, cancer, heart failure, hypertension, cerebral infarction, proteinuria, severe lung damage, eye cataract formation, osteoporosis, emphysema, and renal insufficiency [8,9]. It is worth mentioning that Itai-itai disease, which occurred in Japan in the 1950s, was caused by the long-term intake of cadmium-contaminated rice [10]. On average, weekly Cd accumulation was as high as 3–4 mg kg^{-1} body weight in Japan at that time [11]. Between 1990 and 2015, the average dietary Cd intake of the general population more than doubled in China [12,13]. Therefore, reducing Cd uptake by crops, especially rice, is of great significance to food safety and human health.

The purpose of this review is to explore the mechanism of cadmium uptake and transport and the genetic characteristics of Cd accumulation in rice, and to summarize the research status of genes and QTLs related to cadmium stress and cadmium accumulation in rice. It has important guiding significance for breeding high-quality rice varieties with a low accumulation of Cd in grain and the safe production of rice in mild and moderate Cd-contaminated soil.

2. Toxic Effects of Cadmium Exposure on Rice

Cd stress seriously affects rice germination and growth [2,3,14–18], and it was found that excessive Cd exposure can not only significantly decrease the rice seed germination rate [14], but also cause chlorosis and necrosis in rice plants during the vegetative stage [19,20]. Cd stress causes severe physical and physiological changes in rice plants as it causes a reduction in the length; width; and number of roots, shoots, and leaves. Furthermore, chlorophyll contents, stomatal conductance, and the water use efficiency of rice are also significantly affected [3,17,18,21–23]. Cd also affects the absorption and transport of essential nutrients in rice [15,16,18–20]. Additionally, Cd can be transported to rice grains, reducing their yield, quality, and nutrients [15,16,24–27]. In general, Cd stress inhibits rice growth [18,28–30].

Rice possesses some tolerance mechanisms to cadmium at physiological and molecular levels [31–35]. As root cell walls of the outermost layer have direct contact with the soil solution, this protects the protoplasts against Cd toxicity [36–38]. Furthermore, plants reduce Cd translocation to the shoots by immobilizing Cd in the cell walls and vacuoles of root cells, thus reducing their sensitivity and the harm of Cd to another cellular organelle [39–42]. Several adenosine triphosphate (ATP)-binding cassette (ABC) proteins have been reported to mediate vacuolar compartmentation of Cd-glutathione and/or phytochelatin (PC) conjugates in *Arabidopsis thaliana* [43,44]. Rice *OsPDR5/ABCG43* is likely to encode ABC-type protein functions in Cd extrusion from the cytoplasm [45]. Overexpression of Cd transporter OsHMA3 located in vacuole membranes in rice roots can increase the tolerance of rice to Cd and reduce the accumulation of Cd in grains [46,47]. Exudates of roots contain metal chelators which play a role in the adjustment of the rhizosphere pH and the metal chelating process [48]. Most of the chelated toxic metals inside plants target vacuoles through metal detoxification processes [38,49]. Organic acids secreted from roots, e.g., malate, citrate, etc., are involved in metal uptake, the long-distance transport of metal, and the transport of metal into vacuoles [50,51]. It was found that chelators play a crucial role in keeping Cd in the rice roots and form a barrier in Cd translocation [52].

Cd stress can induce plants to enhance their antioxidant defense system and regulate ion homeostasis to improve their tolerance to Cd [32,53–58]. For example, Cd stress can induce plants to increase the production of glutathione (GSH), abscisic acid (ABA), salicylic acid (SA), jasmonic acid (JA), and nitric oxide (NO) [59–65]. Mitogen-activated protein kinase OsWJUMK1, OsMSRMK2, OsMSRMK3, and OsMAPK2 can affect rice root growth under Cd stress by regulating auxin signal changes [66–69]. Auxin transporter *OsAUX1* has been reported to be involved in root development and the Cd stress response in rice [34]. In addition, the concentration of iron and cadmium was positively correlated during rice seedling growth [70]. It has been reported that increasing the supply of boron, iron, zinc, silicon, or magnesium can reduce the accumulation and toxicity of cadmium in rice [71–76].

In addition, some genes related to Cd stress have been reported in rice (Table 1). *OsHMA9* is a copper efflux protein located in the plasma membrane, which may have a cadmium efflux function to excrete Cd from root cells and reduce Cd accumulation in rice [77]. Knock-out of the low cadmium gene (*LCD*) reduced the accumulation of cadmium and increased the growth of rice under the condition of an excessive cadmium supply, and LCD may be a protein related to cadmium homeostasis [78]. Overexpression of *OsCDT1* can increase the growth of *Arabidopsis thaliana* under cadmium treatment; the cysteine-rich peptide encoded by *OsCDT1* is possibly involved in rice Cd tolerance [79]. *OsCLT1* probably mediates the export of γ-glutamylcysteine and glutathione from plastids to the cytoplasm, which in turn affects As and Cd detoxification in rice [80].

Table 1. Genes of Rice Reported to be Regulated During Cadmium (Cd)-Exposure.

Gene	Chr.	Physical Location (bp)	Gene Name	Function	Reference
OsCDT3	1	4066623–4067218	Encoding a Cys-rich peptide	Cd uptake inhibitor	[79]
Ospdr9	1	24075065–24082181	Multidrug resistance ABC transporter	Redox protection in Cd stress	[81]
OsWJUMK1	1	29398191–29402466	Mitogen-activated protein kinase	Cd signal	[82]
OsHsfA4a	1	31370413–31372729	Heat shock transcription factor gene	Cd tolerance	[83]
OsAUX1	1	36998334–37004685	Auxin transport protein	Root development and Cd stress response	[34]
OsCLT1	1	42086484–42095424	CRT-like transporter 1	Cd tolerance	[80]
OsLCD	1	42162592–42166462	Low cadmium	Cd tolerance and accumulation	[78]
OsZIP1	1	42905566–42907474	Zinc- and iron-regulated transporter	Cd and Zn transport	[84,85]
ricMT	1	43047164–43047861	Metallothionein gene	Cd tolerance	[86]
OsCDT4	2	6078179–6079111	Encoding a Cys-rich peptide	Cd uptake inhibitor	[79]
OsNAAT1	2	11997094–12002633	Nicotinamide aminotransferase gene	Cd accumulation	[87]
CAL1	2	25190487–25191188	defensin-like protein	Cd accumulation in leaf	[88]
OsYSL2	2	26170387–26174970	Metal-nicotinamide transporter	Cd translocation	[89]
OsCd1	3	842577–846408	Major facilitator superfamily	Cd uptake	[90]
OsNramp2	3	5655157–5659147	Natural resistance-associated macrophage protein	Cd transporter, Cd accumulation	[91]
OsMSRMK2	3	9847700–9850473	Mitogen-activated protein kinase	Cd signal	[67]
OsMTI-1b	3	9957335–9958362	Metallothionein-like protein 1B	Cd tolerance	[92]
PEZ1	3	20793053–20799805	Phenol efflux protein	Cd accumulation	[93]
OsCDT1/OsCCX2	3	25613825–25616179	Cation/calcium (Ca) exchanger 2	Cd tolerance and translocation	[79,94]
OsIRT2	3	26276301–26277206	Iron-regulated transporter	Cd and Fe transporter	[95]
OsIRT1	3	26286156–26292023	Iron-regulated transporter	Cd and Fe transporter	[95,96]

Table 1. *Cont.*

Gene	Chr.	Physical Location (bp)	Gene Name	Function	Reference
OsZIP3	4	31078200–31080734	Zinc- and iron-regulated transporter	Cd accumulation	[84]
OsMTP1	5	1675488–1679056	Metal tolerance protein gene	Cd translocation	[97]
OsZIP6	5	3807974–3810752	Zinc- and iron-regulated transporter	Cd transport	[98]
OsCDT5	5	4665325–4667853	Encoding a Cys-rich peptide	Cd uptake inhibitor	[79]
OsZIP7	5	6090801–6094068	Zinc- and iron-regulated transporter	Cd and Zn accumulation	[99]
OsPCS2	6	167367–174319	Plant chelatase synthase 2	Cd tolerance	[100]
OsCDT2	6	2261681–2263972	Encoding a Cys-rich peptide	Cd uptake inhibitor	[79]
OsLCT1	6	22566775–22571982	Low affinity cation transporter	Cd transporter in phloem	[101,102]
OsHMA9	6	27517100–27523604	P-Type Heavy Metal ATPase	Cd efflux	[77]
OsMSRMK3	6	29398191–29402466	Mitogen-activated protein kinase	Cd signal	[82]
OsHMA2	6	29477949–29480905	P-Type Heavy Metal ATPase	Cd and Zn translocation	[103,104]
OsHMA3	7	7405745–7409553	P-Type Heavy Metal ATPase	Sequestration of Cd in root	[46,47,105]
OsNramp5	7	8871436–8878905	Natural resistance-associated macrophage protein	Cd, Mn, and Fe transporters	[106–110]
OsNramp1	7	8966025–8970882	Natural resistance-associated macrophage protein	Cd and Fe transporters	[111–113]
OsABCG43	7	20214025–20218702	ATP-binding cassette transporter	Cd compartmentalization	[45]
OsMAPK2	8	3307520–3310590	Mitogen-activated protein kinase	Cd signal	[68]
OsHIR1	8	19011814–19015998	Heavy metal-induced RING E3 ligase 1	Cd uptake	[114]
SISAP1	9	18760704–18761836	Subspecies indica stress-associated protein gene	Cd tolerance	[115]
OsPCR1	10	826309–824623	Plant cadmium resistance 1	Cd tolerance	[116]
rgMT	11	28827746–28828439	Metallothionein-like protein	Cd tolerance	[117]
RCS1	12	26698650–26703087	Cytosolic cysteine synthase gene	Cd complexation via sulfur	[118]

3. Uptake and Transport Pathway of Cd in Rice

Cadmium is transported from the roots to shoots and then to grains through four steps: (i) uptake by roots; (ii) transportation to shoots through loading to the xylem; (iii) distribution and transportation through nodes; and (iv) transportation to grains through the phloem from leaf blades (Figure 1).

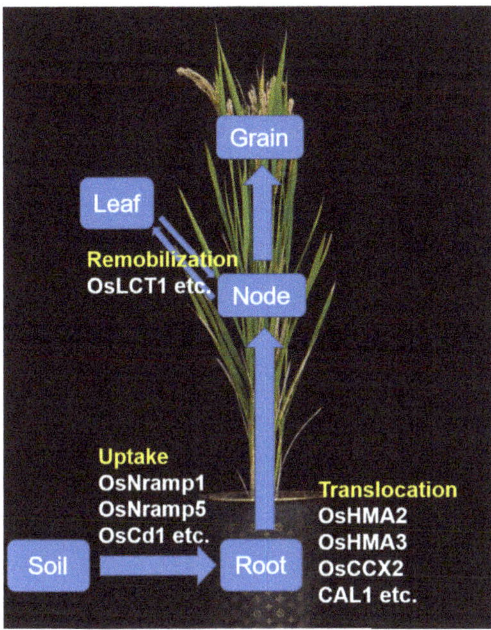

Figure 1. A schematic of cadmium transport from the soil to grains in rice. Cadmium is absorbed from the soil by the roots, and *OsNramp1*, *OsNramp5*, and *OsCd1* mediate this process. *OsHMA3* plays a key role in cadmium segregation to vacuoles in root cells and thus negatively regulates cadmium xylem loading. *OsHMA2*, *OsCCX2*, and *CAL1* regulate cadmium transport to the xylem. *OsLCT1* contributes to cadmium remobilization from leaf blades via the phloem and is likely to play a part in intervascular cadmium transfer at nodes.

3.1. Functional Analysis of Related Genes

Cd can enter rice plants through the uptake mechanism of essential elements such as Mn, Zn, and Fe, etc. [106,107,119]. Fe^{2+} transporters *OsIRT1* and *OsIRT2* display Cd^{2+} influx activity in yeast, which indicates that *OsIRT1* and *OsIRT2* may play a role in cadmium uptake in the root system [95,120]. Overexpression of *OsIRT1* significantly increased the accumulation of Cd in roots and shoots in Murashige & Skoog (MS) medium containing excess Cd, but no obvious phenotype was observed under field conditions, suggesting that *OsIRT1* may be involved in cadmium uptake in roots, but its contribution is largely affected by environmental conditions [96]. Oryza sativa Natural Resistance-Associated Macrophage Protein 5 (OsNramp5), located at the plasma membrane of root cells, was found to be the major transporter of Cd uptake in rice roots, responsible for the transport of Cd from the soil solution to the root cells [106,107]. OsNramp5 is also an Mn transporter, and the knock-out of *OsNramp5* can significantly reduce the uptake and accumulation of cadmium in grains, but also lead to the decrease of growth and yield due to manganese deficiency [107–109]. Recently, Liu et al. [121] located a major QTL, *qGMN7.1*, according to the Mn concentration in the grains of a recombinant inbred line (RILS) crossed between 93–11 (low grain Mn) and PA64s (high grain Mn). Fine mapping delimited *qGMN7.1* to a 49.3 kb region containing *OsNRAMP5*, and sequence variations in the *OsNRAMP5* promoter caused changes in its transcript level and in grain Mn levels. Tang et al. [110] reported that a series of new indica rice lines with low cadmium accumulation were developed by knocking out the metal transporter *OsNramp5* using the CRISPR/CAS9 system. OsNRAMP1, located on the plasma membrane, also exhibits the activity of Cd transport, and participates in the uptake and transport of Cd in root cells [111,112]. OsZIP1, a zinc-regulated/iron-regulated transporter-like protein, expression in yeast can enhance its sensitivity to Cd [84], and the overexpression of *OsZIP6* can increase the Cd uptake in *X. laevis oocytes* [98].

After root absorption, xylem-mediated Cd translocation from the roots to shoots is the main factor determining the cadmium accumulation in shoots [122]. *OsHMA2* and *OsHMA3* were reported to play a role in this process [46,103,123,124]. *OsHMA2* participates in the transport of Cd from the roots to shoots and plays an important role in controlling the distribution of Cd through the phloem to developing tissues [103,104,123]. Compared with wild-type (WT) samples, the Cd concentration in the shoots of an *oshma2* mutant was significantly lower [104]. OsHMA3 plays a role in the vacuolar sequestration of Cd in root cells, the overexpression of *OsHMA3* reduces the Cd load in the xylem and Cd accumulation in shoots, and the functional deficiency of *OsHMA3* results in very high root-to-shoot Cd translocation in rice [46,47,105,125]. Recent reports showed that OsCCX2, a putative cation/calcium (Ca) exchanger, was localized in the plasma membrane and plays an important role in Cd transport by impacting Cd root-to-shoot translocation and the Cd distribution in the shoot tissues, and the knock-out of *OsCCX2* resulted in a significant Cd reduction in the grains [94]. Tan et al. [99] reported that *OsZIP7* plays a key role in xylem-loading in roots and inter-vascular transfer in nodes to deliver Zn and Cd upward in rice.

Nodes are the central organ of Cd transport from the xylem to phloem, and play an important role in Cd transport to grains [126–128]. OsLCT1 is a Cd-efflux transporter on the plasma-membrane involved in phloem Cd transport [101]. *OsLCT1* expression was higher in leaf blades and nodes during the reproductive stage, especially in node I. Compared with wild-type (WT), the Cd concentration in phloem exudates and in grains of *OsLCT1* RNAi plants decreased significantly, although the Cd concentration in xylem sap did not differ. These results suggest that *OsLCT1* in leaf blades functions in Cd remobilization by the phloem, and in node I, *OsLCT1* is likely to play a part in intervascular Cd transfer from enlarged large vascular bundles to diffused vascular bundles, which connect to the panicle [101,102]). The positions of cloned cadmium stress-related genes in rice chromosomes are shown in Figure 2.

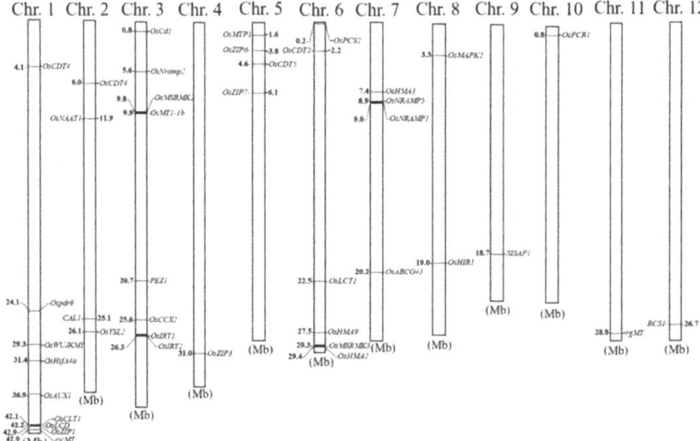

Figure 2. Positions of cloned cadmium stress-related genes in rice chromosomes.

3.2. Location of Related QTLs

Rice varieties show obvious genetic variation in terms of their cadmium accumulation ability, which is a valuable resource for dissecting functional alleles and genetic improvement [19,20,25]. However, only a few quantitative trait loci (QTLs) related to cadmium accumulation in rice have been reported. *OsHMA3*, *CAL1* (Cd Accumulation in Leaf 1), and *OsCd1* are the only Cd-related QTLs cloned so far. OsHMA3 encodes a cadmium transporter located in the vacuole membrane, which transports cadmium into vacuoles for sequestration [105]. Loss of OsHMA3 function significantly increased cadmium transport to rice shoots and grains [101,129]. On the other hand, the overexpression of *OsHMA3* can increase the tolerance of rice to Cd and reduce Cd accumulation in grains [46,105,119]. *CAL1* (cadmium accumulation in leaf 1) was identified and cloned by Luo et al. [88] as a quantitative trait locus (QTL) in rice, which explained 13% of the variation in leaf cadmium concentration in a doubled haploid population. *CAL1* regulates the root-to-shoot translocation of cadmium via the xylem vessels, and knockout mutants of *CAL1* significantly reduced the concentration of cadmium in rice leaves [88]. Yan et al. [90] discovered that the gene *OsCd1* belongs to the major facilitator superfamily through genome-wide association studies (GWAS), which was associated with divergence in rice grain Cd accumulation. Interestingly, the natural variation $OsCd1^{V449}$ in *Japonica*, which is associated with a reduced Cd transport ability and decreased grain Cd accumulation, shows a potential value in low-Cd rice selection [90].

A series of QTLs related to rice varieties that control the Cd concentration in rice have been reported (Table 2). Ishikawa et al. [130] obtained a mapping population consisting of 85 back-cross inbred lines (BIL) from hybridization between a low-cadmium-accumulation variety of *Japonica* rice (Sasanishiki) and a high-cadmium-accumulation variety of *Indica* rice (Habataki). Two QTLs were located on chromosomes 2 and 7, separately, with an increased cadmium concentration in grains. *qGCd7* plays an important role in increasing the cadmium concentration in grains, which can explain 35.5% of phenotypic variation [130]. Kashiwagi et al. [131] identified two QTLs, known as *qcd4–1* and *qcd4–2*, affecting the cadmium concentration in shoots. Sato et al. [132] reported two QTLs controlling the cadmium concentration in brown rice: *qLCdG11* explained 9.4%–12.9% of phenotypic variation and *qLCdG3* explained 8.3%–13.9% of phenotypic variation. Yan et al. [133] constructed an recombinant inbred lines (RIL) population of F7 to identify Cd accumulation and distribution. A total of five main effect QTLs (*scc10* was correlated with Cd accumulation in shoots; *gcc3*, *gcc9*, and *gcc11* with Cd accumulation in grains; and *sgr5* with the Cd distribution ratio in shoots and roots) were detected. Among them, *sgr5* had the greatest effect on the distribution of Cd in grains. Abe et al. [134] used a population consisting of 46 chromosome segment substitution lines (CSSL) to identify eight QTLs related to the grain cadmium content by single-label analysis using ANOVA. The result showed that *qlGCd3* had a high F-test value. A recombinant inbred population derived from Xiang 743/Katy was grown in Cd-polluted fields and used to map the QTLs for Cd accumulation in rice grains, and two QTLs, *qCd-2* and *qCd-7*, were identified in 2014 and 2015 [135]. Liu et al. [136] used 276 accessions with 416 K single nucleotide polymorphisms (SNPs) and performed a genome-wide association analysis of grain Cd concentrations in rice grown in heavily multi-contaminated farmlands, and 17 QTLs were found to be responsible for the grain Cd concentration.

Table 2. Quantitative Trait Loci (QTLs) of Rice Reported to be Regulated during Cadmium (Cd)-Exposure.

Stage	Parent Sources	Population	Marker	Trait	Chr.	QTL	Reference
Seedling stage	Tainan1/Chunjiang06	119 DH, 3651 BC3F3	RFLP	Cd accumulation in leaves	2	*CAL1*	[88]
Seedling stage	Nipponbare/Anjana Dhan	965 F2	SSR	Cd concentration in shoots	7	*OsHMA3*	[105]
Seedling stage	SNU-SG1/Suwon490	91 RIL	124 SSR	Cd concentration in shoots	10	*scc10*	[133]
Seedling stage	Koshihikari/LAC23	46 CSSLs	345 SNP	Cd concentration in shoots	3	*glGCd3*	[134]
Seedling stage	Anjana Dhan/Nipponbare	177 F2	SSR	Root-to-shoot Cd translocation	7	*qCdT7*	[137]
Seedling stage	Badari Dhan/Shwe War	184 F2	141 SSR	Cd concentration in shoots	2,5,11	—	[138]
Seedling stage	JX17/ZYQ8	127 DH	160 RFLP,83 SSR	Shoot/root rate of Cd concentration	3	*qSRR3*	[139]
Seedling stage	JX17/ZYQ8	127 DH	160 RFLP,83 SSR	Cd concentration in roots and shoots	6,7	*qCDS7*, *qCDR6.1*, *qCDR6.2*	[139]
Seedling stage	Azucena/Bala	79 RIL	164 SSR	Cd concentration in leaves	1,3,6	*qCd1*, *qCd3*, *qCd6*,	[140]
Bfore heading	Kasalath/Nipponbare	98 BILs	RFLP and SSR	Cd concentration in leaves and culms	4,11	*qcd4–1*, *qcd4–2*, *qcd11*	[131]
Mature period	Sasanishiki/Habataki	85 BIL	SSR	Cd accumulation in grains	2,7	*qGCd7*	[130]
Mature period	Fukuhibiki/LAC23	126 RIL	454 SNP	Cd accumulation in grains	3,11	*gLCdG3*, *gLCdG11*	[132]
Mature period	SNU-SG1/Suwon490	91 RIL	124 SSR	Cd accumulation in grains	3,5,9,11	*gcc3*, *sgr5*, *gcc9*, *gcc11*	[133]
Mature period	Xiang 743/Katy	115 RIL,	SSR	Cd accumulation in grains	2,7	*qCd-2*, *qCd-7*	[135]
Mature period	Kasalath/Koshihikari	39CSSL	129 RFLP	Cd accumulation in grains	3,6,8	—	[141]
Mature period	Koshihikari/Jarjan	103 BIL	169 SSR	Cd accumulation in grains	7	—	[142]
Mature period	JX17/ZYQ8	127 DH	160 RFLP,83 SSR	Cd accumulation in grains	3,6	*gCdc3*, *gCdc6*	[143]
Mature period		127 rice cultivars	GWAS	Cd accumulation in grains	3	*OsCd1*	[90]
Mature period		378 rice cultivars	GWAS	Cd accumulation in grains	3, 5	*qCd3*, *qCd5.1*, *qCd5.2*	[144]

4. Future Perspectives

Cadmium is a kind of heavy metal that presents extreme biological toxicity. Cd accumulated in rice can enter the food chain, thereby threatening human health [5–7]. Cadmium in rice can be reduced by agronomic practices (including soil amendments, fertilizer management, water management, and tillage management) and bioremediation (including phytoremediation and microbial remediation) [18,145–150]. In addition, understanding the mechanism of cadmium translocation and the factors affecting cadmium accumulation in rice are also important for formulating effective strategies to reduce cadmium accumulation in rice. In recent years, some genes related to cadmium transport in rice have been studied, and significant progress has been made in understanding the mechanism of cadmium uptake and transport. In order to understand the mechanism of cadmium transport in rice, it is necessary to identify more unknown transporters or other molecules.

Biotechnology offers a promising approach to reducing the Cd content in rice grains. Mutations of the *OsNramp5* gene result in obvious decreases in Cd uptake in roots and Cd accumulation in rice grains [106,108,151]. Using the CRISPR/Cas9 gene editing technology to knock out *OsNramp5* in both parental lines, Tang et al. [110] generated a hybrid rice cultivar that accumulated very low levels of Cd in the grain. Another target for gene editing is *OsLCT1*, which is involved in the phloem transport of Cd from the vegetative tissues to the grains [101]. Knockdown of this gene by RNAi reduced the grain Cd concentration by 30%–50% [101]. Overexpression of functional *OsHMA3* in *Nipponbare* decreased Cd translocation and Cd accumulation in rice grains [46,105]. Overexpression of *OsHMA3* is a highly effective method for reducing Cd accumulation in *Indica* rice, and rice grains produced using this approach are almost Cd-free, with little effect on the grain yield or essential micronutrient concentrations [152].

However, commercial transgenic rice is not commonly accepted by the general public and prohibited in many countries. Ishikawa et al. [151] produced three rice mutants by carbon ion-beam irradiation, where cadmium was hardly detected in mutant seeds when planted in cadmium-contaminated paddy fields and there was no significant difference between the mutant and wild-type (WT) in agronomic traits, which could be directly applied to breeding projects. Another possible strategy is marker-assisted breeding, which uses molecular markers to track the genetic composition of rice and bred rice varieties. For example, identifying a low-cadmium-related QTL and then introducing it into high-cadmium cultivars might be a viable approach [122]. However, only a few of QTLs related to cadmium accumulation in rice have been cloned [90,105], and the natural allele variation of grain cadmium accumulation differences among rice varieties has not been fully explored. Further research is necessary to clone more QTLs for controlling grain Cd accumulation, thus providing tools for the marker-assisted molecular breeding of rice cultivars with a low accumulation of Cd in grains.

Author Contributions: J.C. and W.Z. wrote the first draft of the manuscript and organized the tables and figures; L.M and X.F. conceived and supervised their ideas; G.X. and G.Y. reviewed the manuscript. All authors read and approved the final manuscript.

Acknowledgments: This research was financially supported by the National Natural Science Foundation of China (No. 31601286 to L.M.); the CAAS Innovative Team Award to GY's team, Shenzhen Science and Technology Projects (No. JSGG20160608160725473 to Q.Q.); the China Postdoctoral Science Foundation (No. 2016M590160 to L.M.); the Fundamental Research Funds for Central Non-profit Scientific Institution (to G.Y.); the Jiangsu Science Fund for Distinguished Young Scholars (Grant BK20160030 to X.F.); and Dapeng district industry development special funds (KY20180218).

Conflicts of Interest: The authors declare no conflict of interest.

References

1. He, S.Y.; He, Z.L.; Yang, X.E.; Stoffella, P.J.; Baligar, V.C. Soil biogeochemistry, plant physiology, and phytoremediation of cadmium-contaminated soils. *Adv. Agron.* **2015**, *134*, 135–225.
2. Song, W.E.; Chen, S.B.; Liu, J.F.; Chen, L.; Song, N.N.; Li, N.; Liu, B. Variation of Cd concentration in various rice cultivars and derivation of cadmium toxicity thresholds for paddy soil by species-sensitivity distribution. *J. Integr. Agric.* **2015**, *14*, 1845–1854. [CrossRef]
3. Liu, F.; Liu, X.N.; Ding, C.; Wu, L. The dynamic simulation of rice growth parameters under cadmium stress with the assimilation of multi-period spectral indices and crop model. *Field Crop. Res.* **2015**, *183*, 225–234. [CrossRef]
4. Zhao, K.L.; Fu, W.J.; Ye, Z.Q.; Zhang, C.S. Contamination and spatial variation of heavy metals in the soil-rice system in Nanxun County, Southeastern China. *Int. J. Environ. Res. Public Health.* **2015**, *12*, 1577–1594. [CrossRef] [PubMed]
5. Xie, P.P.; Deng, J.W.; Zhang, H.M.; Ma, Y.H.; Cao, D.J.; Ma, R.X.; Liu, R.J.; Liu, C.; Liang, Y.G. Effects of cadmium on bioaccumulation and biochemical stress response in rice (*Oryza sativa* L.). *Ecotox. Environ. Safe.* **2015**, *122*, 392–398. [CrossRef]
6. Xue, D.W.; Jiang, H.; Deng, X.X.; Zhang, X.Q.; Wang, H.; Xu, X.B.; Hu, J.; Zeng, D.L.; Guo, L.B.; Qian, Q. Comparative proteomic analysis provides new insights into cadmium accumulation in rice grain under cadmium stress. *J. Hazard. Mater.* **2014**, *280*, 269–278. [CrossRef] [PubMed]
7. Aziz, R.; Rafiq, M.T.; Li, T.; Liu, D.; He, Z.; Stoffella, P.J.; Sun, K.; Xiaoe, Y. Uptake of cadmium by rice grown on contaminated soils and its bioavailability/toxicity in human cell lines (Caco-2/HL-7702). *J. Agric. Food Chem.* **2015**, *63*, 3599–3608. [CrossRef]
8. Godt, J.; Scheidig, F.; Grosse-Siestrup, C.; Esche, V.; Brandenburg, P.; Reich, A.; Groneberg, D.A. The toxicity of cadmium and resulting hazards for human health. *J. Occup Med. Toxicol.* **2006**, *1*, 22. [CrossRef]
9. Satarug, S.; Baker, J.R.; Urbenjapol, S.; Haswell-Elkins, M.; Reilly, P.E.; Williams, D.J.; Moore, M.R. A global perspective on cadmium pollution and toxicity in non-occupationally exposed population. *Toxicol Lett.* **2003**, *137*, 65–83. [CrossRef]
10. Horiguchi, H.; Teranishi, H.; Niiya, K.; Aoshima, K.; Katoh, T.; Sakuragawa, N.; Kasuya, M. Hypoproduction of erythropoietin contributes to anemia in chronic cadmium intoxication — clinical-study on itai-itai disease in Japan. *Arch. Toxicol.* **1994**, *68*, 632–636. [CrossRef]
11. Tsukahara, T.; Ezaki, T.; Moriguchi, J.; Furuki, K.; Shimbo, S.; Matsuda-Inoguchi, N.; Ikeda, M. Rice as the most influential source of cadmium intake among general Japanese population. *Sci. Total Environ.* **2003**, *305*, 41–51. [CrossRef]
12. Song, Y.; Wang, Y.; Mao, W.; Sui, H.; Yong, L.; Yang, D.; Jiang, D.; Zhang, L.; Gong, Y. Dietary cadmium exposure assessment among the Chinese population. *PLoS ONE* **2017**, *12*, e0177978. [CrossRef] [PubMed]
13. Chen, H.; Yang, X.; Wang, P.; Wang, Z.; Li, M.; Zhao, F.J. Dietary cadmium intake from rice and vegetables and potential health risk: A case study in Xiangtan, southern China. *Sci. Total Environ.* **2018**, *639*, 271–277. [CrossRef] [PubMed]
14. Ahsan, N.; Lee, S.H.; Lee, D.G.; Lee, H.; Lee, S.W.; Bahk, J.D.; Lee, B.H. Physiological and protein profiles alternation of germinating rice seedlings exposed to acute cadmiumtoxicity. *C. R. Biol.* **2007**, *330*, 735–746. [CrossRef] [PubMed]
15. Li, B.; Wang, X.; Qi, X.; Huang, L.; Ye, Z. Identification of rice cultivars with low brown rice mixed cadmium and lead contents and their interactions with the micronutrients iron, zinc, nickel and manganese. *J. Environ. Sci.* **2012**, *24*, 1790–1798. [CrossRef]
16. Li, S.; Yu, J.; Zhu, M.; Zhao, F.; Luan, S. Cadmium impairs ion homeostasis by altering K^+ and Ca^{2+} channel activities in rice root hair cells. *Plant Cell Environ.* **2012**, *35*, 1998–2013. [CrossRef]
17. Wang, Y.; Jiang, X.; Li, K.; Wu, M.; Zhang, R.; Zhang, L.; Chen, G. Photosynthetic responses of *Oryza sativa* L. seedlings to cadmium stress: Physiological, biochemical and ultrastructural analyses. *BioMetals* **2014**, *27*, 389–401. [CrossRef]
18. Kanu, A.S.; Ashraf, U.; Bangura, A.; Yang, D.M.; Ngaujah, A.S.; Tang, X. Cadmium (Cd) Stress in Rice; Phyto-Availability, Toxic Effects, and Mitigation Measures-A Critical Review. *IOSR-JESTFT* **2017**, *11*, 07–23.
19. Liu, J.; Li, K.; Xu, J.; Liang, J.; Lu, X.; Yang, J.; Zhu, Q. Interaction of Cd and five mineral nutrients for uptake and accumulation in different rice cultivars and genotypes. *Field Crop. Res.* **2003**, *83*, 271–281. [CrossRef]

20. Liu, J.G.; Liang, J.S.; Li, K.Q.; Zhang, Z.J.; Yu, B.Y.; Lu, X.L.; Yang, J.C.; Zhu, Q.S. Correlations between cadmium and mineral nutrients in absorption and accumulation in various genotypes of rice under cadmium stress. *Chemosphere* **2003**, *52*, 1467–1473. [CrossRef]
21. Rascio, N.; Dalla Vecchia, F.; La Rocca, N.; Barbato, R.; Pagliano, C.; Raviolo, M.; Gonnelli, C.; Gabbrielli, R. Metal accumulation and damage in rice (*cv.* Vialone nano) seedlings exposed to cadmium. *Environ. Exp. Bot.* **2008**, *62*, 267–278. [CrossRef]
22. Yu, H.; Wang, J.; Fang, W.; Yuan, J.; Yang, Z. Cadmiumaccumulation in different rice cultivars and screening for pollution safe cultivars of rice. *Sci. Total Environ.* **2006**, *370*, 302–309. [CrossRef] [PubMed]
23. Zhang, X.; Gao, H.; Zhang, Z.; Wan, X. Influences of different ion channel inhibitors on the absorption of fluoride in tea plants (Camellia sinesis L.). *Plant Growth Regul.* **2013**, *69*, 99–106.
24. Arao, T.; Ae, N. Genotypic variations in cadmium levels of rice grain. *Soil Sci. Plant Nutr.* **2003**, *49*, 473–479. [CrossRef]
25. He, J.; Zhu, C.; Ren, Y.; Yan, Y.; Jiang, D. Genotypic variation in grain cadmium concentration of lowland rice. *J. Plant Nutr. Soil Sci.* **2006**, *169*, 711–716. [CrossRef]
26. Liu, H.J.; Zhang, J.L.; Christie, P.; Zhang, F.S. Influence of external zinc and phosphorus supply on Cd uptake by rice (*Oryza sativa* L.) seedlings with root surface iron plaque. *Plant Soil.* **2007**, *300*, 105–115. [CrossRef]
27. Rodda, M.S.; Li, G.; Reid, R.J. The timing of grain Cd accumulation in rice plants: The relative importance of remobilisation within the plant and root Cd uptake post flowering. *Plant Soil.* **2011**, *347*, 105–114. [CrossRef]
28. Zhou, H.; Zhou, X.; Zeng, M.; Liao, B.H.; Liu, L.; Yang, W.T.; We, Y.M.; Qiu, Q.Y.; Wang, Y.J. Effects of combined amendments on heavy metal accumulation in rice (*Oryza sativa* L.) planted on contaminated paddy soil. *Ecotoxicol. Environ. Saf.* **2014**, *101*, 226–232. [CrossRef]
29. Mostofa, M.G.; Rahman, A.; Ansary, M.M.U.; Watanabe, A.; Fujita, M.; Tran, L.S.P. Hydrogen sulfide modulates cadmium-induced physiological and biochemical responses to alleviate cadmium toxicity in rice. *Sci. Rep.* **2015**, *5*, 14078. [CrossRef]
30. Rehman, M.Z.; Rizwan, M.; Ghafoor, A.; Naeem, A.; Ali, S.; Sabir, M.; Qayyum, M.F. Effect of inorganic amendments for in situ stabilization of cadmium in contaminated soils and its phyto-availability to wheat and rice under rotation. *Environ. Sci. Pollut. Res.* **2015**, *22*, 16897–16906. [CrossRef]
31. Shah, K.; Nahakpam, S. Heat exposure alters the expression of SOD, POD, APX and CAT isozymes and mitigates low cadmium toxicity in seedlings of sensitive and tolerant rice cultivars. *Plant Physiol. Biochem.* **2012**, *57*, 106–113. [CrossRef] [PubMed]
32. Zhang, C.; Yin, X.; Gao, K.; Ge, Y.; Cheng, W. Non-protein thiols and glutathione S-transferase alleviate Cd stress and reduce root-to-shoot translocation of Cd in rice. *J. Plant Nutr Soil Sci.* **2013**, *176*, 626–633. [CrossRef]
33. Choppala, G.; Saifullah, M.F.; Bolan, N.; Bibi, S.; Iqbal, M.; Rengel, Z.; Kunhikrishnan, A.; Ashwath, N.; Ok, Y.S. Cellular mechanisms in higher plants governing tolerance to cadmium toxicity. *Crit. Rev. Plant Sci.* **2014**, *33*, 374–391. [CrossRef]
34. Yu, C.; Sun, C.; Shen, C.; Wang, S.; Liu, F.; Liu, Y.; Chen, Y.; Li, C.; Qian, Q.; Aryal, B.; et al. The auxin transporter, OsAUX1, is involved in primary root and root hair elongation and in Cd stress responses in rice (*Oryza sativa* L.). *Plant J.* **2015**, *83*, 818–830. [CrossRef] [PubMed]
35. Zhang, X.; Wu, H.; Chen, L.; Liu, L.; Wan, X. Maintenance of mesophyll potassium and regulation of plasma membrane H^+-ATPase are associated with physiological responses of tea plants to drought and subsequent rehydration. *Crop. J.* **2018**, *6*, 611–620. [CrossRef]
36. Parrotta, L.; Guerriero, G.; Sergeant, K.; Cai, G.; Hausman, J.F. Target or barrier? The cell wall of early- and later-diverging plants vs cadmium toxicity: Differences in the response mechanisms. *Front. Plant Sci.* **2015**, *6*, 133. [CrossRef] [PubMed]
37. Loix, C.; Huybrechts, M.; Vangronsveld, J.; Gielen, M.; Keunen, E.; Cuypers, A. Reciprocal Interactions between Cadmium-Induced Cell Wall Responses and Oxidative Stress in Plants. *Front. Plant Sci.* **2017**, *8*, 1867. [CrossRef] [PubMed]
38. Hall, J.L. Cellular mechanisms for heavy metal detoxification and tolerance. *J. Exp. Bot.* **2002**, *53*, 1–11. [CrossRef] [PubMed]
39. Qiu, Q.; Wang, Y.T.; Yang, Z.Y.; Yuan, J.G. Effects of phosphorus supplied in soil on subcellular distribution and chemical forms of cadmium in two Chinese flowering cabbage (Brassica parachinensis L.) cultivars differing in cadmium accumulation. *Food Chem. Toxicol.* **2011**, *49*, 2260–2267. [CrossRef] [PubMed]

40. Wang, X.; Liu, Y.O.; Zeng, G.M.; Chai, L.Y.; Song, X.C.; Min, Z.Y.; Xiao, X. Subcellular distribution and chemical forms of cadmium in *Bechmeria nivea* (L.). Gaud. *Environ. Exp. Bot.* **2008**, *62*, 389–395. [CrossRef]
41. Zhang, J.; Sun, W.; Li, Z.; Liang, Y.; Song, A. Cadmium fate and tolerance in rice cultivars. *Agron. Sustain. Dev.* **2009**, *29*, 483–490. [CrossRef]
42. Fu, X.P.; Dou, C.M.; Chen, Y.X.; Chen, X.C.; Shi, J.Y.; Yu, M.G.; Xu, J. Subcellular distribution and chemical forms of cadmium in *Phytolacca americana* L. *J. Hazard. Mater.* **2011**, *186*, 103–107. [CrossRef] [PubMed]
43. Kim, D.Y.; Bovet, L.; Maeshima, M.; Martinoia, E.; Lee, Y.S. The ABC transporter AtPDR8 is a cadmium extrusion pump conferring heavy metal resistance. *Plant J.* **2007**, *50*, 207–218. [CrossRef] [PubMed]
44. Park, J.; Song, W.Y.; Ko, D.; Eom, Y.; Hansen, T.H.; Schiller, M.; Lee, T.G.; Martinoia, E.; Lee, Y.S. The phytochelatin transporters AtABCC1 and AtABCC2 mediate tolerance to cadmium and mercury. *Plant J.* **2012**, *69*, 278–288. [CrossRef] [PubMed]
45. Oda, K.; Otani, M.; Uraguchi, S.; Akihiro, T.; Fujiwara, T. Rice *ABCG43* is Cd inducible and confers Cd tolerance on yeast. *Biosci. Biotech. Biochem.* **2011**, *75*, 1211–1213. [CrossRef] [PubMed]
46. Sasaki, A.; Yamaji, N.; Ma, J.F. Overexpression of *OsHMA3* enhances Cd tolerance and expression of Zn transporter genes in rice. *J. Exp. Bot.* **2014**, *65*, 6013–6021. [CrossRef] [PubMed]
47. Ueno, D.; Koyama, E.; Yamaji, N.; Ma, J.F. Physiological, genetic, molecular characterization of a high-Cd-accumulating rice cultivar, Jarjan. *J. Exp. Bot.* **2011**, *22*, 2265–2272. [CrossRef] [PubMed]
48. Dong, J.; Mao, W.H.; Zhang, G.P.; Wu, F.B.; Cai, Y. Root excretion and plant tolerance to cadmium toxicity—A review. *Plant Soil Environ.* **2007**, *53*, 193–200. [CrossRef]
49. Haydon, M.J.; Cobbett, C.S. Transporters of ligands for essential metal ions in plants. *New Phytol.* **2007**, *174*, 499–506. [CrossRef] [PubMed]
50. Jabeen, R.; Ahmad, A.; Iqbal, M. Phytoremediation of heavy metals: Physiological and molecular mechanisms. *Bot. Rev.* **2009**, *75*, 339–364. [CrossRef]
51. Verbruggen, N.; Hermans, C.; Schat, H. Molecular mechanisms of metal hyperaccumulation in plants. *New Phytol.* **2009**, *181*, 759–776. [CrossRef] [PubMed]
52. Nocito, F.F.; Lancilli, C.; Dendena, B.; Lucchini, G.; Sacchi, G.A. Cadmium retention in rice roots is influenced by cadmium availability, chelation and translocation. *Plant Cell Environ.* **2011**, *34*, 994–1008. [CrossRef] [PubMed]
53. Hassan, M.J.; Shao, G.; Zhang, G. Influence of cadmium toxicity on growth and antioxidant enzyme activity in rice cultivars with different grain cadmium accumulation. *J. Plant Nutr.* **2005**, *28*, 1259–1270. [CrossRef]
54. Hsu, Y.T.; Kao, C.H. Cadmium-induced oxidative damage in rice leaves is reduced by polyamines. *Plant Soil.* **2007**, *291*, 27–37. [CrossRef]
55. Lin, R.; Wang, X.; Luo, Y.; Du, W.; Guo, H.; Yin, D. Effects of soil cadmium on growth, oxidative stress and antioxidant system in wheat seedlings *Triticum aestivum* L. *Chemosphere* **2007**, *69*, 89–98. [CrossRef]
56. Roychoudhury, A.; Basu, S.; Sengupta, D.N. Antioxidants and stressrelated metabolites in the seedlings of two indica rice varieties exposed to cadmium chloride toxicity. *Acta Physiol. Plant* **2012**, *34*, 835–847. [CrossRef]
57. Shen, G.M.; Zhu, C.; Du, Q.Z.; Shangguan, L.N. Ascorbate-glutathione cycle alteration in cadmium sensitive rice mutant *cadB1*. *Rice Sci.* **2012**, *19*, 185–192. [CrossRef]
58. Srivastava, R.K.; Pandey, P.; Rajpoot, R.; Rani, A.; Dubey, R.S. Cadmium and lead interactive effects on oxidative stress and antioxidative responses in rice seedlings. *Protoplasma* **2014**, *251*, 1047–1065. [CrossRef]
59. Asgher, M.; Khan, M.I.; Anjum, N.A.; Khan, N.A. Minimising toxicity of cadmium in plants–role of plant growth regulators. *Protoplasma* **2015**, *252*, 399–413. [CrossRef]
60. Aina, R.; Labra, M.; Fumagalli, P.; Vannini, C.; Marsoni, M.; Cucchi, U.; Bracale, M.; Sgorbati, S.; Citterio, S. Thiol-peptide level and proteomic changes in response to cadmium toxicity in *Oryza sativa* L. roots. *Environ. Exp. Bot.* **2007**, *59*, 381–392. [CrossRef]
61. Chao, Y.Y.; Chen, C.Y.; Huang, W.D.; Kao, C.H. Salicylic acidmediated hydrogen peroxide accumulation and protection against Cd toxicity in rice leaves. *Plant Soil.* **2010**, *329*, 327–337. [CrossRef]
62. Zhang, C.H.; Ying, G.E. Response of glutathione and glutathione Stransferase in rice seedlings exposed to cadmium stress. *Rice Sci.* **2008**, *15*, 73–76. [CrossRef]
63. Lee, K.; Bae, D.W.; Kim, S.H.; Han, H.J.; Liu, X.; Park, H.C.; Lim, C.O.; Lee, X.Y.; Chung, W.S. Comparative proteomic analysis of the shortterm responses of rice roots and leaves to cadmium. *J. Plant Physiol.* **2010**, *167*, 161–168. [CrossRef] [PubMed]

64. Khavari-Nejad, R.A.; Najafi, F.; Rezaei, M. The influence of cadmium toxicity on some physiological parameters as affected by iron in rice (Oryza Sativa L.) plant. *J. Plant Nutr.* **2014**, *37*, 1202–1213. [CrossRef]
65. Wang, M.Y.; Chen, A.K.; Wong, M.H.; Qiu, R.L.; Cheng, H.; Ye, Z.H. Cadmium accumulation in and tolerance of rice *Oryza sativa* L. varieties with different rates of radial oxygen loss. *Environ. Pollut.* **2011**, *159*, 1730–1736. [CrossRef] [PubMed]
66. Agrawal, G.K.; Rakwal, R.; Yonekura, M.; Kubo, A.; Saji, H. Rapid induction of defense/stress-related proteins in leaves of rice (*Oryza sativa*) seedlings exposed to ozone is preceded by newly phosphorylated proteins and changes in a 66-kDA ERK-typeMAPK. *J. Plant Physiol.* **2002**, *159*, 361–369. [CrossRef]
67. Agrawal, G.K.; Rakwal, R.; Iwahashi, H. Isolation of novel rice (*Oryza sativa* L.) multiple stress responsive MAP kinase gene, *OsMSRMK2*, whose mRNA accumulates rapidly in response to environmental cues. *Biochem. Biophys. Res. Commun.* **2002**, *294*, 1009–1016. [CrossRef]
68. Yeh, C.M.; Hsiao, L.J.H.; Hsiao, H.J. Cadmium activates a mitogenactivated protein kinase gene and MBP kinases in rice. *Plant Cell Physiol.* **2004**, *45*, 1306–1312. [CrossRef]
69. Zhao, F.Y.; Hu, F.; Zhang, S.Y.; Wang, K.; Zhang, C.R.; Liu, T. MAPKs regulate root growth by influencing auxin signaling and cell cyclerelated gene expression in cadmium-stressed rice. *Environ. Sci. Pollut. Res.* **2013**, *20*, 5449–5460. [CrossRef]
70. Wang, X.; Yao, H.; Wong, M.H.; Ye, Z. Dynamic changes in radial oxygen loss and iron plaque formation and their effects on Cd and As accumulation in rice (*Oryza sativa* L.). *Environ. Geochem. Health* **2013**, *35*, 779–788. [CrossRef]
71. Cheng, H.; Wang, M.; Wong, M.H.; Ye, Z. Does radial oxygen loss and iron plaque formation on roots alter Cd and Pb uptake and distribution in rice plant tissues? *Plant Soil.* **2014**, *375*, 137–148. [CrossRef]
72. Shao, J.F.; Che, J.; Yamaji, N.; Shen, R.F.; Ma, J.F. Silicon reduces cadmium accumulation by suppressing expression of transporter genes involved in cadmium uptake and translocation in rice. *J. Exp. Bot.* **2017**, *68*, 5641–5651. [CrossRef]
73. Chen, Z.; Tang, Y.T.; Yao, A.J.; Cao, J.; Wu, Z.H.; Peng, Z.R.; Wang, S.Z.; Xiao, S.; Baker, A.J.M.; Qiu, R.L. Mitigation of Cd accumulation in paddy rice (*Oryza sativa* L.) by Fe fertilization. *Environ. Pollut.* **2017**, *231*, 549–559. [CrossRef] [PubMed]
74. Huang, G.; Ding, C.; Zhou, Z.; Zhang, T.; Wang, X. A tillering application of zinc fertilizer based on basal stabilization reduces Cd accumulation in rice (*Oryza sativa* L.). *Ecotoxicol. Environ. Saf.* **2019**, *167*, 338–344. [CrossRef] [PubMed]
75. Chen, D.; Chen, D.; Xue, R.; Long, J.; Lin, X.; Lin, Y.; Jia, L.; Zeng, R.; Song, Y. Effects of boron, silicon and their interactions on cadmium accumulation and toxicity in rice plants. *J. Hazard. Mater.* **2019**, *367*, 447–455. [CrossRef]
76. Hermans, C.; Conn, S.J.; Chen, J.; Xiao, Q.; Verbruggen, N. An update on magnesium homeostasis mechanisms in plants. *Metallomics* **2013**, *5*, 1170–1183. [CrossRef] [PubMed]
77. Lee, S.; Kim, Y.Y.; Lee, Y.; An, G. Rice P1B-type heavy-metal ATPase, OsHMA9, is a metal efflux protein. *Plant Physiol.* **2007**, *145*, 831–842. [CrossRef]
78. Shimo, H.; Ishimaru, Y.; An, G.; Yamakawa, T.; Nakanishi, H.; Nishizawa, N.K. *Low cadmium* (*LCD*), a novel gene related to cadmium tolerance and accumulation in rice. *J. Exp Bot.* **2011**, *62*, 5727–5734. [CrossRef]
79. Kuramata, M.; Masuya, S.; Takahashi, Y.; Kitagawa, E.; Inoue, C.; Ishikawa, S.; Youssefian, S.; Kusano, T. Novel cysteine-rich peptides from Digitaria ciliaris and Oryza sativa enhance tolerance to cadmium by limiting its cellular accumulation. *Plant Cell Physiol.* **2009**, *50*, 106–117. [CrossRef]
80. Yang, J.; Gao, M.X.; Hu, H.; Ding, X.M.; Lin, H.W.; Wang, L.; Xu, J.M.; Mao, C.Z.; Zhao, F.J.; Wu, Z.C. OsCLT1, a CRT-like transporter 1, is required for glutathione homeostasis and arsenic tolerance in rice. *New Phytol.* **2016**, *211*, 658–670. [CrossRef]
81. Moons, A. *Ospdr9*, which encodes a PDR-type ABC transporter, is induced by heavy metals, hypoxic stress and redox perturbations in rice roots. *FEBS Lett.* **2003**, *553*, 370–376. [CrossRef]
82. Agrawal, G.K.; Agrawal, S.K.; Shibato, J.; Iwahashi, H.; Rakwal, R. Novel rice MAP kinases *OsMSRMK3* and *OsWJUMK1* involved in encountering diverse environmental stresses and developmental regulation. *Biochem. Biophys. Res. Commun.* **2003**, *300*, 775–783. [CrossRef]
83. Shim, D.; Jae-Ung, H.; Lee, J.; Lee, S.; Choi, Y.; An, G.; Martinoia, E.; Lee, Y. Orthologs of the class A4 heat shock transcription factor HsfA4a confer cadmium tolerance in wheat and rice. *Plant Cell.* **2009**, *21*, 4031–4043. [CrossRef] [PubMed]

84. Ramesh, S.A.; Shin, R.; Eide, D.J.; Schachtman, D.P. Differential metal selectivity and gene expression of two zinc transporters from rice. *Plant Physiol.* **2003**, *133*, 126–134. [CrossRef] [PubMed]
85. Chou, T.S.; Chao, Y.Y.; Huang, W.D.; Hong, C.Y.; Kao, C.H. Effect of magnesium deficiency on antioxidant status and cadmium toxicity in rice seedlings. *J. Plant Physiol.* **2011**, *168*, 1021–1030. [CrossRef] [PubMed]
86. Yu, L.H.; Umeda, M.; Liu, J.Y.; Zhao, N.M.; Uchimiya, H. A novel MT gene of rice plants is strongly expressed in the node portion of the stem. *Gene* **1998**, *206*, 29–35. [CrossRef]
87. Cheng, L.; Wang, F.; Shou, H.; Huang, F.; Zheng, L.; He, F.; Li, J.; Zhao, F.J.; Ueno, D.; Ma, J.F.; et al. Mutation in nicotianamine aminotransferase stimulated the Fe(II) acquisition system and led to iron accumulation in rice. *Plant Physiol.* **2007**, *145*, 1647–1657. [CrossRef]
88. Luo, J.S.; Huang, J.; Zeng, D.L.; Peng, J.S.; Zhang, G.B.; Ma, H.L.; Guan, Y.; Yi, H.Y.; Fu, Y.L.; Han, B.; et al. A defensin-like protein drives cadmium efflux and allocation in rice. *Nat. Commun.* **2018**, *9*, 645. [CrossRef]
89. Masuda, H.; Ishimaru, Y.; Aung, M.S.; Kobayashi, T.; Kakei, Y.; Takahashi, M.; Higuchi, K.; Nakanishi, H.; Nishizawa, N.K. Iron biofortification in rice by the introduction of multiple genes involved in iron nutrition. *Sci. Rep.* **2012**, *2*, 543. [CrossRef]
90. Yan, H.; Xu, W.; Xie, J.; Gao, Y.; Wu, L.; Sun, L.; Feng, L.; Chen, X.; Zhang, T.; Dai, C.; et al. Variation of a major facilitator superfamily gene contributes to differential cadmium accumulation between rice subspecies. *Nat. Commun.* **2019**, *10*, 2562. [CrossRef]
91. Zhao, J.; Yang, W.; Zhang, S.; Yang, T.; Liu, Q.; Dong, J.; Fu, H.; Mao, X.; Liu, B. Genome-wide association study and candidate gene analysis of rice cadmium accumulation in grain in a diverse rice collection. *Rice* **2018**, *11*, 61. [CrossRef] [PubMed]
92. Ansarypour, Z.; Shahpiri, A. Heterologous expression of a rice metallothionein isoform (*OsMTI-1b*) in *Saccharomyces cerevisiae* enhances cadmium, hydrogen peroxide and ethanol tolerance. *Braz. J. Microbiol.* **2017**, *48*, 537–543. [CrossRef] [PubMed]
93. Ishimaru, Y.; Kakei, Y.; Shimo, H.; Bashir, K.; Sato, Y.; Sato, Y.; Uozumi, N.; Nakanishi, H.; Nishizawa, N.K. A rice phenolic efflux transporter is essential for solubilizing precipitated apoplasmic iron in the plant stele. *J. Biol. Chem.* **2011**, *286*, 24649–24655. [CrossRef] [PubMed]
94. Hao, X.; Zeng, M.; Wang, J.; Zeng, Z.; Dai, J.; Xie, Z.; Yang, Y.; Tian, L.; Chen, L.; Li, D. A Node-Expressed Transporter OsCCX2 Is Involved in Grain Cadmium Accumulation of Rice. *Front. Plant Sci.* **2018**, *9*, 476. [CrossRef] [PubMed]
95. Nakanishi, H.; Ogawa, I.; Ishimaru, Y.; Mori, S.; Nishizawa, N.K. Iron deficiency enhances cadmium uptake and translocation mediated by the Fe^{2+} transporters OsIRT1 and OsIRT2 in rice. *Soil Sci. Plant Nutr.* **2006**, *52*, 464–469. [CrossRef]
96. Lee, S.; An, G. Over-expression of *OsIRT1* leads to increased iron and zinc accumulations in rice. *Plant Cell Environ.* **2009**, *32*, 408–416. [CrossRef]
97. Yuan, L.; Yang, S.; Liu, B.; Zhang, M.; Wu, K. Molecular characterization of a rice metal tolerance protein, OsMTP1. *Plant Cell Rep.* **2012**, *31*, 67–79. [CrossRef]
98. Kavitha, P.G.; Kuruvilla, S.; Mathew, M.K. Functional characterization of a transition metal ion transporter, OsZIP6 from rice (*Oryza sativa* L.). *Plant Physiol. Biochem.* **2015**, *97*, 165–174.
99. Tan, L.; Zhu, Y.; Fan, T.; Peng, C.; Wang, J.; Sun, L.; Chen, C. *OsZIP7* functions in xylem loading in roots and inter-vascular transfer in nodes to deliver Zn/Cd to grain in rice. *Biochem. Biophys. Res. Commun.* **2019**, *512*, 112–118. [CrossRef]
100. Das, N.; Bhattacharya, S.; Bhattacharyya, S.; Maiti, M. Identification of alternatively spliced transcripts of rice phytochelatin synthase 2 gene *OsPCS2* involved in mitigation of cadmium and arsenic stresses. *Plant Mol. Biol.* **2017**, *94*, 167–183. [CrossRef]
101. Uraguchi, S.; Kamiya, T.; Sakamoto, T.; Kasai, K.; Sato, Y.; Nagamura, Y.; Yoshida, A.; Kyozuka, J.; Ishikawa, S.; Fujiwara, T. Low-affinity cation transporter (*OsLCT1*) regulates cadmium transport into rice grains. *Proc. Natl. Acad Sci.* **2011**, *108*, 20959–20964. [CrossRef] [PubMed]
102. Uraguchi, S.; Kamiya, T.; Clemens, S.; Fujiwara, T. Characterization of OsLCT1, a cadmium transporter from indica rice *Oryza sativa*. *Physiol. Plant* **2014**, *151*, 339–347. [CrossRef] [PubMed]
103. Takahashi, R.; Ishimaru, Y.; Shimo, H.; Ogo, Y.; Senoura, T.; Nishizawa, N.K.; Nakanishi, H. The OsHMA2 transporter is involved in root-to-shoot translocation of Zn and Cd in rice. *Plant Cell Environ.* **2012**, *35*, 1948–1957. [CrossRef] [PubMed]

104. Yamaji, N.; Xia, J.; Mitani-Ueno, N.; Yokosho, K.; Ma, J.F. Preferential delivery of zinc to developing tissues in rice is mediated by P-type heavy metal ATPase OsHMA2. *Plant Physiol.* **2013**, *162*, 927–939. [CrossRef] [PubMed]
105. Ueno, D.; Yamaji, N.; Kono, I.; Huang, C.F.T.; Yano, M.; Ma, J.F. Gene limiting cadmium accumulation in rice. *Proc. Natl. Acad Sci. USA* **2010**, *107*, 16500–16505. [CrossRef]
106. Ishimaru, Y.; Takahashi, R.; Bashir, K.; Shimo, H.; Senoura, T.; Sugimoto, K.; Ono, K.; Yano, M.; Ishikawa, S.; Arao, T.; et al. Characterizing the role of rice NRAMP5 in manganese, iron and cadmium transport. *Sci. Rep.* **2012**, *2*, 286. [CrossRef] [PubMed]
107. Sasaki, A.; Yamaji, N.; Yokosho, K.; Ma, J.F. Nramp5 is a major transporter responsible for manganese and cadmium uptake in rice. *Plant Cell.* **2012**, *24*, 2155–2167. [CrossRef]
108. Yang, M.; Zhang, Y.Y.; Zhang, L.; Hu, J.; Zhang, X.; Lu, K.; Dong, H.; Wang, D.; Zhao, F.J.; Huang, C.F.; et al. OsNRAMP5 contributes to manganese translocation and distribution in rice shoots. *J. Exp. Bot.* **2014**, *65*, 4849–4861. [CrossRef]
109. Takahashi, R.; Ishimaru, Y.; Shimo, H.; Bashir, K.; Senoura, T.; Sugimoto, K.; Ono, K.; Suzui, N.; Kawachi, N.; Ishii, S.; et al. From laboratory to field: *OsNRAMP5*-knockdown rice is a promising candidate for Cd phytoremediation in paddy fields. *PLoS ONE* **2014**, *9*, e98816. [CrossRef]
110. Tang, L.; Mao, B.; Li, Y.; Lv, Q.; Zhang, L.; Chen, C.; He, H.; Wang, W.; Zeng, X.; Shao, Y.; et al. Knockout of *OsNramp5* using the CRISPR/Cas9 system produces low Cd-accumulating indica rice without compromising yield. *Sci. Rep.* **2017**, *7*, 14438. [CrossRef]
111. Takahashi, R.; Ishimaru, Y.; Nakanishi, H.; Nishizawa, N.K. Role of the iron transporter OsNRAMP1 in cadmium uptake and accumulation in rice. *Plant Signal. Behav.* **2011**, *6*, 1813–1816. [CrossRef] [PubMed]
112. Takahashi, R.; Ishimaru, Y.; Senoura, T.; Shimo, H.; Ishikawa, S.; Arao, T.; Nakanishi, H.; Nishizawa, N.K. The OsNRAMP1 iron transporter is involved in Cd accumulation in rice. *J. Exp. Bot.* **2011**, *62*, 4843–4850. [CrossRef] [PubMed]
113. Tiwari, M.; Sharma, D.; Dwivedi, S.; Singh, M.; Tripathi, R.D.; Trivedi, P.K. Expression in Arabidopsis and cellular localization reveal involvement of rice NRAMP, OsNRAMP1, in arsenic transport and tolerance. *Plant Cell Environ.* **2014**, *37*, 140–152. [CrossRef] [PubMed]
114. Lim, S.D.; Hwang, J.G.; Han, A.R.; Park, Y.C.; Lee, C.; Ok, Y.S.; Jang, C.S. Positive regulation of rice RING E3 ligase OsHIR1 in arsenic and cadmium uptakes. *Plant Mol. Biol.* **2014**, *85*, 365–379. [CrossRef] [PubMed]
115. 125Mukhopadhyay, A.; Vij, S.; Tyagi, A.K. Overexpression of a zinc-finger protein gene from rice confers tolerance to cold, dehydration, and salt stress in transgenic tobacco. *Proc. Natl. Acad. Sci. USA* **2004**, *101*, 6309–6314. [CrossRef] [PubMed]
116. Wang, F.J.; Wang, M.; Liu, Z.P.; Shi, Y.; Han, T.Q.; Ye, Y.Y.; Gong, N.; Sun, J.W.; Zhu, C. Different responses of low grain-Cd-accumulating and high grain-Cd-accumulating rice cultivars to Cd stress. *Plant Mol. Biol.* **2015**, *96*, 261–269.
117. Jin, S.; Cheng, Y.; Guan, Q.; Liu, D.; Takano, T.; Liu, S. A metallothionein-like protein of rice (rgMT) functions in *E. coli* and its gene expression is induced by abiotic stresses. *Biotechnol. Lett.* **2006**, *28*, 1749–1753.
118. Harada, E.; Choi, Y.E.; Tsuchisaka, A.; Obata, H.; Sano, H. Transgenic tobacco plants expressing a rice cysteine synthase gene are tolerant to toxic levels of cadmium. *J. Plant Physiol.* **2001**, *158*, 655–661. [CrossRef]
119. Lu, L.L.; Tian, S.K.; Yang, X.E.; Li, T.Q.; He, Z.L. Cadmium uptake and xylem loading are active processes in the hyperaccumulator Sedum alfredii. *J. Plant Physiol.* **2009**, *166*, 579–587. [CrossRef]
120. Ishimaru, Y.; Suzuki, M.; Tsukamoto, T.; Suzuki, K.; Nakazono, M.; Kobayashi, T.; Wada, Y.; Watanabe, S.; Matsuhashi, S.; Takahashi, M.; et al. Rice plants take up iron as an Fe^{3+}-phytosiderophore and as Fe^{2+}. *Plant J.* **2006**, *45*, 335–346. [CrossRef]
121. Liu, C.; Chen, G.; Li, Y.; Peng, Y.; Zhang, A.; Hong, K.; Jiang, H.; Ruan, B.; Zhang, B.; Yang, S.; et al. Characterization of a major QTL for manganese accumulation in rice grain. *Sci. Rep.* **2017**, *7*, 17704. [CrossRef] [PubMed]
122. Uraguchi, S.; Fujiwara, T. Cadmium transport and tolerance in rice: Perspectives for reducing grain cadmium accumulation. *Rice* **2012**, *5*, 1–8. [CrossRef] [PubMed]
123. Satoh-Nagasawa, N.; Mori, M.; Nakazawa, N.; Kawamoto, T.; Nagato, Y.; Sakurai, K.; Takahashi, H.; Watanabe, A.; Akagi, H. Mutations in rice (*Oryza sativa*) heavy metal ATPase 2 (*OsHMA2*) restrict the translocation of zinc and cadmium. *Plant Cell Physiol.* **2012**, *53*, 213–224. [CrossRef] [PubMed]

124. Satoh-Nagasawa, N.; Mori, M.; Sakurai, K.; Takahashi, H.; Watanabe, A.; Akagi, H. Functional relationship heavy metal P-type ATPases (*OsHMA2* and *OsHMA3*) of rice (*Oryza sativa*) using RNAi. *Plant Biotechnol.* **2013**, *30*, 511–515. [CrossRef]
125. Miyadate, H.; Adachi, S.; Hiraizumi, A.; Tezuka, K.; Nakazawa, N.; Kawamoto, T.; Katou, K.; Kodama, I.; Sakurai, K.; Takahashi, H.; et al. OsHMA3, a P1B-type of ATPase affects root-to-shoot cadmium translocation in rice by mediating efflux into vacuoles. *New Phytol.* **2011**, *189*, 190–199. [CrossRef] [PubMed]
126. Tanaka, K.; Fujimaki, S.; Fujiwara, T.; Yoneyama, T.; Hayashi, H. Quantitative estimation of the contribution of the phloem in cadmium transport to grains in rice plants (Oryza sativa L.). *Soil Sci. Plant Nutr.* **2007**, *53*, 72–77. [CrossRef]
127. Wu, Z.C.; Zhao, X.H.; Sun, X.C.; Tan, Q.L.; Tang, Y.F.; Nie, Z.J.; Hu, C.X. Xylem transport and gene expression play decisive roles in cadmium accumulation in shoots of two oilseed rape cultivars (Brassica napus). *Chemosphere* **2015**, *119*, 1217–1223. [CrossRef]
128. Fujimaki, S.; Suzui, N.; Ishioka, N.S.; Kawachi, N.; Ito, S.; Chino, M.; Nakamura, S. Tracing cadmium from culture to spikelet: Noninvasive imaging and quantitative characterization of absorption, transport, and accumulation of cadmium in an intact rice plant. *Plant Physiol.* **2010**, *152*, 1796–1806. [CrossRef]
129. Yan, J.; Wang, P.; Wang, P.; Yang, M.; Lian, X.; Tang, Z.; Huang, C.F.; Salt, D.E.; Zhao, F.J. A loss-of-function allele of *OsHMA3* associated with high cadmium accumulation in shoots and grain of Japonica rice cultivars. *Plant Cell Environ.* **2016**, *39*, 1941–1954. [CrossRef]
130. Ishikawa, S.; Abe, T.; Kuramata, M.; Yamaguchi, M.; Ando, T.; Yamamoto, T.; Yano, M. A major quantitative trait locus for increasing cadmium-specific concentration in rice grain is located on the short arm of chromosome 7. *J. Exp. Bot.* **2010**, *61*, 923–934. [CrossRef]
131. Kashiwagi, T.; Shindoh, K.; Hirotsu, N.; Ishimaru, K. Evidence for separate translocation pathways in determining cadmium accumulation in grain and aerial plant parts in rice. *BMC Plant Biol.* **2009**, *9*, 8. [CrossRef] [PubMed]
132. Sato, H.; Shirasawa, S.; Maeda, H.; Nakagomi, K.; Kaji, R.; Ohta, H.; Yamaguchi, M.; Takeshi, N. Analysis of QTL for lowering cadmium concentration in rice grains from 'LAC- 23'. *Breed. Sci.* **2011**, *61*, 196–200.
133. Yan, Y.F.; Lestari, P.; Lee, K.J.; Kim, M.Y.; Lee, S.H.; Lee, B.W. Identification of quantitative trait loci for cadmi um accumulation and distribution in rice (Oryza sativa). *Genome* **2013**, *56*, 227–232. [CrossRef] [PubMed]
134. Abe, T.; Nonoue, Y.; Ono, N.; Omoteno, M.; Kuramata, M.; Fukuoka, S.; Yamamoto, T.; Yano, M.; Ishikawa, S. Detection of QTLs to reduce cadmium content in rice grains using LAC23/Koshihikari chromosome segment substitution lines. *Breed. Sci.* **2013**, *63*, 284–291.
135. Liu, W.; Pan, X.; Li, Y.; Duan, Y.; Min, J.; Liu, S.; Liu, L.; Sheng, X.; Li, X. Identification of QTLs and Validation of *qCd-2* Associated with Grain Cadmium Concentrations in Rice. *Rice Sci.* **2019**, *26*, 42–49. [CrossRef]
136. Liu, X.; Chen, S.; Chen, M.; Zheng, G.; Peng, Y.; Shi, X.; Qin, P.; Xu, X.; Teng, S. Association Study Reveals Genetic Loci Responsible for Arsenic, Cadmium and Lead Accumulation in Rice Grain in Contaminated Farmlands. *Front. Plant Sci.* **2019**, *10*, 61. [CrossRef] [PubMed]
137. Ueno, D.; Koyama, E.; Kono, I.; Ando, T.; Yano, M.; Ma, J.F. Identification of a novel major quantitative trait locus controlling distribution of Cd between roots and shoots in rice. *Plant Cell Physiol.* **2009**, *50*, 2223–2233. [CrossRef] [PubMed]
138. Ueno, D.; Kono, I.; Yokosho, K.; Ando, T.; Yano, M.; Ma, J.F. A major quantitative trait locus controlling cadmium translocation in rice (*Oryza sativa*). *New Phytol.* **2009**, *182*, 644–653. [CrossRef]
139. Xue, D.; Chen, M.; Zhang, G. Mapping of QTLs associated with cadmium tolerance and accumulation during seedling stage in rice (*Oryza sativa* L.). *Euphytica* **2009**, *165*, 587. [CrossRef]
140. Norton, G.J.; Deacon, C.M.; Xiong, L.Z.; Huang, S.Y.; Meharg, A.A.; Price, A.H. Genetic mapping of the rice ionome in leaves and grain: Identification of QTLs for 17 elements including arsenic, cadmium, iron and selenium. *Plant Soil* **2010**, *329*, 139–153. [CrossRef]
141. Ishikawa, S.; Ae, N.; Yano, M. Chromosomal regions with quantitative trait loci controlling cadmium concentration in brown rice (*Oryza sativa*). *New Phytol.* **2005**, *168*, 345–350. [CrossRef] [PubMed]
142. Abe, T.; Taguchi-Shiobara, F.; Kojima, Y.; Ebitani, T.; Kuramata, M.; Yamamoto, T.; Yano, M.; Ishikawa, S. Detection of a QTL for accumulating Cd in rice that enables efficient Cd phytoextraction from soil. *Breed. Sci.* **2011**, *61*, 43–51. [CrossRef]
143. Zhang, X.Q.; Zhang, G.P.; Guo, L.B.; Wang, H.Z.; Zeng, D.L.; Dong, G.J.; Qian, Q.; Xue, D.W. Identification of quantitative trait loci for Cd and Zn concentrations of brown rice grown in Cd-polluted soils. *Euphytica* **2011**, *180*, 173–179. [CrossRef]

144. Huang, Y.; Sun, C.; Min, J.; Chen, Y.; Tong, C.; Bao, J. Association Mapping of Quantitative Trait Loci for Mineral Element Contents in Whole Grain Rice (*Oryza sativa* L.). *J. Agric. Food Chem.* **2015**, *63*, 10885–10892. [CrossRef] [PubMed]
145. Guo, G.; Zhou, Q.; Ma, L.Q. Availability and assessment of fixing additives for the in-situ remediation of heavy metal contaminated soils: A review. *Environ. Monit. Assess.* **2006**, *116*, 513–528. [CrossRef] [PubMed]
146. Madejón, E.; de Mora, A.P.; Felipe, E.; Burgos, P.; Cabrera, F. Soil amendments reduce trace element solubility in a contaminated soil and allow regrowth of natural vegetation. *Environ. Pollut.* **2006**, *139*, 40–52. [CrossRef]
147. Yan, Y.; Zhou, Y.Q.; Liang, C.H. Evaluation of phosphate fertilizers for the immobilization of Cd in contaminated soils. *PLoS ONE* **2015**, *10*, e0124022. [CrossRef] [PubMed]
148. Hu, P.J.; Ouyang, Y.N.; Wu, L.H.; Shen, L.B.; Luo, Y.M.; Christie, P. Effects of water management on arsenic and cadmium speciation and accumulation in an upland rice cultivar. *J. Environ. Sci.* **2015**, *27*, 225–231. [CrossRef]
149. Honma, T.; Ohba, H.; Kaneko-Kadokura, A.; Makino, T.; Nakamura, K.; Katou, H. Optimal soil Eh, pH, and water management for simultaneously minimizing arsenic and cadmium concentrations in rice grains. *Environ. Sci. Technol.* **2016**, *50*, 4178–4185. [CrossRef]
150. Yu, L.L.; Zhu, J.Y.; Huang, Q.Q.; Su, D.C.; Jiang, R.F.; Li, H.F. Application of a rotation system to oilseed rape and rice fields in Cd-contaminated agricultural land to ensure food safety. *Ecotox. Environ. Safe.* **2014**, *108*, 287–293. [CrossRef]
151. Ishikawa, S.; Ishimaru, Y.; Igura, M.; Kuramata, M.; Abe, T.; Senoura, T.; Hase, Y.; Arao, T.; Nishizawa, N.K.; Nakanishi, H. Ion-beam irradiation, gene identification, and marker-assisted breeding in the development of low-cadmium rice. *Proc. Natl. Acad. Sci. USA* **2012**, *109*, 19166–19171. [CrossRef] [PubMed]
152. Lu, C.; Zhang, L.; Tang, Z.; Huang, X.Y.; Ma, J.F.; Zhao, F.J. Producing cadmium-free Indica rice by overexpressing *OsHMA3*. *Environ. Int.* **2019**, *126*, 619–626. [CrossRef] [PubMed]

© 2019 by the authors. Licensee MDPI, Basel, Switzerland. This article is an open access article distributed under the terms and conditions of the Creative Commons Attribution (CC BY) license (http://creativecommons.org/licenses/by/4.0/).

Article

Hormesis in Plants: The Role of Oxidative Stress, Auxins and Photosynthesis in Corn Treated with Cd or Pb

Eugeniusz Małkowski [1,*], Krzysztof Sitko [1,*], Michał Szopiński [1], Żaneta Gieroń [1], Marta Pogrzeba [2], Hazem M. Kalaji [3] and Paulina Zieleźnik-Rusinowska [1]

1 Plant Ecophysiology Team, Faculty of Natural Sciences, Institute of Biology, Biotechnology and Environmental Protection, University of Silesia in Katowice, 40-032 Katowice, Poland; mszopinski@us.edu.pl (M.S.); zgieron@us.edu.pl (Ż.G.); pzieleznik@us.edu.pl (P.Z.-R.)

2 Institute for Ecology of Industrial Areas, 40-844 Katowice, Poland; m.pogrzeba@ietu.pl

3 Department of Plant Physiology, Institute of Biology, Warsaw University of Life Sciences WULS-SGGW, 02-776 Warsaw, Poland; hazem@kalaji.pl

* Correspondence: eugeniusz.malkowski@us.edu.pl (E.M.); krzysztof.sitko@us.edu.pl (K.S.)

Received: 30 December 2019; Accepted: 16 March 2020; Published: 19 March 2020

Abstract: Hormesis, which describes the stimulatory effect of low doses of toxic substances on growth, is a well-known phenomenon in the plant and animal kingdoms. However, the mechanisms that are involved in this phenomenon are still poorly understood. We performed preliminary studies on corn coleoptile sections, which showed a positive correlation between the stimulation of growth by Cd or Pb and an increase in the auxin and H_2O_2 content in the coleoptile sections. Subsequently, we grew corn seedlings in hydroponic culture and tested a wide range of Cd or Pb concentrations in order to determine hormetic growth stimulation. In these seedlings the gas exchange and the chlorophyll *a* fluorescence, as well as the content of chlorophyll, flavonol, auxin and hydrogen peroxide, were measured. We found that during the hormetic stimulation of growth, the response of the photosynthetic apparatus to Cd and Pb differed significantly. While the application of Cd mostly caused a decrease in various photosynthetic parameters, the application of Pb stimulated some of them. Nevertheless, we discovered that the common features of the hormetic stimulation of shoot growth by heavy metals are an increase in the auxin and flavonol content and the maintenance of hydrogen peroxide at the same level as the control plants.

Keywords: hormesis; growth; photosynthesis; chlorophyll *a* fluorescence; cadmium; lead

1. Introduction

Hormesis is a phenomenon that is defined as the stimulatory effect of low doses of toxic substances, e.g., heavy metals, on a single biological parameter of a given organism [1]. The hormetic effect is described by a reversed U-shaped biphasic curve in which a low dose of a toxic substance has a stimulatory effect; however, when the dose is increased, the toxic effect starts to be visible [2–4]. Hormesis is considered to be a universal phenomenon that is common in nature, which is independent of the type of stressor, the organism in which it occurs or the physiological process [2–4]. This notion is supported by numerous studies that have been carried out on a wide range of organisms (from microorganisms through plants to mammals) [5,6]. Some studies have suggested that hormesis represents an evolutionary adaptive response to environmental factors that interfere with homeostasis [7,8].

Growth, oxidative stress and/or photosynthetic activity are the physiological parameters that are most frequently examined by scientists who are interested in hormesis in plants [3,4,8–11]. Many scientists have connected the stimulation of plant growth during hormesis with a low level of oxidative stress [3,7,8,10,12]. Their results show that reactive oxygen species (ROS) are undoubtedly involved in the hormetic effect. However, there is a dearth of information that could explain the mechanisms that underlie the hormesis phenomenon [6].

Cd and Pb are heavy metals that are commonly known for their lack of any biological functions in plants [13] and for their toxic effects on plant growth and development [14–17]. Pb is mostly accumulated in roots, which negatively affects their growth and cell division. This metal is characterized by a low translocation from the root to shoot [18]. As a result, the shoot growth is inhibited to a lesser degree than that of the root [14,19,20]. By contrast, Cd is easily transported from the root to the aboveground parts of a plant and thereby exerts a negative influence on the whole plant [15,19]. Both metals cause a disturbance in the proper functioning of the root, which negatively affects the mineral nutrition of plants [14,15,21]. Moreover, Cd and Pb induce oxidative stress in plants and also affect photosynthesis, water relations and hormonal balance [15,21,22].

Coleoptile sections of the Poaceae family, which are excised from etiolated seedlings, are a model object in plant elongation growth studies because their cells do not undergo division but only elongate [23–25]. Coleoptile sections are frequently used to determine the influence of various factors on the elongation growth, e.g., heavy metals [26–29]. It was found that the Pb or Cd that was added to an incubation solution inhibited both the endogenous and exogenous auxin-induced growth of coleoptile sections [27–29]. On the other hand, Małkowski et al. [20] observed that the elongation growth of coleoptile sections that had been excised from corn seedlings growing for 24 h in the presence of Pb at 100 or 1000 μM, and then incubated in a control medium with auxin (IAA), was 47% or 69% higher, respectively, compared to the growth of the sections that had been cut from seedlings that had not been treated with the heavy metal. This phenomenon seems to be similar to hormetic stimulation of growth, although its mechanism is not known.

The growth of coleoptile sections is induced by indole-3-acetic acid (IAA). This applies to both endogenous growth and growth that is stimulated by exogenous auxin [25,30,31]. Schopfer et al. [32] documented that the presence of oxygen reactive species (ROS) (e.g., H_2O_2) in the cell wall is necessary to promote the auxin-induced growth of coleoptile sections and that the removal of reactive oxygen species (ROS) inhibited elongation growth.

The toxic effects of heavy metals such as Pb or Cd induce oxidative stress in plants, which is associated with the production of ROS [15,21,22]. Therefore, it was hypothesized that the hormetic effect on the growth of the coleoptile sections excised from seedlings that had been treated with Pb that was observed by Małkowski et al. [20] was associated with the induction of oxidative stress in the seedlings and the accumulation of ROS (e.g., H_2O_2). The coleoptile sections contained higher concentrations of ROS, and therefore, after the exogenous IAA was administered, showed a higher elongation growth (hormesis). To verify this hypothesis, the main goal of the study was to determine:

– whether the hormetic effect of elongation growth of the coleoptile sections cut from corn seedlings that had previously been treated with Pb that was observed by Małkowski et al. [20] would also occur in the case of Cd-treated plants

– whether stimulating the elongation growth of coleoptile sections is correlated with higher H_2O_2 and IAA content in the sections

– whether the relationships between stimulating growth (hormesis) and the content of H_2O_2 and/or IAA that was observed during studies conducted with coleoptile sections will be confirmed in studies with whole corn seedlings

– whether the changes that were observed in the seedling shoots are related to changes in photosynthesis and transpiration rates

2. Results

2.1. Experiment with the Coleoptile Sections

2.1.1. Influence of Cd and Pb on the Elongation Growth of Corn Coleoptile Sections

A significant stimulation of elongation growth was observed in the coleoptile sections that had been excised from the corn seedlings that had been treated with 1000 μM Cd or Pb and incubated for 24 h in the control medium (APW) without heavy metals. As a result, there was a three-fold and two-fold increase in elongation growth for the Cd- and Pb-treated plants, respectively, compared to the control. The stimulation of the growth of the sections that had been excised from the seedlings that had been treated with 100 μM Cd or Pb was lower compared to 1000 μM, but it was still significantly higher than in the control (Figure 1a,b).

2.1.2. Influence of Cd and Pb on the Auxin Concentration in Corn Coleoptile Sections

In the coleoptile sections that had been excised from the control seedlings, the auxin concentration was approximately 0.33 μmol g^{-1} FW (Figure 1c,d). A significant increase in the auxin concentration was observed in the coleoptile sections that had been excised from the seedlings growing for 24 h in the medium with 1000 μM Cd or Pb. In the plants that had been treated with Cd, the auxin content in the coleoptile sections was more than two-fold higher (124%) (Figure 1c), whereas in the sections that had been excised from the Pb-treated plants, it was 79% higher (Figure 1d) compared to the control. In the sections that had been excised from the seedlings that had been treated with 100 μM Cd or Pb, the increase in the auxin concentration was still significantly higher compared to the control, 43% and 21% for Cd and Pb, respectively (Figure 1c,d).

2.1.3. Influence of Cd and Pb on the Hydrogen Peroxide Concentration in Corn Coleoptile Sections

The H_2O_2 concentration in the coleoptile sections that had been excised from the control seedlings was 0.07 μmol g^{-1} FW (Figure 1e,f). The concentration of 10 μM did not result in significant changes in the H_2O_2 concentration (for Cd) (Figure 1e) or increased concentration of hydrogen peroxide in coleoptile sections slightly (37% for Pb) (Figure 1f). While treating the seedlings with Cd or Pb at concentrations of 100 and 1000 μM considerably increased the H_2O_2 content in the excised coleoptile sections by 57% compared to the control (Figure 1e,f).

2.2. Experiment with the Corn Seedlings

The experiments were conducted on 14-day-old seedlings that had been treated with Cd or Pb for the last four days.

2.2.1. Influence of Cd and Pb on the Elongation Growth of Corn Shoots

In the seedlings that were treated with Cd for four days, a statistically significant stimulation of growth was observed only at concentrations of 2.5 and 10 μM (Figure 2a). A significant inhibition of growth was measured for the plants treated with Cd at a concentration of 100 μM compared to the control (Figure 2a). A significant stimulation of growth of corn shoots treated with Pb was observed for concentrations of 1, 5 and 10 μM, although there was significant inhibition for the highest tested heavy metal (HM) concentration (Figure 2b). It is worth noting that the highest level of growth stimulation was measured for concentrations of 10 μM and 5 μM of Cd and Pb, respectively (Figure 2a,b).

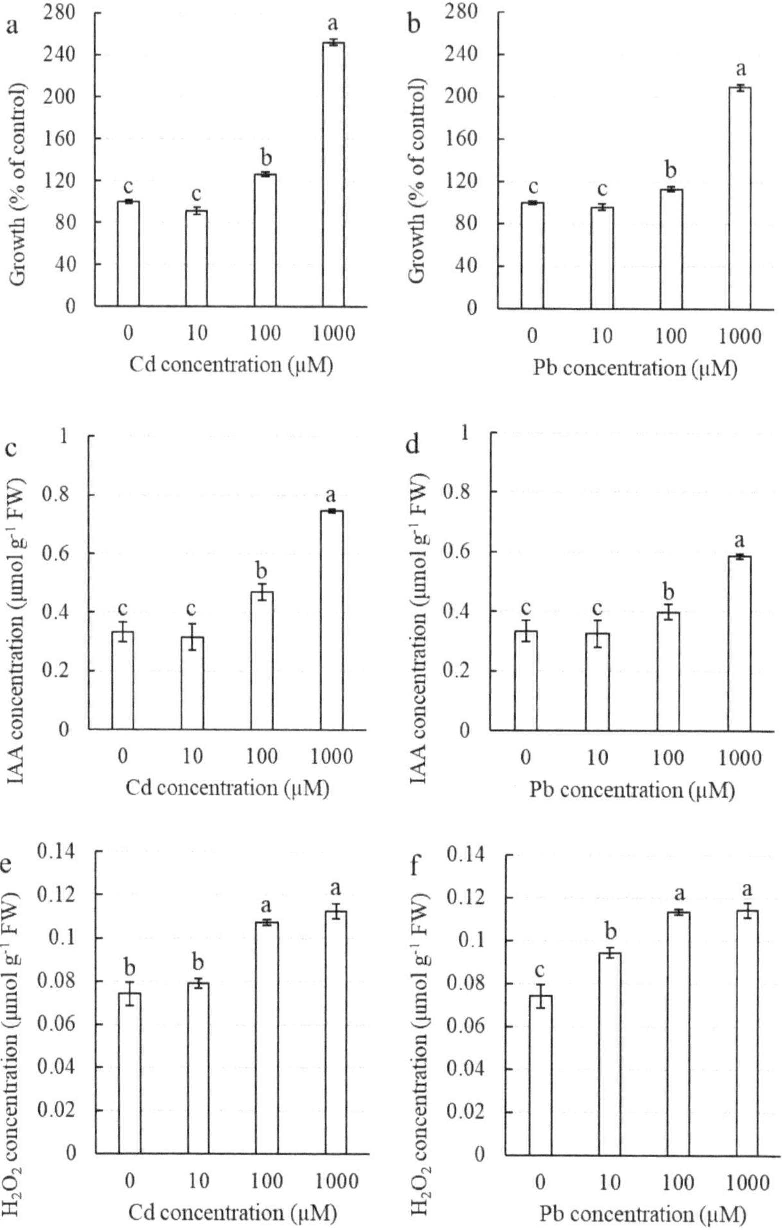

Figure 1. Physiological characteristics of the corn coleoptile sections that had been excised from the seedlings that had been treated with Cd or Pb for 24 h. The growth of sections was measured after incubation in APW for 24 h—(**a**) Cd and (**b**) Pb; concentration of IAA—(**c**) Cd and (**d**) Pb; concentration of H_2O_2—(**e**) Cd and (**f**) Pb. The values are the means ± SE (n = 3). Means followed by the same letter are not significantly different from each other using the LSD test ($p < 0.05$).

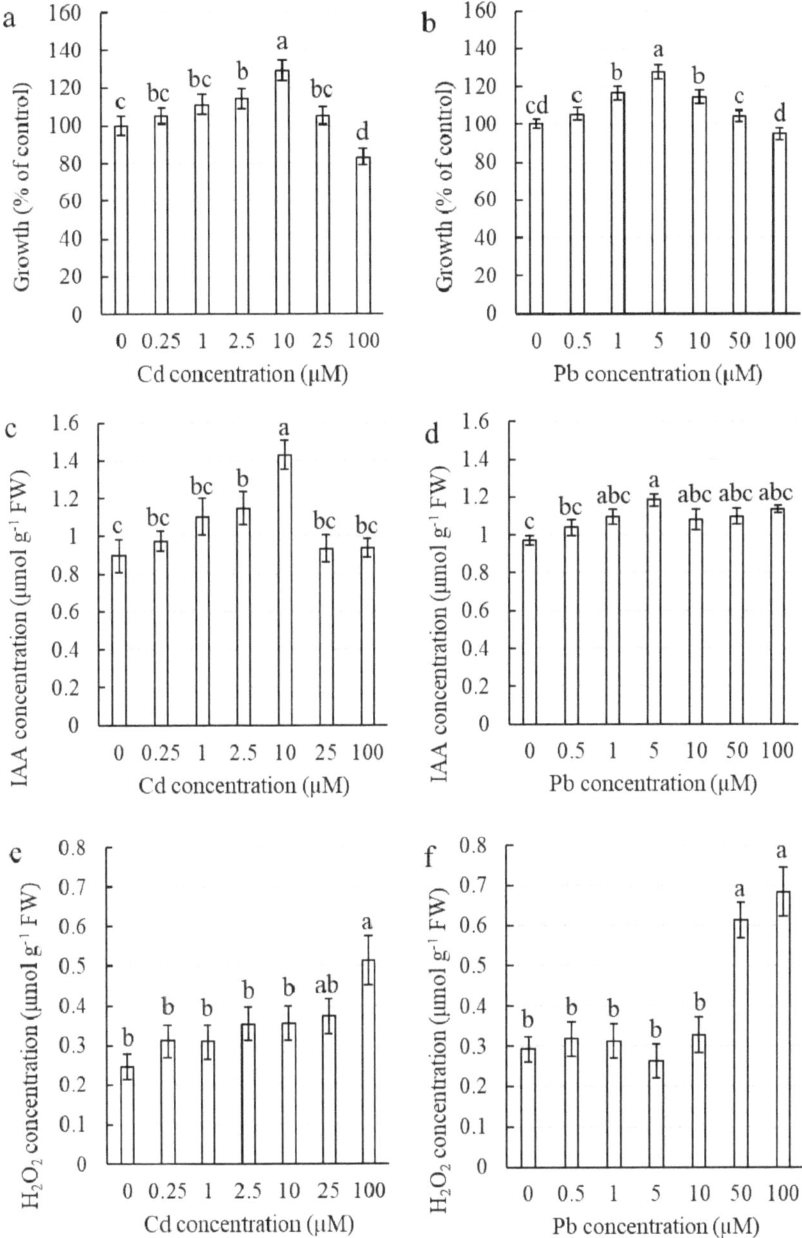

Figure 2. Physiological characteristics of the corn shoots treated with Cd or Pb for four days. Growth of shoots—(**a**) Cd, (**b**) Pb; Auxin concentration in the leaves—(**c**) Cd, (**d**) Pb; H_2O_2 concentration in the leaves—(**e**) Cd, (**f**) Pb. The values are the means ± SE ($n = 8$, except for growth where $n = 18$). Means followed by the same letter are not significantly different from each other using the LSD test ($p < 0.05$).

2.2.2. Influence of Cd and Pb on Auxin and Hydrogen Peroxide in Corn Leaves

The highest and statistically significant (compared to the control) content of auxin was measured for corn leaves in plants treated with Cd at the concentration of 2.5 and 10 μM (Figure 2c). In the plants that were treated with Pb at a concentration of 5 μM, there was a considerably higher content of IAA compared to the control, which was the only significant difference among all of the tested Pb concentrations (Figure 2d).

In the plants treated with Pb, there was a statistically significant increase in H_2O_2 content in the leaves for the 50 and 100 μM concentrations, whereas in the Cd-treated plants, there was a considerably higher content of H_2O_2 compared to the control only for a concentration of 100 μM of Cd (Figure 2e,f).

2.2.3. The Toxic Effect of Cd and Pb on the Photosynthetic Apparatus

To highlight the toxic effect of Cd and Pb on photosystem II, the curves of the relative variable fluorescence (ΔV_t) were plotted (Figure 3). ΔV_t was calculated as the subtraction between the fluorescence curves that registered for the plants treated with the specific concentrations of HM and the averaged fluorescence values from the control plants ($\Delta V_t = ((F_t - F_0)/F_v) - V_{control}$). The toxic effect of Cd was clearly visible in the time-course of the ΔVt curves. Only for a concentration of 0.25 μM Cd was the ΔVt curve comparable to the control (Figure 3a). At low concentrations, Cd seemed to mainly inhibit the activity of the FNR (Ferredoxin-NADP+ Reductase) complex (high ΔH and ΔG steps). As the concentration of Cd was increased, the characteristic peaks that might indicate damage to the individual elements of the electron-transport chain, such as the oxygen evolving complex (Figure 3a), started to appear (ΔK, ΔJ and ΔI).

The plants treated with Pb were characterized by a much milder time-course of the ΔV_t curves (Figure 3b). The greatest changes were observed for the ΔH and G step, which could be correlated with damage to the FNR. The concentrations of 1 μM and 5 μM of Pb resulted in ΔH peaks that were below the control, which may indicate a higher efficiency of the final electron acceptors in PSI compared to the control (Figure 3b). It is noteworthy that the hormetic effect of the growth of the seedlings treated with Cd was found simultaneously with a significantly high decrease in the performance of the photosystems, whereas the activity of the photosynthetic apparatus of the Pb hormetic plants was comparable to the control.

The phenomenological pipeline models of energy fluxes through the leaf cross sections of the corn plants treated with different concentrations of Cd or Pb are presented in Figure 4 (Cd) and Figure 5 (Pb). The plants treated with Cd at a concentration of 0.25 μM were characterized by the highest values of the parameters that described all of the fluxes, but they did not differ significantly compared to the control (Figure 4). With an increase in the Cd concentration in the medium, there was an increase in the toxic effect of this metal on photosynthetic apparatus. At a concentration of 25 μM Cd, the electron transport through PSII was inhibited to 57% of the control and for 100 μM of Cd, it was inhibited to 50% of the control values. The percentage of active reaction centers in 100 μM of Cd decreased to 65% of the control value (Figure 4).

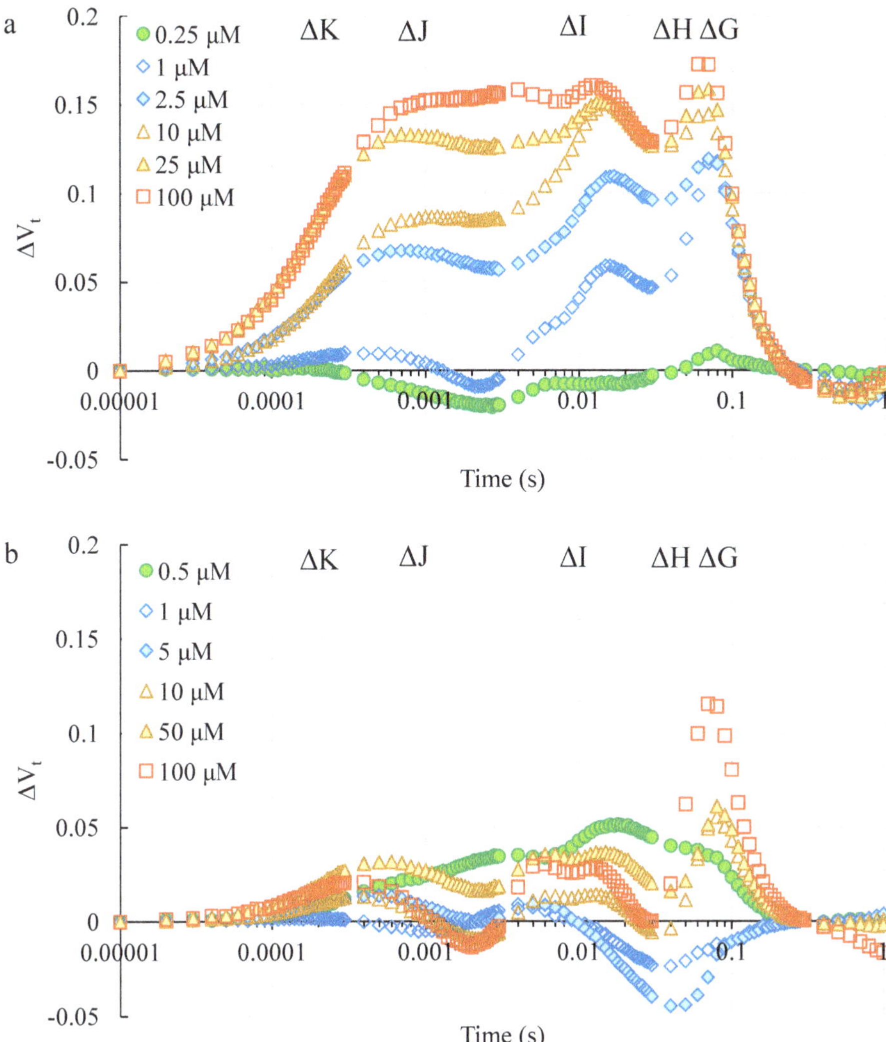

Figure 3. The effect of Cd (**a**) and Pb (**b**) on the relative variable fluorescence of the chlorophyll *a* ($\Delta V_t = ((F_t - F_0)/F_v) - V_{control}$) of the studied corn leaves. For the ΔV_t analysis, the fluorescence of the leaves of the control plant was the reference and equaled 0. The values are the means ($n = 15$).

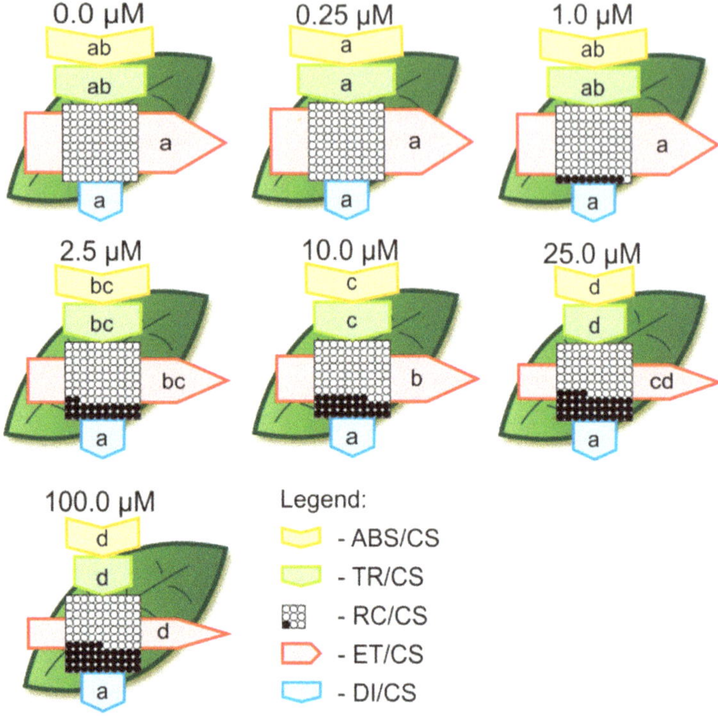

Figure 4. Leaf models showing the phenomenological energy fluxes per the excited cross sections (CS) of corn leaves treated with different Cd concentrations. Each relative value is the mean ($n = 15$) and is represented by the size of the correct parameters (arrows). Letters in the arrows correspond to the statistical significance (LSD test, $p < 0.05$). ABS/CS—absorption flux per CS approximated; TR/CS—trapped energy flux per CS; ET/CS—electron transport flux per CS; RC/CS—% of active/inactive reaction centers as circles inscribed in the square (white = active, black = inactive); DI/CS—dissipated energy flux per CS.

Treatment with Cd had no impact on dissipated energy through the cross sections of the leaves. A significant inhibition of energy flux parameters in plants treated with Pb was only observed at concentrations of 50 and 100 μM (Figure 5). These concentrations also caused a decrease in the dissipated energy through cross section. The limitation of electron transport at a concentration of 100 μM of Pb was also observed, but only by 10% compared to the control. The inhibition of reaction center activity at this concentration was determined to be 77% of the control (Figure 5).

Treatment with Cd caused a decrease in the chlorophyll content in the corn leaves (Figure 6a). As a result, the chlorophyll content varied from 75% to 80% of the control for the concentrations from 10 μM to 100 μM of Cd. In contrast to Cd, most of the investigated Pb concentrations caused a significant increase in the chlorophyll content in the corn leaves (Figure 6b). For both of the investigated heavy metals, there was an increase in the flavonol content in the leaves with an increase in metal concentration in the hydroponic medium (Figure 6c,d). It is noteworthy that the higher content of flavonols in the leaves was related with a hormetic increase in growth for both metals, whereas an increase of the chlorophyll content at a concentration that caused hormesis was only observed for the corn treated with Pb.

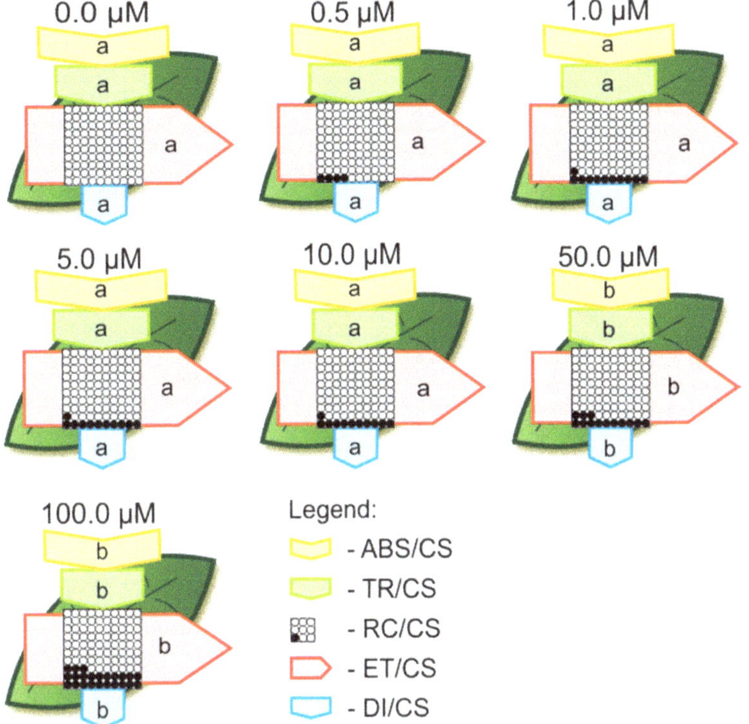

Figure 5. Leaf models showing the phenomenological the energy fluxes per excited cross sections (CS) of the corn leaves treated with different Pb concentrations. Each relative value is the mean (n = 15) and is represented by the size of the correct parameters (arrows). Letters in the arrows correspond to the statistical significance (LSD test, $p < 0.05$). ABS/CS—absorption flux per CS approximated; TR/CS—trapped energy flux per CS; ET/CS—electron transport flux per CS; RC/CS—% of active/inactive reaction centers as circles inscribed in the square (white = active, black = inactive); DI/CS—dissipated energy flux per CS.

There was a stimulation of the photosynthetic rate by Cd only at a concentration of 0.25 µM, whereas at concentrations from 2.5 µM to 100 µM of Cd, there was a significant decrease compared to the control (Figure 7a). At the lowest investigated concentration, Cd considerably stimulated the transpiration rate and stomatal conductance, but the higher concentrations decreased these parameters compared to the control (Figure 7c,d). It is worth noting that the increase in growth of the corn shoots was measured at the concentrations of Cd at which all of the gas exchange parameters were significantly lower compared to the control.

In contrast to Cd, there was a significant stimulation of the photosynthetic rate for Pb for the range of concentrations from 0.5 to 10 µM, whereas at the highest investigated concentration, the photosynthetic rate did not differ compared to the control (Figure 7b). Moreover, the transpiration rate of corn treated with Pb was characterized by a similar relation as the photosynthetic rate (Figure 7d). No significant increase was only observed for the stomatal conductance compared to the control (Figure 7e). Interestingly, there was an increase in the photosynthetic and transpiration rate that was caused by the Pb treatment for the concentrations at which an increase in the growth of the corn shoots was also documented.

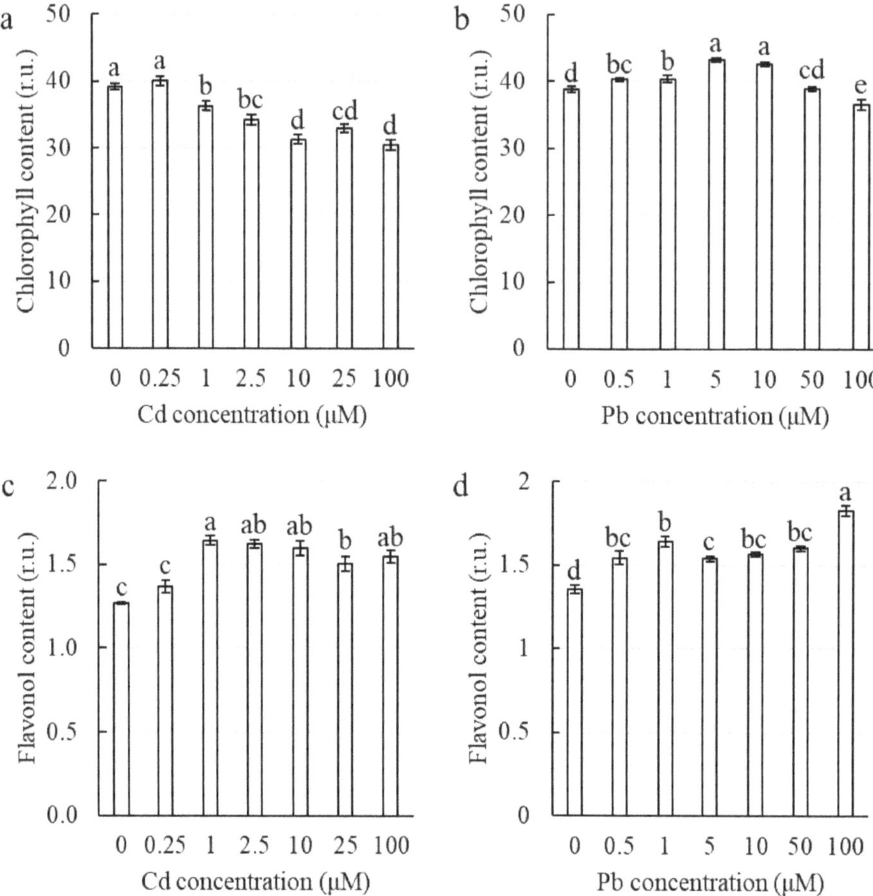

Figure 6. The effect of Cd and Pb on the chlorophyll and flavonol content in the corn leaves. Cd (**a**,**c**) and Pb (**b**,**d**). The values are the means ± SE (n = 15). Means followed by the same letter are not significantly different from each other using the LSD test ($p < 0.05$).

On the basis of the principal component analysis (PCA), the positive correlation between the auxin concentration and hormetic growth stimulation by both HMs was demonstrated (Figure S1, Tables S1 and S2). PCA made it possible to isolate each concentration as a separate group characterized by unique plant physiological status. What is particularly noteworthy is that plants showed hormetic growth stimulation formed groups apart from both control and high concentrations of both heavy metals (Figure S1).

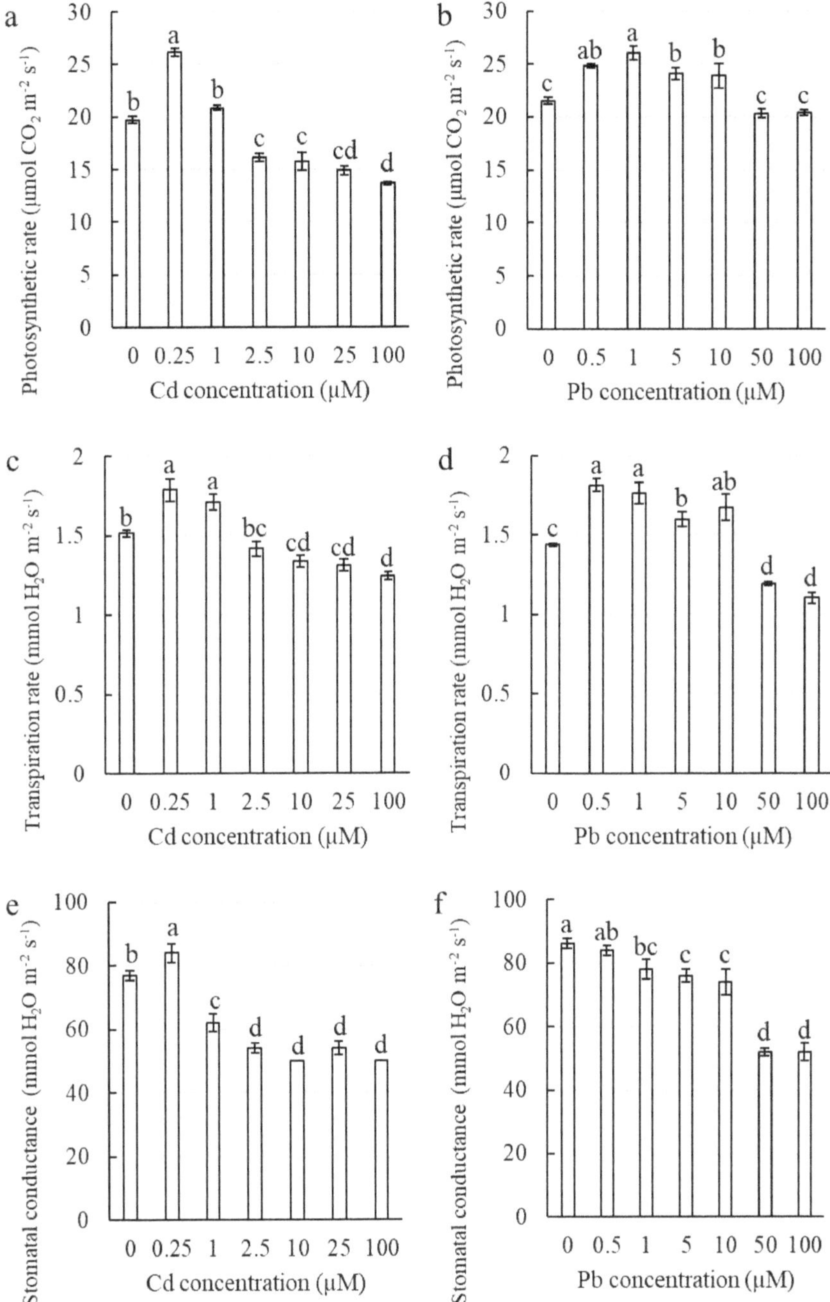

Figure 7. The effect of Cd and Pb on the photosynthetic and transpiration rate and stomatal conductance in the corn leaves. Cd (**a**,**c**,**e**) and Pb (**b**,**d**,**f**). The values are the means ± SE (n = 15). Means followed by the same letter are not significantly different from each other using LSD test ($p < 0.05$).

3. Discussion

It has been well documented that incubating corn coleoptile sections in a medium with Pb or Cd inhibits the elongation growth of the sections [26–29]. In our preliminary study, however, the corn coleoptile sections that had been excised from seedlings that had been treated with Pb and incubated in a medium without Pb grew substantially better than the coleoptile sections that had been excised from the control seedlings [20]. In the current study, we showed that stimulating the growth of coleoptile sections by pretreating the seedlings with Pb or Cd is the result of the higher content of IAA and H_2O_2 in these sections (Figure 1). This finding is in agreement with the data presented by Schopfer et al. [32], who documented that the presence of oxygen reactive species (ROS) (e.g., H_2O_2) in the cell wall is necessary to promote the auxin-induced growth of coleoptile sections. Most investigations have shown that treating plants with heavy metals decreases the content of IAA in plants, see [33] and literature therein [34–39]. In the current study, we found that Pb and Cd increased the content of IAA in the corn coleoptile sections, but only at the concentrations at which stimulation of elongation growth was observed (Figure 1). Hence, we can conclude that for an increase in IAA and H_2O_2 content is necessary for the elongation growth stimulation by HMs.

Because as a model object, coleoptile sections have a simple structure and are a short-lived organ [40], we conducted other experiments with corn seedlings. The main goal of the experiments with the seedlings was to determine whether an increase in both the IAA and H_2O_2 content in the shoots is necessary to induce the hormetic effect on growth.

Many investigations have been conducted to study the effects of heavy metal toxicity on plants. However, fewer works have addressed the stimulatory effect of sub-toxic levels of HMs—the phenomenon of hormesis. However, hormesis has been gaining more and more interest in recent years [2–4,41]. The stimulatory effect of low concentrations of Cd on plant growth was reported for *Lonicera japonica* by Jia et al. [8,42] and for *Brassica juncea* by Seth et al. [43]. Moreover, it was reported that Pb also had a stimulatory effect on the growth of corn [44] and *Arabis paniculata* [45]. In our research, we observed a significant increase in the growth of the shoots of the corn by 29% and 27%, respectively, compared to the control when treated with 10 μM Cd and 5 μM Pb (Figure 2a,b).

The majority of the experiments that have been conducted that have focused on the effect of Cd and Pb on plants have reported an increase in the reactive oxygen species (ROS) level as a response to HM treatment that was accompanied by an inhibition of growth [46–52]. However, in our research, which was conducted on corn shoots treated with Cd or Pb at the concentrations for which growth stimulation was found, there was no significant change in the H_2O_2 content compared to the control (Figure 2e,f). Lin et al. [53] and Jia et al. [8] also showed no significant increase in the level of oxidative stress in plants in which growth stimulation was observed under a low Cd treatment compared to the untreated plants. These results contradict the results that were obtained in the current study for coleoptile sections for which an increase in the H_2O_2 content was always necessary to induce hormesis (Figure 1).

Phytohormones such as auxins have been found to play an important role in plant tolerance and the alleviation of the stress that is induced by HMs [54]. It has been suggested that the cellular level of auxins increases under a low level of abiotic stress, which stimulates vegetative growth, thus leading to the hormetic effect [3,55]. Although Elobeid and Polle ([33] and literature therein) reported that a high concentration of Cd in *Glycine max* inhibited auxin biosynthesis and reduced growth, whereas exposure to a low concentration of Cd stimulated auxin biosynthesis. Other studies have shown that the application of exogenous auxin improves a plant's protection against HM toxicity and can reverse the growth inhibition that is caused by heavy metals [54]. Liphadzi et al. [56] documented that the addition of exogenous IAA caused a significant increase in the biomass of the roots and stems of *Helianthus annuus* plants grown in soil that had been moderately contaminated with Pb compared to the untreated plants. We found that treatment with Cd and Pb at a concentration of 10 μM and 5 μM, respectively, at which we observed a hormetic effect, significantly increased the content of IAA in the corn shoots compared to the untreated plants (Figure 2c,d). The same trend was observed in our experiments with the coleoptile sections in which the induction of elongation growth was always connected with a higher IAA content (Figure 1). Based on our results, we suggest that the increase in auxin content in response to subtoxic levels of HMs such as Cd and Pb plays a key role in the phenomenon of hormesis in plants.

Figlioli et al. [44] found that all of the investigated concentrations of Pb (from 10 μM to 1000 μM) significantly stimulated the growth (both the length of the seedlings and their fresh weight) of corn plants in cultivated garden soil. However, the maximum quantum efficiency of PSII did not differ significantly, whereas the chlorophyll content was significantly higher compared to the control, but only for plants treated with 1000 μM of Pb [44]. A similar effect of Pb was observed on *Pisum sativum* by Rodriguez et al. [57] and on corn by Nyitrai et al. [58]. An increase in the chlorophyll content that was caused by Pb was also confirmed in *Populus* × *canescenc* trees [59] and *Arabis paniculata* [45]. The results presented in the current study also confirmed that the phenomenon of hormesis, which is caused by Pb, was positively correlated with a higher chlorophyll content and that there was no significant difference in the PSII yield compared to the control (Figures 3b, 5 and 6b). On the other hand, improvements in the content of photosynthetic and accessory pigments (such as chlorophyll a, chlorophyll b, total chlorophyll and/or carotenoids) were observed in the leaves of different plant species after Cd exposure [4]. Conversely, González et al. [11] observed a decrease in the chlorophyll content and PSII performance in barley plants, which simultaneously showed an increase in growth that was caused by Cd treatment. Generally, the stimulation of growth by Cd is positively correlated with a significant decrease in the total chlorophyll content and activity of the photosystems [11,43]. The results presented in this paper also confirmed this relationship, which may additionally prove the high toxic effect of Cd on these parameters (Figures 3a, 4 and 6a).

There is a dearth of data on the effect of both Cd and Pb on the content of flavonols during hormesis. However, Zhang et al. [60] found that the addition of exogenous flavonols diminished the toxic effect of Pb on growth and the ROS level in *Arabidopsis thaliana*. They also discovered that flavonols increased the activities of the antioxidant enzymes and detoxified the ROS, which prevented oxidative damage [60]. Moreover, Cetin et al. [61] described the dose-dependent stimulation of the synthesis of flavonols in grape cell suspension cultures that had been treated with Cd. These results are in agreement with the data obtained in our work, which suggests that the corn plants maintained the H_2O_2 content at the level of the control due to a higher content of flavonols, which seems to be a part of the hormesis mechanism (Figure 6c,d). In this work, we showed that corn treated with Cd or Pb and that was characterized by hormetic growth stimulation also had a significant increase in the flavonol content in the leaves for both heavy metal treatments for the first time.

There are many papers that discuss the toxic effect of Cd or Pb on the gas exchange in plants [4,62–67], but there is a paucity of published works that describe the gas exchange parameters in relation to the hormetic stimulation of growth. Ban et al. [68] showed that a low dose of Pb stimulated growth, the photosynthetic rate, transpiration and stomatal conductance in corn. In the shoots of many plant species, the hormesis that is caused by Cd is associated with an increase in the photosynthesis activity, which improves plant development [4,41]. In our study, the hormetic stimulation of growth that was caused by Pb was correlated with increased photosynthetic and transpiration rates, while the hormetic stimulation of growth that was caused by Cd led to a significant decrease of all of the investigated plant gas exchange parameters (Figure 7). Based on these results, we can suggest that stimulating growth by HMs during hormesis does not have to be correlated with an increase in CO_2 assimilation and transpiration.

To summarize, we confirmed that hormesis is a complex process, which is still not fully understood or investigated. In this work, we present the effect of two heavy metals (Pb and Cd), which are characterized by different physicochemical properties, on the hormetic stimulation of corn shoot growth. Based on our results, we conclude that there are common features that seem to be necessary to induce the hormetic stimulation of shoot growth by heavy metals. They are an increase in the auxin and flavonol content and the maintenance of hydrogen peroxide at the level of the control

(Figure 8). We would like to stress that an increase in the IAA and flavonol content have been proposed as the key factors in hormesis here for the first time. In general, the literature on metals and hormesis is dominated by studies on the role of oxidative stress and phytohormones such as abscisic acid, ethylene or jasmonates, with little information on the role of IAA and flavonols [3,4,6,41,54]. The results presented here also suggest that an increase in the photosynthesis and transpiration rates are not necessary for stimulating hormetic growth. To sum up, the presented results shed new light on the mechanism of hormesis and can open new research directions, which should encompass analyses of the gene expression and phytohormone crosstalk.

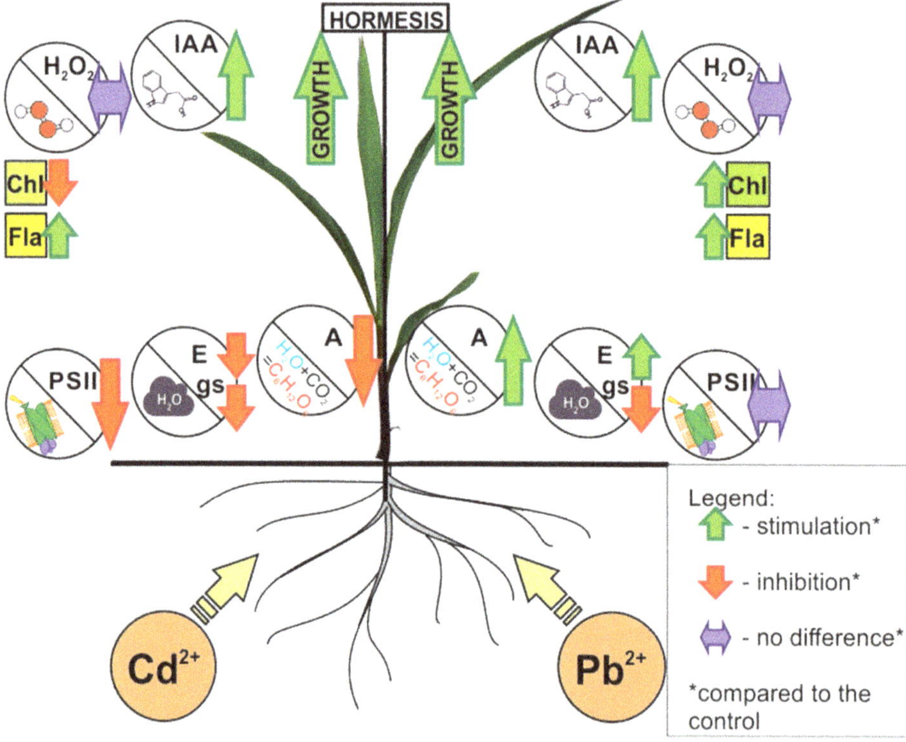

Figure 8. Model summarizing the differences in the corn plant response during hormetic growth stimulation after treatment with low doses of Pb or Cd. It was documented that the shoot growth stimulation for both metals was correlated with an increase in the IAA and flavonol content in the shoots. This increase in the content of both substances was accompanied by the maintenance of the H_2O_2 content at the level of the control. In conclusion, an increase in the IAA content and a lack of oxidative stress seem to play a key role in the hormetic stimulation of shoot growth by Pb and Cd. Abbreviation list: A—photosynthetic rate; Chl—chlorophyll content; E—transpiration rate; Fla—flavonol content; gs—stomatal conductance; H_2O_2—hydrogen peroxide content; IAA—auxin content; PSII—photosystem II performance and quantum yield.

4. Materials and Methods

4.1. Plant Material

In the experiment with coleoptiles sections the caryopsis of *Zea mays* L. cv. 'KOSMO 230' were used. First, they were soaked in tap water for 2 h; then, the seeds were sown in plastic trays that had been lined with moistened cellulose sheets. The trays were placed in an incubator (MIR-533, SANYO, Moriguchi, Japan) and left for three days in the dark for seed germination at a temperature of 27 ± 0.5 °C, ≈100% humidity and were watered as needed.

The caryopsis of cv. 'LOKATA' were used for the experiment with corn seedlings, because the seeds of cv. 'KOSMO 230' were no longer available. A preliminary study showed that the coleoptile sections that had been excised from the seedlings of cv. 'LOKATA' that had been treated with Pb or Cd reacted in a way that was similar to the sections that had been excised from the seedlings of cv. 'KOSMO 230'. As a result, the caryopsis of cv. 'LOKATA' were germinated in the same manner as the cv. 'KOSMO 230' seeds.

4.2. Hydroponic Cultures

In the experiment with coleoptiles, the three-day-old etiolated *Z. mays* L. cv. 'KOSMO 230' seedlings with equal coleoptile lengths (1–1.5 cm) were transferred into hydroponic cultures (100 seedlings/2.5 l) with artificial pond water (APW) composed of the following: 1 mM KCl, 0.1 mM NaCl and 0.1 mM $CaCl_2$ dissolved in deionized water [14]. The initial pH of APW was established at 5.8 ± 0.1. The following treatments were applied APW (control), APW + 10 μM, APW + 100 μM and APW + 1000 μM of $CdCl_2$ or $PbCl_2$. The seedlings were cultivated for the next 24 h in the incubator in conditions similar to those described in 4.1. From four-day-old etiolated corn seedlings, 1 cm long coleoptile sections were excised with blade razor starting 3 mm from the tip; next, the first leaf was removed using a dissecting needle. The concentrations of H_2O_2 and auxin was measured in the prepared sections, or they were incubated for 24 h in the APW for the measurements of elongation growth.

In the experiment with seedlings, the three-day-old etiolated corn seedlings cv. 'Lokata' with only the primary root and coleoptile length of 3.5 ± 0.5 cm were transferred into hydroponic cultures, which were carried out in plastic containers (nine seedlings/container) that were filled with 2850 mL (315 mL/seedling) of a nutrient solution. A Hoagland solution [69] was used in the hydroponic cultures with the initial pH established at 5.9 ± 0.05. The seedlings were grown in a greenhouse under artificial light using sodium lamps (HPS), a photoperiod of 16/8 h day/night and an average energy of light of 250 μmoL E m^{-2} s^{-1}. The temperature in the greenhouse was 20 ± 1 °C and the air humidity was 30 ± 5%. During the first seven days of the experiment, medium was changed twice. After seven days of plant cultivation the medium was changed into a Hoagland solution (control) or a Hoagland solution with 0.25, 1, 2.5, 10, 25 or 100 μM $CdCl_2$ or 0.5, 1, 5, 10, 50 or 100 μM of $PbCl_2$. The concentrations of the metals were selected based on preliminary studies.

4.3. Growth Measurement

In the experiment with coleoptiles, a column of 10 coleoptile sections was placed on a stainless-steel wire to preserve their natural vertical orientation, and the length of the column was measured with an accuracy to 0.1 mm using calipers. Next, the column was introduced into a measuring cylinder that had been filled with 100 mL APW (10 mL/segment; this volume prevents the medium pH to be changed by the coleoptile sections) with an initial pH 5.8–6.0 and incubated in the dark for the next 24 h at room temperature. After 24 h, the length of the column was measured once again. The endogenous growth of the coleoptile sections was assessed as the difference between the length of the sections of a column after 24 h of incubation and the length of the sections of a column at the start of the experiment. The measurements were repeated twice.

In the experiment with corn seedlings, the length of the shoots was measured after seven days of cultivation on control solution by measuring from the first node to the end of the longest leaf. Then, after the four days of growth in the medium with HMs, the length of the shoots was measured again. The growth of the shoots was assessed as the difference between the length of the shoots before and the length of shoots after four days of metal treatment. From each container, the smallest seedling was rejected and the remaining eight seedlings were used for the other measurements. Because Pb easily precipitates in the presence of phosphate ions, 2 mM of $NH_4H_2PO_4$ were not administrated to the medium with $PbCl_2$ [17,18,70,71]. All of the investigated physiological parameters were measured on the fourth day of the HM treatment on second fully developed leaf counting from the bottom of the seedling.

4.4. Measurement of the Hydrogen Peroxide and Indole Compound Content in the Corn Coleoptile Sections and Leaves

The hydrogen peroxide content in the corn coleoptile sections and leaves was measured according to the method of Bouazizi et al. [72]. Fresh leaf tissues (150 mg) were homogenized in 1.5 mL of 0.1% trichloroacetic acid (TCA). The homogenate was centrifuged at 12,000× *g* for 15 min and 0.5 mL of the supernatant was added to a 0.5 mL potassium phosphate buffer (10 mM, pH 7.0) and 1 mL potassium iodide (KI) (1 M). The absorbance was measured at 390 nm and the content of H_2O_2 was determined using a standard curve.

The content of the indole compounds in the corn coleoptile sections and leaves was measured according to the method of Wójcikowska et al. [73]. Fresh leaf tissues (150 mg) were homogenized in 1.5 mL of 10 × PBS (phosphate buffer solution). The homogenate was centrifuged at 15,000× *g* for 25 min and 0.3 mL of the supernatant was added to 0.025 mL of orthophosphate acid (0.01 M) and 1.2 mL of Salkowski's reagent (150 mL H_2SO_4; 250 mL ddH_2O; 7.5 mL 0.5 M $FeCl_3 \times 6H_2O$). The absorbance was measured at 530 nm and the content of indole compound was determined using a standard curve. It should be emphasized that Wójcikowska et al. [73] showed a direct correlation between the content of the indole compounds and indolyl-3-acetic acid (IAA) in plant tissue, and therefore, the term auxin (IAA) content is used in the current study. A spectrophotometer (SPECORD® 250, Analytik Jena, Jena, Germany) was used to determine the hydrogen peroxide and auxin content in the plant samples.

4.5. Measurements of the Photosynthetic Characteristics, Transpiration and Pigment Content

All of the measurements were performed on the second fully developed leaf counting from the bottom of the seedling for eight plants from each treatment after four days of HM treatment. The chlorophyll *a* fluorescence (ChlF) was measured using a Plant Efficiency Analyzer (PocketPEA fluorimeter, Hansatech Instruments Ltd., King's Lynn, England). Before the measurement, each selected leaf was adapted in the dark for 30 min using dedicated leaf clips. After adaptation, a saturating light pulse of 3500 μmol photons m^{-2} s^{-1} was applied for 1 s, which closed all of the reaction centers, and then the fluorescence parameters were measured. The measurements were performed without damaging the plant material.

The plant gas exchange parameters such as the net photosynthetic rate (A), stomatal conductance (gs) and transpiration rate (E) were measured on the same leaves as the ChlF. The measurements were performed at the end of experiment using an infrared gas analyzer with a special narrow chamber (LCpro+, ADC Bioscientific, Hoddesdon, UK) under controlled climate conditions (T = 21 °C, PAR = 1500 μmol m^{-2} s^{-1}). The measurements were performed at noon.

The content of chlorophyll and flavonols was measured using a Dualex sensor (Force-A, Orsay, France). The pigment content was measured on the same leaves as for the ChlF and gas exchange measurements. The measurements were performed without damaging the plant material. The pigment content was calculated on the basis of the fluorescence and light transmission measurement with a 5-mm-diameter probe. More details on pigment content index measurements are presented by Cerovic et al. [74]

4.6. Statistical Analysis

The results are shown as the means ± SE. The statistically significant differences among mean values were determined using a one-way ANOVA and the post hoc Fischer LSD test ($p < 0.05$). The statistical analysis was performed using Statistica v.13.1 (Dell Inc., Round Rock, TX, USA). The principal components analysis (PCA) was used to identify the dominant groups of the factors that characterized physiological status of plants treated with Cd or Pb. The pipeline models of the energy fluxes through a leaf's cross section were done using CorelDRAW X6 (Corel Corp., Ottawa, ON, Canada).

Supplementary Materials: Supplementary materials can be found at http://www.mdpi.com/1422-0067/21/6/2099/s1.

Author Contributions: E.M. and K.S. conceived and designed the research. K.S., M.S., Ż.G. and P.Z.-R. conducted the experiments. E.M., K.S., M.S., Ż.G., M.P., H.M.K. and P.Z.-R. analyzed the data. E.M., K.S. and M.S. wrote the first draft of the manuscript, which was then edited by all of the authors. All authors have read and agreed to the published version of the manuscript.

Funding: This research received no external funding.

Acknowledgments: The authors would like to thank Michele L. Simmons, B.A., the University of Silesia, Katowice, Poland, for improving the English style.

Conflicts of Interest: The authors declare no conflicts of interest.

Abbreviations

A	photosynthetic rate
ABS/CS	absorption flux per CS
APW	artificial pond water
ChlF	chlorophyll a fluorescence
CS	excited cross section of leaf
DI/CS	dissipation energy flux per CS
ΔVt	control; variable fluorescence
E	transpiration rate
ET/CS	electron transport per CS
FNR	Ferredoxin-NADP+ Reductase
F0	minimal fluorescence, when all PS II RCs are open (at t = 0)
Fm	maximal fluorescence, when all PS II RCs are closed
Fv	maximal variable fluorescence
Ft	fluorescence at time t
gs	stomatal conductance
HM	heavy metal
IAA	auxin
OJIP	transient chlorophyll a fluorescence rise induced during a dark-to-strong light transition, where O is equivalent to F0 and P is for peak equivalent to Fm
PAR	photosynthetic active radiation
PSI	photosystem I
PSII	photosystem II
RC	reaction center of PSII
RC/CS	% of active reaction centers per CS compared to the control
ROS	reactive oxygen species
TR/CS	trapped energy flux per CS

References

1. Kendig, E.L.; Le, H.H.; Belcher, S.M. Defining hormesis: Evaluation of a complex concentration response phenomenon. *Int. J. Toxicol.* **2010**, *29*, 235–246. [CrossRef] [PubMed]
2. Agathokleous, E.; Kitao, M.; Calabrese, E.J. Hormesis: A compelling platform for sophisticated plant science. *Trends Plant Sci.* **2019**, *24*, 318–327. [CrossRef] [PubMed]
3. Shahid, M.; Niazi, N.K.; Rinklebe, J.; Bundschuh, J.; Dumat, C.; Pinelli, E. Trace elements-induced phytohormesis: A critical review and mechanistic interpretation. *Crit. Rev. Environ. Sci. Technol.* **2019**, 1–32. [CrossRef]
4. Carvalho, M.E.A.; Castro, P.R.C.; Azevedo, R.A. Hormesis in plants under Cd exposure: From toxic to beneficial element? *J. Hazard. Mater.* **2020**, *384*, 121434. [CrossRef] [PubMed]

5. Calabrese, E.J. Hormesis is central to toxicology, pharmacology and risk assessment. *Hum. Exp. Toxicol.* **2010**, *29*, 249–261. [CrossRef] [PubMed]
6. Poschenrieder, C.; Cabot, C.; Martos, S.; Gallego, B.; Barcelo, J. Do toxic ions induce hormesis in plants? *Plant Sci.* **2013**, *212*, 15–25. [CrossRef]
7. Costantini, D.; Metcalfe, N.B.; Monaghan, P. Ecological processes in a hormetic framework. *Ecol. Lett.* **2010**, *13*, 1435–1447. [CrossRef]
8. Jia, L.; He, X.; Chen, W.; Liu, Z.; Huang, Y.; Yu, S. Hormesis phenomena under Cd stress in a hyperaccumulator—*Lonicera japonica* Thunb. *Ecotoxicology* **2013**, *22*, 476–485. [CrossRef]
9. Shah, K.; Kumar, R.G.; Verma, S.; Dubey, R.S. Effect of cadmium on lipid peroxidation, superoxide anion generation and activities of antioxidant enzymes in growing rice seedlings. *Plant Sci.* **2001**, *161*, 1135–1144. [CrossRef]
10. Patnaik, A.R.; Achary, V.M.M.; Panda, B.B. Chromium (VI)-induced hormesis and genotoxicity are mediated through oxidative stress in root cells of *Allium cepa* L. *Plant Growth Regul.* **2013**, *71*, 157–170. [CrossRef]
11. González, C.I.; Maine, M.A.; Cazenave, J.; Sanchez, G.C.; Benavides, M.P. Physiological and biochemical responses of *Eichhornia crassipes* exposed to Cr (III). *Environ. Sci. Pollut. Res.* **2015**, *22*, 3739–3747. [CrossRef] [PubMed]
12. Luna-López, A.; González-Puertos, V.Y.; López-Diazguerrero, N.E.; Königsberg, M. New considerations on hormetic response against oxidative stress. *J. Cell Commun. Signal.* **2014**, *8*, 323–331. [CrossRef] [PubMed]
13. Mengel, K.; Kirkby, E.A. *Principles of Plant Nutrition*, 5th ed.; Springer: Dordrecht, The Netherlands; Kluwer Academic Publishers: Kluwer, The Netherlands, 2001.
14. Małkowski, E.; Kita, A.; Galas, W.; Karcz, W.; Kuperberg, J.M. Lead distribution in corn seedlings (*Zea mays* L.) and its effect on growth and the concentrations of potassium and calcium. *Plant Growth Regul.* **2002**, *37*, 69–76. [CrossRef]
15. Vassilev, A.; Perez-Sanz, A.; Semane, B.; Carleer, R.; Vangronsveld, J. Cadmium accumulation and tolerance of two *Salix* genotypes hydroponically grown in presence of cadmium. *J. Plant Nutr.* **2005**, *28*, 2159–2177. [CrossRef]
16. Kopittke, P.M.; Asher, C.J.; Kopittke, R.A.; Menzies, N.W. Toxic effects of Pb^{2+} on growth of cowpea (*Vigna unguiculata*). *Environ. Pollut.* **2007**, *150*, 280–287. [CrossRef]
17. Kopittke, P.M.; Blamey, F.P.C.; Asher, C.J.; Menzies, N.W. Trace metal phytotoxicity in solution culture: A review. *J. Exp. Bot.* **2010**, *61*, 945–954. [CrossRef]
18. Małkowski, E.; Sitko, K.; Zieleźnik-Rusinowska, P.; Gieroń, Ż.; Szopiński, M. Heavy Metal Toxicity: Physiological Implications of Metal Toxicity in Plants. In *Plant Metallomics and Functional Omics*; Springer: Berlin/Heidelberg, Germany, 2019; pp. 253–301.
19. Seregin, I.V.; Ivanov, V.B. Physiological aspects of cadmium and lead toxic effects on higher plants. *Russ. J. Plant Physiol.* **2001**, *48*, 523–544. [CrossRef]
20. Małkowski, E.; Burdach, Z.; Karcz, W. Effect of lead ions on the growth of intact seedlings and coleoptile segments of corn (*Zea mays* L.). In Proceedings of the International Scientific Meeting Ecophysiological Aspects of Plant Responses to Stress Factors, Cracow, Poland, 12–14 June 1997; pp. 235–239.
21. Ismail, S.; Khan, F.; Iqbal, M.Z. Phytoremediation: Assessing tolerance of tree species against heavy metal (Pb and Cd) toxicity. *Pak. J. Bot.* **2013**, *45*, 2181–2186.
22. Sharma, P.; Dubey, R.S. Lead toxicity in plants. *Braz. J. Plant Physiol.* **2005**, *17*, 35–52. [CrossRef]
23. Edelmann, H.G. Lateral redistribution of auxin is not the means for gravitropic differential growth of coleoptiles: A new model. *Physiol. Plant.* **2001**, *112*, 119–126. [CrossRef]
24. Carpita, N.C.; Defernez, M.; Findlay, K.; Wells, B.; Shoue, D.A.; Catchpole, G.; Wilson, R.H.; McCann, M.C. Cell wall architecture of the elongating maize coleoptile. *Plant Physiol.* **2001**, *127*, 551–565. [CrossRef] [PubMed]
25. Karcz, W.; Burdach, Z. A comparison of the effects of IAA and 4-Cl-IAA on growth, proton secretion and membrane potential in maize coleoptile segments. *J. Exp. Bot.* **2002**, *53*, 1089–1098. [CrossRef] [PubMed]
26. Lane, S.D.; Martin, E.S.; Garrod, J.F. Lead toxicity effects on indole-3-ylacetic acid-induced cell elongation. *Planta* **1978**, *144*, 79–84. [CrossRef] [PubMed]
27. Małkowski, E.; Stolarek, J.; Karcz, W. Toxic effect of Pb^{2+} ions on extension growth of cereal plants. *Pol. J. Environ. Stud.* **1996**, *5*, 41–45.

28. Munzuroglu, O.; Geckil, H. Effects of metals on seed germination, root elongation, and coleoptile and hypocotyl growth in *Triticum aestivum* and *Cucumis sativus*. *Arch. Environ. Contam. Toxicol.* **2002**, *43*, 203–213. [CrossRef]
29. Kurtyka, R.; Małkowski, E.; Burdach, Z.; Kita, A.; Karcz, W. Interactive effects of temperature and heavy metals (Cd, Pb) on the elongation growth in maize coleoptiles. *Comptes Rendus Biol.* **2012**, *335*, 292–299. [CrossRef]
30. Iino, M. Kinetic modelling of phototropism in maize coleoptiles. *Planta* **1987**, *171*, 110–126. [CrossRef]
31. Kutschera, U.; Bergfeld, R.; Schopfer, P. Cooperation of epidermis and inner tissues in auxin-mediated growth of maize coleoptiles. *Planta* **1987**, *170*, 168–180. [CrossRef]
32. Schopfer, P.; Liszkay, A.; Bechtold, M.; Frahry, G.; Wagner, A. Evidence that hydroxyl radicals mediate auxin-induced extension growth. *Planta* **2002**, *214*, 821–828. [CrossRef]
33. Elobeid, M.; Polle, A. Interference of heavy metal toxicity with auxin physiology. In *Metal Toxicity in Plants: Perception, Signaling and Remediation*; Springer: Berlin/Heidelberg, Germany, 2012; pp. 249–259.
34. Hu, Y.F.; Zhou, G.; Na, X.F.; Yang, L.; Nan, W.B.; Liu, X.; Zhang, Y.Q.; Li, J.L.; Bi, Y.R. Cadmium interferes with maintenance of auxin homeostasis in *Arabidopsis* seedlings. *J. Plant Physiol.* **2013**, *170*, 965–975. [CrossRef]
35. Wang, R.; Wang, J.; Zhao, L.; Yang, S.; Song, Y. Impact of heavy metal stresses on the growth and auxin homeostasis of *Arabidopsis* seedlings. *Biometals* **2015**, *28*, 123–132. [CrossRef] [PubMed]
36. Yu, C.; Sun, C.; Shen, C.; Wang, S.; Liu, F.; Liu, Y.; Chen, Y.; Li, C.; Qian, Q.; Aryal, B. The auxin transporter, Os AUX 1, is involved in primary root and root hair elongation and in Cd stress responses in rice (*Oryza sativa* L.). *Plant J.* **2015**, *83*, 818–830. [CrossRef] [PubMed]
37. Yuan, H.-M.; Huang, X. Inhibition of root meristem growth by cadmium involves nitric oxide-mediated repression of auxin accumulation and signalling in *Arabidopsis*. *Plant Cell Environ.* **2016**, *39*, 120–135. [CrossRef] [PubMed]
38. Zhan, Y.; Zhang, C.; Zheng, Q.; Huang, Z.; Yu, C. Cadmium stress inhibits the growth of primary roots by interfering auxin homeostasis in *Sorghum bicolor* seedlings. *J. Plant Biol.* **2017**, *60*, 593–603. [CrossRef]
39. Luo, Y.; Wei, Y.; Sun, S.; Wang, J.; Wang, W.; Han, D.; Shao, H.; Jia, H.; Fu, Y. Selenium modulates the level of auxin to alleviate the toxicity of cadmium in tobacco. *Int. J. Mol. Sci.* **2019**, *20*, 3772. [CrossRef]
40. Bennetzen, J.L.; Hake, S.C. *Handbook of Corn: Its Biology*; Springer Science + Business Media LLC: New York, NY, USA, 2009.
41. Muszyńska, E.; Labudda, M. Dual role of metallic trace elements in stress biology—From negative to beneficial impact on plants. *Int. J. Mol. Sci.* **2019**, *20*, 3117. [CrossRef]
42. Jia, L.; Liu, Z.; Chen, W.; Ye, Y.; Yu, S.; He, X. Hormesis Effects Induced by Cadmium on Growth and Photosynthetic Performance in a Hyperaccumulator, *Lonicera japonica* Thunb. *J. Plant Growth Regul.* **2015**, *34*, 13–21. [CrossRef]
43. Seth, C.S.; Chaturvedi, P.K.; Misra, V. The role of phytochelatins and antioxidants in tolerance to Cd accumulation in *Brassica juncea* L. *Ecotoxicol. Environ. Saf.* **2008**, *71*, 76–85. [CrossRef]
44. Figlioli, F.; Sorrentino, M.C.; Memoli, V.; Arena, C.; Maisto, G.; Giordano, S.; Capozzi, F.; Spagnuolo, V. Overall plant responses to Cd and Pb metal stress in maize: Growth pattern, ultrastructure, and photosynthetic activity. *Environ. Sci. Pollut. Res.* **2019**, *26*, 1781–1790. [CrossRef]
45. Tang, Y.-T.; Qiu, R.-L.; Zeng, X.-W.; Ying, R.-R.; Yu, F.-M.; Zhou, X.-Y. Lead, zinc, cadmium hyperaccumulation and growth stimulation in *Arabis paniculata* Franch. *Environ. Exp. Bot.* **2009**, *66*, 126–134. [CrossRef]
46. Sidhu, G.P.S.; Singh, H.P.; Batish, D.R.; Kohli, R.K. Effect of lead on oxidative status, antioxidative response and metal accumulation in *Coronopus didymus*. *Plant Physiol. Biochem.* **2016**, *105*, 290–296. [CrossRef] [PubMed]
47. Wu, Z.; Yin, X.; Bañuelos, G.S.; Lin, Z.-Q.; Liu, Y.; Li, M.; Yuan, L. Indications of selenium protection against cadmium and lead toxicity in oilseed rape (*Brassica napus* L.). *Front. Plant Sci.* **2016**, *7*, 1875. [CrossRef] [PubMed]
48. Rui, H.; Chen, C.; Zhang, X.; Shen, Z.; Zhang, F. Cd-induced oxidative stress and lignification in the roots of two *Vicia sativa* L. varieties with different Cd tolerances. *J. Hazard. Mater.* **2016**, *301*, 304–313. [CrossRef] [PubMed]
49. Alyemeni, M.N.; Ahanger, M.A.; Wijaya, L.; Alam, P.; Bhardwaj, R.; Ahmad, P. Selenium mitigates cadmium-induced oxidative stress in tomato (*Solanum lycopersicum* L.) plants by modulating chlorophyll fluorescence, osmolyte accumulation, and antioxidant system. *Protoplasma* **2018**, *255*, 459–469. [CrossRef]

50. Wei, T.; Lv, X.; Jia, H.; Hua, L.; Xu, H.; Zhou, R.; Zhao, J.; Ren, X.; Guo, J. Effects of salicylic acid, Fe (II) and plant growth-promoting bacteria on Cd accumulation and toxicity alleviation of Cd tolerant and sensitive tomato genotypes. *J. Environ. Manag.* **2018**, *214*, 164–171. [CrossRef]
51. El-Banna, M.F.; Mosa, A.; Gao, B.; Yin, X.; Ahmad, Z.; Wang, H. Sorption of lead ions onto oxidized bagasse-biochar mitigates Pb-induced oxidative stress on hydroponically grown chicory: Experimental observations and mechanisms. *Chemosphere* **2018**, *208*, 887–898. [CrossRef]
52. Khan, M.M.; Islam, E.; Irem, S.; Akhtar, K.; Ashraf, M.Y.; Iqbal, J.; Liu, D. Pb-induced phytotoxicity in para grass (*Brachiaria mutica*) and Castorbean (*Ricinus communis* L.): Antioxidant and ultrastructural studies. *Chemosphere* **2018**, *200*, 257–265. [CrossRef]
53. Lin, R.; Wang, X.; Luo, Y.; Du, W.; Guo, H.; Yin, D. Effects of soil cadmium on growth, oxidative stress and antioxidant system in wheat seedlings (*Triticum aestivum* L.). *Chemosphere* **2007**, *69*, 89–98. [CrossRef]
54. Bücker-Neto, L.; Paiva, A.L.S.; Machado, R.D.; Arenhart, R.A.; Margis-Pinheiro, M. Interactions between plant hormones and heavy metals responses. *Genet. Mol. Biol.* **2017**, *40*, 373–386. [CrossRef]
55. Wani, S.H.; Kumar, V.; Shriram, V.; Sah, S.K. Phytohormones and their metabolic engineering for abiotic stress tolerance in crop plants. *Crop J.* **2016**, *4*, 162–176. [CrossRef]
56. Liphadzi, M.S.; Kirkham, M.B.; Paulsen, G.M. Auxin-enhanced root growth for phytoremediation of sewage-sludge amended soil. *Environ. Technol.* **2006**, *27*, 695–704. [CrossRef] [PubMed]
57. Rodriguez, E.; da Conceição Santos, M.; Azevedo, R.; Correia, C.; Moutinho-Pereira, J.; de Oliveira, J.M.P.F.; Dias, M.C. Photosynthesis light-independent reactions are sensitive biomarkers to monitor lead phytotoxicity in a Pb-tolerant *Pisum sativum* cultivar. *Environ. Sci. Pollut. Res.* **2015**, *22*, 574–585. [CrossRef]
58. Nyitrai, P.; Bóka, K.; Gáspár, L.; Sárvári, É.; Lenti, K.; Keresztes, Á. Characterization of the stimulating effect of low-dose stressors in maize and bean seedlings. *J. Plant Physiol.* **2003**, *160*, 1175–1183. [CrossRef] [PubMed]
59. Szuba, A.; Karliński, L.; Krzesłowska, M.; Hazubska-Przybył, T. Inoculation with a Pb-tolerant strain of *Paxillus involutus* improves growth and Pb tolerance of *Populus×canescens* under in vitro conditions. *Plant Soil* **2017**, *412*, 253–266. [CrossRef]
60. Zhang, X.; Yang, H.; Cui, Z. Alleviating effect and mechanism of flavonols in *Arabidopsis* resistance under Pb–HBCD stress. *ACS Sustain. Chem. Eng.* **2017**, *5*, 11034–11041. [CrossRef]
61. Cetin, E.S.; Babalik, Z.; Hallac-Turk, F.; Gokturk-Baydar, N. The effects of cadmium chloride on secondary metabolite production in *Vitis vinifera* cv. cell suspension cultures. *Biol. Res.* **2014**, *47*, 47. [CrossRef] [PubMed]
62. Romanowska, E.; Igamberdiev, A.U.; Parys, E.; Gardeström, P. Stimulation of respiration by Pb^{2+} in detached leaves and mitochondria of C3 and C4 plants. *Physiol. Plant.* **2002**, *116*, 148–154. [CrossRef]
63. He, J.-Y.; Ren, Y.-F.; Zhu, C.; Yan, Y.-P.; Jiang, D.-A. Effect of Cd on growth, photosynthetic gas exchange, and chlorophyll fluorescence of wild and Cd-sensitive mutant rice. *Photosynthetica* **2008**, *46*, 466. [CrossRef]
64. Dias, M.C.; Monteiro, C.; Moutinho-Pereira, J.; Correia, C.; Gonçalves, B.; Santos, C. Cadmium toxicity affects photosynthesis and plant growth at different levels. *Acta Physiol. Plant.* **2013**, *35*, 1281–1289. [CrossRef]
65. Per, T.S.; Khan, S.; Asgher, M.; Bano, B.; Khan, N.A. Photosynthetic and growth responses of two mustard cultivars differing in phytocystatin activity under cadmium stress. *Photosynthetica* **2016**, *54*, 491–501. [CrossRef]
66. Silva, S.; Pinto, G.; Santos, C. Low doses of Pb affected *Lactuca sativa* photosynthetic performance. *Photosynthetica* **2017**, *55*, 50–57. [CrossRef]
67. Szopiński, M.; Sitko, K.; Rusinowski, S.; Corso, M.; Hermans, C.R.M.; Verbruggen, N.; Małkowski, E. Toxic effects of Cd and Zn on the photosynthetic apparatus of the *Arabidopsis halleri* and *Arabidopsis arenosa* pseudo-metallophytes. *Front. Plant Sci.* **2019**, *10*, 748. [CrossRef] [PubMed]
68. Ban, Y.; Xu, Z.; Yang, Y.; Zhamg, H.; Chen, H.; Tang, M. Effect of dark septate endophytic fungus *Gaeumannomyces cylindrosporus* on plant growth, photosynthesis and Pb tolerance of maize (*Zea mays* L.). *Pedosphere* **2017**, *27*, 283–292. [CrossRef]
69. Bloom, A.J. Mineral nutrition. In *Plant Physiology and Development*, 6th ed.; Taiz, L., Zeiger, E., Møller, I.M., Murphy, A., Eds.; Sinauer Associates, Inc.: Sunderalnd, MA, USA, 2015; pp. 119–142.
70. Antosiewicz, D.M. Study of calcium-dependent lead-tolerance on plants differing in their level of Ca-deficiency tolerance. *Environ. Pollut.* **2005**, *134*, 23–34. [CrossRef]

71. Brunet, J.; Varrault, G.; Zuily-Fodil, Y.; Repellin, A. Accumulation of lead in the roots of grass pea (*Lathyrus sativus* L.) plants triggers systemic variation in gene expression in the shoots. *Chemosphere* **2009**, *77*, 1113–1120. [CrossRef]
72. Bouazizi, H.; Jouili, H.; Geitmann, A.; El Ferjani, E. Copper toxicity in expanding leaves of *Phaseolus vulgaris* L.: Antioxidant enzyme response and nutrient element uptake. *Ecotoxicol. Environ. Saf.* **2010**, *73*, 1304–1308. [CrossRef]
73. Wójcikowska, B.; Jaskóła, K.; Gąsiorek, P.; Meus, M.; Nowak, K.; Gaj, M.D. LEAFY COTYLEDON2 (LEC2) promotes embryogenic induction in somatic tissues of *Arabidopsis*, via YUCCA-mediated auxin biosynthesis. *Planta* **2013**, *238*, 425–440. [CrossRef]
74. Cerovic, Z.G.; Masdoumier, G.; Ben Ghozlen, N.; Latouche, G. A new optical leaf-clip meter for simultaneous non-destructive assessment of leaf chlorophyll and epidermal flavonoids. *Physiol. Plant.* **2012**, *146*, 251–260. [CrossRef]

© 2020 by the authors. Licensee MDPI, Basel, Switzerland. This article is an open access article distributed under the terms and conditions of the Creative Commons Attribution (CC BY) license (http://creativecommons.org/licenses/by/4.0/).

Article

Full-Length Transcriptome Assembly of Italian Ryegrass Root Integrated with RNA-Seq to Identify Genes in Response to Plant Cadmium Stress

Zhaoyang Hu, Yufei Zhang, Yue He, Qingqing Cao, Ting Zhang, Laiqing Lou * and Qingsheng Cai

College of Life Sciences, Nanjing Agricultural University, Nanjing 210095, China; 2016216002@njau.edu.cn (Z.H.); 2018116014@njau.edu.cn (Y.Z.); 2018116015@njau.edu.cn (Y.H.); 2016116018@njau.edu.cn (Q.C.); 2016116017@njau.edu.cn (T.Z.); qscai@njau.edu.cn (Q.C.)

* Correspondence: loulq@njau.edu.cn; Tel.: +86-25-84396408

Received: 22 December 2019; Accepted: 4 February 2020; Published: 6 February 2020

Abstract: Cadmium (Cd) is a toxic heavy metal element. It is relatively easily absorbed by plants and enters the food chain, resulting in human exposure to Cd. Italian ryegrass (*Lolium multiflorum* Lam.), an important forage cultivated widely in temperate regions worldwide, has the potential to be used in phytoremediation. However, genes regulating Cd translocation and accumulation in this species are not fully understood. Here, we optimized PacBio ISO-seq and integrated it with RNA-seq to construct a de novo full-length transcriptomic database for an un-sequenced autotetraploid species. With the database, we identified 2367 differentially expressed genes (DEGs) and profiled the molecular regulatory pathways of Italian ryegrass with Gene Ontology (GO) and Kyoto Encyclopedia of Genes and Genomes (KEGG) analysis in response to Cd stress. Overexpression of a DEG *LmAUX1* in *Arabidopsis thaliana* significantly enhanced plant Cd concentration. We also unveiled the complexity of alternative splicing (AS) with a genome-free strategy. We reconstructed full-length UniTransModels using the reference transcriptome, and 29.76% of full-length models had more than one isoform. Taken together, the results enhanced our understanding of the genetic diversity and complexity of Italian ryegrass under Cd stress and provided valuable genetic resources for its gene identification and molecular breeding.

Keywords: alternative splicing; cadmium; Italian ryegrass root; transcriptome; cadmium; *LmAUX1*

1. Introduction

The genus *Lolium* L., belonging to the *Poaceae* family, is native to Europe, North Africa, and temperate Asia and has been introduced to almost all temperate regions in the world [1]. Italian ryegrass (*L. multiflorum* Lam.) and perennial ryegrass (*L. perenne* L.) are two important species, and both of them are valuable forage grasses. Because of their desirable characteristics, such as high yields, tolerance to grazing, rapid establishment, and high palatability and digestibility for ruminant animals, they have been cultivated through hybrid or molecular breeding in the last hundred years [2]. Moreover, molecular breeding has played an increasingly important role in breeding programs of late [3]. In many plants, it has shown promise in increasing yield as well as resistance to multiple biotic and abiotic stresses. However, molecular breeding depends on the availability of a reference genome. Until now, in species of the *Lolium* genus, only a draft genome of perennial ryegrass (*L. perenne* L.) was reported [4]. Additionally, there are still gaps in the genome, which limits the understanding of the molecular regulation mechanism, especially for closely related species, such as Italian ryegrass, so it provides little reference value. Italian ryegrass, also known as annual ryegrass, is used for turf, forage, and quick cover in the event of erosion [5], and it also has some new uses, such as lignocellulosic ethanol conversion programs [6] and phytoremediation [7].

Cadmium (Cd) is a widespread heavy metal element in nature, is highly toxic to almost all plants and animals, and is without any known biological functions [8]. Cd uptake by plants from Cd-contaminated soils happens relatively easily, resulting in human Cd exposure through the food chain [9]. Soil Cd pollution has drawn public attention due to its direct adverse impact on plant growth and the uncertainties regarding the safety of produce [10]. Many previous studies have shown that Cd exposure may cause a series of damages to plants [11], such as lipid peroxidation, enzyme inactivation, an overproduction of reactive oxygen species (ROSs), a disturbed assimilation of essential elements (for example, copper, iron, and zinc), activated programmed cell death (PCD), and membrane and DNA damage [12,13]. Additionally, these damages eventually lead to various toxic phenotypes, such as leaf yellowing, growth reduction, plant withering, or even death [14]. In recent years, common plant mechanisms, in response to Cd stress, have been revealed, such as Cd transporters, Cd cellular compartmentation, and chelation [15]. In *Sedum alfredii*, a hyperaccumulator, an apoplastic pathway mediated by abscisic acid (ABA), regulated the uptake of Cd [16]; in rice, several transporters participating in the absorption, transport, and distribution of Cd have been characterized, including OsHMA3 [8], OsNramp5 [17], and OsCAL1 [18]; in *Arabidopsis thaliana*, the ABC-type transporter AtABCC3 induced by Cd increased plant Cd tolerance mediated by phytochelatin [19].

RNA-seq, as a high-throughput next-generation sequencing (NGS) technique, has become an indispensable tool for transcriptome-wide analysis of differential gene expression and gene regulatory networks. It has recently been widely used for researching plants in response to various stresses, including Cd stress, in such plants as switchgrass [20], rice [21], and *S. alfredii* Hance [22]. However, the limitations of RNA-seq, i.e., short read lengths and amplification biases, restrict researchers to accurately obtain full-length transcripts and differential splicing of mRNA [23], particularly for polyploid species or species lacking a high-quality reference transcriptome [24]. As a third-generation sequencing technique, single-molecule real-time (SMRT) sequencing can overcome these limitations by generating long reads without further assembly. It is highly suitable for isoform discovery, in de novo transcriptome analysis [25]. Isoform sequencing (ISO-seq) from PacBio dependent on the SMRT sequencing platform has been used to analyze full-length transcriptomes in sorghum [23], maize [25], rice [26], *Pteris vittata* [27], and pepper [28].

Roots act as the first barrier to Cd exposure for plants grown in Cd-contaminated soils. Identifying Cd-inducible genes in Italian ryegrass roots can deepen our understanding of molecular regulatory pathways in response to Cd stress. In this study, we constructed a de novo full-length transcriptomic database for Italian ryegrass root with the ISO-seq technique. Integrated with RNA-seq, we identified the differentially expressed genes (DEGs) and analyzed the regulated networks related to Cd tolerance and translation. The function of a DEG *LmAUX1* was validated in *A. thaliana*. The overexpression of this gene significantly decreases plant Cd tolerance and enhances plant Cd concentration. Thus, it is possible that this gene could be used for regulating Cd accumulation in Italian ryegrass.

2. Results

2.1. Assembly of the Full-Length Reference Transcriptomic Database of Italian Ryegrass

Roots are the first organ to contact Cd stress. To identify gene comportments that foremost regulated plant Cd uptake and tolerance, we treated Italian ryegrass with 50 μmol L^{-1} Cd for six hours. Metal concentrations of Italian ryegrass roots were significant affected by Cd (Figure 1). The concentrations of Zn, Fe, Mn, and Cu were significantly increased, and the Cd concentration reached 291.60 mg k^{-1}. To profile the transcriptomic information, total RNA of Italian ryegrass roots was extracted after a six-hour treatment with Cd, and three control and three treated samples were then combined into one sample in equal amounts. The poly(A) RNA was enriched from the mixed RNA sample. Subsequently, according to the results of the pre-experiment, the length of Italian ryegrass mRNA was mostly enriched in a range less than 4 kb (Figure S1). We prepared two overlapping cDNA libraries with insert fragments of 1–4 kb and 3–10 kb, and 20 cycles LD-PCR cycles for a 3–10 kb library were amplified (1–4 kb library, 16 cycles). A library was sequenced by a SMRT cell with the PacBio Sequel platform (Pacific Biosciences, Menlo Park, CA, USA). In summary, we obtained 415,985 and 386,341 reads of inserts for the 1–4 kb library and 3–10 kb library, respectively. Furthermore, 69.06% and 40.47% of them were full-length reads, which had poly(A) tail signals, 5′ adaptor sequences, and 3′ adapter sequences. The average full-length non-chimeric (FLNC) reads were 1711 and 3540 base pairs (bps) for the 1–4 kb library and 3–10 kb library, respectively (Table S1). Usually, the FLNC reads of each cDNA library contain repetitive isoforms. According to the isoform-level clustering algorithm (ICE) analysis, the FLNC reads were aligned, and similarity sequences were assigned to a cluster. Each cluster was identified as a uniform isoform. The isoform sequences were then polished and integrated with non-full-length non-chimeric reads. Isoforms with a predicted accuracy of >99% were considered high-quality isoforms, and others were regarded as low-quality. The average lengths of consensus isoform reads were modified to 1756 and 3653 bps in the 1–4 kb and 3–10 kb libraries, respectively (Table S2).

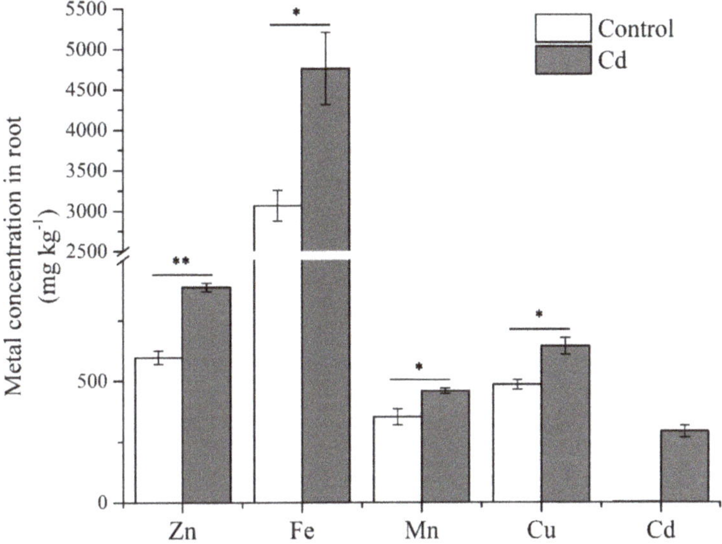

Figure 1. Metal concentrations of Italian ryegrass roots. Data are means ±SE (n = 3). Asterisks indicate significant differences (* $p < 0.05$ and ** $p < 0.01$ by Duncan's test).

To further improve the accuracy of the PacBio ISO-seq data, we sequenced six samples of the Italian ryegrass root with the same treatment as PacBio ISO-seq on the Illumina Hiseq X Ten platform (Illumina, San Diego, CA, USA) with 150 bp pair-ends. In total, each of RNA-seq samples generated more than 39,340,473 clean reads (Table S3). The low-quality isoforms generated by ICE proofread were corrected with RNA-seq short reads using LSC software (http://augroup.org/LSC/LSC/), and the high-quality isoforms and corrective low-quality isoforms were then combined as whole full-length isoforms. In standard ISO-seq analysis, a transcript may generate different isoforms, and these isoforms may be assigned to different libraries, so we removed redundant isoforms using CD-HIT-EST software (http://www.bioinformatics.org/cd-hit/). Finally, the non-redundant high-quality transcripts were obtained and considered the reference transcriptome with a size of 340 Mb.

2.2. Annotation of the Full Length Reference Transcriptome

All isoforms of the reference transcriptome were aligned to the protein and nucleotide databases including NCBI non-redundant proteins (NR), NCBI non-redundant nucleotides (NT), Swiss-Prot, Gene Ontology (GO), Kyoto Encyclopedia of Genes and Genomes (KEGG), and KOG. As shown in Table 1, in total, 146,545 isoforms were annotated; there were 123,344 isoforms annotated in the GO database and 72,725 isoforms annotated in the KEGG database. The best harbored isoforms in the reference transcriptome were in the NR database, and 145,825 isoforms hit in the NR database.

Table 1. Information of function annotation.

Database	Annotated Number	Length 0–<1000	Length 1000–<2000	Length 2000–<3000	Length 3000–<6000	Length ≥6000
GO	123,344	5759	59,645	21,204	33,834	2902
KEGG	72,725	3579	38,770	11,890	16,983	1503
KOG	81,350	3319	41,539	13,234	21,480	1778
NR	145,825	7126	72,604	24,372	38,526	3197
NT	139,068	6697	68,837	23,560	36,922	3052
Swiss-Prot	130,374	5837	64,719	22,006	34,869	2943
All	146,545	7269	72,943	24,448	38,678	3207

Overall, there were 30,797 isoforms mapped in at least three databases, and 46,902 isoforms hit four databases (KOG, NR, KEGG, and Swiss-Prot) (Figure 2A). For GO analysis, the isoforms were annotated in three categories—biological process (BP, 352,144 isoforms), cellular component (CC, 353,049 isoforms), and molecular function (MF, 156,073 isoforms). Within the functional classifications, cellular process (GO:0009987, 81,045 isoforms) and metabolic process (GO:0008152, 74,167 isoforms) were the two most functional terms in BP; cell (GO:0005623, 79,012 isoforms) and cell part (GO:0044464, 78,846 isoforms) were the two most functional terms in CC. Catalytic activity (GO:0003824, 63,137 isoforms) and binding (GO:0005488, 70,972 isoforms) were the most abundant functional terms in MF (Figure 2B). To further profile the pathways that all isoforms take part in, the KEGG Orthology identifiers (KOs) that annotated isoforms enriched in the KEGG database were classified in five metabolic pathways (Hierarchy 1). In cellular process KOs, the isoforms were mainly associated with transport and catabolism (6254 isoforms, 8.6%) pathways. There were 2269 isoforms (3.1%) that took part in environmental information processing KOs. In genetic information processing KOs, there were 20,447 isoforms (28.1%) that participated, and they were mainly focused on folding, sorting, and degradation pathways. More than 43.0% (31,292) isoforms hit for metabolism KOs, which were involved in many pathways, such as carbohydrate metabolism, amino acid metabolism, and global and overview maps. Further, less than 5.4% (3891) of isoforms were categorized into organismal system KOs and were involved in environmental adaptation pathways (Figure 2C).

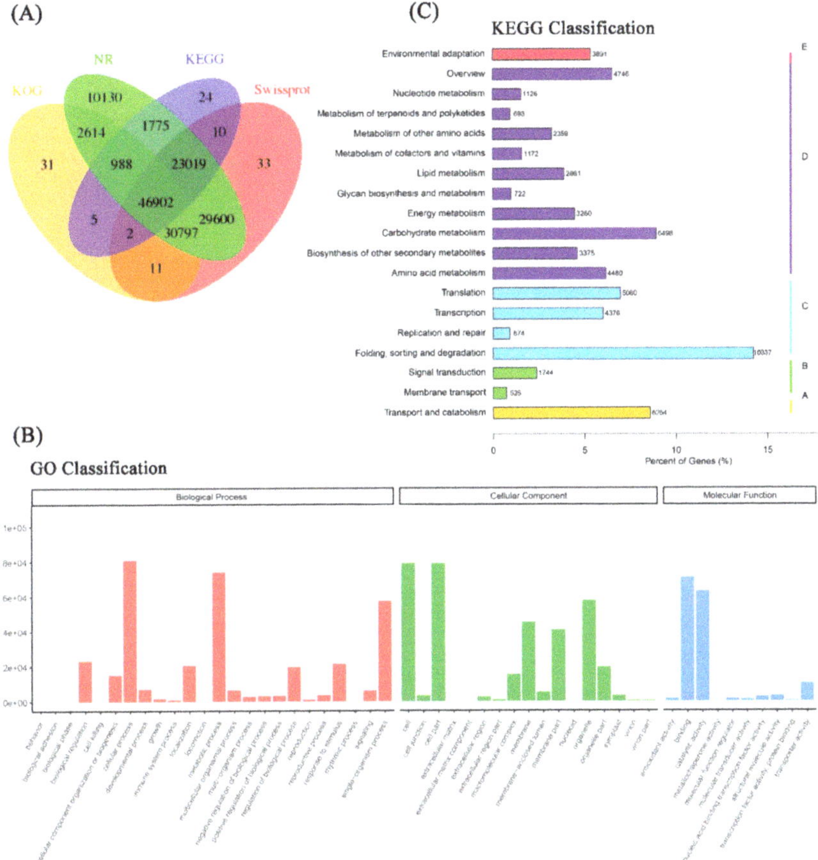

Figure 2. Functional annotation of the full-length reference transcriptome. (**A**) Venn diagram of all isoforms from the reference transcriptome hits of KOG, non-redundant proteins (NR), Kyoto Encyclopedia of Genes and Genomes (KEGG), and SwissProt databases. (**B**) Gene ontology classification of all identified isoforms from the reference transcriptome. (**C**) KEGG classification of all identified isoforms from the reference transcriptome.

2.3. Identification and Functional Profiles of Differentially Expressed Genes

The filtered clean reads from Illumina RNA-seq were mapped to the reference transcriptome generated by PacBio ISO-seq. Of all the reads, 68.73–71.43% were mapped to the reference transcriptome, 9.22–10.29% were uniquely mapped reads, and 98.71–90.78% were multiple aligned reads (Table 2). The expression levels of all isoforms were calculated by RESM software and shown as FPKM values. With the edgeR package, the different expression levels of the isoforms were assessed according to the threshold (FDR values < 0.001). A total of 2367 differentially expressed genes (DEGs) were obtained in response to Cd stress, 1944 DEGs were significantly upregulated, 423 DEGs were significantly downregulated (Figure 3), and relative expression levels of 20 random selected DEGs were further verified using qRT-PCR (Figure S2). Specific primers were designed with Primer Premier 5.0 (PREMIER Biosoft International, Palo Alto, CA, USA) (Table S6).

Table 2. Information of RNA-seq clean reads mapped with the reference transcriptome.

Sample Name	Total Reads	Mapped Reads (%)	Unique Mapped Reads (%)	Multi Mapped Reads (%)
Control-1	96,605,492	66,392,568 (68.73%)	6,831,848 (10.29%)	59,560,720 (89.71%)
Control-2	97,407,032	67,836,176 (69.64%)	6,790,096 (10.01%)	61,046,080 (89.99%)
Control-3	78,680,946	56,203,632 (71.43%)	5,183,088 (9.22%)	51,020,544 (90.78%)
Cd-1	102,165,716	71,120,258 (69.61%)	7,203,654 (10.13%)	63,916,604 (89.87%)
Cd-2	88,241,902	61,102,730 (69.24%)	6,008,134 (9.83%)	55,094,596 (90.17%)
Cd-3	89,834,568	63,034,550 (70.17%)	6,166,310 (9.78%)	56,868,240 (90.22%)

Total Reads: number of clean reads; Unique Mapped Reads (ratio): number of one read mapped to unique site (%); Multi Mapped Reads (ratio): number of one read mapped to multiple sites (%).

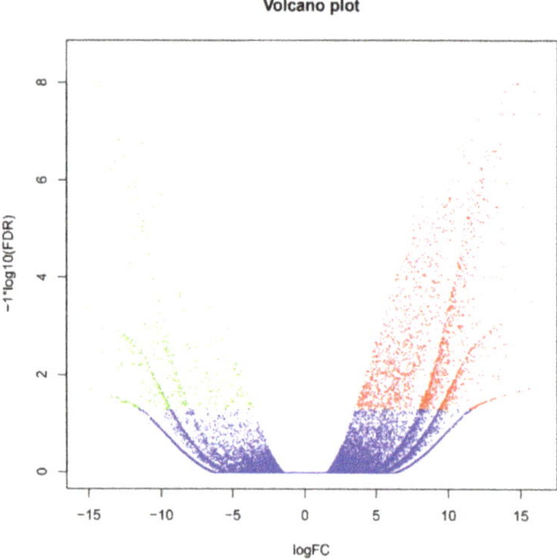

Figure 3. Distribution and expression levels of differentially expressed genes (DEGs) in Italian ryegrass roots exposed or not to cadmium (Cd). The x-axis represents the $\log_2$ [Fold Change] values under the mean normalized expression of all isoforms (y-axis). Red dots indicate the upregulated DEGs, and green dots represent the down regulated DEGs.

To characterize the functions of the DEGs, GO enrichment analysis was performed to categorize potential functions (Supplementary Materials Table S1). The DEGs significantly annotated in the top 20 terms in the BP, MF, and CC categories are shown in Tables 3–5. For the BP category, the four most enriched GO terms were associated with oxidation reduction processes (GO:0055114), response to heat (GO:0009408), protein folding (GO:0006457), and response to oxidative stress (GO:0006979). The two terms involved in unfolded protein binding (GO:0051082) and heme binding (GO:0020037) were assigned to the MF category. The extracellular regions (GO:0005576) and vacuoles (GO:0005773) were the two most important terms in the CC category. The DEGs were also subject to KEGG analysis, and 1094 DEGs were assigned to 205 pathways (Supplementary Materials Table S2). The DEGs significantly enriched in the top 20 significant KOs are shown in Figure 4. The highly enriched pathways of DEGs were protein processing in the endoplasmic reticulum (ko04141) and antigen processing and presentation (ko04612). The endocytosis (ko04144) and glutathione metabolism (ko00480) also appeared in the main KEGG pathways of the DEGs.

Table 3. Gene Ontology (GO) analysis of DEGs significantly annotated in the biological process (BP).

NO.	GO.ID	Term	Significant
1	GO:0055114	oxidation-reduction process	221
2	GO:0006457	Protein folding	103
3	GO:0009813	Flavonoid biosynthetic process	38
4	GO:0042744	Hydrogen peroxide catabolic process	19
5	GO:0052696	Flavonoid glucuronidation	32
6	GO:0009992	Cellular water homeostasis	17
7	GO:0015793	Glycerol transport	17
8	GO:0006833	Water transport	18
9	GO:0006979	Response to oxidative stress	82
10	GO:0010041	Response to iron(III) ion	14
11	GO:0006098	pentose-phosphate shunt	8
12	GO:0009414	Response to water deprivation	14
13	GO:0009651	Response to salt stress	34
14	GO:0006024	Glycosaminoglycan biosynthetic process	5
15	GO:0006065	UDP-glucuronate biosynthetic process	5
16	GO:0009408	Response to heat	154
17	GO:0044550	Secondary metabolite biosynthetic process	41
18	GO:0019521	D-gluconate metabolic process	7
19	GO:0006559	L-phenylalanine catabolic process	17
20	GO:0006857	Oligopeptide transport	1

Table 4. GO analysis of DEGs significantly annotated in the molecular function (MF).

NO.	GO.ID	Term	Significant
1	GO:0051082	Unfolded protein binding	65
2	GO:0020037	Heme binding	52
3	GO:0003924	GTPase activity	36
4	GO:0015250	Water channel activity	17
5	GO:0015254	Glycerol channel activity	17
6	GO:0004601	Peroxidase activity	23
7	GO:0005525	GTP binding	39
8	GO:0005200	Structural constituent of cytoskeleton	20
9	GO:0004197	cysteine-type endopeptidase activity	15
10	GO:0080043	quercetin_3-O-glucosyltransferase activity	27
11	GO:0080044	quercetin_7-O-glucosyltransferase activity	27
12	GO:0033760	2′-deoxymugineic-acid_2′-dioxygenase activity	14
13	GO:0003979	UDP-glucose_6-dehydrogenase activity	5
14	GO:0004623	phospholipase_A2 activity	7
15	GO:0019904	protein domain specific binding	6
16	GO:0004497	monooxygenase activity	36
17	GO:0016709	oxidoreductase activity, acting on paired donors, with incorporation or reduction	17
18	GO:0045548	Phenylalanine ammonia-lyase activity	17
19	GO:0004616	phosphogluconate dehydrogenase (decarboxylating) activity	8
20	GO:0003700	Transcription factor activity, sequence-specific DNA binding	19

Table 5. GO analysis of DEGs significantly annotated in the cellular component (CC).

NO.	GO.ID	Term	Significant
1	GO:0005576	Extracellular region	79
2	GO:0005615	Extracellular space	15
3	GO:0005764	lysosome	15
4	GO:0005773	vacuole	67
5	GO:0005788	Endoplasmic reticulum lumen	21
6	GO:0009505	plant-type cell wall	24
7	GO:0009941	Chloroplast envelope	12
8	GO:0005874	microtubule	17
9	GO:0048046	apoplast	24
10	GO:0009570	Chloroplast stroma	1
11	GO:0046658	Anchored component of plasma membrane	4
12	GO:0005774	Vacuolar membrane	25
13	GO:0009535	Chloroplast thylakoid membrane	5
14	GO:0005777	peroxisome	3
15	GO:0009506	plasmodesma	37
16	GO:0005730	nucleolus	8
17	GO:0048226	Casparian strip	5
18	GO:0008287	Protein serine/threonine phosphatase complex	1
19	GO:0005618	Cell wall	41
20	GO:0031201	SNARE complex	2

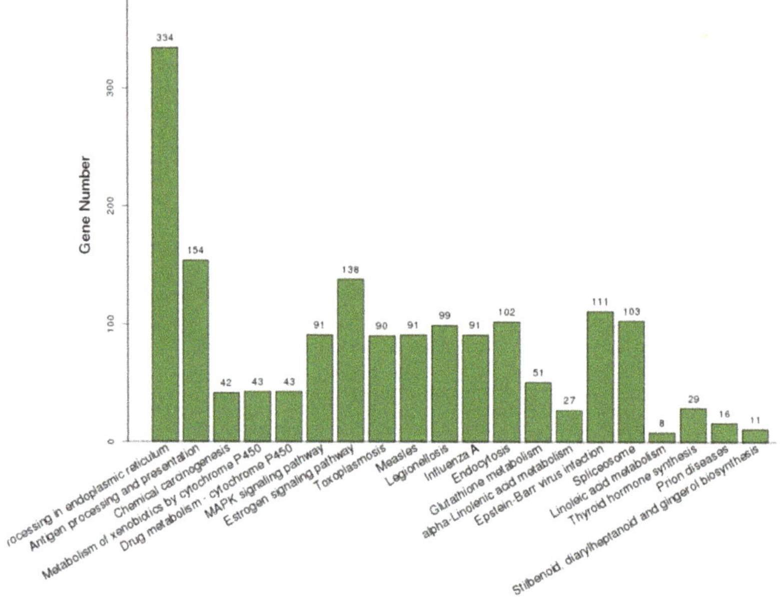

Figure 4. KEGG pathway enrichment analysis based on the DEGs significantly enriched in KOs. The number of above each bar was the number of DEGs enriched in each KO.

2.4. Functional Validation of LmAUX1 in Response to Cd

To confirm whether these putative genes were associated in plant Cd tolerance, we cloned a significantly downregulated gene, according to the blastp results, and named it *LmAUX1*. We hypothesized that it may affect plant Cd uptake and distribution. The predicted CDS of this gene (Multiflorum_1-4k_c11285_f6p6_2149) was isolated and over-expressed in *A. thaliana* (Col-0 and *aux1-7* mutant).

The FPKM values obtained from RNA-seq and relative expression levels verified by qRT-PCR of the DEG (Multiflorum_1-4k_c11285_f6p6_2149) are shown in Figure 5A,B, respectively. According to the analysis of the phylogenetic tree, it was a homologue gene of *AUX1* in *A. thaliana*, *Brachypodium distachyon*, *Zea mays*, and rice (Figure 5C). Multiple transgenic lines of *A. thaliana* (Col-0 ecotype and *aux1-7* mutant) were generated by the T-DNA-inserted fragment harboring *35S:lmAUX1*. Two independent transgenic lines appeared as Cd-tolerance trails, respectively. The T3 generations of transgenic lines, the *aux1-7* mutant, and Col-0 were treated with and without 50 μM $CdCl_2$ in 1/2 MS medium (Figure 6A and Figure S3A), and the results clearly showed that the relative length of root elongation in the transgenic lines was significantly lower than those of the *aux1-7* mutant and Col-0 (Figure 6B and Figure S3B). Col-0, the *aux1-7* mutant, and the overexpression lines were grown in soils irrigated with water or with a 100 or 500 μM $CdCl_2$ solution, respectively (Figure 7A), and the shoot Cd concentration of transgenic lines was significant increased compared to the *aux1-7* mutant (Figure 7B) and Col-0 (Figure S4). The shoot Zn, Mn, and Cu concentrations of transgenic lines was not significant affected compared to the *aux1-7* mutant and Col-0, but the Fe concentration of the transgenic lines was significant changed compared to the *aux1-7* mutant (Figure S5) and Col-0 (Figure S6), except *aux1-7* and its overexpression lines (L1 and L2) treated with 500 μM Cd.

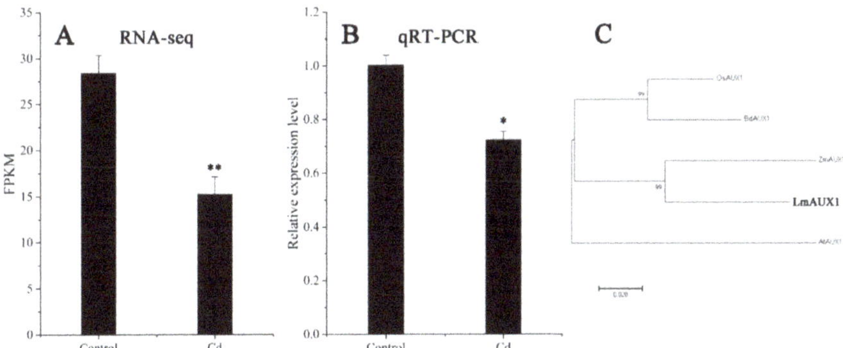

Figure 5. Expression level and phylogenetic tree analysis of *AUX1*. (**A**) Expression level of *LmAUX1* calculated by RNA-seq. Data are means ± SE (n = 3). Asterisks indicate significant differences (** $p < 0.01$ by Duncan's test). (**B**) Relative expression level of *LmAUX1* verified by qRT-PCR. Data are means ± SE (n = 4). Asterisks indicate significant differences (* $p < 0.05$ by Duncan's test). (**C**) Phylogenetic tree of AUX1 proteins: the phylogenetic tree was constructed from an alignment of AUX1 proteins from five different plant species by neighbor-joining methods with bootstrapping analysis (1000 replicates), AtAUX1 (*Arabidopsis thaliana*, AT2G38120), BdAXU1 (*Brachypodium distachyon*, Bradi2g55340), LmAUX1 (*Lolium multiflorum*, Multiflorum_1-4k_c11285_f6p6_2149), OsAUX1 (*Oryza sativa*, Os01g63770), and ZmAXU1 (*Zea mays*, PWZ17958).

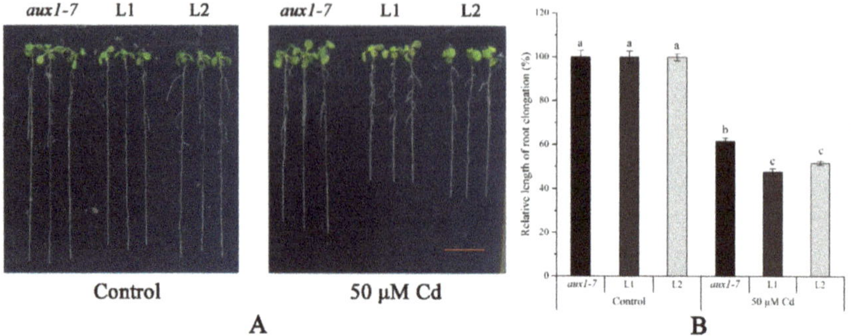

Figure 6. The Cd tolerance experiment on plates of *LmAUX1*. *LmAUX1* was transformed into the *Arabidopsis thaliana aux1-7* mutant with the *CaMV35S* promoter, and two lines (L1 and L2) were used for the experiment. (**A**) Phenotype of plants grown in 1/2 MS medium with and without 50 μM Cd for 4–5 d. Scale bar = 1 cm. (**B**) Relative length of root elongation. Data are means ± SE (n ≥ 15); different letters above the bars were significantly different at $p < 0.05$ (Duncan's test).

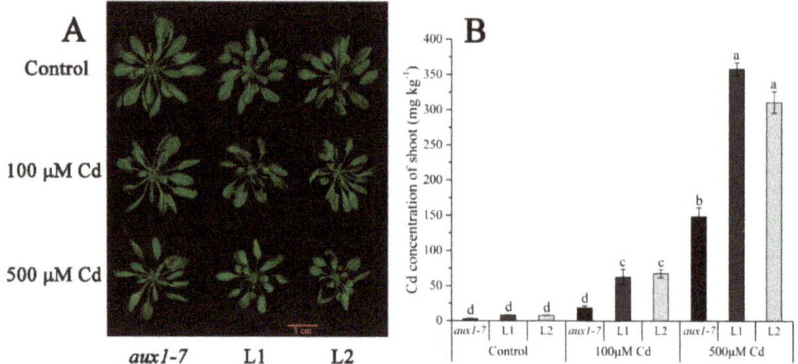

Figure 7. Cd translocated experiment in soil of *LmAUX1*. *LmAUX1* was transformed into the *Arabidopsis thaliana aux1-7* mutant with the *CaMV35S* promoter, and two lines (L1 and L2) were used for the experiment. (**A**) Phenotype of plants grown in soils irrigated with 0, 100, or 500 μM Cd solution for 3–4 weeks. (**B**) The Cd concentration of shoot. Data are means ± SE (n = 3); different letters above the bars are significantly different at $p < 0.05$ (Duncan's test).

2.5. Alternative Splicing Identification

The inherent advantage of PacBio ISO-seq makes it possible to understand the complexity of alternative splicing (AS) at the whole transcriptome scale even without a reference genome. In this study, using the full-length reference transcriptomic database, we reconstructed the unique models of all transcripts by cogent software, named full-length UniTransModels. Transcript isoforms of Italian ryegrass were clustered by mapping individual ISO-seq consensus isoforms back to the reconstructed full-length UniTransModels. In summary, a total of 29.76% of the UniTransModels had more than one isoform, and almost half of the UniTransModels (14.33%) had two isoforms; there were still some UniTransModels with more than 10 isoforms (797, 1.12%) (Figure 8A). The different types of AS events were assessed based on the UniTransModels as the references instead of canonical genome-based AS events. Retained introns (RIs) were the majority of AS events, and together with alternative 5′ end or 3′ end AS events, these three types of AS events constituted more than 90% of detected events (Figure 8B).

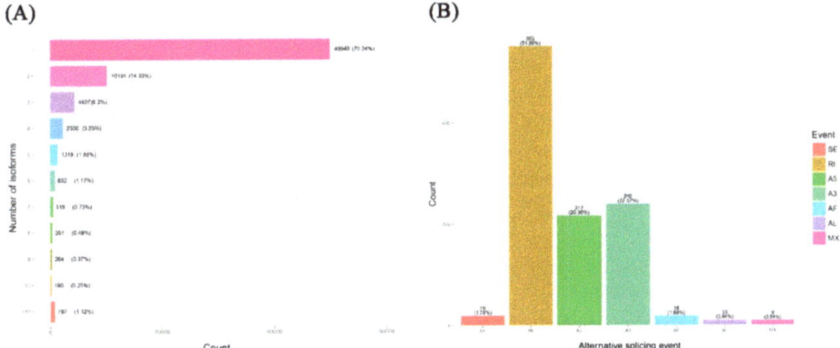

Figure 8. Alternative splicing (AS) analysis of the *Lolium multiflorum* full-length transcriptome using ISO-seq. (**A**) Distribution of isoform numbers for UniTransModels. (**B**) Numbers of different AS events detected in full-length transcriptome. SE: skipping exon; RI: retained intron; A5: alternative 5′ splice-site; A3: alternative 3′ splice-site; AF: alternative first exon; AL: alternative last exon; MX: mutually exclusive exons.

By aligning Illumina short reads to transcript models (UniTransModels), we could further validate the reliability of isoforms detected by the pipeline without a reference genome, and different AS events of some isoforms in UniTransModels in control and Cd stress treatments are illustrated in Figure S7.

3. Discussion

In the past decade, next-generation sequencing (NGS) technology has been widely used for genetic discovery and research [29], and the well-known short-read RNA-seq has become an integral part in evaluating whole RNA expression patterns [30]. However, the sequenced fragments of NGS are usually between 50 and 300 bp, and this was insufficient in the accurate reconstruction and reliable expression estimation of transcript variants, especially for species with low-quality genomes or even without reference genomes [31]. The studies of the species that lack reliable genetic information were seriously restricted in the precise understanding of the genetic diversity and whole expression patterns. Single-molecule long-read sequencing, which was a third sequencing technique, can identify the full structure of individual transcripts by full-length sequencing, which can uniquely reveal the complexity of the transcriptome [23,25].

Italian ryegrass is one of the most widely used forage species in the world [32]. For now, forage grasses are not only used for feedstuff and soil stability enhancer, but also potentially used for biofuel and bioremediation activities [5]. Herbaceous crops may play an important role in sustainable bioenergy feedstock in the future [33]. Italian ryegrass can provide substantial biomass feedstocks due to its high yields and rapid growth. A report indicated that Italian ryegrass was a high-potential bioethanol resource with a higher yield of bioethanol (84.6%) compared to switchgrass (77.7%) and silvergrass (64.3%) [6]. Molecular breeding in crops has shown promise in increasing yield and multiple stress in a changing climate and environment, and it may improve the desirable characteristics in the breeding of Italian ryegrass with molecular biology methods. Transcriptomic analyses are important for understanding gene function. The current knowledge about ryegrass is mainly dependent on gene expression sequenced NGS technologies [34,35]. As a result, the fragmentation of RNA information cannot fully characterize the complexity and diversity of the transcriptome.

In this study, we constructed the transcriptome of Italian ryegrass root using PacBio ISO-seq. We aimed to obtain full and accurate transcriptome information about ryegrass so as to unveil the various responsive networks under Cd stress. Given our pre-experiment, we constructed two overlapped libraries (inserted sizes of 1–4 kb and 3–10 kb), rather than the normal libraries, i.e., 1–4 kb and 4–10 kb libraries. In summary, we obtained 8.12 and 5.52 GB of raw data, respectively

(Table S4). According to the SMRT standard protocol procedures, the raw data were then further filtered (Table S5 and Figure S8). The results of quality control indicated that, instead of using the multiple size-fractionated libraries method [25], shortages in the sequencing data between 3 and 4 kb were supplemented, and the overlapped strategy avoided the biases that shorter fragments of a unique library sequence were regnant (Figure S9). With further classification, clustering, polishing, and comprehensive filter procedures as well as integration with Illumina RNA-seq data, we obtained a 340 Mb full length reference transcriptome, and 146,545 isoforms were annotated with different protein and nucleotide databases. The various functions of Italian ryegrass transcripts were thus unveiled.

RNA-seq is one of the most well-known transcriptomic research pipelines. It is useful for understanding the expressed patterns and for identifying genes, be they of model or non-model species [36]. Cadmium is a toxic heavy metal element, causing serious harm to almost all plants, animals, and humans [8]. Understanding the regulatory mechanisms of genes controlled under Cd may contribute to precise breeding with safer forage or new uses of Italian ryegrass in phytoremediation. In this work, nine-day Italian ryegrass seedlings were treated with and without Cd treatment. After six hours, the concentrations of heavy metals were significant affected by Cd. The mRNAs of six Italian ryegrass roots with and without Cd treatment were sequenced with RNA-seq. Over 60 GB of raw data were obtained. Heat map analysis indicated a correlation of all samples, which were clustered as the control group and the Cd group (Figure S10). The data were aligned with the reference transcriptome generated by PacBio ISO-seq, and a large number of DEGs were generated, which provided valuable information on the transcriptional regulation of genes in responses to Cd stress in Italian ryegrass.

Gene Ontology (GO) enrichment analysis enhanced our understanding of DEGs. The DEGs of biological processes involved oxidation–reduction processes, protein folding, and responses to oxidative stress, which was present in the highest proportion. Cadmium can lead to a rise in ROSs and disturb redox homeostasis [37]. Overproduction of ROS can generate oxidative stress, causing protein carbonylation, which can lead to misfolded or inactivated proteins [38]. In addition, some DEGs are involved in responses to heat terms. Heat Shock Proteins (HSPs) play a crucial role in plant tolerance to multiple stress [39]. Some HSP/chaperone genes have been reported to contribute to the positive enhancement of plant Cd tolerance [20,40,41]. HSPs can also act as molecular chaperones to process the folding, assembly, translocation, and degradation of matched proteins [42]. For example, the HSP chaperones are responsible for the folding of some nascent proteins to alleviate endoplasmic reticulum (ER) stress [43], including kinases [44], transcription factors [45], and transporters [46]. Previous transcriptomic research on Cd-treated plants has also emphasized that these biological processes all serve as important functions in adaptation to biotic and abiotic stresses, but these are not limited to Cd stress [21,47]. In addition, some DEGs have been classified as contributing to cellular components, such as extracellular regions, vacuoles, and cell walls in common with other RNA-seq analyses [48]. Additionally, few Cd-responsive genes in Italian ryegrass have been cloned and characterized. Therefore, a list of DEGs as candidate genes could be helpful for further studies (Supplementary Materials Table S3).

In previous studies, the homeostasis of auxin was found to be disturbed by Cd in *A. thaliana* [49,50]. Additionally, our previous study also showed that Cd interfered with the auxin homeostasis (Figures S11 and S12), and the loss of function of the auxin transporter enhanced plant Cd tolerance (Figure S13). We transformed the *lmAUX1* into the *aux1-7* mutant, and the tropic movement of roots was regained (Figure S14). In the Cd tolerance on plates, the relative length of root elongation was significant inhibited. Additionally, soil experiments showed that the Cd concentration of shoots was significantly increased. These results indicate that *LmAUX1* can regulate plant Cd tolerance and accumulation. In Italian ryegrass, the relative expression level of *lmAUX1* was downregulated under Cd stress, so it is suggested that Italian ryegrass enhanced self-tolerance by decreasing the expression level of *LmAUX1*.

Alternative splicing (AS) is an important post-transcriptional regulatory mechanism for the diversity of transcriptomes and proteomes [51]. Recently, the versatile roles of AS in plants have been further clarified, such as the salt stress tolerance in *A. thaliana* and the mineral nutrient homeostasis in rice. In *A. thaliana*, the U1 spliceosomal protein, AtU1A, is responsible for recognizing the 5′ splice site in the initial step of pre-mRNA splicing. The 5′ splice site in the *Arabidopsis aconitase* (*ACO1*) pre-mRNA can be sequentially recognized by the U1 snRNA that is associated with the AtU1A protein. Splicing of *ACO1* pre-mRNA is important for maintaining proper ROS levels in the cell and for contributing to salt tolerance [52]. Ser/Arg (SR)-rich proteins interact with pre-mRNA sequences and splicing factors during spliceosome assembly in order to perform essential functions in constitutive and AS. In rice, three SR protein-encoding genes (*SR40*, *SCL57*, and *SCL25*) regulate P uptake and remobilization between leaves and shoots of rice [53]. However, these studies have mainly focused on model plants using NGS techniques. For polyploid species or un-sequenced species, it is difficult to accurately identify transcript isoforms [24]. PacBio ISO-seq is a more reliable and superior strategy to understand the whole transcriptomes. It can directly generate accurate AS isoforms, rather than short read sequencing, to annotate a novel whole-genome assembly [54]. Here, we characterized AS in autotetraploid Italian ryegrass using PacBio ISO-seq. In common with results with respect to cotton [24], rice [26], strawberry [29], and switchgrass [55], the retained intron events contributed the most AS events. In addition, the percentage of unique isoforms is greater than those of other species; it appeared that polyploidy also played an important role in increasing the transcript diversity of Italian ryegrass [56]. Our results only scratch the surface of the AS in Italian ryegrass under Cd stress without the reference genome, and we expect to reveal more AS transcript-harboring genes under multiple stresses in the future.

4. Materials and Methods

4.1. Plant Materials, Growth Conditions, and Treatment

A Cd-tolerant ryegrass cultivar ("IdyII") was chosen for this study [57]. The Italian ryegrass seeds were treated with 50% H_2SO_4 for 5 min, washed with deionized water thoroughly, subsequently disinfected with 30% NaClO for 15 min, and then rinsed with deionized water again. After that, the seeds were preliminarily germinated on a net floating on deionized water at 25 °C in the dark. Three days later, the germinated seeds were cultivated in a greenhouse with a day/night temperature of 25/20 °C, 65 ± 5% relative humidity, and a 12 h light/dark cycle with 150 μmol m^{-2} s^{-1} light intensity for two days. The seedlings with a uniform height of about 4 cm were transferred to a 1 L black plastic beaker filled with a 1/4 strength modified Hoagland solution with the following composition: 1.5 mmol L^{-1} KNO_3, 1.0 mmol L^{-1} $Ca(NO_3){\cdot}4H_2O$, 0.5 mmol L^{-1} $NH_4H_2PO_4$, 0.25 mmol L^{-1} $MgSO_4{\cdot}H_2O$, 50 μmol L^{-1} KCl, 25 μmol L^{-1} H_3BO_3, 2.0 μmol L^{-1} $MnSO_4{\cdot}H_2O$, 2.0 μmol L^{-1} $ZnSO_4{\cdot}7H_2O$, 0.5 μmol L^{-1} $CuSO_4{\cdot}5H_2O$, 0.5 μmol L^{-1} $(NH4)_6MO_7O_{24}{\cdot}4H_2O$, and 20 μmol L^{-1} Fe(II)-EDTA-Na_2. The pH value was adjusted to 6.5, and the nutrient solution was renewed every 3 d. Seedlings were continuously cultivated in the greenhouse. After pre-culture for 9 d, plants were treated with and without 50 μmol L^{-1} Cd supplied as $CdCl_2$ for 6 h. There were three biological replicates for the assays. A black beaker containing seven seedlings represented a biological replicate. From each beaker, two seedlings used for PacBio ISO-seq and Illumina RNA-seq were chosen, respectively. Other seedlings were used for the determination of metal concentrations.

4.2. RNA Isolation, RNA-seq, and PacBio Full-Length ISO-SeqLibrary Preparation and Sequencing

Total RNAs of root samples were extracted using TRIzol (Invitrogen™, Life Technologies, Carlsbad CA, USA), and residual DNA was removed with RNase-free DNase I (New England BioLabs, Ipswich, MA, USA) according to the manufacturer's instructions. The total RNA was quantified and assessed with an Agilent Bioanalyzer 2100 (Agilent Technologies, Palo Alto, CA, USA), and the RNA integrity numbers (RIN) of all samples were greater than 9.0. Total RNA samples were sent to Berry Genomics (http://www.berrygenomics.com) for sequencing.

The RNA-seq libraries were constructed using the TruSeq RNA Sample Prep Kit (Illumina, San Diego, CA, USA) as per the manufacturer's instructions. The libraries were sequenced using Illumina Hiseq X Ten (Illumina, San Diego, CA, USA) as 150 bp paired-end reads.

For the PacBio full-length ISO-seq libraries, two equal mixed RNAs (2 μg, mixed with equal three controls and three-treatment total RNAs) were used for 1–4 and 3–10 kb cDNA library construction, respectively. Poly(A) RNA was isolated using the Poly(A) Purist^{TM} Kit (Ambion, Inc., Austin, TX, USA)). RNA was reverse-transcribed into cDNA using the SMARTer^{TM} PCR cDNA Synthesis Kit (Clontech Laboratories Inc., Palo Alto, California, USA). In PCR amplification, for 1–4 kb library construction, the LD-PCR included 16 cycles, and for 3–10 kb library, the LD-PCR included 20 cycles. The 1–4 and 3–10 kb cDNA fragments were generated by BluePippin size selection (Sage Science, Beverly, MA, USA). The size selected cDNA products were applied to constructed SMRTbell Template libraries (SMRTBell Template Prep Kit, Pacific Biosciences, Menlo Park, CA, USA). Each library used one SMRT cell sequenced on a PacBio sequel platform (Pacific Biosciences, Menlo Park, CA, USA).

4.3. Assembly of Reference Transcriptomic Database

Raw short reads of Illumina RNA-seq were filtered by removing reads containing adaptors, reads containing poly-N (>10%), and low-quality reads. Q20, Q30, GC-contents, and sequence duplication levels of the clean data were calculated. Clean reads were used for next analysis.

The standard protocol of ISO-seq (SMRT Analysis 2.3) was used to process the raw PacBio full-length ISO-seq data. Raw long reads were filtered, and the reads for which length was no more than 50 bp and of which accuracy was less than 0.75 were removed. The reads of inserts (ROIs) were obtained from the circular consensus sequences (CCS). After examining for 5′ and 3′ adaptors and poly(A) signals, full-length and non-full-length cDNA reads were defined, and chimeric reads were removed, which included sequencing primers. According to the isoform-level clustering algorithm (ICE), the full-length non-chimeric reads were aligned, and similarity sequences were assigned to a cluster. Each cluster was identified as a uniform isoform. Non-full-length cDNA reads were then applied to polish each cluster. The isoform sequences (with a predicted accuracy of >99%) were considered high-quality isoforms, and others were regarded as low-quality isoforms. The Illumina short reads were used to improve the low-quality isoform accuracies by LSC (LSC 2.0, http://augroup.org/LSC/LSC)). The modified low-quality isoforms and high-quality isoforms were combined as high-quality full-length transcripts. Finally, the redundancies were moved using CD-HIT-EST (CD-HIT-EST 4.6, http://www.bioinformatics.org/cd-hit, parameterization: c = 0.99, T = 6, G = 1, U = 10, s = 0.999, d = 40, and p = 1) to obtain non-redundant high-quality transcripts. The transcripts were considered the reference transcriptome and used for further analysis.

4.4. Annotation of Gene Function

To understand the functions of the reference transcriptome, the protein-coding sequences (CDS) of all transcripts were predicted by TransDecoder (Transdecoder 3.0.0, https://github.com/TransDecoder) at the default setting. The coding likelihood scores of six open reading frames (ORFs) with a length of coding protein sequences of more than 100 amino acids (aa) in a transcript were counted, and the highest score of ORFs was retained as the CDS. If a candidate ORF was found fully encapsulated by the coordinates of another candidate ORF, the longer one was reported [58]. All predicted CDSs were aligned to protein and nucleotide databases (NCBI non-redundant protein (NR), NCBI non-redundant nucleotide (NT), Swiss-Prot, Gene Ontology (GO), KEGG, and KOG) using blast.

4.5. Quantification of Gene Expression Levels

All clean reads of Illumina RNA-seq corresponding to each pooled sample were mapped against the assembled reference transcriptome using bowtie 2 (V2.1.0). The FPMK (fragments per Kilobase Million) values of each isoform were computed for each pooled sample and normalized using the RPKM (RSEM 1.2.15) [59].

4.6. Identification and Function Assessment of DEGs

EdgeR package (edgeR 3.14.0) was used to identify differentially expressed genes (DEGs) between the control and Cd treatment. Significant DEGs were identified with the threshold of FDR < 0.001 and $\log_2$ (FoldChange) > 1.

The GO enrichment analysis was implemented to assess the functions of DEGs using the topGO R package (topGO 2.24.0). The KEGG pathways were applied to enrich the DEGs to different biochemical metabolic pathways and signal transduction pathways with KOBAS software (KOBAS 2.0).

4.7. Validation of DEGs with qRT-PCR

The relative expression levels of 20 randomly selected DEGs were validated with qRT-PCR. New total RNAs were isolated from the plants re-cultivated as were previous samples used for sequencing. The cDNA was synthesized with the HiScript® Q RT SuperMix for qPCR (Vazyme, #R223-1, Nanjing, China). The qRT-PCR was carried out using the ChamQTM SYBR® Master Mix (Vazyme, #Q311-02/03, Nanjing, China) in the QuantStudio 5 Real-Time PCR System (Applied Biosystems, Foster City, CA, USA). All samples were normalized with the reference gene *LmeEF1A(s)*, calculated using the $2^{-\Delta\Delta Ct}$ method, and shown as mean ±SE. Four biological replicates were performed.

4.8. Vector Construction and Ectopic Overexpression of LmAUX1 in Arabidopsis Thaliana

To reveal the functions controlled by the genes of Italian ryegrass response to Cd stress, a DEG (Multiflorum_1-4k_c11285_f6p6_2149) was selected for functional characterization. A blastp search in the National Center for Biotechnology Information (NCBI) with predicted protein coding sequences of this gene as the default setting indicated that it coded an auxin transporter. Thus, it was named *LmAUX1*. The RNA-seq result showed that the relative expression level of this gene was significantly decreased in response to Cd stress, and this was validated with qRT-PCR. The phylogenetic tree, based on amino acid sequences of predicted protein coding sequences and AUX1 amino acid sequences of several other plant species, was constructed using MEGA 7 software (Molecular Evolutionary Genetics Analysis 7.0 software). Bootstrapping analysis was processed as the neighbor-joining method (1000 replicates).

The predicted CDSs of *LmAUX1* were cloned into the pCAMBIA1305-eGFP vector with the *CaMV35s* promoter. The constructed vector was chemically transformed into the *Agrobacterium tumefaciens* strain "EHA105," and the target gene was then transformed into *A. thaliana* (*LmAUX1* was transformed into both the Col-0 ecotype and the *aux1-7* mutant, respectively) by the floral dip method. Positive transgenic lines were screened with 25 mg L^{-1} hygromycin on the half-strength Murashige and Skoog solid medium supplied with 1% sucrose and 0.8% agar (pH 5.8) (1/2 MS medium). More than 15 transgenic lines were obtained, and two T3 lines of them were used for functional characterization. The *LmAUX1* overexpression lines transformed into *aux1-7* mutants named L1 and L2, and *LmAUX1* overexpression lines were transformed into Col-0 ecotypes named OE1 and OE2.

For Cd tolerance experiments on plates, seeds were surface-sterilized with 10% NaClO for 15 min and plated on 1/2 MS medium. The seeds were vernalized for two days in darkness at 4 °C and then cultivated in a growth chamber with a 16/8 h light/dark cycle and a temperature of 23/20 °C for the day/night cycle. After 4–5 days, seedlings were transferred to 1/2 MS medium with and without 50 μM $CdCl_2$. After five days, the seedlings were photographed, and the length of primary root elongation was measured using the ImageJ software. The experiments were repeated twice.

For Cd transferred experiments in soil, Col-0, *aux1-7*, and *LmAUX1* overexpression lines (L1, L2, OE1, and OE2) were grown in soil under a long-day photoperiod (16/8 h light/dark) for two weeks. Subsequently, the plants were grown on soils irrigated with a 0, 100, or 500 μM Cd solution every two or three day for 3–4 weeks.

4.9. Determination of Metal Concentrations

The samples were dried at 70 °C for 3 d and then subjected to with mixed acid (HNO3 + HClO4 (85:15, *v/v*)) using a DigiBlock ED54-iTouch Digester (LabTech, Beijing, China). The samples were completely digested at 170 °C until they turned white; subsequently, the residual acid was volatilized, and all samples were re-dissolved in 10 mL 2.5% HNO3. The concentrations of metals were determined using am inductively coupled plasma optical emission spectrometer (ICP-OES, Perkin Elmer Optima 8000).

4.10. Microscopic Imaging of GUS Staining and Fluorescence of DII-VENUS

GUS staining was carried out according to the method reported in Hu et al. [50]. GUS activity in primary root apices in 5-day-old *DR5::GUS* seedlings treated with 0, 25, 50, 100 μM Cd for 3–4 days on 1/2 MS plates. GUS-stained images were observed under a stereo microscope (ZEISS Stemi 2000-C) and photographed by a CCD camera (Canon PowerShot A620).

Confocal microscopy was performed using a confocal laser scanning microscope (PerkinElmer, Waltham, MA, USA, UltraVIEW® VoX) according to the manufacturer's instructions, excitation and emission wavelengths were 488 to 520 nm for DII-VENUS. Auxin signaling level in primary root apices in five-day-old *Arabidopsis thaliana DII-VENUS* seedlings treated with 0, 25, 50, 100 μM Cd for 3–4 days on 1/2 MS plates.

4.11. Alternative Splicing Identification

The reference transcriptome was re-constructed using Cogent (coding genome reconstruction tool, Cogent v1.4), and a set of transcripts models was then generated and named UniTransModels [60]. All isoforms of the reference transcriptome (transcripts before Cogent reconstruction) were then aligned to UniTransModels with GMAP software (GMAP v2017-06-20). At the same time, a gff file was generated, including the coordinate information of isoforms with respect to the UniTransModels. The transcripts were mapped to the UniTransModels and used for the splicing junctions assessed. The same splicing junctions of the transcripts were collapsed together, and these transcripts with different splicing junctions were identified as transcriptional isoforms of UniTransModels. Additionally, alternative splicing (AS) events were detected with the SUPPA pipeline (https://bitbucket.org/regulatorygenomicsupf/suppa/src/master/) using default settings [61].

The AS events were shown by sashimi plots generated using the IGV browser (IGV 2.5) with bam files and a gff file [62]. The bam files were obtained by mapping the Illumina short reads against UniTransModels using TopHat (TopHat 2.0.4). The FASTA file of the UniTransModels was then imported into IGV as a reference, and the corresponding mapped bam files and isoform gff files were loaded into IGV as tracks. After that, the sashimi plots were obtained from the bam file track. In the opened sashimi plot, the red peaks and junctions are from the Illumina bam file and the blue isoform bars are from the gff file.

4.12. Accession Numbers

The raw sequence data sequenced in this study have been deposited in the Genome Sequence Archive in the BIG Data Centre, Beijing Institute of Genomics (BIG), the Chinese Academy of Sciences (GSA, http://bigd.big.ac.cn/gsa), under accession number CRA001799.

5. Conclusions

In summary, according to the analyses of PacBio long read ISO-seq and Illumina short read RNA-seq data, we constructed a high-quality Italian ryegrass reference transcriptome and unveiled a comprehensive picture of the regulatory networks in Italian ryegrass under Cd stress. We then characterized a DEG (*LmAUX1*) expressed in *A. thaliana*, which significantly enhanced plant Cd accumulation. It is possible that this gene can be used to breed a high-Cd-accumulation Italian ryegrass cultivar. This work deepens our understanding of the mechanism of Italian ryegrass in response to Cd stress and contributes to its gene identification and molecular breeding.

Supplementary Materials: Supplementary materials can be found at http://www.mdpi.com/1422-0067/21/3/1067/s1.

Author Contributions: Conceptualization: Z.H., Q.C., and L.L.; Methodology: Z.H.; Formal Analysis: Z.H.; Investigation: Z.H., Y.Z., Y.H., Q.C., and T.Z.; Data Curation: Z.H.; Writing—Review and Editing: Z.H.; Supervision: Q.C.; Funding Acquisition: Q.C. All authors have read and agreed to the published version of the manuscript.

Funding: This research was funded by the Jiangsu Science and Technology Support Program for Social Development, grant number BE2014709.

Conflicts of Interest: The authors declare no conflict of interest.

References

1. Charmet, G.; Balfourier, F.; Chatard, V. Taxonomic relationships and interspecific hybridization in the genus *Lolium* (grasses). *Genet. Resour. Crop Evolut.* **1996**, *43*, 319–327. [CrossRef]
2. Aguirre, A.A.; Studer, B.; Frei, U.L.; übberstedt, T. Prospects for hybrid breeding in bioenergy grasses. *BioEnergy Res.* **2012**, *5*, 10–19. [CrossRef]
3. Fè, D.; Ashraf, B.; Byrne, S.; Czaban, A.; Roulund, N.; Lenk, I.; Asp, T. Prospects for introducing genomic selection info forage grass breeding. In *EGF at 50: The Future of European Grasslands. In Proceedings of the 25th General Meeting of the European Grassland Federation, Aberystwyth, Wales, 7–11 September 2014*; Centre for Agriculture and Biosciences International: Wallingford, UK, 2014; pp. 830–832. Available online: https://www.cabdirect.org/cabdirect/abstract/20143369178 (accessed on 7 September 2014).
4. Byrne, S.L.; Nagy, I.; Pfeifer, M.; Armstead, I.; Swain, S.; Studer, B.; Mayer, K.; Campbell, J.D.; Czaban, A.; Hentrup, S.; et al. A synteny-based draft genome sequence of the forage grass *Lolium perenne*. *Plant J.* **2015**, *84*, 816–826. [CrossRef] [PubMed]
5. Inoue, M.; Stewart, A.; Cai, H.W. Italian ryegrass. In *Genetics, Genomics and Breeding of Forage Crops*; Kole, C., Yamada, T., Cai, H.W., Eds.; CRC Press: Boca Raton, FL, USA, 2013; pp. 36–57. [CrossRef]
6. Yasuda, M.; Takenouchi, Y.; Nitta, Y.; Ishii, Y.; Ohta, K. Italian ryegrass (*Lolium multiflorum* Lam) as a high-potential bio-ethanol resource. *BioEnergy Res.* **2015**, *8*, 1303–1309. [CrossRef]
7. Zhu, T.; Fu, D.; Yang, F. Effect of saponin on the phytoextraction of Pb, Cd and Zn from soil using Italian ryegrass. *Bull. Environ. Contam. Toxicol.* **2015**, *94*, 129–133. [CrossRef] [PubMed]
8. Daisei, U.; Naoki, Y.; Izumi, K.; Feng, H.C.; Tsuyu, A.; Masahiro, Y.; Ma, J.F. Gene limiting cadmium accumulation in rice. *Proc. Natl. Acad. Sci. USA* **2010**, *107*, 16500–16505. [CrossRef]
9. Clemens, S.; Ma, J.F. Toxic heavy metal and metalloid accumulation in crop plants and foods. *Annu. Rev. Plant Biol.* **2016**, *67*, 489–512. [CrossRef]
10. Wang, P.; Chen, H.; Kopittke, P.M.; Zhao, F.J. Cadmium contamination in agricultural soils of China and the impact on food safety. *Environ. Pollut.* **2019**, *249*, 1038–1048. [CrossRef]
11. Rani, A.; Kumar, A.; Lal, A.; Pant, M. Cellular mechanisms of cadmium-induced toxicity: A review. *Int. J. Environ. Health Res.* **2014**, *24*, 378–399. [CrossRef]
12. Zhang, L.; Pei, Y.; Wang, H.; Jin, Z.; Liu, Z.; Qiao, Z.; Fang, H.; Zhang, Y. Hydrogen sulfide alleviates cadmium-induced cell death through restraining ROS accumulation in roots of *Brassica rapa* L. Ssp. *Pekinensis. Oxid. Med. Cell. Longev.* **2015**, *2015*, 804603. [CrossRef]
13. Gupta, D.K.; Pena, L.B.; Romero-Puertas, M.C.; Hernandez, A.; Inouhe, M.; Sandalio, L.M. NADPH oxidases differentially regulate ROS metabolism and nutrient uptake under cadmium toxicity. *Plant Cell Environ.* **2016**, *40*, 509–526. [CrossRef] [PubMed]
14. Rizwan, M.; Ali, S.; Abbas, T.; Zia-Ur-Rehman, M.; Hannan, F.; Keller, C.; Al-Wabel, M.I.; Ok, Y.S. Cadmium minimization in wheat: A critical review. *Ecotoxicol. Environ. Saf.* **2016**, *130*, 43–53. [CrossRef] [PubMed]
15. Lin, Y.; Aarts, M.G.M. The molecular mechanism of zinc and cadmium stress response in plants. *Cell. Mol. Life Sci.* **2012**, *69*, 3187–3206. [CrossRef]
16. Tao, Q.; Jupa, R.; Liu, Y.; Luo, J.; Li, J.; Kováč, J.; Ki, B.; Li, Q.; Wu, K.; Liang, Y.; et al. Abscisic acid-mediated modifications of radial apoplastic transport pathway play a key role in cadmium uptake in hyperaccumulator *Sedum alfredii*. *Plant Cell Environ.* **2019**, *42*, 1425–1440. [CrossRef] [PubMed]

17. Ishikawa, S.; Ishimaru, Y.; Igura, M.; Kuramata, M.; Abe, T.; Senoura, T.; Hase, Y.; Arao, T.; Nishizawa, N.K.; Nakanishi, H. Ion-beam irradiation, gene identification, and marker-assisted breeding in the development of low-cadmium rice. *Proc. Natl. Acad. Sci. USA* **2012**, *109*, 19166–19171. [CrossRef] [PubMed]
18. Luo, J.S.; Huang, J.; Zeng, D.L.; Peng, J.S.; Zhang, G.B.; Ma, H.L.; Guan, Y.; Yi, H.Y.; Fu, Y.L.; Han, B.; et al. A defensin-like protein drives cadmium efflux and allocation in rice. *Nat. Commun.* **2018**, *9*, 645. [CrossRef]
19. Brunetti, P.; Zanella, L.; De Paolis, A.; Di Litta, D.; Cecchetti, V.; Falasca, G.; Barbieri, M.; Altamura, M.M.; Costantino, P.; Cardarelli, M. Cadmium-inducible expression of the ABC-type transporter AtABCC3 increases phytochelatin-mediated cadmium tolerance in *Arabidopsis*. *J. Exp. Bot.* **2015**, *66*, 3815–3829. [CrossRef]
20. Song, G.; Yuan, S.; Wen, X.; Xie, Z.; Lou, L.; Hu, B.; Cai, Q.; Xu, B. Transcriptome analysis of Cd-treated switchgrass root revealed novel transcripts and the importance of HSF/HSP network in switchgrass Cd tolerance. *Plant Cell Rep.* **2018**, *37*, 1485–1497. [CrossRef]
21. He, F.; Liu, Q.; Zheng, L.; Cui, Y.; Shen, Z.; Zheng, L. RNA-Seq analysis of rice roots reveals the involvement of post-transcriptional regulation in response to cadmium stress. *Front. Plant Sci.* **2015**, *6*, 1136. [CrossRef]
22. Gao, J.; Sun, L.; Yang, X.; Liu, J.X. Transcriptomic analysis of cadmium stress response in the heavy metal hyperaccumulator *Sedum alfredii* Hance. *PLoS ONE* **2014**, *8*, e64643. [CrossRef]
23. Abdel-Ghany, S.E.; Hamilton, M.; Jacobi, J.L.; Ngam, P.; Devitt, N.; Schilkey, F.; Ben-Hur, A.; Reddy, A.S. A survey of the sorghum transcriptome using single-molecule long reads. *Nat. Commun.* **2016**, *7*, 11706. [CrossRef] [PubMed]
24. Wang, M.; Wang, P.; Liang, F.; Ye, Z.; Li, J.; Shen, C.; Pei, L.; Wang, F.; Hu, J.; Tu, L.; et al. A global survey of alternative splicing in allopolyploid cotton: Landscape, complexity and regulation. *New Phytol.* **2018**, *217*, 163–178. [CrossRef] [PubMed]
25. Wang, B.; Tseng, E.; Regulski, M.; Clark, T.A.; Hon, T.; Jiao, Y.; Lu, Z.; Olson, A.; Stein, J.C.; Ware, D. Unveiling the complexity of the maize transcriptome by single-molecule long-read sequencing. *Nat. Commun.* **2016**, *7*, 11708. [CrossRef] [PubMed]
26. Zhang, G.; Sun, M.; Wang, J.; Lei, M.; Li, C.; Zhao, D.; Huang, J.; Li, W.; Li, S.; Li, J.; et al. PacBio full-length cDNA sequencing integrated with RNA-seq reads drastically improves the discovery of splicing transcripts in rice. *Plant J.* **2018**, *97*, 296–305. [CrossRef]
27. Yan, H.; Gao, Y.; Wu, L.; Wang, L.; Zhang, T.; Dai, C.; Xu, W.; Feng, L.; Ma, M.; Zhu, Y.G.; et al. Potential use of the *Pteris vittata* arsenic hyperaccumulation-regulation network for phytoremediation. *J. Hazard. Mater.* **2019**, *368*, 386–396. [CrossRef]
28. Wang, J.; Deng, Y.; Zhou, Y.; Liu, D.; Yu, H.; Zhou, Y.; Lv, J.; Ou, L.; Li, X.; Ma, Y.; et al. Full-length mRNA sequencing and gene expression profiling reveal broad involvement of natural antisense transcript gene pairs in pepper development and response to stresses. *Plant J.* **2019**. [CrossRef]
29. Yuan, H.; Yu, H.; Huang, T.; Shen, X.; Xia, J.; Pang, F.; Wang, J.; Zhao, M. The complexity of the *Fragaria x ananassa* (octoploid) transcriptome by single-molecule long-read sequencing. *Hortic. Res.* **2019**, *6*, 46. [CrossRef]
30. Anvar, S.Y.; Allard, G.; Tseng, E.; Sheynkman, G.M.; de Klerk, E.; Vermaat, M.; Yin, R.H.; Johansson, H.E.; Ariyurek, Y.; den Dunnen, J.T.; et al. Full-length mRNA sequencing uncovers a widespread coupling between transcription initiation and mRNA processing. *Genome Biol.* **2018**, *19*, 46. [CrossRef]
31. Steijger, T.; Abril, J.F.; Engstrom, P.G.; Kokocinski, F.; Hubbard, T.; Guigo, R.; Harrow, J.; Bertone, P. Assessment of transcript reconstruction methods for RNA-seq. *Nat. Methods* **2013**, *10*, 1177–1184. [CrossRef]
32. Hirata, M.; Cai, H.; Inoue, M.; Yuyama, N.; Miura, Y.; Komatsu, T.; Takamizo, T.; Fujimori, M. Development of simple sequence repeat (SSR) markers and construction of an SSR-based linkage map in Italian ryegrass (*Lolium multiflorum* Lam.). *Theor. Appl. Genet.* **2006**, *113*, 270–279. [CrossRef]
33. Lee, D.K.; Aberle, E.; Anderson, E.K.; Anderson, W.; Baldwin, B.S.; Baltensperger, D.; Barrett, M.; Blumenthal, J.; Bonos, S.; Bouton, J. Biomass production of herbaceous energy crops in the United States: Field trial results and yield potential maps from the multiyear regional feedstock partnership. *GCB Bioenergy* **2018**, *10*, 698–716. [CrossRef]
34. Duhoux, A.; Carrere, S.; Gouzy, J.; Bonin, L.; Delye, C. RNA-Seq analysis of rye-grass transcriptomic response to an herbicide inhibiting acetolactate-synthase identifies transcripts linked to non-target-site-based resistance. *Plant Mol. Biol.* **2015**, *87*, 473–487. [CrossRef] [PubMed]

35. Knorst, V.; Byrne, S.; Yates, S.; Asp, T.; Widmer, F.; Studer, B.; Kolliker, R. Pooled DNA sequencing to identify SNPs associated with a major QTL for bacterial wilt resistance in Italian ryegrass (*Lolium multiflorum* Lam.). *Theor. Appl. Genet.* **2019**, *132*, 947–958. [CrossRef] [PubMed]
36. Jain, M. A next-generation approach to the characterization of a non-model plant transcriptome. *Curr. Sci.* **2011**, *101*, 1435–1439.
37. Chmielowska-Bąk, J.; Gzyl, J.A.; Rucińska-Sobkowiak, R.; Arasimowicz-Jelonek, M.; Deckert, J. The new insights into cadmium sensing. *Front. Plant Sci.* **2014**, *5*, 245. [CrossRef]
38. Braconi, D.; Bernardini, G.; Santucci, A. Linking protein oxidation to environmental pollutants: Redox proteomic approaches. *J. Proteom.* **2011**, *74*, 2324–2337. [CrossRef]
39. Wang, W.; Vinocur, B.; Shoseyov, O.; Altman, A. Role of plant heat-shock proteins and molecular chaperones in the abiotic stress response. *Trends Plant Sci.* **2004**, *9*, 244–252. [CrossRef]
40. Shim, D.; Hwang, J.U.; Lee, J.; Lee, S.; Choi, Y.; An, G.; Martinoia, E.; Lee, Y. Orthologs of the class A4 heat shock transcription factor *HsfA4a* confer cadmium tolerance in wheat and rice. *Plant Cell* **2009**, *21*, 4031–4043. [CrossRef]
41. Cai, S.Y.; Zhang, Y.; Xu, Y.P.; Qi, Z.Y.; Li, M.Q.; Ahammed, G.J.; Xia, X.J.; Shi, K.; Zhou, Y.H.; Reiter, R.J. *HsfA1a* upregulates melatonin biosynthesis to confer cadmium tolerance in tomato plants. *J. Pineal Res.* **2017**, *62*, e12387. [CrossRef]
42. Tamás, M.; Sharma, S.; Ibstedt, S.; Jacobson, T.; Christen, P. Heavy metals and metalloids as a cause for protein misfolding and aggregation. *Biomolecules* **2014**, *4*, 252–267. [CrossRef]
43. Liu, L.; Cui, F.; Li, Q.; Yin, B.; Zhang, H.; Lin, B.; Wu, Y.; Xia, R.; Tang, S.; Xie, Q. The endoplasmic reticulum-associated degradation is necessary for plant salt tolerance. *Cell Res.* **2011**, *21*, 957. [CrossRef]
44. Piper, P.; Truman, A.; Millson, S.; Nuttall, J. Hsp90 chaperone control over transcriptional regulation by the yeast Slt2(Mpk1)p and human ERK5 mitogen-activated protein kinases (MAPKs). *Biochem. Soc. Trans.* **2006**, *34*, 783–785. [CrossRef]
45. Rodriguez, F.; Arsène-Ploetze, F.; Rist, W.; Rüdiger, S.; Schneider-Mergener, J.; Mayer, M.P.; Bukau, B. Molecular basis for regulation of the heat shock transcription factor σ32 by the DnaK and DnaJ chaperones. *Mol. Cell* **2008**, *32*, 347–358. [CrossRef] [PubMed]
46. Kimura, T.; Kambe, T. The functions of metallothionein and ZIP and ZnT transporters: An overview and perspective. *Int. J. Mol. Sci.* **2016**, *17*, 336. [CrossRef]
47. Chen, C.; Cao, Q.; Jiang, Q.; Li, J.; Yu, R.; Shi, G. Comparative transcriptome analysis reveals gene network regulating cadmium uptake and translocation in peanut roots under iron deficiency. *BMC Plant Biol.* **2019**, *19*, 35. [CrossRef]
48. Lin, C.Y.; Trinh, N.N.; Fu, S.F.; Hsiung, Y.C.; Chia, L.C.; Lin, C.W.; Huang, H.J. Comparison of early transcriptome responses to copper and cadmium in rice roots. *Plant Mol. Biol.* **2013**, *81*, 507–522. [CrossRef]
49. Yuan, H.M.; Huang, X. Inhibition of root meristem growth by cadmium involves nitric oxide-mediated repression of auxin accumulation and signalling in *Arabidopsis*. *Plant Cell Environ.* **2016**, *39*, 120–135. [CrossRef]
50. Hu, Y.F.; Zhou, G.; Na, X.F.; Yang, L.; Nan, W.B.; Liu, X.; Zhang, Y.Q.; Li, J.L.; Bi, Y.R. Cadmium interferes with maintenance of auxin homeostasis in *Arabidopsis* seedlings. *J. Plant Physiol.* **2013**, *170*, 965–975. [CrossRef]
51. Marquez, Y.; Brown, J.W.; Simpson, C.; Barta, A.; Kalyna, M. Transcriptome survey reveals increased complexity of the alternative splicing landscape. *Arabidopsis Genome Res.* **2012**, *22*, 1184–1195. [CrossRef]
52. Gu, J.; Xia, Z.; Luo, Y.; Jiang, X.; Qian, B.; Xie, H.; Zhu, J.K.; Xiong, L.; Zhu, J.; Wang, Z.Y. Spliceosomal protein U1A is involved in alternative splicing and salt stress tolerance in *Arabidopsis thaliana*. *Nucleic Acids Res.* **2018**, *46*, 1777–1792. [CrossRef]
53. Dong, C.; He, F.; Berkowitz, O.; Liu, J.; Cao, P.; Tang, M.; Shi, H.; Wang, W.; Li, Q.; Shen, Z.; et al. Alternative splicing plays a critical role in maintaining mineral nutrient homeostasis in rice (*Oryza sativa*). *Plant Cell* **2018**, *30*, 2267–2285. [CrossRef]
54. Li, J.; Harata-Lee, Y.; Denton, M.D.; Feng, Q.; Rathjen, J.R.; Qu, Z.; Adelson, D.L. Long read reference genome-free reconstruction of a full-length transcriptome from *Astragalus membranaceus* reveals transcript variants involved in bioactive compound biosynthesis. *Cell Discov.* **2017**, *3*, 17031. [CrossRef]
55. Zuo, C.; Blow, M.; Sreedasyam, A.; Kuo, R.C.; Ramamoorthy, G.K.; Torres-Jerez, I.; Li, G.; Wang, M.; Dilworth, D.; Barry, K.; et al. Revealing the transcriptomic complexity of switchgrass by PacBio long-read sequencing. *Biotechnol. Biofuels* **2018**, *11*, 170. [CrossRef]

56. Yoo, M.J.; Liu, X.; Pires, J.C.; Soltis, P.S.; Soltis, D.E. Nonadditive gene expression in polyploids. *Annu. Rev. Genet.* **2014**, *48*, 485–517. [CrossRef]
57. Fang, Z.; Lou, L.; Tai, Z.; Wang, Y.; Yang, L.; Hu, Z.; Cai, Q. Comparative study of Cd uptake and tolerance of two Italian ryegrass (*Lolium multiflorum*) cultivars. *Peer J.* **2017**, *5*, e3621. [CrossRef]
58. Tian, J.; Feng, S.; Liu, Y.; Zhao, L.; Tian, L.; Hu, Y.; Yang, T.; Wei, A. Single-molecule long-read sequencing of *Zanthoxylum bungeanum* maxim. Transcriptome: Identification of aroma-related genes. *Forests* **2018**, *9*, 765. [CrossRef]
59. Zhou, Y.; Tang, Q.; Wu, M.; Mou, D.; Liu, H.; Wang, S.; Zhang, C.; Ding, L.; Luo, J. Comparative transcriptomics provides novel insights into the mechanisms of selenium tolerance in the hyperaccumulator plant *Cardamine hupingshanensis*. *Sci. Rep.* **2018**, *8*, 2789. [CrossRef]
60. Wu, T.D.; Reeder, J.; Lawrence, M.; Becker, G.; Brauer, M.J. GMAP and GSNAP for genomic sequence alignment: Enhancements to speed, accuracy, and functionality. In *Statistical Genomics*; Humana Press: New York, NY, USA, 2016; pp. 283–334. [CrossRef]
61. Alamancos, G.P.; Pagès, A.; Trincado, J.L.; Bellora, N.; Eyras, E. Leveraging transcript quantification for fast computation of alternative splicing profiles. *RNA* **2015**, *21*, 1521–1531. [CrossRef]
62. Thorvaldsdóttir, H.; Robinson, J.T.; Mesirov, J.P. Integrative Genomics Viewer (IGV): High-performance genomics data visualization and exploration. *Brief. Bioinform.* **2013**, *14*, 178–192. [CrossRef]

© 2020 by the authors. Licensee MDPI, Basel, Switzerland. This article is an open access article distributed under the terms and conditions of the Creative Commons Attribution (CC BY) license (http://creativecommons.org/licenses/by/4.0/).

International Journal of
Molecular Sciences

Article

Comparative Transcriptome Analysis of the Molecular Mechanism of the Hairy Roots of *Brassica campestris* L. in Response to Cadmium Stress

Yaping Sun, Qianyun Lu, Yushen Cao, Menghua Wang, Xiyu Cheng * and Qiong Yan *

College of Life Sciences and Bioengineering, School of Science, Beijing Jiaotong University, Beijing 100044, China; 17121607@bjtu.edu.cn (Y.S.); 15121580@bjtu.edu.cn (Q.L.); 16121632@bjtu.edu.cn (Y.C.); 18121618@bjtu.edu.cn (M.W.)

* Correspondence: xycheng@bjtu.edu.cn (X.C.); qyan@bjtu.edu.cn (Q.Y.); Tel.: +86-138-1027-4418 (X.C.); +86-138-1093-8871 (Q.Y.)

Received: 18 October 2019; Accepted: 20 December 2019; Published: 26 December 2019

Abstract: *Brassica campestris* L., a hyperaccumulator of cadmium (Cd), is considered a candidate plant for efficient phytoremediation. The hairy roots of *Brassica campestris* L are chosen here as a model plant system to investigate the response mechanism of *Brassica campestris* L. to Cd stress. High-throughput sequencing technology is used to identify genes related to Cd tolerance. A total of 2394 differentially expressed genes (DEGs) are identified by RNA-Seq analysis, among which 1564 genes are up-regulated, and 830 genes are down-regulated. Data from the gene ontology (GO) analysis indicate that DEGs are mainly involved in metabolic processes. Glutathione metabolism, in which glutathione synthetase and glutathione *S*-transferase are closely related to Cd stress, is identified in the Kyoto Encyclopedia of Genes and Genomes (KEGG) analysis. A Western blot shows that glutathione synthetase and glutathione *S*-transferase are involved in Cd tolerance. These results provide a preliminary understanding of the Cd tolerance mechanism of *Brassica campestris* L. and are, hence, of particular importance to the future development of an efficient phytoremediation process based on hairy root cultures, genetic modification, and the subsequent regeneration of the whole plant.

Keywords: *Brassica campestris* L.; cadmium; glutathione synthetase; glutathione *S*-transferase; transcriptome

1. Introduction

Cadmium (Cd) is one of the most toxic and most common heavy metals, causing serious pollution to farmland and active transfer in soil-plant systems [1–12]. Cd intake may lead to tissue inflammation, fibrosis, and even various cancers in humans [13–17], necessitating the development of an efficient strategy to eliminate Cd from soil. Phytoremediation is an effective and low-cost technique for removing Cd from contaminated soil [18–20]. Currently, many studies in this area have been carried out using whole plants, which have a limited lifespan and must be replaced and re-established after each experiment. It is also difficult to evaluate the effects of various factors such as light, temperature, soil pH, and rhizosphere microorganisms in different batch experiments [21,22]. Hairy roots, which possess all the morphological and physiological characteristics of normal roots and allow indefinite propagation, have been proven as a convenient experimental tool for investigating the interactions between plant cells and metal ions [22–25]. The hairy roots of *Thlaspi caerulescens* were chosen to study oxidative stress and the response of the antioxidant defense system under Cd stress [26]. The hairy roots of *Trifolium repens* was used as a sensitive tool for monitoring and assessing Cd contamination in the environment [27]. Notably, further genetic modification via the Ri plasmid of *Agrobacterium rhizogenes* is quite straightforward for the introduction of enhanced Cd accumulation traits [23]. Given these attractive characteristics of the hairy root culture process, it has been chosen as a model plant system to investigate the Cd tolerance of hyperaccumulating plants in this work.

Studies on the Cd tolerance mechanisms of hyperaccumulating plants are of great importance [28]. It was reported that glutathione (GSH) is an important non-enzymatic antioxidant in cells which can effectively scavenge free radicals produced by Cd stress and can reduce peroxidative damage [29]. During this process, GSH synthetase (GSHB), glutathione *S*-transferase (GST) and glutathione reductase (GSR) also play an important role in the intracellular detoxification of intracellular xenobiotics and toxic compounds [30–32]. Three basic helix-loop-helix transcription factors (FER-like Deficiency Induced Transcripition Factor (*FIT*), *AtbHLH38*, and *AtbHLH39*) of *Arabidopsis* were reported to be involved in the plant's response to Cd stress, and the tolerance of transgenic *Arabidopsis* to Cd was significantly improved [33]. Moreover, an RNA-Seq method, which can be used to analyze gene expression, has been successfully developed to further investigate the molecular mechanism of response to various biotic and abiotic stresses in various plants [8,24,34–37]. Data on transcriptome differences indicate that the difference in the iron-deficient transporter gene in response to Cd stress was found to be related to Cd uptake and redistribution in both *Solanum nigrum* L. and *Solanum torvum* L. [38]. Another study of comparative transcriptomics showed that the transporter gene NcNramp1 had higher expression in the roots and shoots of the elite Cd accumulator Ganges in response to Cd stress [39]. The gene NcNramp1, which is one of the major transporters involved in Cd hyperaccumulation in *Noccaea caerulescens*, appears to be the main cause of high Cd accumulation by increasing *Noccaea caerulescens* gene expression [39].

Notably, among different candidates for phytoremediation of Cd pollution, *Brassica* species are considered promising hyperaccumulators [40,41]. A previous study reported that most of the metal transporter genes (74.8%, 202/270) responded to Cd stress, suggesting that at least some of them are involved in Cd uptake and translocation in *Brassica napus* [42]. miR158 may be required for *Brassica napus* tolerance to Cd by decreasing BnRH24 levels [43]. A total of 84 ATP binding cassette (ABC) genes of *Brassica napus* were up-regulated under Cd stress [44]. The response mechanism of the hairy roots of the *Brassica* species to Cd stress is of particular significance in the development of an efficient phytoremediation process based on hairy root cultures, genetic modification, and subsequent regeneration of the whole plant. However, studies on the absorption, accumulation, and tolerance of Cd in the hairy roots of *Brassica campestris* L. are still limited, and the molecular mechanism remains unclear.

The hairy roots of *Brassica campestris* L. could survive in a medium with the addition of 100–200 μM $CdCl_2$ [45,46], as seen in our previous studies. The high Cd accumulation in the hairy roots exceeded 1000 mg/kg [47], which is significantly higher than the threshold level of 100 mg Cd/kg for a Cd-hyperaccumulating plant [46] and comparable to those (204–7408 mg/kg) obtained in whole Cd-hyperaccumulating plants [48,49]. Established hairy roots of *Brassica campestris* L. have been chosen as a model system for our further studies on the plant's response under Cd stress. Transcriptome sequencing by RNA-Seq is performed to investigate the gene transcription pattern and the related molecular mechanism of hairy roots exposed to Cd. Moreover, the function of some distinguished genes such as GSHB and GST also are verified, providing important information for future genetic modification of hairy roots and regeneration of the whole plant with enhanced abilities.

2. Results

2.1. An Overview of the mRNA of *Brassica campestris* L. under Cadmium Stress

To investigate the response of the Cd hyperaccumulator *Brassica campestris* L. to Cd stress, we analyzed and compared mRNA expression profiles of Cd-treated and untreated hairy roots of *Brassica campestris* L. Six samples were measured using high-throughput sequencing, and the average data volume of one sample was 6.72 GB (Table 1). Clean reads were obtained from raw reads by removing reads with poor contaminants, low mass, or a high N content of unknown bases. The Q20 value of all clean reads of samples exceeded 98%, and the error rate was less than 1% (Table 1), indicating that the quality of the sequencing met the requirements of subsequent analysis. There were no AT or GC separations in the sequencing report, indicating that sequencing was stable (Figure S1). Overall, if the proportion of low quality (quality < 20) bases was low, the sequencing quality was good (Figure S2). After obtaining clean reads, the clean reads were aligned to the reference genome sequence by Hierarchical Indexing for Spliced Alignment of Transcripts (HISAT). The average ratio of each sample reached 64.97% (Table S1), and the uniform ratio between samples indicated that the clean reads data between samples were comparable. To conclude, the transcriptome sequencing results were credible.

Table 1. Summary of sequence reads for six RNA samples including three replicate control treatments (C1–C3) and three samples of cadmium treatment (T1–T3).

Sample	Raw Reads (Mb)	Clean Reads (Mb)	Clean Bases (Gb)	Clean Reads Q20 (%)	Clean Reads Q30 (%)	Clean Reads Rate (%)
C 1	53.34	44.28	6.64	98.26	94.45	83.01
C 2	54.83	45.37	6.81	98.37	94.75	82.74
C 3	56.47	45.28	6.79	98.39	94.82	80.19
T 1	54.85	45.03	6.75	98.34	94.67	82.09
T 2	54.77	44.14	6.62	98.34	94.69	80.6
T 3	54.88	44.71	6.71	98.28	94.52	81.48

According to the untreated and treated samples, the similarity heat map shows that the correlations among control replicates C1, C2, and C3 were relatively high; the correlations among replicates within the Cd treatment T1, T2, and T3 were comparatively high. This reveals that Cd treatment caused a large difference in the transcriptome of hairy roots of *Brassica campestris* L. (Figure 1A). Through the framed genetic tree (Figure 1B), it is further illustrated that the closer the expression profile, the higher the correlation, which indicates that Cd stress greatly changed the expression profile of the hairy roots of *Brassica campestris* L. Above all, the results of transcriptome sequencing were reliable.

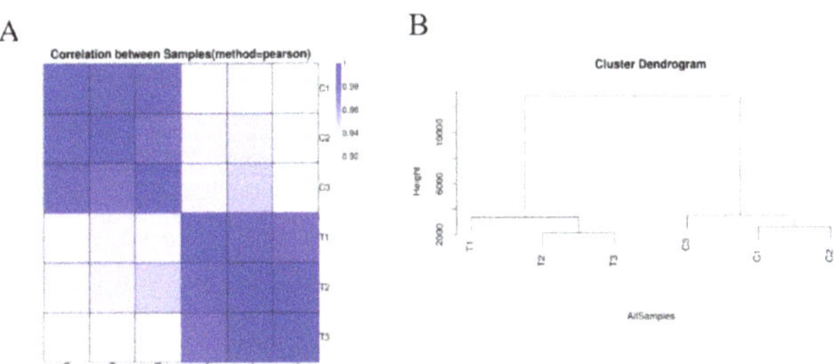

Figure 1. Sample quality test. (**A**) Heatmap of Pearson correlation between samples. The x- and y-axes represent each sample. (**B**) Hierarchical clustering between all and samples. C1, untreated hairy roots group number 1; C2, untreated hairy roots group number 2; C3, untreated hairy roots group number 3; T1, hairy roots treated with 200 μM Cd for 1 d, group number 1; T2, hairy roots treated with 200 μM Cd for 1 d, group number 2; T3, hairy roots treated with 200 μM Cd for 1 d, group number 3.

2.2. Analysis of Differentially Expressed Genes (DEGs)

A total of 35,385 genes were detected by sequencing C and T samples. Among these genes, 30,873 genes were found in both C and T samples, while 2490 and 2022 genes were observed in C and T, respectively (Figure S3). There was a significant difference in unigenes between T and C ($p < 0.05$), of which 1564 unigenes were up-regulated and 830 were down-regulated after $CdCl_2$ treatment (Figure 2.).

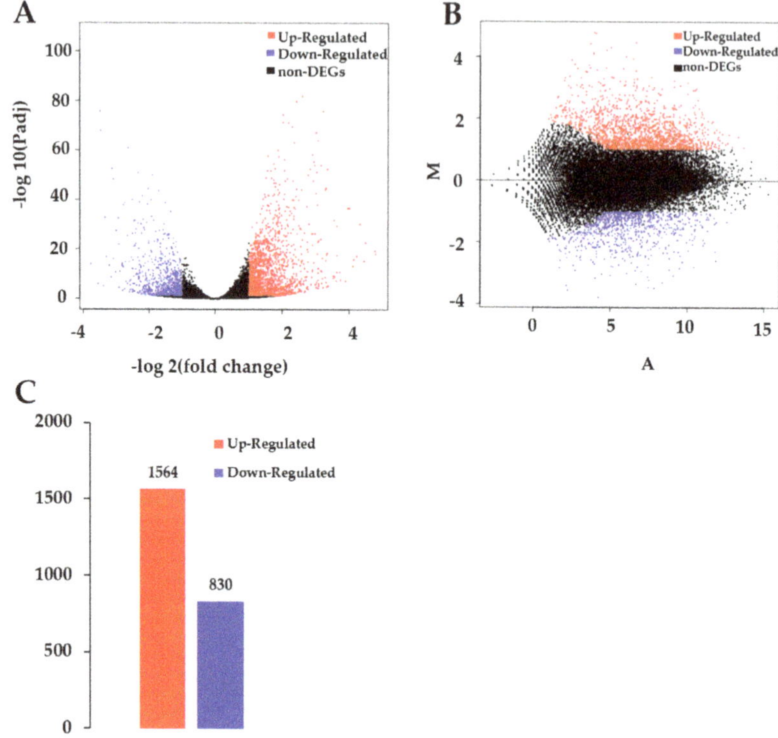

Figure 2. Analysis of differentially expressed genes (DEGs) between control and treatment groups. (**A**) Volcano plot of DEGs. The x-axis represents the log2-transformed fold change. The y-axis represents the -log10-transformed significance. (**B**) Minus-versus-add (MA) plot of DEGs. The x-axis represents value A (log2-transformed mean expression level). The y-axis represents value M (log2-transformed fold change). Red points represent up regulated DEGs. Blue points represent down regulated DEGs. Black points represent non-DEGs. (**C**) Summary of DEGs. The x-axis represents the compared samples. The y-axis represents DEG numbers. Red color represents up-regulated DEGs. Blue color represents down-regulated DEGs.

2.3. Gene Ontology (GO) Functional Analysis of DEGs

Following the identification of DEGs, we analyzed their functions using GO; we classified and enriched their GO function. Detailed proportions of the GO annotation for DEGs are shown in Figure 3, indicating that molecular functions, biological processes, and cellular components were well represented. The GO functional classification results show that the DEGs were mainly enriched in 47 GO terms (Figure 3). It was found that the biological functions of the DEGs in the hairy roots of *Brassica campestris* L. were enriched in cell stimulation responding to Cd stress. Among the most abundant genes in hairy roots were those enriched in response to stimulus (43.6%), an organic substance (15.9%), an endogenous stimulus (14.0%), a chemical (21.2%), a carbohydrate (4.6%), stress (21.2%), oxygen-containing (6.4%) and hormone (11.8%) (Table S2).

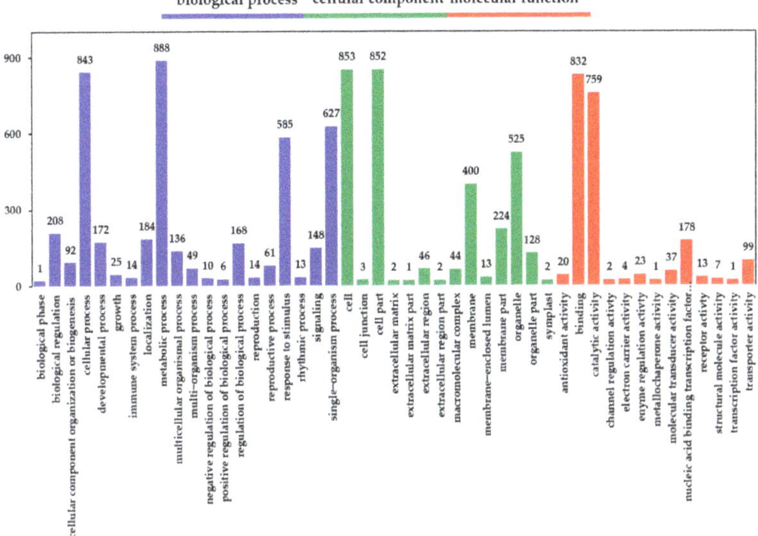

Figure 3. Gene ontology (GO) classification of DEGs. The x-axis represents the GO term. The y-axis represents the number of DEGs.

2.4. Pathway Functional Analysis Results

Different genes in the hairy roots coordinate with each other to perform their biological functions, and pathway-based analysis helps to further understand the biological functions of genes. We mapped the DEGs to the reference canonical pathways in the Kyoto Encyclopedia of Genes and Genomes (KEGG) database to further identify the active metabolic pathways involved in the responses to Cd. The pathway classification results are shown in Figure 4. First, we divided the differentially expressed genes into the following six categories: metabolism, organic systems, environmental information processing, genetic information processing, cellular processes, and human diseases. Then, we further divided the six categories into 20 subcategories. We listed the pathway and the top six differential gene enrichments, as shown in Table S3. It can be seen from the table that Cd stress caused changes in the hairy root metabolic pathway, secondary metabolites, plant–pathogen interactions, plant hormone signal transduction, starch and carbohydrate metabolism, and RNA transport in *Brassica campestris* L. Altogether, there were 433 significant differentially expressed genes involved in the metabolic signaling pathway, accounting for 24.41% of the total number of differentially expressed genes in the signaling pathway (Table S3).

2.5. Real-Time PCR Confirmation of the Gene Expression

To verify the reliability of sequencing, eight differentially expressed genes were randomly selected for RT-PCR validation. Real-time PCR results showed that the transcription levels of these eight genes were consistent with the results of transcriptome sequencing (Figure 5, Table 2). Using this, we can conclude that the results of transcriptome sequencing can represent differences in transcriptome levels in the hairy roots of *Brassica campestris* L. caused by Cd stress. We found that the Gene ID 103874543 involved in the synthesis of glutathione was up-regulated.

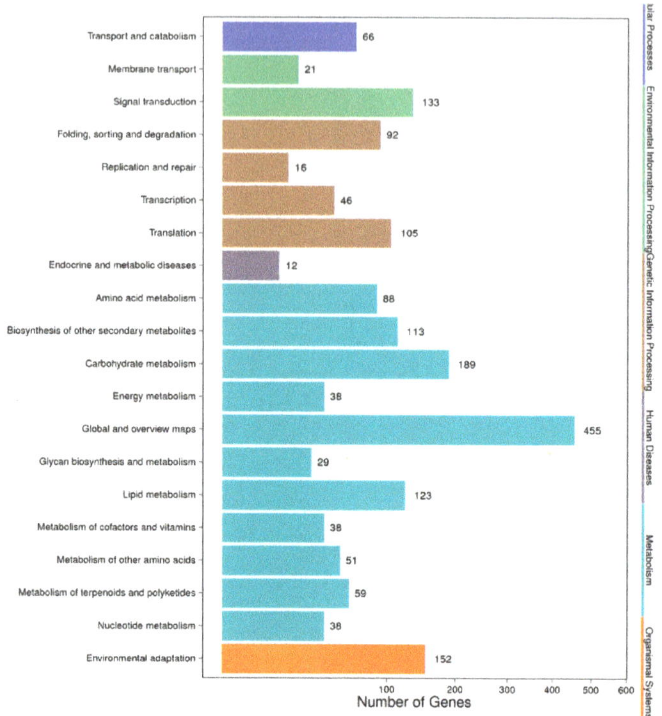

Figure 4. Pathway classification of DEGs. The x-axis represents the number of DEGs. The y-axis represents the pathway name.

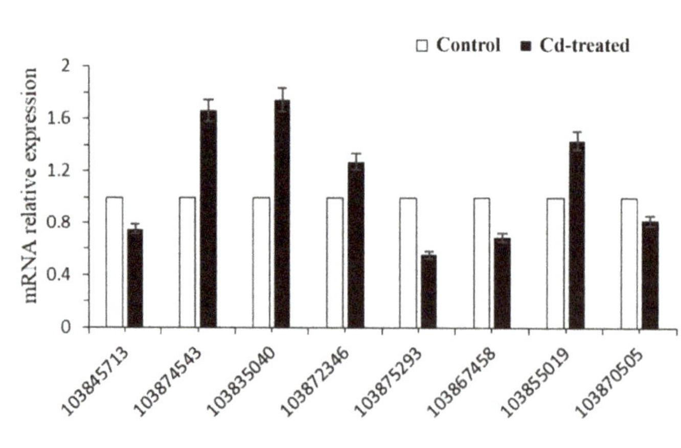

Figure 5. Relative expression of significant differentially expressed genes in hairy root cultures of *Brassica campestris* L.

Table 2. Differential expression of eight target genes (relative to reference genes) quantified by qPCR in control and Cd-treated hairy root samples.

Gene ID	*Nr* Description	C-Expression	T-Expression	log2FoldChange (T/C)	*p*-Value
103845713	peroxidase N	1547.49	311.39	−2.31	4.20×10^{-10}
103874543	glutathione synthetase	406.26	1117.96	1.46	1.86×10^{-16}
103835040	calcium-transporting ATPase 2	402.56	974.08	1.27	3.25×10^{-16}
103872346	ABC transporter G family member 40	532.77	4004.34	2.91	3.07×10^{-62}
103875293	aquaporin TIP1-2	2508.98	1039.61	−1.27	3.03×10^{-18}
103867458	aquaporin NIP2-1	2308.78	703.43	−1.71	5.19×10^{-47}
103855019	protein NRT1/ PTR FAMILY 2.11	584.22	2262.11	1.95	3.82×10^{-34}
103870505	ras-related protein RABC2b	16.88	1.44	−3.55	4.90×10^{-8}

2.6. Identification of Key Gene/Protein and Expression Validation

The GO and KEGG analysis of transcriptome sequencing showed that the GSH metabolic pathway plays an important role in regulating the Cd enrichment ability and defense ability of rapeseed hairy roots. Therefore, we continue to study the GSH metabolic pathway. Using the KEGG database, we obtained the following signal path map (Figure S4). By analyzing key genes in the GSH metabolic signaling pathway, significant up-regulation of GSHB and GST in transcriptome sequencing results caught our attention, so we selected validation of GSHB and GST protein expression levels.

Western Blot indicated that the expression of GSHB at 200 μM Cd stress was 1.85-fold higher than that of the control group, and the expression level of GST was 2.57-fold higher than that of the control group. The amount of actin protein was stable across the treatments (Figure 6). This result further demonstrated that, not only the mRNA, but also the proteins of GSHB and GST were highly upregulated in the hairy roots of *Brassica campestris* L.

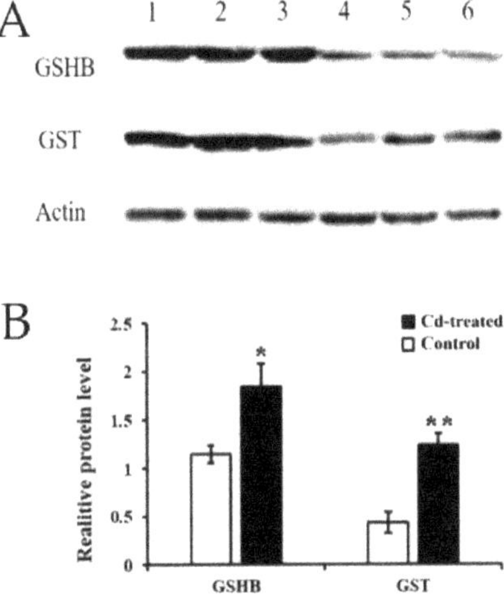

Figure 6. Relative expression of target genes in the hairy roots of *Brassica campestris* L. (**A**) Expression of GST and GSH protein. Lines 1, 2, 3 Cd treatment (200 μM for 24 h); 4, 5, 6: control group; (**B**) Grayscale analysis. (* $p < 0.05$, ** $p < 0.01$).

3. Discussion

It is particularly important to investigate the Cd tolerance and the related molecular mechanisms of the hairy roots of *Brassica campestris* L. [47,50,51]. Our previous studies showed that the high Cd accumulation in the hairy roots of *Brassica campestris* L. reached more than 1000 mg/kg [47]. The growth of hairy roots of *Brassica campestris* L. was significantly inhibited when it grow in the medium with $CdCl_2$ of 200 μM for 24 h [47]. Under this condition, the hairy roots indicated obvious browning and decaying. Additionally, a high Malondialdehyde (MDA) content and microscopic observation of the hairy roots indicated that cell membrane damage and apoptosis appeared, which increased our interest to further investigate the response molecular mechanism using RNA-Seq to perform transcriptome sequencing.

Results indicate that the hairy roots of *Brassica campestris* L. may improve their Cd tolerance and response to the Cd stress as follows. First, the synthesis of some antioxidants (e.g., GSH and AsA) in the hairy roots may be enhanced by up-regulating expression of the related genes (Figure 7). The KEEG analysis and Western Blot test indicated that the GSHB genes were significantly up-regulated and the GSHB content was improved, hence definitely benefiting the corresponding GSH synthesis (Figure 6 and Figure S4). When considering GSH as one of the main antioxidants in plant cells [52,53], it, therefore, is expected to enhance the scavenging of reactive oxygen species (ROS) [54].

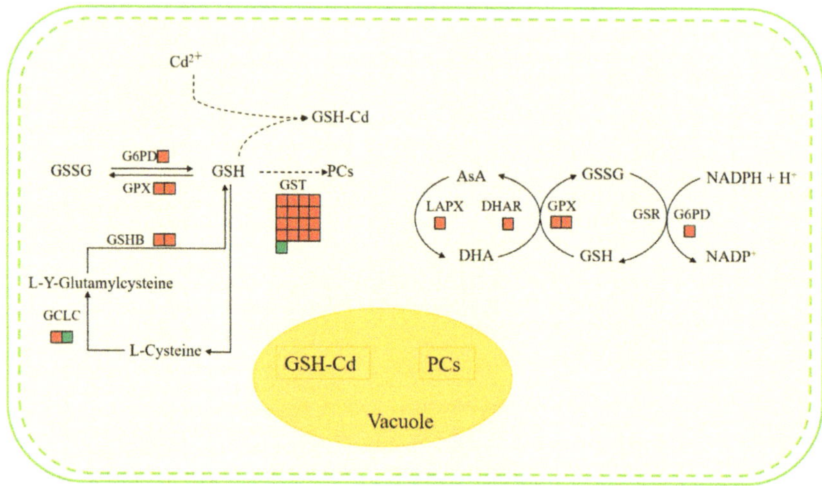

Figure 7. Main changes in *Brassica campestris* L. metabolism after Cd exposure. Red boxes indicate an increased abundance in Cd-treated plants compared with controls, and green boxes indicate a reduced abundance in Cd-treated plants compared with controls. Abbreviations: GSHB, glutamine synthase; GSH, glutathione; GSSG, oxidized glutathione; GST, glutathione-S-transferase; GPX, glutathione peroxidase; GCLC, glutamate-cysteine ligase catalytic subunit; GSH-Cd, glutathione-cadmium complex; PCs, phytochelatins; AsA, ascorbate; MDH, malate dehydrogenase; MDHA, monodehydroascorbate; LAPX, L-ascorbate peroxidase; G6PD, glucose-6-phosphate 1-dehydrogenase; GSR, glutathione reductase.

Second, detoxification of the hairy roots from Cd stress could be further enhanced by up-regulating the expression of the genes involved in other GSH metabolism aside from synthesis catalyzed by GSHB (Figure 7). The reaction of scavenging ROS can be catalyzed by the enzymes including GPX or GST with GSH as the substrate. Described in the Results section, the GST and GPX (103830386, 103862090) genes in the glutathione metabolic pathway were up-regulated and the enhanced GST synthesis also was verified by a Western Blot test (Figure 6 and Figure S4). Conversely, the up-regulated G6PD (103836988) gene in the Cd-treated hairy roots (Figure S4) benefitted the synthesis of G6PD, which catalyzed glucose 6-phosphate to produce a co-enzyme (i.e., NADPH) of GSR and help to regenerate GSH from GSSG [55]. Previous studies also observed that the expression of genes related to some key enzymes of GSH metabolism, including ROS scavenging (e.g., GPX and GST) and GSH regeneration from GSSG (e.g., GSR), increased in response to different heavy metal stress (e.g., Ni, Cd etc.) [55–57].

Third, the hairy roots of *Brassica campestris* L. may improve their tolerance to Cd by metal-binding and/or metal-chelating reactions by GSH and/or phytochelatins (PCs). As a major non-enzymatic antioxidant, GSH also may be involved in cell defense against toxicants via binding them or their metabolites with catalysis of GST [53,58]. Previous studies have shown that high levels of expression of most GST genes result in higher GST activity, which increases maize root tolerance to abiotic stress [59]. Data from the GO analysis, regarding the GST enzyme gene in this work, indicate that GST mainly plays a role in the toxin metabolic process and the response to metal ions (Figures S4 and S5). Alternately, PCs may play an important role in the detoxification of heavy metals in this process [60]. Data analyses indicate that the expression of the glutamate-cysteine ligase catalytic subunit (103861360), which participates in cysteine and methionine metabolism following GSH synthesis and synthesis of subsequent chelating agents (e.g., PCs) with GSH as a precursor, were up-regulated in plants under Cd stress (Figure 7). Previous studies also found that a large number of PCs were observed in rice plants growing under heavy metal stress [53,61]. The functional *Vicia sativa* Phytochelatin Synthase 1 (PCS1) homolog is overexpressed in *Arabidopsis thaliana* [62]. The expression levels of GSHB, GSR, PCs, and other chelating peptide synthesis-related enzymes in hyperaccumulators under heavy metal stress were significantly increased [63–67].

To summarize, all these factors could provide an integrated protection for hairy roots from Cd stress. Together with previous studies, results in this work present a better understanding of the regulation network of the hairy roots of *Brassica campestris* L. under Cd stress, hence providing important information for further genetic modification and subsequent regeneration of the whole plant with improved accumulation abilities.

4. Materials and Methods

4.1. Plant Material L and Cadmium Stress Treatment

The hairy roots of *Brassica campestris* L. used in this study were previously induced by infection of the seedlings with *Agrobacterium rhizogenes* in our laboratory [45]. The hairy roots were maintained in Murashige and Skoog medium (with vitamins and sugar) medium (Coolaber, Beijing, China) based on previous methods [45]. Briefly, 0.5 g of fresh hairy roots of *Brassica campestris* L. were placed in a 100 mL pyridoxine flask containing 50 mL Murashige and Skoog medium (with vitamins and sugar) liquid medium (NH_4NO_3 1650 mg/L, $CaCl_2$ 332.2 mg/L, $MgSO_4$ 180.7 mg/L, KNO_3 1900 mg/L, KH_2PO_4 170 mg/L, H_3BO_3 6.2 mg/L, $CoCl_2{\cdot}6H_2O$ 0.025 mg/L, $CuSO_4{\cdot}5H_2O$ 0.025 mg/L, FeNaEDTA 36.7 mg/L, $MnSO_4{\cdot}H_2O$ 16.9 mg/L, $Na_2MoO_4{\cdot}2H_2O$ 0.25 mg/L, KI 0.83 mg/L, $ZnSO_4{\cdot}7H_2O$ 8.6 mg/L, glycine 2 mg/L, moy-inositol 100 mg/L, nicotinic acid 0.5 mg/L, pyridoxine HCl 0.5 mg/L, thiamine HCl 0.1 mg/L) and incubated at 100 r/min and 28 °C for 5 d. Our previous results showed that the growth of hairy roots of *Brassica campestris* L. was significantly inhibited when grown in a medium with 200 μM $CdCl_2$ for 24 h [47]. Therefore, this treatment condition was chosen for further response mechanism study in this work. The Cd-treated hairy roots of *Brassica campestris* L. were collected and rinsed three times with deionized water. The samples were then dried and stored at −80 °C. Untreated and Cd-treated samples were labeled C and T, respectively. There were three replicates for each treatment.

4.2. Construction and Sequencing of mRNA Libraries

Total RNA was extracted using a TRIzol reagent and was treated with DNase to remove genomic DNA contamination. The eukaryotic mRNA was enriched with Oligo(dT) magnetic beads, the interrupter reagent was used to break the mRNA into short fragments in a Thermomixer (Eppendorf, Hamburg, Germany) to synthesize a strand of cDNA using the interrupted mRNA as a template and, then, the preparation was performed. A two-strand synthesis reaction system synthesized a double-strand cDNA and then used a kit for purification and recovery, cohesive end repair, the addition of base "A" to the 3′ end of the cDNA, and the attachment of a linker. Subsequently, the fragment size was chosen, followed by PCR amplification. A suitable library was inspected with an Agilent 2100 Bioanalyzer and an ABI StepOne Plus Real-Time PCR System. Then, we performed paired-end sequencing on an Illumina Hiseq4000 (BGI, Shenzhen, China) following the manufacturer's protocol. The six RNA libraries consisted of three control libraries and three Cd-treated libraries.

4.3. Sequence and Primary Analysis

We used the Illumina paired-end RNA-seq approach to sequence the hairy roots of *Brassica campestris* L. transcriptomes, each producing 6 Gb of multiple data, resulting in a total of 40.32 Gb of sequences. Prior to assembly, low-quality readings were removed, including reads containing sequencing adaptors, sequencing primers, and nucleotides with a quality score below 20. The original sequence data were submitted to the NCBI Database under registration number PRJNA543954.

4.4. RNA-Seq Reads Mapping

We compared the readings of the different samples to the *Brassica campestris* L. (Field mustard) ID: 229 cabbage canola reference genomic sequence using the HISAT2 software. The alignment process can be divided into the following three parts: (1) aligning of reads to a known transcriptome (optional); (2) aligning of the aligned pairs of reads on the reference genome; (3) unpaired read segments are aligned to the reference genome. Reads aligned to the specified reference genome were called mapped reads, and subsequent information analysis was performed based on mapped reads.

4.5. Transcript Abundance Estimation and Differential Expression Testing

The results of the HISAT2 [68] alignment to the reference genome were submitted to htseq-count (v0.6.0) [69] for processing, and the read count of each transcript was obtained. Single gene expression levels were calculated using readings per kilobase read (RPKM), which eliminated the effects of gene length and sequencing levels during the calculation of gene expression, making the samples comparable. Then, we used DEGseq for analysis, using the normalization method of quantiles, fold change ≥2, and p-value < 0.05 as the threshold for determining whether the gene was differentially expressed to obtain DEGs.

4.6. Gene Annotation, Classification, and Metabolic Pathway Analysis

To study the functional partitioning of DEGs in the hairy roots of *Brassica campestris* L. under Cd stress, we used GO and KEGG for further annotation, classification, and metabolic pathway analysis [70–72]. First, a gene ontology (GO) enrichment analysis of differentially expressed genes was performed by the clusterProfiler R package in which the gene length deviation was corrected. A GO term with a corrected p-value of less than 0.05 was considered to be significantly enriched by differentially expressed genes. The KEGG pathway was retrieved from the KEGG web server (http://www.genome.jp/kegg/). The clusterProfiler R package was used to test the statistical enrichment analysis of differentially expressed genes in the KEGG pathway.

4.7. Quantitative Real-Time PCR Validation

Eight DEGs were randomly selected for RT-PCR validation. The primer sequences and reference genes of these genes are listed in Table S4. Total RNA (0.2 μg) from each root sample was reverse transcribed using a PrimScript® RT Kit (Takara, Beijing, China) and random primers according to the manufacturer's instructions. Quantitative PCR reactions were performed in a 20 μL reaction volume using a Promega GoTaq® qPCR Master Mix Real Time PCR kit (Takara, Beijing, China) according to the manufacturer's instructions. The reaction was carried out on anSLAN-90P (Hongshi Medical

Technology Co., LTD., Shanghai, China). Each biological replica was technically replicated three times. Relative RNA expression of the selected genes, which is the expression of these genes relative to an internal reference gene, was used as an indicator of the genes' expression in different samples [73–75]. The relative expression levels of the selected genes were calculated using the 2-ΔΔCT method and the probable ubiquitin-conjugating enzyme E2 21 (BnUBC21) gene was used as a reference gene to correct gene expression [76]. Three replicates were performed for each sample.

4.8. Western Blot Analysis

The extraction of total hairy root protein was carried out using a Plant Protein Extraction Kit (CoWin Biosciences, Beijing, China). The extracted proteins were assayed by a Pierce™ BCA Protein Assay Kit (Thermo Fisher Scientific, Massachusetts MA, USA). Standard Western blots were performed. Immunoblotting was carried out by using the following primary antibodies: GSH-S (Agrisera AB, Vännäs, Swedish), GST class-phi (Agrisera AB, Vännäs, Swedish), and β-actin (CoWin Biosciences, Beijing, China). After incubation with the secondary antibodies, the signal was developed by chemiluminescence and autoradiography. Densitometric analysis was performed using ImageJ software (National Institutes of Health, Bethesda, NY, USA).

4.9. Statistical Analysis

All the experimental data were obtained with three or more repetitions. The data obtained from the experiment were analyzed using IBM's SPSS 20 software by one-way analysis of variance (ANOVA). Statistical analysis was calculated by Duncan's method ($p < 0.05$).

5. Conclusions

During this study, 2364 DEGs were discovered under Cd stress in the hairy roots of *Brassica campestris* L. based on RNA-Seq analysis. These genes were mainly involved in transcription-related processes, defense, stress responses, and transport processes in the response of *Brassica campestris* L. to Cd stress. Furthermore, data from Western blot tests indicated that the signaling pathway for glutathione synthesis and metabolism played an important role in the response to Cd stress. These results provide valuable information for enhancing our understanding of the heavy metal tolerance of *Brassica campestris* L. and the corresponding molecular mechanism. It is expected that the Cd-accumulating ability will be further improved by combining transgenic modification of the hairy roots, such as over-expression of the genes involved in Cd hyperaccumulation with subsequent plant regeneration, hence making the whole regenerated plant of *Brassica campestris* L. more promising for future application in phytoremediation.

Supplementary Materials: Supplementary materials can be found at http://www.mdpi.com/1422-0067/21/1/180/s1. Figure S1: Distribution of base composition on clean reads. X axis represents base position along reads. Y axis represents base content percentage; Figure S2: Distribution of base quality on clean reads. X axis represents base positions along reads. Y axis represents base quality value; Figure S3: Venn diagram analysis; Figure S4: Glutathione metabolism in *Brasssica campestris* L; Figure S5: Molecular Function with GST unigene; Table S1: Summary of Genome Mapping; Table S2: The results of GO terms enrichment in transgenic *Brasssica campestris* L. hairy roots; Table S3: The results of pathway enrichment in *Brasssica campestris* L. hairy roots; Table S4: Real Time PCR primer.

Author Contributions: Q.Y. planned and designed the research. X.C. reviewed and revised the article. Y.S. and Q.L. contributed to sample collection, performed experiments and data collection. Y.S. analyzed the data, interpreted results, and wrote the manuscript. Y.C., M.W. and Y.S. provided suggestions in drafting the manuscript and edited the manuscript. All authors have read and agreed to the published version of the manuscript.

Funding: This study was financially funded by Fundamental Research Funds for the Central Universities (grant number 2014JBM118; grant number 2017JBM073).

Acknowledgments: The authors thank the Fundamental Research Funds for the Central Universities (NO 2014JBM118, 2017JBM073).

Conflicts of Interest: The authors declare that the research was conducted in the absence of any commercial or financial relationships that could be construed as a potential conflict of interest.

Abbreviations

GO	Gene Ontology
KEGG	Kyoto Encyclopedia of Genes and Genomes
DEGs	Differentially expressed genes
GSHB	Glutathione synthetase
GSH	Glutathione
GST	Glutathione S-transferase
GSSG	Oxidized glutathione
GPX	Glutathione peroxidase
GCLC	Glutamate-cysteine ligase catalytic subunit
GSH-Cd	Glutathione-cadmium complex
PCs	Phytochelatins
AsA	Ascorbate
MDH	Malate dehydrogenase
MDHA	Monodehydroascorbate
LAPX	L-ascorbate peroxidase
G6PD	Glucose-6-phosphate 1-dehydrogenase
GSR	Glutathione reductase
ROS	Reactive oxygen species
Cd	Cadmium

References

1. Han, Y.L.; Yuan, H.Y.; Huang, S.Z.; Guo, Z.; Xia, B.; Gu, J. Cadmium tolerance and accumulation by two species of Iris. *Ecotoxicology* **2007**, *16*, 557–563. [CrossRef] [PubMed]
2. Yamaguchi, H.; Fukuoka, H.; Arao, T.; Ohyama, A.; Nunome, T.; Miyatake, K.; Negoro, S. Gene expression analysis in cadmium-stressed roots of a low cadmium-accumulating solanaceous plant, *Solanum torvum*. *J. Exp. Bot.* **2010**, *61*, 423–437. [CrossRef] [PubMed]
3. Zhang, M.; Liu, X.C.; Yuan, L.Y.; Wu, K.Q.; Duan, J.; Wang, X.L.; Yang, L.X. Transcriptional profiling in cadmium-treated rice seedling roots using suppressive subtractive hybridization. *Plant Physiol. Biochem.* **2012**, *50*, 79–86. [CrossRef] [PubMed]
4. Kim, S.; Arora, M.; Fernandez, C.; Landero, J.; Caruso, J.; Chen, A. Lead, mercury, and cadmium exposure and attention deficit hyperactivity disorder in children. *Environ. Res.* **2013**, *126*, 105–110. [CrossRef]
5. Ninkov, M.; Popov Aleksandrov, A.; Demenesku, J.; Mirkov, I.; Mileusnic, D.; Petrovic, A.; Grigorov, I.; Zolotarevski, L.; Tolinacki, M.; Kataranovski, D.; et al. Toxicity of oral cadmium intake: Impact on gut immunity. *Toxicol. Lett.* **2015**, *237*, 89–99. [CrossRef]
6. Wang, Y.; Xu, L.; Shen, H.; Wang, J.J.; Liu, W.; Zhu, X.W.; Wang, R.H.; Sun, X.C.; Liu, L.W. Metabolomic analysis with GC-MS to reveal potential metabolites and biological pathways involved in Pb & Cd stress response of radish roots. *Sci. Rep. UK* **2015**, *5*. [CrossRef]
7. Wang, Y.; Chen, L.; Gao, Y.; Zhang, Y.; Wang, C.; Zhou, Y.; Hu, Y.; Shi, R.; Tian, Y. Effects of prenatal exposure to cadmium on neurodevelopment of infants in Shandong, China. *Environ. Pollut.* **2016**, *211*, 67–73. [CrossRef]
8. Yue, R.Q.; Lu, C.X.; Qi, J.S.; Han, X.H.; Yan, S.F.; Guo, S.L.; Liu, L.; Fu, X.L.; Chen, N.N.; Yin, H.Y.; et al. Transcriptome Analysis of Cadmium-Treated Roots in Maize (*Zea mays* L.). *Front. Plant Sci.* **2016**, *7*. [CrossRef]
9. Buha, A.; Matovic, V.; Antonijevic, B.; Bulat, Z.; Curcic, M.; Renieri, E.A.; Tsatsakis, A.M.; Schweitzer, A.; Wallace, D. Overview of Cadmium Thyroid Disrupting Effects and Mechanisms. *Int. J. Mol. Sci.* **2018**, *19*, 1501. [CrossRef]
10. Gustin, K.; Tofail, F.; Vahter, M.; Kippler, M. Cadmium exposure and cognitive abilities and behavior at 10years of age: A prospective cohort study. *Environ. Int.* **2018**, *113*, 259–268. [CrossRef]
11. de Souza Reis, I.N.R.; Alves de Oliveira, J.; Ventrella, M.C.; Otoni, W.C.; Marinato, C.S.; Paiva de Matos, L. Involvement of glutathione metabolism in *Eichhornia crassipes* tolerance to arsenic. *Plant Biol.* **2019**. [CrossRef] [PubMed]

12. Xu, Z.G.; Dong, M.; Peng, X.Y.; Ku, W.Z.; Zhao, Y.L.; Yang, G.Y. New insight into the molecular basis of cadmium stress responses of wild paper mulberry plant by transcriptome analysis. *Ecotoxicol. Environ. Saf.* **2019**, *171*, 301–312. [CrossRef] [PubMed]
13. Moon, S.H.; Lee, C.-M.; Nam, M.J. Cytoprotective effects of taxifolin against cadmium-induced apoptosis in human keratinocytes. *Hum. Exp. Toxicol.* **2019**, *38*, 096032711984694. [CrossRef] [PubMed]
14. Xia, L.; Chen, S.; Dahms, H.-U.; Ying, X.; Peng, X.J.E. Cadmium induced oxidative damage and apoptosis in the hepatopancreas of Meretrix meretrix. *Ecotoxicology* **2016**, *25*, 959–969. [CrossRef] [PubMed]
15. Jablonska, E.; Socha, K.; Reszka, E.; Wieczorek, E.; Skokowski, J.; Kalinowski, L.; Fendler, W.; Seroczynska, B.; Wozniak, M.; Borawska, M.H.; et al. Cadmium, arsenic, selenium and iron-Implications for tumor progression in breast cancer. *Environ. Toxicol. Pharm.* **2017**, *53*, 151–157. [CrossRef] [PubMed]
16. Fujiki, K.; Inamura, H.; Sugaya, T.; Matsuoka, M. Blockade of ALK4/5 signaling suppresses cadmium- and erastin-induced cell death in renal proximal tubular epithelial cells via distinct signaling mechanisms. *Cell Death Differ.* **2019**. [CrossRef]
17. Wallace, D.R.; Spandidos, D.A.; Tsatsakis, A.; Schweitzer, A.; Djordjevic, V.; Djordjevic, A.B. Potential interaction of cadmium chloride with pancreatic mitochondria: Implications for pancreatic cancer. *Int. J. Mol. Med.* **2019**, *44*, 145–156. [CrossRef]
18. Feng, N.X.; Yu, J.; Zhao, H.M.; Cheng, Y.T.; Mo, C.H.; Cai, Q.Y.; Li, Y.W.; Li, H.; Wong, M.H. Efficient phytoremediation of organic contaminants in soils using plant-endophyte partnerships. *Sci. Total Environ.* **2017**, *583*, 352–368. [CrossRef]
19. Sarwar, N.; Imran, M.; Shaheen, M.R.; Ishaque, W.; Kamran, M.A.; Matloob, A.; Rehim, A.; Hussain, S. Phytoremediation strategies for soils contaminated with heavy metals: Modifications and future perspectives. *Chemosphere* **2017**, *171*, 710–721. [CrossRef]
20. Hussain, I.; Aleti, G.; Naidu, R.; Puschenreiter, M.; Mahmood, Q.; Rahman, M.M.; Wang, F.; Shaheen, S.; Syed, J.H.; Reichenauer, T.G. Microbe and plant assisted-remediation of organic xenobiotics and its enhancement by genetically modified organisms and recombinant technology: A review. *Sci. Total Environ.* **2018**, *628–629*, 1582–1599. [CrossRef]
21. Rofkar, J.R.; Dwyer, D.F. Effects of light regime, temperature, and plant age on uptake of arsenic by *Spartina pectinata* and *Carex stricta*. *Int. J. Phytoremediation* **2011**, *13*, 528–537. [CrossRef] [PubMed]
22. Braga, R.R.; dos Santos, J.B.; Zanuncio, J.C.; Bibiano, C.S.; Ferreira, E.A.; Oliveira, M.C.; Silva, D.V.; Serrão, J.E. Effect of growing *Brachiria brizantha* on phytoremediation of picloram under different pH environments. *Ecol. Eng.* **2016**, *94*, 102–106. [CrossRef]
23. Al-Shalabi, Z.; Doran, P.M. Metal uptake and nanoparticle synthesis in hairy root cultures. *Adv. Biochem. Eng. Biotechnol.* **2013**, *134*, 135–153. [CrossRef] [PubMed]
24. Gao, J.; Zhang, Y.Z.; Lu, C.L.; Peng, H.; Luo, M.; Li, G.K.; Shen, Y.O.; Ding, H.P.; Zhang, Z.M.; Pan, G.T.; et al. The development dynamics of the maize root transcriptome responsive to heavy metal Pb pollution. *Biochem. Biophys. Res. Commun.* **2015**, *458*, 287–293. [CrossRef]
25. Ibanez, S.; Talano, M.; Ontanon, O.; Suman, J.; Medina, M.I.; Macek, T.; Agostini, E. Transgenic plants and hairy roots: Exploiting the potential of plant species to remediate contaminants. *New Biotechnol.* **2016**, *33*, 625–635. [CrossRef]
26. Rengasamy, B. Cadmium tolerance and antioxidative defenses in hairy roots of the cadmium hyperaccumulator, *Thlaspi caerulescens*. *Biotechnol. Bioeng.* **2003**, *83*, 158–167.
27. Shi, H.P.; Zhu, Y.F.; Wang, Y.L.; Tsang, P.K. Effect of cadmium on cytogenetic toxicity in hairy roots of *Wedelia trilobata* L. and their alleviation by exogenous CaCl2. *Environ. Sci. Pollut. Res. Int.* **2014**, *21*, 1436–1443. [CrossRef]
28. Zhao, J.; Xia, B.; Meng, Y.; Yang, Z.; Pan, L.; Zhou, M.; Zhang, X. Transcriptome Analysis to Shed Light on the Molecular Mechanisms of Early Responses to Cadmium in Roots and Leaves of King Grass (*Pennisetum americanum* x *P. purpureum*). *Int. J. Mol. Sci.* **2019**, *20*. [CrossRef]
29. Alscher, R.G. Biosynthesis and antioxidant function of glutathione in plants. *Physiol. Plant* **1989**, *77*, 457–464. [CrossRef]
30. Chronopoulou, E.; Georgakis, N.; Nianiou-Obeidat, I.; Madesis, P.; Perperopoulou, F.; Pouliou, F.; Vasilopoulou, E.; Ioannou, E.; Ataya, F.S.; Labrou, N.E. Plant Glutathione Transferases in Abiotic Stress Response and Herbicide Resistance. In *Glutathione in Plant Growth, Development, and Stress Tolerance;*

Hossain, M.A., Mostofa, M.G., Diaz-Vivancos, P., Burritt, D.J., Fujita, M., Tran, L.-S.P., Eds.; Springer International Publishing: Cham, Germany, 2017; pp. 215–233. [CrossRef]
31. Nianiou-Obeidat, I.; Madesis, P.; Kissoudis, C.; Voulgari, G.; Chronopoulou, E.; Tsaftaris, A.; Labrou, N.E. Plant glutathione transferase-mediated stress tolerance: Functions and biotechnological applications. *Plant Cell Rep.* **2017**, *36*, 791–805. [CrossRef]
32. Zhang, D.J.; Li, C.X. Genotypic differences and glutathione metabolism response in wheat exposed to copper. *Environ. Exp. Bot* **2019**, *157*, 250–259. [CrossRef]
33. Wu, H.; Chen, C.; Du, J.; Liu, H.; Cui, Y.; Zhang, Y.; He, Y.; Wang, Y.; Chu, C.; Feng, Z.; et al. Co-Overexpression FIT with AtbHLH38 or AtbHLH39 in Arabidopsis-Enhanced Cadmium Tolerance via Increased Cadmium Sequestration in Roots and Improved Iron Homeostasis of Shoots. *Plant Physiol.* **2012**, *158*, 790. [CrossRef] [PubMed]
34. Lin, C.Y.; Trinh, N.N.; Fu, S.F.; Hsiung, Y.C.; Chia, L.C.; Lin, C.W.; Huang, H.J. Comparison of early transcriptome responses to copper and cadmium in rice roots. *Plant Mol. Biol.* **2013**, *81*, 507–522. [CrossRef] [PubMed]
35. Ma, J.; Cai, H.M.; He, C.W.; Zhang, W.J.; Wang, L.J. A hemicellulose-bound form of silicon inhibits cadmium ion uptake in rice (*Oryza sativa*) cells. *New Phytol.* **2015**, *206*, 1063–1074. [CrossRef]
36. Shi, X.; Sun, H.J.; Chen, Y.T.; Pan, H.W.; Wang, S.F. Transcriptome Sequencing and Expression Analysis of Cadmium (Cd) Transport and Detoxification Related Genes in Cd-Accumulating Salix integra. *Front. Plant Sci.* **2016**, *7*. [CrossRef]
37. Theriault, G.; Michael, P.; Nkongolo, K. Comprehensive Transcriptome Analysis of Response to Nickel Stress in White Birch (*Betula papyrifera*). *PLoS ONE* **2016**, *11*, e0153762. [CrossRef]
38. Xu, J.; Sun, J.H.; Du, L.G.; Liu, X.J. Comparative transcriptome analysis of cadmium responses in *Solanum nigrum* and *Solanum torvum*. *New Phytol.* **2012**, *196*, 110–124. [CrossRef]
39. Milner, M.J.; Mitani-Ueno, N.; Yamaji, N.; Yokosho, K.; Craft, E.; Fei, Z.J.; Ebbs, S.; Zambrano, M.C.; Ma, J.F.; Kochian, L.V. Root and shoot transcriptome analysis of two ecotypes of *Noccaea caerulescens* uncovers the role of NcNramp1 in Cd hyperaccumulation. *Plant J.* **2014**, *78*, 398–410. [CrossRef]
40. Shen, G.; Niu, J.; Deng, Z. Abscisic acid treatment alleviates cadmium toxicity in purple flowering stalk (*Brassica campestris* L. ssp. chinensis var. purpurea Hort.) seedlings. *Plant Physiol. Biochem.* **2017**, *118*, 471–478. [CrossRef]
41. Li, J.T.; Gurajala, H.K.; Wu, L.H.; van der Ent, A.; Qiu, R.L.; Baker, A.J.M.; Tang, Y.T.; Yang, X.E.; Shu, W.S. Hyperaccumulator Plants from China: A Synthesis of the Current State of Knowledge. *Environ. Sci. Technol.* **2018**, *52*, 11980–11994. [CrossRef]
42. Zhang, X.D.; Meng, J.G.; Zhao, K.X.; Chen, X.; Yang, Z.M. Annotation and characterization of Cd-responsive metal transporter genes in rapeseed (*Brassica napus*). *Biometals* **2018**, *31*, 107–121. [CrossRef] [PubMed]
43. Zhang, X.D.; Sun, J.Y.; You, Y.Y.; Song, J.B.; Yang, Z.M. Identification of Cd-responsive RNA helicase genes and expression of a putative BnRH 24 mediated by miR158 in canola (*Brassica napus*). *Ecotoxicol. Environ. Saf.* **2018**, *157*, 159–168. [CrossRef] [PubMed]
44. Zhang, X.D.; Zhao, K.X.; Yang, Z.M. Identification of genomic ATP binding cassette (ABC) transporter genes and Cd-responsive ABCs in *Brassica napus*. *Gene* **2018**, *664*, 139–151. [CrossRef] [PubMed]
45. Liu, J.Y. *Research on Agrobacterium Rhizogenes Mediated IRI l Gene Transformation in Cd Hypemccumulator Brassica campestris L*; Beijing Jiaotong University: Beijing, China, 2014.
46. Liu, H.; Zhao, H.; Wu, L.; Liu, A.; Zhao, F.J.; Xu, W. Heavy metal ATPase 3 (HMA3) confers cadmium hypertolerance on the cadmium/zinc hyperaccumulator *Sedum plumbizincicola*. *New Phytol.* **2017**, *215*, 687–698. [CrossRef] [PubMed]
47. Lu, Y.; Cao, Y.; Chen, Y.; Yan, Q. The physiological response and iron and potassium contents in the hairy roots of *Brassica rape* L. under cadmium stress. *Chin. J. Appl. Environ. Biol.* **2018**, *24*, 8. [CrossRef]
48. Šimonová, E.; Henselová, M.; Masarovičová, E.; Kohanová, J. Comparison of tolerance of *Brassica juncea* and *Vigna radiata* to cadmium. *Biol. Plant.* **2007**, *51*, 488–492. [CrossRef]
49. Van Engelen, D.L.; Sharpe-Pedler, R.C.; Moorhead, K.K. Effect of chelating agents and solubility of cadmium complexes on uptake from soil by *Brassica juncea*. *Chemosphere* **2007**, *68*, 401–408. [CrossRef]
50. Li, H.; Han, X.; Qiu, W.; Xu, D.; Wang, Y.; Yu, M.; Hu, X.; Zhuo, R. Identification and expression analysis of the GDSL esterase/lipase family genes, and the characterization of SaGLIP8 in *Sedum alfredii* Hance under cadmium stress. *PeerJ* **2019**, *7*. [CrossRef]

51. Zhang, J.; Zhu, Q.; Yu, H.; Li, L.; Zhang, G.; Chen, X.; Jiang, M.; Tan, M. Comprehensive Analysis of the Cadmium Tolerance of Abscisic Acid-, Stress- and Ripening-Induced Proteins (ASRs) in Maize. *Int. J. Mol. Sci.* **2019**, *20*, 133. [CrossRef]
52. Wang, C.; Zhang, S.H.; Wang, P.F.; Qian, J.; Hou, J.; Zhang, W.J.; Lu, J. Excess Zn alters the nutrient uptake and induces the antioxidative responses in submerged plant *Hydrilla verticillata* (L.f.) Royle. *Chemosphere* **2009**, *76*, 938–945. [CrossRef]
53. Geng, A.; Wang, X.; Wu, L.; Wang, F.; Wu, Z.; Yang, H.; Chen, Y.; Wen, D.; Liu, X. Silicon improves growth and alleviates oxidative stress in rice seedlings (*Oryza sativa* L.) by strengthening antioxidant defense and enhancing protein metabolism under arsanilic acid exposure. *Ecotoxicol. Environ. Saf.* **2018**, *158*, 266–273. [CrossRef] [PubMed]
54. Fu, Y.; Yang, Y.; Chen, S.; Ning, N.; Hu, H. Arabidopsis IAR4 Modulates Primary Root Growth Under Salt Stress Through ROS-Mediated Modulation of Auxin Distribution. *Front. Plant Sci.* **2019**, *10*, 522. [CrossRef] [PubMed]
55. Borges, K.L.R.; Salvato, F.; Loziuk, P.L.; Muddiman, D.C.; Azevedo, R.A. Quantitative proteomic analysis of tomato genotypes with differential cadmium tolerance. *Environ. Sci. Pollut. Res. Int.* **2019**. [CrossRef] [PubMed]
56. Dixit, P.; Mukherjee, P.K.; Ramachandran, V.; Eapen, S. Glutathione transferase from Trichoderma virens enhances cadmium tolerance without enhancing its accumulation in transgenic Nicotiana tabacum. *PLoS ONE* **2011**, *6*, e16360. [CrossRef] [PubMed]
57. Zhou, Q.; Guo, J.J.; He, C.T.; Shen, C.; Huang, Y.Y.; Chen, J.X.; Guo, J.H.; Yuan, J.G.; Yang, Z.Y. Comparative Transcriptome Analysis between Low- and High-Cadmium-Accumulating Genotypes of Pakchoi (*Brassica chinensis* L.) in Response to Cadmium Stress. *Environ. Sci. Technol.* **2016**, *50*, 6485–6494. [CrossRef] [PubMed]
58. Lyubenova, L.; Nehnevajova, E.; Herzig, R.; Schroder, P. Response of antioxidant enzymes in *Nicotiana tabacum* clones during phytoextraction of heavy metals. *Environ. Sci. Pollut. Res. Int.* **2009**, *16*, 573–581. [CrossRef]
59. Li, D.; Xu, L.; Pang, S.; Liu, Z.; Wang, K.; Wang, C. Variable Levels of Glutathione S-Transferases Are Responsible for the Differential Tolerance to Metolachlor between Maize (*Zea mays*) Shoots and Roots. *J. Agric. Food Chem.* **2017**, *65*, 39–44. [CrossRef]
60. Jacquart, A.; Brayner, R.; El Hage Chahine, J.M.; Ha-Duong, N.T. Cd(2+) and Pb(2+) complexation by glutathione and the phytochelatins. *Chem. Biol. Interact.* **2017**, *267*, 2–10. [CrossRef]
61. Gupta, C.K.; Singh, B. Uninhibited biosynthesis and release of phytosiderophores in the presence of heavy metal (HM) favors HM remediation. *Environ. Sci. Pollut. Res. Int.* **2017**, *24*, 9407–9416. [CrossRef]
62. Zhang, X.; Rui, H.; Zhang, F.; Hu, Z.; Xia, Y.; Shen, Z. Overexpression of a Functional Vicia sativa PCS1 Homolog Increases Cadmium Tolerance and Phytochelatins Synthesis in Arabidopsis. *Front. Plant Sci.* **2018**, *9*, 107. [CrossRef]
63. Foyer, C.H.; Souriau, N.; Perret, S.; Lelandais, M.; Kunert, K.J.; Pruvost, C.; Jouanin, L. Overexpression of glutathione reductase but not glutathione synthetase leads to increases in antioxidant capacity and resistance to photoinhibition in poplar trees. *Plant Physiol.* **1995**, *109*, 1047–1057. [CrossRef] [PubMed]
64. Guo, J.; Dai, X.; Xu, W.; Ma, M. Overexpressing GSH1 and AsPCS1 simultaneously increases the tolerance and accumulation of cadmium and arsenic in *Arabidopsis thaliana*. *Chemosphere* **2008**, *72*, 1020–1026. [CrossRef] [PubMed]
65. Gill, S.S.; Tuteja, N. Reactive oxygen species and antioxidant machinery in abiotic stress tolerance in crop plants. *Plant Physiol. Biochem.* **2010**, *48*, 909–930. [CrossRef] [PubMed]
66. Cheng, M.C.; Ko, K.; Chang, W.L.; Kuo, W.C.; Chen, G.H.; Lin, T.P. Increased glutathione contributes to stress tolerance and global translational changes in Arabidopsis. *Plant J.* **2015**, *83*, 926–939. [CrossRef]
67. Kuluev, B.R.; Berezhneva, Z.A.; Mikhaylova, E.V.; Postrigan, B.N.; Knyazev, A.V. Productivity and Stress-Tolerance of Transgenic Tobacco Plants with a Constitutive Expression of the Rapeseed Glutathione Synthetase Gene BnGSH. *Russ. J. Genet. Appl. Res.* **2018**, *8*, 190–196. [CrossRef]
68. Kim, D.; Langmead, B.; Salzberg, S.L. HISAT: A fast spliced aligner with low memory requirements. *Nat. Methods* **2015**, *12*, 357. [CrossRef]
69. Robinson, M.D.; McCarthy, D.J.; Smyth, G.K. edgeR: A Bioconductor package for differential expression analysis of digital gene expression data. *Bioinformatics* **2010**, *26*, 139–140. [CrossRef]

70. Foissac, S.; Sammeth, M. ASTALAVISTA: Dynamic and flexible analysis of alternative splicing events in custom gene datasets. *Nucleic Acids Res.* **2007**, *35*, W297–W299. [CrossRef]
71. Sammeth, M.; Foissac, S.; Guigo, R. A general definition and nomenclature for alternative splicing events. *PLoS Comput. Biol.* **2008**, *4*, e1000147. [CrossRef]
72. Sammeth, M. Complete Alternative Splicing Events Are Bubbles in Splicing Graphs. *J. Comput. Biol.* **2009**, *16*, 1117–1140. [CrossRef]
73. Pfaffl, M.W. A new mathematical model for relative quantification in real-time RT-PCR. *Nucleic Acids Res.* **2001**, *29*, e45. [CrossRef] [PubMed]
74. Pfaffl, M.W.; Horgan, G.W.; Dempfle, L. Relative expression software tool (REST) for group-wise comparison and statistical analysis of relative expression results in real-time PCR. *Nucleic Acids Res.* **2002**, *30*, e36. [CrossRef] [PubMed]
75. Qiu, H.; Durand, K.; Rabinovitch-Chable, H.; Rigaud, M.; Gazaille, V.; Clavere, P.; Sturtz, F.G. Gene expression of HIF-1alpha and XRCC4 measured in human samples by real-time RT-PCR using the sigmoidal curve-fitting method. *Biotechniques* **2007**, *42*, 355–362. [CrossRef] [PubMed]
76. Chen, X.; Truksa, M.; Shah, S.; Weselake, R.J. A survey of quantitative real-time polymerase chain reaction internal reference genes for expression studies in *Brassica napus*. *Anal. Biochem.* **2010**, *405*, 138–140. [CrossRef] [PubMed]

© 2019 by the authors. Licensee MDPI, Basel, Switzerland. This article is an open access article distributed under the terms and conditions of the Creative Commons Attribution (CC BY) license (http://creativecommons.org/licenses/by/4.0/).

Article

Isolation and Characterization of Copper- and Zinc-Binding Metallothioneins from the Marine Alga *Ulva compressa* (Chlorophyta)

Antonio Zúñiga [1,2], Daniel Laporte [1,*], Alberto González [1], Melissa Gómez [1], Claudio A. Sáez [2,3] and Alejandra Moenne [1,*]

1 Laboratory of Marine Biotechnology, Faculty of Chemistry and Biology, University of Santiago of Chile, Alameda 3363, Santiago 9170022, Chile; antonio.zt@gmail.com (A.Z.); alberto.gonzalezfi@usach.cl (A.G.); melissa.gomez@usach.cl (M.G.)

2 HUB AMBIENTAL UPLA, Vicerrectoría de Investigación, Postgrado e Innovación, University of Playa Ancha, Avenida Carvallo 270, Valparaíso 2340000, Chile; claudio.saez@upla.cl

3 Laboratory of Aquatic Environmental Research, Center of Advances Studies, University of Playa Ancha, Traslaviña 450, Viña del Mar 2520000, Chile

* Correspondence: daniel.laporte@usach.cl (D.L.); alejandra.moenne@usach.cl (A.M.)

Received: 23 September 2019; Accepted: 22 December 2019; Published: 25 December 2019

Abstract: In this work, transcripts encoding three metallothioneins from *Ulva compressa* (UcMTs) were amplified: The 5′and 3′ UTRs by RACE-PCR, and the open reading frames (ORFs) by PCR. Transcripts encoding UcMT1.1 (*Crassostrea*-like), UcMT2 (*Mytilus*-like), and UcMT3 (*Dreissena*-like) showed a 5′UTR of 61, 71, and 65 nucleotides and a 3′UTR of 418, 235, and 193 nucleotides, respectively. UcMT1.1 ORF encodes a protein of 81 amino acids (MW 8.2 KDa) with 25 cysteines (29.4%), arranged as three motifs CC and nine motifs CXC; UcMT2 ORF encode a protein of 90 amino acids (9.05 kDa) with 27 cysteines (30%), arranged as three motifs CC, nine motifs CXC, and one motif CXXC; UcMT3 encode a protein of 139 amino acids (13.4 kDa) with 34 cysteines (24%), arranged as seven motifs CC and seven motifs CXC. UcMT1 and UcMT2 were more similar among each other, showing 60% similarity in amino acids; UcMT3 showed only 31% similarity with UcMT1 and UcMT2. In addition, UcMTs displayed structural similarity with MTs of marine invertebrates MTs and the terrestrial invertebrate *Caenorhabtidis elegans* MTs, but not with MTs from red or brown macroalgae. The ORFs fused with GST were expressed in bacteria allowing copper accumulation, mainly in MT1 and MT2, and zinc, in the case of the three MTs. Thus, the three MTs allowed copper and zinc accumulation in vivo. UcMTs may play a role in copper and zinc accumulation in *U. compressa*.

Keywords: copper and zinc; expression in bacteria; metal accumulation; metallothioneins; marine alga; *Ulva compressa*

1. Introduction

Metallothioneins (MTs) are low molecular weight proteins, of around 10 kDa, that are rich in cysteine residues allowing the binding of divalent or monovalent metal ions such as Zn^{2+}, Cd^{2+}, Pb^{2+}, Hg^{2+}, Cu^{1+}, Ag^{1+}, among others [1–3]. MTs participate in metal accumulation and detoxification in vertebrates, invertebrates, plants, algae, and bacteria [1–3]. Cysteine residues in MTs are usually arranged as CC, CXC, and/or CXXC motifs and they correspond to around 30% of amino acids. In vertebrates, cysteine residues in MTs are contained in two domains, α and β, that are separated by a linker of variable sizes. Vertebrate MTs are rich in glycine and alanine amino acids, ranging from 10% to 20% residues [2] and invertebrate and plant MTs can contain histidine and aromatic residues [4,5]. MTs were first discovered in horse kidney, although they have been isolated from kidney and liver of other mammals [6]. In mammals, such as humans and mouse, there are four MT isoforms and the linker is constituted by three amino acids [2]. In fish, such as in rainbow trout, there are two MTs and the linker is formed by four amino acids [7]. In invertebrates, there are mainly two MTs, and the cysteine-rich domains are separated by a linker of four to seven amino acids and they have been described in organisms such as the nematode *Caenorhabtidis elegans* [8], the gastropod *Arianta arbustorum* [9], the snail *Helix pomatia* [10,11], the mollusks *Crassostrea virginica* [12], *Mytilus edulis* [13], and *Dreissena polymorpha* [14] and in the equinodern *Strongylocentrotus purpuratus* [15]. The yeast *Saccharomyces cerevisiae* displays two MTs, CUP-1 and CRS5, and the linker is constituted by three amino acids [16,17]. The cyanobacteria *Synechococcus* sp. showed a single MT, SmtA, and the linker that separates the two cysteine-rich domains is constituted by 15 amino acids [2]. The first MT isolated in plants was wheat germ MT, the first cloned MT in plants was *Mimulus guttatus MT*, and the first cloned MT from marine algae was *Fucus vesiculosus* MT [18–20]. Plant MTs can be classified in four types corresponding to type 1, 2, 3, and 4 [3]. The types 1, 2, and 3 MTs have cysteine-rich domains separated by a linker of around 40 amino acids and a linker of about 15 amino acids, and they differ in the arrangement of cysteines in cysteine-rich domains [3]. Type 4 MTs have mainly a linker of around 15 amino acids [3]. In *Arabidopsis thaliana,* there are seven MTs corresponding to MT1a, MT1c, MT2a, MT2b, MT3, MT4a, and MT4b and in *Populus trichcarpa x deltoides* there are six MTs corresponding to MT1a, MT1b, MT1a, MT1b, MT3a, and MT3b [21,22]. In addition, the study of algal genomes has shown that the brown macroalgae (Phaeophyceae) *Fucus serratus*, *Ectocarpus siliculosus*, and *Sargassum binderi*, and the red macroalgae (Rodophyceae) *Chondrus crispus* and *Euchema denticulatum* encode a single MT [3]. Red and brown algae MTs are related among each other considering the arrangement of cysteines [3]. Until now, no MTs have been cloned or characterized in green macroalgae (Chlorophyceae).

The green macroalga *Ulva compressa* is the dominant species in copper-polluted coastal areas of northern Chile and in other parts of the world [23,24]. It has been shown that the alga collected in the field accumulate copper in its tissue [23]. Furthermore, the alga cultivated in vitro with 2.5, 5, 7.5, and 10 µM for 0 to 12 days showed a linear accumulation of intracellular copper with increasing concentrations of the metal reaching a maximal accumulation of 620 µg g^{-1} of dry weight (DW), at day 12 with 10 µM copper [25]. In addition, *U. compressa* extrudes copper ions to the extracellular medium reaching a maximal concentration at intracellular level of around 900 µg g^{-1} of DW [24]. In contrast, the green alga *Ulva fasciata* cultivated with 0.3 µM copper for 14 days reached an intracellular level of copper of 2000 µg g^{-1} of DW suggesting that this alga may not extrude copper ions to the culture medium [26]. On the other hand, the level of MT transcripts in *U. compressa* cultivated with 7.5 and 10 µM of copper increased from days 3 to 12 [24]. Thus, it is possible that accumulation of intracellular copper is mediated by MTs in *U. compressa.* In this sense, it has been shown that *Arabidopsis thaliana* plants deficient in MT1a accumulate 30% less copper in the shoots than control plants [26] and *A. thaliana* mutants deleted in four MTs accumulate 45% less copper in the shoots [27]. In addition, rat fibroblasts having a deletion of MTI and MTII accumulate less copper than control cells and is subjected to an increase in oxidative stress [28]. Thus, it is possible that *U. compressa* can accumulate copper mediated by MTs. A transcriptomic analysis performed with the alga cultivated with 10 µM copper for 0 and 24 h allowed the identification of 7 potential MTs in *U. compressa* and their levels increased from days 3 to 12 of metal exposure [29].

In this work, we cloned three putative *U. compressa* MTs previously described as *Crassostrea*-like, *Mytilus*-like, and *Dreissena*-like MTs [26], which were renamed UcMT1, UcMT2, and UcMT3, respectively. In this work, the ORF of each UcMT was cloned and expressed in the bacteria *E. coli* as a fusion protein with a glutahione-S-transferase (GST) allowing accumulation of copper and zinc in vivo. Thus, the marine alga *U. compressa* may accumulate intracellular copper and zinc through UcMTs.

2. Results

2.1. Sequences of Transcripts Encoding UcMTs

Total RNA and mRNAs were isolated from *U. compressa* cultivated with 10 μM copper for three days. The 5′and 3′untranslated regions (UTR) of transcripts encoding three UcMTs were amplified using RACE-PCR technique, as well as the open reading frame (ORF) using conventional PCR. UcMT1.1 transcript (formerly *Crassostrea*-like *mt*) and protein are described in Table 1 (Figure 1). It is important to mention that two other transcripts that were closely related with UcMT1.1 were isolated, corresponding to UcMT1.2 and UcMT1.3; UcMT1.2 showed a deletion of 48 nucleotides after G in position 471 of the 3′UTR region of UcMT1.1, and UcMT1.3 displayed the same deletion mentioned before and a deletion of 58 nucleotides after C in position 607 in the 3′UTR of UcMT1.1 (Figure 1). UcMT2 transcript (formerly *Mytilus*-like *mt*) and protein is described in Table 1 (Figure 2) as well as UcMT3 transcript (formerly *Dreissena*-like *mt*) and protein (Figure 3). The linker region of UcMT1.1, UcMT2, and UcMT2 correspond to, 9, 9, and 23 amino acids, respectively (Figures 1–3).

Figure 1. Complete cDNA and amino acid sequences of metallothionein UcMT1.1 (*Crassostrea*-like) from *U. compressa*. Initiation and termination codons are highlighted in bold; * indicate stop codon, cysteines arranged as CXC and CC motifs are underlined in the amino acid sequence of UcMT1. UcMT1.2 presents a deletion of 48 nucleotides located after the G in position 471 in the 3′UTR of MT1.1, and the deletion is underlined. UcMT3 present the deletion previously mentioned, and a deletion located after the C in position 607 in the 3′UTR of UcMT1.1, and the deletion is underlined.

Table 1. Characteristics of *U. compressa* metallothioneins (UcMTs) transcripts and proteins. For transcripts: Nucleotides (nt); for proteins amino acids (aa) and cysteines (cys).

	Transcript (nt)	3′ UTR (nt)	ORF (nt)	5′ UTR (nt)	Protein (aa)	MW (kDa)	Total Cys	% Cys	Cys Motifs
UcMT1.1	726	61	246	418	81	8.2	25	29.4	3CC, 9CXC
UcMT2	580	71	273	236	90	9.05	27	30	3CC, 9CXC, 1CXXC
UcMT3	679	65	420	194	139	13.4	34	24	7CC, 7CXC

```
-71   AACCGCTGTCAACAACCGCTGTCAACACACCAACAACTCGCCAGTTCTCATATAAA  -16
-15   GCATCATCTGAAGACATGGACTGCCGTTGCGACAAGGCATGCTGCCAATCCGCGGGG  42
                     M  D  C  R  C  D  K  A  C  C  Q  S  A  G

43    GAGTGCACCTGCGAGGCTGGATGCCCTTGCCATCAGGAAGGCGGTTGCTGCACTCCT  99
      E  C  T  C  E  A  G  C  P  C  H  Q  E  G  G  C  C  T  P

100   GAGACCTGCACTTGCTCAGTGTCCAAGAAGTGTCTCGACACCTGCGCCTGCAAGACC  156
      E  T  C  T  C  S  V  S  K  K  C  L  D  T  C  A  C  K  T

157   ACCGATGAGGGCTGTCCTTGCTCACCCGCGAAGGATTGCCCGTGCGGTACCGCCGAG  213
      T  D  E  G  C  P  C  S  P  A  K  D  C  P  C  G  T  A  E

214   TGCGGCTGCGCCTGCTGCAATGATGGCTGCGACTGTACAGCCTGTCCGGGTTGTGTG  270
      C  G  C  A  C  C  N  D  G  C  D  C  T  A  C  P  G  C  V

271   TAGCTCGTTCGTAGAGCTGTCTGAAGATCGTCGGCTATGTCGTTACGCAGACGCCAG  327
      *

328   CATGGTCGACTAAGGACGCTGCTTGTTAGGCCTCAGTGGTTGACAGCAACTTCTGCT  384
385   TTCCAGAGATGTAATATGTCTTTTGTTCAATACCATGGTATTGACTATACAGTGGAG  441
442   TAGGCATGCTGGCGGGCACGGGCTGTCTGCGATGATTCGTCGAGATGGTTTGTAACT  498
499   CGGGTGAAGCC                                                509
```

Figure 2. Complete cDNA and amino acid sequences of metallothionein UcMT2 (*Mytilus*-like) from the marine alga *U. compressa*. Initiation and termination codons are highlighted in bold; * indicate stop codon, cysteines arranged as CXC, CC, and CXXC motifs are underlined in the amino acid sequence of UcMT2.

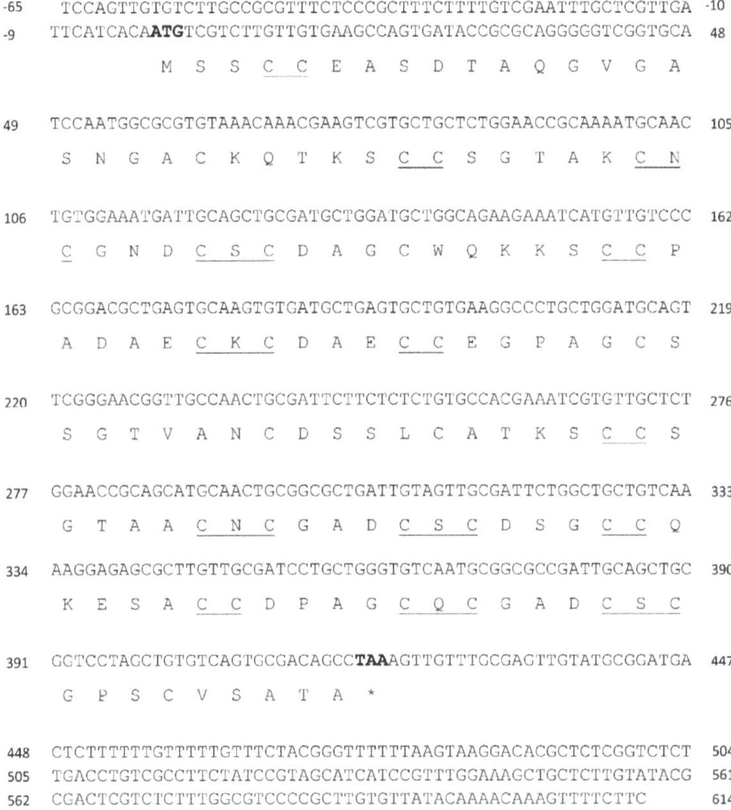

Figure 3. Complete cDNA and amino acid sequences of metallothionein UcMT3 (*Dreissena*-like) from the marine alga *U. compressa*. Initiation and termination codons are highlighted in bold; * indicate stop codon, cysteines arranged as CXC and CC motifs are underlined in the amino acid sequence of UcMT3.

2.2. Similarities in Amino Acids of UcMTs and Hierarchical Clustering of MTs

UcMT1.1 and UcMT2 were more closely related among each other and shared 52.2% identity and 60% similarity in amino acids; indeed, both sequences contained arrangement of cysteines corresponding to 3 CC and 9 CXC motifs, but UcMT2 also contained a CXXC motif (Figure 4A). In contrast, UcMT3 shared only 30% similarity in amino acids with MT1.1 and MT2 and contained arrangements of cysteines corresponding to 7 CC and 7 CXC (Figure 4A). UcMT3 showed an extra N-terminal sequence of 10 amino acids, an internal additional sequence of 19 amino acids, and an extra C-terminal sequence of 8 amino acids, compared with UcMT1.1 and UcMT2 (Figure 4A). Thus, UcMT1 and UcMT2 may have derived from UcMT3 by deletions of initial, internal, and terminal nucleotide sequences.

The hierarchical clustering of vertebrate, invertebrate, and plants MTs constructed with 237 protein sequences (including the 3 UcMTs) demonstrated that UcMT1.1 and UcMT2 grouped mainly with marine crustacean MTs, such as those of the lobster *Homarus americanus* and the crabs *Carcinus maena* and *Scylla serrata* (Figure S1). In addition, UcMT3 clustered with the nematode *C. elegans* MTs, as well as with MTs of marine equinoderms MTs such as those of *Sterechinus neumayeri*, *Strongylocentrus purpuratus*, and *Paracentrotus livudus* (Figure S1). On the other hand, UcMTs grouped as a different clade with MTs from Rodophyceae and Phaeophyceae (Figure 4B). Thus, *U. compressa* MTs are more closely related with marine invertebrate MTs and the terrestrial invertebrate *C. elegans* MTs, and not to MTs from other marine macroalgae.

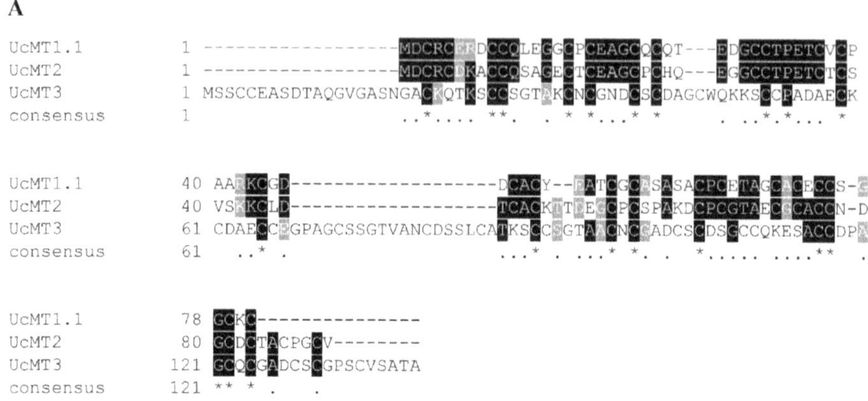

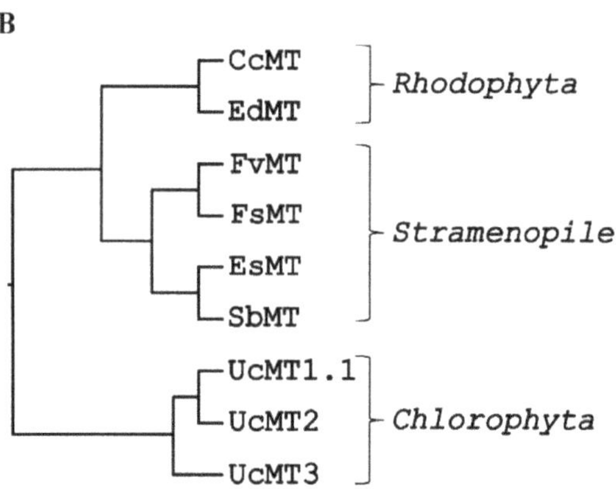

Figure 4. Alignment of amino acid sequences of metallotioneins (MTs) UcMT1.1, UcMT2, and UcMT3 from the marine alga *U. compressa* (**A**). Identical amino acids are indicated in black and similar amino acids are indicated in gray. * indicates identical amino acids and · similar amino acids. Hierarchical clustering of the amino acid sequences of UcMTs and MTs found in other marine macroalgae (**B**).

2.3. Expression of UcMTs-GST in Bacteria and Detection of GST-Tag

The ORFs of UcMTs were cloned in an *E. coli* expression vector, which allows the expression of MTs fused with the enzyme glutathione-S-transferase (GST) from the platyhelminthe *Schistosoma japonicum* (26 kDa), an enzyme that contain a single cysteine in the N-terminal domain, and do not bind metals. After 1 to 12 h of culture, the induction of protein expression with IPTG allows the visualization of increasing levels of UcMT1.1-GST (34.2 kDa. Figure 5A), UcMT2-GST (35.05 KDa, Figure 5B), and UcMT3-GST (39.4 KDa, Figure 5C). These proteins were detected by Western blot using an antibody prepared against *S. japonicum* GST indicating that the overexpressed proteins correspond to UcMTs fused with GST (Figure 5C).

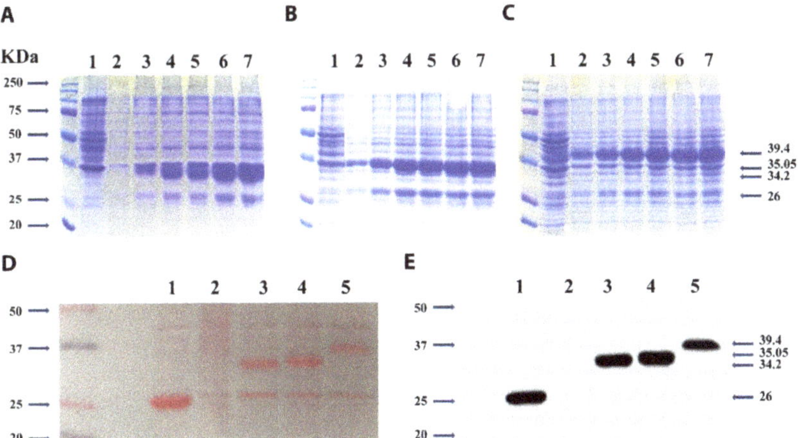

Figure 5. Visualization of protein extracts of bacteria overexpressing UcMT1.1-GST (**A**), UcMT2-GST (**B**), and UcMT3-GST (**C**) obtained from bacteria cultivated without IPTG (lane 1) or with 0.5 mM IPTG for 1, 3, 6, 9, and 12 h (lanes 2 to 7). Visualization of proteins bacterial extracts overexpressing UcMTs-GST transferred to a nitrocellulose membrane stained with Ponceau red dye (**D**) or incubated with anti-GST antibody and revealed by chemiluminescence (**E**). Proteins of extracts overexpressing GST (lane 1), not overexpressing GST (lane 2), and overexpressing UcMT1.1-GST (lane 3), UcMT2-GST (lane 4), and UcMT3-GST (lane 5) for 6 h. Arrows (left side) indicate molecular weights of standard proteins and arrows (right side) indicate molecular weights of UcMT3-GST (39.4 kDa), UcMT2-GST (35.05 kDa), UcMT1 (34.2 kDa), and GST (26 kDa).

2.4. UcMTs-GST-Mediated Accumulation of Copper or Zinc In Vivo

Transformed bacteria expressing MTs-GST were cultivated with 0.5 mM IPTG for 30 min, and with 1 mM copper and IPTG for 6 h. Control bacteria expressing only GST accumulate 0.35 μg of copper mg^{-1} of dry weight (DW), whereas those expressing UcMT1.1-GST, UcMT2-GST, and UcMT3-GST accumulate 1.8, 1.7, and 1.4 times more copper than the control, respectively (Figure 6A). On the other hand, transformed bacteria were cultivated with 0.5 mM IPTG for 30 min and, with 1 mM zinc and IPTG for 6 h. Control bacteria accumulate 0.49 μg mg^{-1} of zinc mg^{-1} of DW whereas those expressing UcMT1.1-GST, UcMT2-GST, and UcMT3-GST accumulate 4,1, 3.8, and 3.4 times more zinc than the control, respectively; although these increases were not significantly different among each other (Figure 6A). Thus, expression of UcMTs-GST mediates accumulation of copper and zinc in vivo.

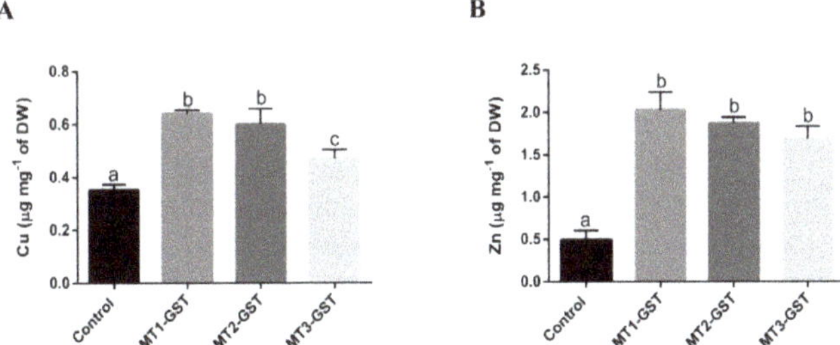

Figure 6. Level of copper (**A**) and zinc (**B**) in *E. coli* cultivated without IPTG for 6 h (control) and with 0.5 mM IPTG for 6 h and overexpressing UcMT1.1, UcMT2, and UcMT3. Levels of copper and zinc are expressed in micrograms per gram of bacterial dry weight (DW). Bars represent mean values of three independent experiments ± SD. Letters indicate significant differences among mean values ($p < 0.05$).

3. Discussion

In this work, we isolated the complete sequence of transcripts encoding three MTs from the marine alga *U. compressa*. These transcripts encode UcMT1.1, UcMT2, and UcMT3 with a MW of 8.3, 9.05, and 13.4 kDa, respectively, and containing 24–30% of cysteines. It is important to mention that UcMT1.1 and UcMT2 seemed to be structurally more related among each other than to UcMT3. UcMT1.1 and UcMT2 showed deletions in amino acids sequences compared with MT3 at the initial, internal, and terminal part of the protein compared with UcMT3. Thus, UcMT2 and UcMT3 genes may have derived from UcMT3 gene. In the case of UcMT1, it seems that more than a single copy exists in *U. compressa* genome since three different transcripts of UcMT1 were isolated corresponding to UcMT1.1, UcMT1.2 and UcMT1.3. In this sense, it has been shown that the increase in number of copies of Cup1 MT gene in the yeast *S. cereviciae* allowed an increased tolerance to copper [30]. In addition, several copies of domains α and β of MTs exist in the mollusk *C. gigas* [12]. Thus, it is not surprising that UcMT1 may be present in multiple copies in *U. compressa* genome, but the latter remained to be confirmed. The sequencing of *U. compressa* genome is already in course.

Interestingly, UcMTs resemble marine invertebrate MTs. Mollusk MTs are constituted by 73–75 amino acids and contain 20–21 cysteine residues (28%), arranged mainly as CXC and CC motifs [5]. UcMTs are longer than mollusk MTs since they are constituted by 81, 90, and 139 amino acids; although the content of cysteines is similar to mollusk MTs (28–30%). However, mollusk MTs reported until now do not contain histidine or aromatic residues [5]. In addition, MT from marine crustacean and equinoderms showed cysteines (29–30%) arranged as CC and CXC and they do not contain histidine or aromatic aminoacids [25,26]. It is important to mention that we showed that UcMT1.1 contains a tyrosine, UcMT2 a histidine, and UcMT3 a tryptophan. In this sense, the two MTs found in the worm *C. elegans* are constituted by 75 amino acids and contain 19 cysteines (25%), arranged as CXC and CC, and contain a tyrosine and histidine residues [8]. Thus, UcMTs are more similar to *C. elegans* MTs but longer and contain a higher percentage of cysteines (28–30%). This indicates that UcMTs are unique MTs, and their sequences are more closely related with MTs of marine invertebrates and *C. elegans* MTs. It is not surprising that UcMTs showed similarities to marine invertebrate MTs since *F. vesiculosus* MT is also more related to mollusk MTs than to plant MTs [20]. *F. vesiculosus* MT showed a linker of 14 amino acids, which is longer than the linker of vertebrates MTs (4 aa) and the linkers of plant MTs (4 and 40 aa) [20]. The linker of UcMTs has a size of 9–13 amino acids that is more closely related with *F. vesiculosus* and invertebrate MTs linker size.

In addition, the sequence of UcMTs is different from the MT cloned from *F. vesiculosus*, which is constituted by 67 amino acids (6.9 kDa) and contains 16 cysteine residues (24%) arranged as CXC, but not as CC [3]. The only structural similarity of *Fucus mt* is that this transcript showed a short 5′UTR of 64 nucleotides and a long 3′UTR of 960 nucleotides [8]. Furthermore, MTs identified in the genomes of Rodophyceae *C. cripus* and *E. denticulatum* are constituted by 69 to 71 amino acids and contain 12 to 14 cysteines (20%) arranged mainly as CXC, but not as CC [2]. Furthermore, UcMTs grouped in a different clade compared with MTs from other marine macroalgae. Thus, UcMTs of the green macroalga *U. compressa* are distantly related with those found in red and brown macroalgae.

On the other hand, we showed that UcMTs-GST mediates the accumulation of copper and zinc in bacteria. In particular, UcMT1.1 and UcMT2, which are structurally more closely related, allowed higher accumulation of copper compared with UcMT3. Moreover, the three MTs allowed accumulation of zinc with similar efficiencies among them. Thus, the three UcMTs differentially bind copper and similarly bind zinc in vivo but their affinity for copper and zinc need to be further investigated. In this sense, it has been demonstrated that MTs are either more Zn- or Cu-binding thioneins, preference associated with cysteine arrangements and the nature of the other amino acids that constitute the MT [4,31]. Likewise, mouse MT1, MT2, and MT3 are more Zn-thionein and, in contrast, yeast cup-1, *Drosophila* MntA and MntB, and mouse MT4, are more Cu-thioneins [4,31]. Thus, the nature of UcMTs regarding their affinity for copper or zinc need to be further investigated.

It is now clearly established that green and red macroalgae are more closely related among each other, and with terrestrial plants, than with brown macroalgae [32–34]. Green and red macroalgae belong to the kingdom Plantae, as terrestrial plants [35] whereas brown macroalgae belong to the kingdom Chromalveolta [34]. The latter is based on the observations that green and red algae contain a plastid that derive from a single event of endosymbiosis by a cyanobacteria, whereas brown algae plastids derive from a secondary or tertiary endosymbiosis event of green or red microalgae [32,33]. Thus, it is unexpected that the three UcMTs of the green macroalga *U. compressa* are not closely related with other marine alga MTs, in particular to red macroalgae MTs. In contrast, the major similarity is with MTs of marine invertebrate and *C. elegans* MTs. Considering that marine algae appeared on earth around a billion years ago, and marine invertebrates emerged around 500 million years ago [34,36] and, moreover, that UcMTs are longer than MTs of marine invertebrates, it is then possible that marine invertebrates have acquired MTs genes from marine green macroalgae by horizontal gene transfer; however, this hypothesis need to be further investigated. Furthermore, it is possible that *U. compressa* contain additional MTs that differentially bind copper, zinc, and other heavy metals, as it has been predicted in [23], but the latter need to be further analyzed. In this sense, it has been shown that the equinoderm *Paracentrotus livudus* exhibits 7 MTs [35] suggesting that the four other potential UcMTs can be functional MTs. Thus, additional UcMTs may exist in the genome of *U. compressa* and these MTs will be cloned and characterized in the future.

In conclusion, we showed that transcripts encoding three MTs were cloned and sequenced; they encode unique MTs with homology with marine invertebrate and *C. elegans* MTs. UcMTs expressed in bacteria allowed copper and zinc accumulation in vivo. Thus, it is likely that these UcMTs may participate in copper and zinc accumulation in the marine alga *U. compressa*.

4. Materials and Methods

4.1. Sampling of Algae and Water Collection

The green macroalga *U. compressa* was collected during spring 2017 from the high intertidal zone at Cachagüa Beach (32°34′ S), a site in central Chile with no history of metal pollution [23]. Algal samples were transported to the laboratory in sealed plastic bags inside a cool-box. In the laboratory, material was rinsed three times with filtered seawater, cleaned manually, and sonicated three times for 2 min using an ultrasound bath (HiLab Innovation Systems, model SK221OHP) to remove epiphytes. The algae were maintained in aerated seawater under an irradiance of 50 μmoles m^{-2} s^{-1} on a photoperiod of 12 h:12 h light:dark cycle, at 14 °C for 4 days, prior to experimentation. Seawater was obtained from Quintay (33°12′ S), a pristine site in central Chile, filtered through 0.45 and 0.2 μm pore size membranes and stored in darkness at 4 °C.

4.2. Algal Culture and RNA Extraction

U. compressa (1 g of FT) was cultivated in 100 mL of filtrated seawater containing 10 μM $CuCl_2$ (Merck, Darmstat, Germany) for 3 days. The alga was washed with 10 mM Tris-50 mM EDTA pH 7.0, in order to eliminate copper and other metals bound to cell walls [37].

4.3. Purification of Total RNAs and mRNAs for RACE-PCR

Total RNAs were extracted as described in [38]. *U. compressa* (150 mg of FT) was frozen in liquid nitrogen and pulverized in a mortar. One mL Trizol reagent (Invitrogen, Carlsbad, CA, USA) was added and the alga was homogenized with a pestle until thawing. The mixture was centrifuged at 12,000× *g* for 10 min at 4 °C, and the supernatant was recovered. Chloroform (200 μL) was added and the mixture was vortexed for 10 s and left at room temperature for 3 min. The mixture was centrifuged at 12,000× *g* for 15 min at 4 °C, and the aqueous phase was recovered. Isopropanol (500 μL) was added and the solution incubated for 10 min at room temperature. The solution was centrifuged at 12,000× *g* for 10 min at 4 °C, and the supernatant removed. The pellet was washed twice with 1 mL of 75%

ethanol, gently vortexed, and centrifuged at 7000× *g* for 5 min at 4 °C. The ethanol phase was removed, the pellet dried for 15 min at room temperature, dissolved in 50 μL of ultrapure water treated with DEPC (water-DEPC), and incubated for 10 min at 60 °C.

Total RNAs were quantified using Nanodrop spectrophotometer (Tecan, Zürich, Switzerland); the integrity was verified by agarose gel electrophoresis and stored at −80 °C. Messenger RNAs were purified from 100 μg of total RNA using NucleoTrap mRNA minikit (Macherey-Nagel, Düren, Germany), mRNAs were eluted in 25 μL of water-DEPC, and normally 1 μg of mRNAs was obtained from 100 μg of total RNAs.

4.4. Amplification of 5′RACE cDNAs

The amplification of cDNAs was performed using MMLV reverse transcriptase kit (Promega, Madison, WI, USA). To this end, 4.1 μL of purified mRNAs (150 ng) were mixed with 2.5 μL of primer 1 for MT1 (5′GTAGCAGGCACAGTCGTCA3′), primer 2 for MT2 (5′AGGCCTAACAAGCAGCGTCC3′), or primer 3 for MT3 (5′CCGAGAGCGTGTCCTTACTT3′) at 10 μM, and 8.4 μL of water-DEPC were added to complete a final volume of 15 μL. The mixture was denatured at 65 °C for 5 min and cooled on ice for 1 min. Then, 0.7 μL of RNase inhibitor (10 U μL^{-1}), 5 μL of MMLV buffer 5×, 1.5 μL of dNTPs mixture (10 mM of each dNTP), 1 μL de MMLV reverse transcriptase (200 U μL^{-1}), and 1.8 μL of water- DEPC were added to complete a final volume of 25 μL. The mixture was sequentially incubated at 42 °C for 50 min, at 50 °C for 10 min, at 55 °C for 5 min, and at 70 °C for 10 min in order to inactivate MMLV reverse transcriptase. cDNAs were treated with RNAse H, purified using GEL/PCR purification mini kit (Favorgen Biotech Corp., Changzhi, Taiwan), and eluted in 40 μL of water-DEPC.

Purified cDNAs (14.5 μL) were incubated with 0.5 μL of terminal transferase (TdT), 20 U μL^{-1}, New England Biolabs, Ipswich, MA, USA), 1 μL of dCTP, 2 μL of $CoCl_2$ (2.5 mM), and 2 μL of TdT buffer 10×, to complete a final volume of 20 μL. The mixture was incubated at 37 °C, and at 75 °C for 20 min to inactivate TdT in order to obtain cDNAs having a 3′ C-tail.

The first round of amplification was performed using PCR kit (Favorgene, London, UK) and 1 μL of C-tailed cDNAs mixed with 10 μL of PCR mixture, 0.6 μL of 5′RACE adapter primer (5′GGCCACGCGTCGACTAGTACGGGIIGGGIIGGGIIG3′), and with 0.6 μL of primer 4 for MT1 (5′GGCATACGCACGTCTCGGG3′), primer 5 for MT2 (5′CTGCGTAACGACATAGCCGA3′), or primer 6 for MT3 (5′GCAGCCAGAATCGCAACTAC3′), at 10 μM, and 6.8 μL of water-DEPC to complete a final volume of 20 μL. The mixture was incubated at 95 °C for 3 min and subjected to 40 cycles of denaturation at 95 °C for 5 s, hybridization at 63 °C for 10 s and amplification at 72 °C for 15 s, using a real-time thermocycler RotorGene 6000. The second round of amplification was performed with the mixture diluted 100 times in distilled water-DEPC, and 14.4 μL of the cDNAs were mixed with 10 μL of PCR mixture, 0.6 μL of abridged universal adaptor primer (AUAP) (5′GGCCACGCGTCGACTAGTAC3′) and primer 7 for MT1, (5′GCAACCATCTTCGGTTTGGC3′), primer 8 for MT2, (5′ATCCTTCGCGGGTGAGCAAG3′), and primer 9 for MT3 (5′CACAGTTGCATTCTGCGGTT3′) at 10 μM and with 6.8 μL of water-DEPC to complete a final volume of 20 μL. The amplification was performed using 40 cycles of amplification mentioned before. The amplified fragments were analyzed in a 2% agarose gel, stained with SYBR green (Invitrogen, Carlsbad, CA, USA), and visualized on a UV trans-illuminator.

4.5. Amplification of 3′RACE cDNAs

Initial cDNAs were obtained using MMLV reverse transcriptase kit (Promega, Madison, WI, USA). To this end, 6.8 μL of mRNAs (250 ng) were mixed with 5 μL of 3′RACE adapter primer with an oligo-dT tail (5′GGCCACGCGTCGACTAGTACTTTTTTTTTTTTTTTTT3′) at 10 μM and with 3.2 μL of water-DEPC to complete a final volume of 15 μL. The mixture was denatured at 65 °C for 5 min and then cooled on ice for 1 min. Then, 0.7 μL of RNAse inhibitor (10 U μL^{-1}), 5 μL of MMLV buffer 5×, 1.5 μL of dNTPs mixture (10 mM of each dNTP), 1 μL de MMLV reverse transcriptase (200 U μL^{-1}), and 1.8 μL of water- DEPC were added to complete a final volume of 25 μL. The mixture was incubated at 42 °C for 1 h and at 70 °C for 10 min to inactivate MMLV reverse transcriptase and diluted to 50 μL with water-DEPC. cDNAs were treated with RNAse H, purified using GEL/PCR purification mini kit (Favorgen Biotech Corp., Changzhi, Taiwan), and eluted in 40 μL of water-DEPC.

The first round of amplification was performed using PCR kit (Favorgene, London, UK) and 1 μL of cDNAs were mixed with 0.6 μL of AUAP (5′GGCCACGCGTCGACTAGTAC3′), 0.6 μL of primer 10 for MT1 (5′CAGTGCCAAACCGAAGATGG3′), primer 11 for MT2 (5′GATGAGGGCTGTCCTTGCTC3′), or primer 12 for MT3 (5′AGTGTGATGCTGAGTGCTGT3′) at 10 μM and 6.8 μL of water-DEPC to complete a final volume of 20 μL. The mixture was incubated at 95 °C for 3 min and subjected to 40 cycles of denaturation at 95 °C for 5 s, hybridization at 63 °C for 10 s, and amplification at 72 °C for 15 s, using a real-time thermocycler RotorGene 6000. The second round of amplification was performed with the mixture diluted 100 times in distilled water-DEPC, and 14.4 μL of the cDNAs were mixed with 10 μL of PCR mixture, 0.6 μL of AUAP forward (5′GGCCACGCGTCGACTAGTAC3′) and with primer 13 for MT1 (5′GGTTGCAAGTGCTAGCTGAC3′), primer 14 for MT2 (5′GCTTGTTAGGCCTCAGTGGT3′), or primer 15 for MT3 (5′TGTCAGTGCGACAGCCTAA3′) and with 6.8 μL of water-DEPC to complete a final volume of 20 μL. The amplification was performed using 40 cycles of amplification mentioned before. The amplified fragments were analyzed in a 2% agarose gel, stained with SYBR green (Invitrogen, Carlsbad, CA, USA) and visualized on a UV trans-illuminator.

4.6. Amplification of UcMT ORFs

MT1.1 ORF was amplified using primer 16 forward (5′ATGGACTGCCGTTGCG3′) and primer 17 reverse (5′GCACTTGCAACCGCCAGAGC3′); MT2 ORF was amplified using primer 18 forward (5′ATGAACTGCTGTTGCGA3′) and primer 19 reverse (5′GACACAGCCCGGACAGGC3′); and MT3 ORF was amplified using primer 20 forward (5′ATGTCGTCTTGTTGTGAAGC3′) and primer 21 reverse (5′GGCTGTCGCACTGACACAG3′). The mixture was incubated at 95 °C for 2 min and subjected to 35 cycles of denaturation at 95 °C for 15 s, hybridization at 65 °C for UcMT1.1, 63 °C for MT2 and 62 °C for UcMT3, during 10 s, and amplification at 72 °C for 15 s, using a real-time thermocycler RotorGene 6000.

4.7. Cloning of UcMTs 5′and 3′UTRs, and UcMT ORFs in pGEM-T Vector

The 5′RACE and 3′RACE amplification fragments obtained in the second PCR (see above) and those of UcMTs ORFs were subjected to electrophoresis in 2% agarose gel. The piece of agarose gel containing the stained fragments was removed from the gel and placed in Eppendorf tubes. The amplified fragments were eluted from agarose using Gel/PCR purification kit (Favorgen, London, UK), recovered in 50 μL of water-DEPC and stored at 4 °C. Amplified fragments were ligated with the cloning vector pGEM-T easy (Promega, Madison, WI, USA) and transformed in *E. coli* competent cells One Shot TOP 10 (Invitrogen, Carlsbad, CA, USA). Transformed *E. coli* cells were cultivated in 10 mL of LB medium (10 g tryptone, 5 g yeast extract, and 100 g NaCl in 1 L of distilled water) supplemented with 100 μg mL^{-1} of ampicillin. The culture was centrifuged at 3,000 x g for 5 min in a centrifuge model Nuwind (Nuaire, Plymouth, MN, USA). Transformed pGEM-T vectors were purified from the bacterial pellet using Wizard Plus SV Miniprep DNA Purification System (Promega, Madison, WI, USA). To check cloning of 5′UTR fragments in pGEM-T, primers AUAP and primers 7, 8, and 9 were used. To check cloning of 3′UTR in pGEM-T, primer AUAP and primers 13, 14, and 15 were used. PCR conditions were identical to those used for the amplification of 5′and 3′RACE ends (mentioned above). To amplify the ORFs, primers 16 and 17 for MT1, primers 18 and 19 for MT2, and primers 20 and 21 for MT3 were used and to verify the insertion PCR conditions were those used to amplify ORFs (mentioned above). Cloned fragments were sequenced using an ABI3730XL (Macrogen, Seoul, Korea).

4.8. Cloning of UcMTs ORFS in pGEX Expression Vector

UcMTs were synthesized and subjected to codon optimization for expression in *E. coli* by Genscript (Piscataway, NJ, USA) and then ligated to the expression vector pGEX-5X-1 (Genscript) which allowed fusion of UcMTs ORFS with the enzyme glutahione-S-transferase (GST) from the platyhelminthes *Schistosoma japonicum* (26 Kda), an enzyme containing a single cysteine that does not bind metals; the fusion proteins were named UcMT1.1-GST, UcMT2-GST, and UcMT3-GST. The recombinant vectors were sequenced by Genscript to verify the correct insertion of complete ORFs.

4.9. Transformation of Expression Vectors in Bacteria

The recombinant expression vectors were transformed in competent *E. coli* strain BL21 (DE3) (Sigma-Aldrich, Saint Louis, MO, USA). To this end, 200 μL of competent cells BL21 (DE3) were incubated with 50 ng of recombinant expression vector containing UcMTs-GST. Then, 800 μL of SOC medium (2% (*w*/*v*) tryptone, 0.5% (*w*/*v*) of yeast extract, 10 mM NaCl, 2.5 mM KCl, 10 mM $MgCl_2$, 10 mm $MgSO_4$, and 20 mM glucose) were added, and cultivated at 37 °C for 45 min. An aliquot of 200 μL was cultured on solid LB medium containing 100 μg L^{-1} of carbenicillin in a Petri dish overnight. Individual colonies were selected for each UcMT, cultivated with LB medium, and stored at −80 °C in LB medium containing 15% glycerol.

4.10. Expression of UcMTs-GST in Bacteria

Recombinant *E. coli* were cultured 15 mL of LB medium until OD_{600}= 0.6, 0.5 mM of isopropyl-β-D-1thiogalactopyranoside (IPTG) was added, and samples of 1 mL were obtained after 1, 3, 6, 9, and 12 h of culture at 37 °C. The samples were centrifuged at 7000× *g* for 10 min, washed with PBS pH 7.4 (10 mM Na_2HPO_4, 1.8 mM KH_2PO_4, 140 mM NaCl, 2.7 mM KCl), and centrifuged again in similar conditions. Pellets were suspended in 50 μL of protein loading buffer 2× (125 mM Tris HCl pH=6.8, 4% (*w*/*v*) SDS, 20% (*v*/*v*) glycerol, 10% (*v*/*v*) β-mercaptoethanol, 0.004% (*w*/*v*) bromophenol blue), and heated at 95 °C for 5 min. A sample of 10 μL was analyzed in a denaturant polyacrylamide gel (12%) and proteins were stained with Coomassie blue staining solution (25% (*v*/*v*) methanol, 5% (*v*/*v*) acetic acid, and 0.1% (*w*/*v*) of Comassie blue G-250).

4.11. Purification of UcMTs-GST by Affinity Column

A sample of 2 mL of *E. coli* transformed with expression vectors containing a UcMT-GST were added to 100 mL LB medium containing 100 μg mL^{-1} ampicillin and cultured at 37 °C overnight. A sample of 10 mL was added to 1 L of LB medium containing 100 μg mL^{-1} ampicillin and 0.5 mM IPTG, in quadruplicate, until OD_{600}= 0.6 (aprox. 2.5 h). The 4 L of culture were centrifuged at 6000× *g* for 10 min. The pellet was washed twice with 300 mL of PBS and bacteria were suspended in 10 mL of PBS containing 5 mM dithiotreitol (DTT) and 1 tablet of protease inhibitor cocktail (Roche, Manheim, Germany). The bacterial suspension was sonicated for 20 s, with 20 s of pause, for 5 min. The suspension was centrifuged at 6000 rpm for 10 min and supernatant was recovered. Protein concentration was determined using Bradford method [39] and adjusted with PBS-5 mM DTT (PBS-DTT) to 1 mg mL^{-1}. UcMTs-GST were purified by HPLC using GSTrap HP (General Electric, Uppsala, Sweden) at 5 bars of pressure, washed with PBS-DTT, and eluted with 3 mL of buffer 50 mL Tris-HCl-10 mM GSH. Normally, 1–2 mg of purified UcMT-GST was obtained from 4 L of bacterial culture and proteins were quantified using Bradford method [40].

4.12. Detection of UcMTs-GST with Anti-GST Antibody

Transformed bacteria were cultured in 100 mL of LB medium until OD_{600} = 0.6, 0.5 mM IPTG was added and the mixture incubated for 6 h. The culture was centrifuged at 6000× *g* for 5 min, the pellet suspended in 5 mL of buffer PBS, and sonicated for 20 s, with 20 s of pause, for 5 min. Proteins (20 μg) were separated by electrophoresis in a denaturant 12% polyacrylamide gel and transferred to a nitrocellulose membrane for 10 min using Trans Blot Turbo apparatus (BioRad, Hercules, CA, USA). The membrane was stained with Ponceau Red dye and washed with 10 mL of distilled water. The membrane was incubated in 10 mL TTBS (20 mM Tris-HCl pH 7.5, 0.1 mM NaCl, 0.05% Tween-20) containing 5% skim milk for 1 h, washed twice with 10 mL TTBS for 15 min, incubated with 10 mL TTBS containing 3% skim milk and the antibody anti-GST (Sigma-Aldrich, St Louis, MO, USA) diluted 5000 times, and washed four times in TTBS for 15 min. The membrane was incubated in TTBS containing 3% skim milk and the secondary antibody prepared against rabbit IgG coupled to hoseradish peroxidase (Agrisera, Vännas, Sweden) diluted 2000 times, for 1 h, and washed four times with TTBS for 15 min. Proteins were detected using ECL Western Blotting System kit (Amersham, Buckinghamshire, UK) and revealed using a C-Digits chemiluminiscence Western blot scanner Li-Cor (Lincoln, NE, USA) and Image Studio Digits software version 4.0 Li-Cor.

4.13. Quantification of Copper and Zinc in Bacteria Expressing UcMTs-GST

Recombinant bacteria were cultured in 100 mL of LB medium containing 100 mg mL^{-1} of carbenecillin until DO_{600} = 0.6, with 0.5 mM IPTG for 30 min, and with 1 mM $CuSO_4$ or 1 mM $ZnCl_2$ and IPTG for 6 h. Bacterial pellets showed a weight of 26–42 mg for copper cultures and 46–62 mg for zinc cultures. Pellets were dried at 60 °C for 48 h, suspended in 5 mL of 60% (*v*/*v*) HNO_3, and incubated at 85 °C for 2 h. The solutions were filtered through 0.22 μm MCE filters (TCL, Santiago, Chile) and analyzed by flame atomic emission spectrophotometry using an atomic emission spectrophotometer ThermoFisher (Waltham, MA, USA).

4.14. Hierarchical Clustering of UcMTs

Amino acid sequences corresponding to MTs of different animal and plant species (234 in total) were selected from revised SwissProt repository of the UniprotKB database (www.uniprot.org). Alignment of these sequences was performed with Clustal W software with default setting. This alignment was used to generate the phylogenetic reconstruction to represent a hierarchical clustering using UPGMA algorithm based on distance. Phylogenetic and hierarchical clustering analyses were conducted using MEGA software version X [39].

4.15. Statistical Analyses

Statistical analyses were performed with the Prism 6 statistical package (Graph Pad software Inc., San Diego, CA, USA). Following confirmation of normality and homogeneity of variance, significant differences among treatments were determined by two-way ANOVA and Tukey's multiple comparison post-hoc test, at a 95% confidence interval.

Supplementary Materials: Supplementary materials can be found at http://www.mdpi.com/1422-0067/21/1/153/s1.

Author Contributions: A.Z. amplified and cloned 5′and 3′UTR and ORFs of transcripts encoding UcMTs, expressed UcMTs-GST in bacteria, and analyzed copper and zinc accumulation. A.G. and M.G. purified UcMTs-GST. D.L. guided A.Z., did hierarchical clustering of MTs, Western hybridization of UcMTs-GST, and UV-VIS spectra of MTs-GST. C.A.S. revised the manuscript and participate in discussions, and A.M. designed experimental work and wrote the manuscript. All authors have read and agreed to the published version of the manuscript.

Funding: This work was financed by Fondecyt 1160013 to A.M.

Acknowledgments: We thank Luis Lemus and Cristián Vera who helped with UV-VIS spectra analyses.

Conflicts of Interest: The authors declare no conflict of interest.

Abbreviations

GSH	glutathione
GST	glutathione-S-transferase
MTs	metallothioneins
PCs	phytochelatins
UV	ultraviolet

References

1. Cobbett, C.; Goldsbrough, P. Phytochelatins and metallothioneins: Role in heavy metal detoxification and homeostasis. *Ann. Rev. Plant. Biol.* **2002**, *53*, 159–182. [CrossRef] [PubMed]
2. Blindauer, C.A.; Leszczyszyn, O.I. Metallothioneins: Unparalleled diversity in structures and functions for metal ions homeostasis and more. *Nat. Prod. Rep.* **2010**, *27*, 720–741. [CrossRef] [PubMed]
3. Leszczyszyn, O.I.; Imam, H.T.; Blindauer, C.A. Diversity and distribution of plant metallothioneins: A review of structure, properties and functions. *Metallomics* **2013**, *5*, 1146–1169. [CrossRef] [PubMed]
4. Palacios, O.; Atrian, S.; Capdevila, M. Zn- and Cu-thioneins: A functional classification for metallothioneins. *J. Biol. Inorg. Chem.* **2011**, *16*, 991–1009. [CrossRef] [PubMed]
5. Isani, G.; Carpene, E. Metallothioneins, unconventional proteins from unconventional animals: A long journey from nematodes to mammals. *Biomolecules* **2014**, *4*, 435–457. [CrossRef] [PubMed]
6. Margoshes, M.; Vallee, B.L. A cadmium protein from equine kidney cortex. *J. Am. Chem. Soc.* **1957**, *79*, 4813–4814. [CrossRef]
7. Kille, P.; Lee, W.E.; Darke, B.M.; Winge, D.R.; Dameron, C.T.; Stephens, P.E.; Kay, J. Sequestration of cadmium and copper by recombinant rainbow trout and human metallothioneins and by a chimeric (mermaid and fishman) proteins with interchanged domains. *J. Biol. Chem.* **1992**, *267*, 8042–8049.
8. Freedman, J.H.; Slice, L.W.; Dixon, D.; Fire, A.; Rubin, S.S. The novel metallothionein genes of *Caenorhabditis elegans*. *J. Biol. Chem.* **1993**, *268*, 2554–2564.
9. Berger, B.; Hunziker, P.E.; Hauer, C.R.; Birchler, N.; Dallinger, R. Mass spectrometry and amino acid sequencing of two cadmium-binding metallothionein isoforms from the terrestrial gastropod *Arianta arbustorum*. *Biochem. J.* **1995**, *311*, 951–957. [CrossRef]
10. Berger, B.; Dallinger, R.; Gehrig, P.; Hunziker, P.E. Primary structure of a copper-binding metallothionein from the mantle tissue of the terrestrial gastropod *Helix pomatia* L. *Biochem. J.* **1997**, *328*, 219–224. [CrossRef]
11. Palacios, O.; Pérez-Rafael, S.; Pagani, A.; Dallinger, R.; Atrian, S.; Capdevila, M. Cognate and noncognate metal ion coordination in metal-specific metallothioneins: The *Helix pomatia* system as a model. *J. Biol. Inorg. Chem.* **2014**, *19*, 923–935. [CrossRef] [PubMed]
12. Jenny, M.J.; Ringwood, A.H.; Schey, K.; Warr, G.W.; Chapman, R.W. Diversity of metallothioneins in the American oyster, *Crassostrea virginica*, revealed by transcriptomic and proteomic approaches. *Eur. J. Biochem.* **2004**, *271*, 1702–1712. [CrossRef] [PubMed]
13. Leignel, V.; Laulier, M. Isolation and characterization of *Mytilus edulis* metallothionein genes. *Comp. Biochem. Physiol.* **2006**, *142*, 12–18. [CrossRef] [PubMed]
14. Engelke, J.; Hildebrandt, A. cDNA cloning and cadmium-induced expression of metallothionein mRNA in the zebra mussel *Dreissena polymporpha*. *Biochem. Cell Biol.* **1999**, *77*, 237–241. [CrossRef]
15. Nemer, M.; Thornton, R.D.; Stuebing, E.W.; Harlow, P. Structure, spatial and temporal expression of two sea urchin metallothionein genes, SpMTB and SpMTA. *J. Biol. Chem.* **1991**, *266*, 6586–6593.
16. Butt, T.R.; Sternberg, E.J.; Gorman, J.A.; Clark, P.; Hamer, D.; Rosenberg, M.; Crooke, S. Copper metallothionein of yeast, structure of the gene, and regulation of expression. *Proc. Natl. Acad. Sci. USA* **1984**, *81*, 3332–3336. [CrossRef]
17. Cizewski-Culotta, V.; Howard, W.; Liu, X. CRS5 encode a metallothionein-like protein in *Saccharomyces cerevisiae*. *J. Biol. Chem.* **1994**, *269*, 25295–25302.
18. Lane, B.; Kajioka, R.; Kennedy, T. The wheat germ E_c protein is a zinc-containing metallothionein. *Biochem. Cell Biol.* **1987**, *65*, 1001–1005. [CrossRef]
19. De Miranda, J.R.; Thomas, M.A.; Thurman, D.A.; Tomsett, A.B. Metallothionein genes from the flowering plant *Mimulus guttatus*. *FEBS Lett.* **1990**, *260*, 277–280. [CrossRef]
20. Morris, C.A.; Nicolaus, B.; Sampson, V.; Harwood, J.L.; Kille, P. Identification and characterization of a recombinant metallothionein protein from a marine alga, *Fucus vesiculosus*. *Biochem. J.* **1999**, *338*, 553–560. [CrossRef]
21. Kohler, A.; Blaudez, D.; Chalot, M.; Martin, F. Cloning and expression of multiple metallothioneins from hybrid poplar. *New Phytol.* **2004**, *164*, 83–93. [CrossRef]
22. Guo, W.J.; Meetam, M.; Goldsbrough, P.B. Examining the specific contribution of individual *Arabidopsis* metallothioneins to copper distribution and metal tolerance. *Plant Physiol.* **2008**, *146*, 1697–1706. [CrossRef] [PubMed]

23. Ratkevicius, N.; Correa, J.A.; Moenne, A. Copper accumulation, synthesis of ascorbate and activation of ascorbate peroxidase in *Enteromorpha compressa* (L.) Grev. (Chlorophyta) from heavy metal-enriched environments in northern Chile. *Plant Cell Environ.* **2003**, *26*, 1599–1608. [CrossRef]
24. Seeliger, U.; Cordazzo, C. Field and experimental evaluation of *Enteromorpha* sp. as a quali-quantitative monitoring organism for copper and mercury in estuaries. *Environ. Pollut. Ser.* **1982**, *29*, 197–206. [CrossRef]
25. Navarrete, A.; González, A.; Gómez, M.; Contreras, R.A.; Díaz, P.; Lobos, G.; Brown, M.T.; Sáez, C.A.; Moenne, A. Copper excess detoxification is mediated by a coordinated and complementary induction of glutahione, phytochelatins and metallothioneins in the green seaweed *Ulva compressa*. *Plant Physiol. Biochem.* **2019**, *135*, 423–431. [CrossRef]
26. Geddie, A.W.; Hall, S.G. The effect of salinity and alkalinity on growth and accumulation of copper and zinc in the Chlorophyta *Ulva fasciata*. *Ecotoxicol. Environ. Saf.* **2019**, 203–209. [CrossRef]
27. Benatti, M.R.; Yookongkaev, N.; Meetam, M.; Guo, W.J.; Punyasuk, N.; AbuQamar, S.; Goldsbrough, P. Metallothionein deficiency impacts copper accumulation and redistribution in leaves and seeds of Arabidopsis. *New Phytol.* **2014**, *202*, 940–951. [CrossRef]
28. Tapia, L.; González-Agüero, M.; Cisternas, M.F.; Suazo, M.; Cambiazo, V.; Uauy, R.; González, M. Metallothionein is crucial for safe intracellular copper storage and cell survival at normal and supra-physiological exposure levels. *Biochem. J.* **2004**, *378*, 617–624. [CrossRef]
29. Laporte, D.; Valdés, N.; González, A.; Sáez, C.A.; Zúñiga, A.; Navarrete, A.; Meneses, C.; Moenne, A. Copper-induced overexpression of genes encoding antioxidant system enzymes and metallothioneins involve activation of CaMs, CPKs and MEK1/2 in the marine alga *Ulva compressa*. *Aquat. Toxicol.* **2016**, *177*, 433–440. [CrossRef]
30. Bi, X.W.; Kong, F.; Hu, H.Y.; Viu, X. Role of glutathione in detoxification of copper cadmium in yeast cells having different abilities to express Cup1 protein. *Toxicol. Mech. Meth.* **2007**, *17*, 6. [CrossRef]
31. Valls, M.; Bofill, R.; González-Duarte, R.; González-Duarte, P.; Capdevila, M.; Atrian, S. A new insight into metallothionein (MT) classification and evolution. *J. Biol. Chem.* **2001**, *276*, 32835–32843. [CrossRef] [PubMed]
32. Li, S.; Nosenko, D.; Hackett, J.D.; Bhattacharya, D. Phylogenomic analysis identifies red algal genes of endosymbiotic origin in the Chromalveolata. *Mol. Biol. Evol.* **2006**, *23*, 663–674. [CrossRef] [PubMed]
33. Chan, C.X.; Gross, J.; Yoon, H.S.; Bhattacharya, D. Red and green alga monophyly and extensive gene sharing found in a rich repertoire of red algal genes. *Curr. Biol.* **2011**, *21*, 328–333. [CrossRef] [PubMed]
34. Yoon, H.S.; Hackett, J.D.; Ciniglia, C.; Pinto, G.; Bhattacharya, D. A molecular time for the origin of photosynthetic eukaryotes. *Mol. Biol. Evol.* **2004**, *21*, 809–818. [CrossRef]
35. Ragusa, M.A.; Nicosia, A.; Costa, S.; Cittita, A.; Ginaguzza, F. Metallothionein gene family in the sea urchin *Paracentrotus lividus*: Gene structure, differential expression and phylogenetic analysis. *Int. J. Mol. Sci.* **2017**, *18*, 812.
36. Cary, G.A.; Hinman, V.F. Echinoderm development and evolution in the post-genomic era. *Dev. Biol.* **2017**, *427*, 203–2011. [CrossRef]
37. Hassler, C.S.; Slaveykova, V.I.; Wilkinson, K.J. Discriminating between intra- and extracellular metals using chemical extractions. *Limnol. Oceanogr. Methods* **2004**, *2*, 237–242. [CrossRef]
38. Chomczynski, P.; Sacchi, N. Single-step method of RNA isolation by acid-guanidinium-thiocyanate-phenol-chloroform extraction. *Anal. Biochem.* **1987**, *162*, 156–159. [CrossRef]
39. Stecher, K.; Kniaz, L.M.; Tamurak, K. MEGA X: Molecular evolutionary genetics analysis across computing platform. *Mol. Biol. Evol.* **2018**, *35*, 1547–1549.
40. Bradford, M.M. A rapid and sensitive method for the quantification of microgram quantities of protein utilizing the principle of protein-dye binding. *Anal. Biochem.* **1976**, *72*, 248–254. [CrossRef]

© 2019 by the authors. Licensee MDPI, Basel, Switzerland. This article is an open access article distributed under the terms and conditions of the Creative Commons Attribution (CC BY) license (http://creativecommons.org/licenses/by/4.0/).

Review

Advances in the Mechanisms of Plant Tolerance to Manganese Toxicity

Jifu Li [1,2], Yidan Jia [1,2], Rongshu Dong [1,3], Rui Huang [1,3], Pandao Liu [1,3], Xinyong Li [1,3], Zhiyong Wang [2], Guodao Liu [1,3] and Zhijian Chen [1,2,3,*]

1 Institute of Tropical Crop Genetic Resources, Chinese Academy of Tropical Agriculture Sciences, Haikou 571101, China; li_jifu@163.com (J.L.); YidanJia@163.com (Y.J.); dongrongshu@126.com (R.D.); bluesing@126.com (R.H.); liupandao@foxmail.com (P.L.); lixy051985@163.com (X.L.); liuguodao2008@163.com (G.L.)

2 Institute of Tropical Agriculture and Forestry, Hainan University, Haikou 571101, China; wangzhiyong7989@163.com

3 Key Laboratory of Crop Gene Resources and Germplasm Enhancement in Southern China, Ministry of Agriculture, Hainan 571101, China

* Correspondence: jchen@scau.edu.cn; Tel.: +86-0898-2330-0496

Received: 22 September 2019; Accepted: 12 October 2019; Published: 14 October 2019

Abstract: Manganese (Mn) is an essential element for plant growth due to its participation in a series of physiological and metabolic processes. Mn is also considered a heavy metal that causes phytotoxicity when present in excess, disrupting photosynthesis and enzyme activity in plants. Thus, Mn toxicity is a major constraint limiting plant growth and production, especially in acid soils. To cope with Mn toxicity, plants have evolved a wide range of adaptive strategies to improve their growth under this stress. Mn tolerance mechanisms include activation of the antioxidant system, regulation of Mn uptake and homeostasis, and compartmentalization of Mn into subcellular compartments (e.g., vacuoles, endoplasmic reticulum, Golgi apparatus, and cell walls). In this regard, numerous genes are involved in specific pathways controlling Mn detoxification. Here, we summarize the recent advances in the mechanisms of Mn toxicity tolerance in plants and highlight the roles of genes responsible for Mn uptake, translocation, and distribution, contributing to Mn detoxification. We hope this review will provide a comprehensive understanding of the adaptive strategies of plants to Mn toxicity through gene regulation, which will aid in breeding crop varieties with Mn tolerance via genetic improvement approaches, enhancing the yield and quality of crops.

Keywords: manganese toxicity; Mn detoxification; tolerance mechanism; gene function; subcellular compartment

1. Introduction

Manganese (Mn) is the second most prevalent trace element in the Earth's crust after iron (Fe), and is widely distributed in soils, sediments, and other biological materials [1]. In soils, Mn is present in a wide range of oxidation states, including Mn(II), Mn(III), Mn(IV), Mn(VI), and Mn(VII) [2]. Among the oxidized forms of Mn, divalent Mn(II) is the most soluble species in soils and is also the most available form of Mn for plant acquisition. The solubility of Mn is strongly influenced by soil pH and redox conditions [1,3]. At neutral or higher soil pH, Mn(III) and Mn(IV) are the predominant and insoluble forms of Mn. However, in poorly drained acid soils with pH levels below 5.0 and a reducing environment, oxidized Mn is easily reduced to divalent Mn [4]. Thus, the available Mn in soils is variable and generally ranges from 450 to 4000 mg per kilogram [3]. For example, the concentration of Mn varies between 40 and 1681 mg per kilogram in farmland soils across mainland China [5], while the concentration of Mn in the agricultural soils of central Greece is from 685 to 1307 mg per kilogram [6].

Mn is an example of a transition element that is required for humans, animals, and plants. For most plants, Mn is absolutely necessary at low levels of 20–40 mg per kilogram dry weight [7,8]. Mn is involved in a variety of metabolic processes, including photosynthesis, respiration, fatty acid and protein synthesis, as well as enzyme activation. For example, Mn is an indispensable constitutive element in the Mn cluster structure of the oxygen-evolving complex in photosystem II (PSII) that participates in the water-splitting process, providing necessary electrons for photosynthesis [9,10]. Mn acts as an important cofactor of various enzymes, including superoxide dismutase (MnSOD), catalase (MnCAT), decarboxylases of the tricarboxylic acid (TCA) cycle, and RNA polymerases [8,11]. In addition, Mn is required for multiple steps in the biosynthesis of secondary metabolites, such as lignins, flavonoids, cinnamic acid, and acyl lipids [12].

Despite its necessity, Mn is also considered one of the heavy metals that can be harmful to plants at excessive levels. When the Mn concentration in the aboveground tissues of plants reaches 150 mg per kilogram dry weight, Mn toxicity can generally occur, especially for plants growing in acid soils [13,14]. Many previous studies demonstrate that Mn toxicity can disrupt various physiological processes in plant cells, such as triggering oxidative stress, inhibiting enzyme activity, impeding chlorophyll biosynthesis and photosynthesis, and preventing the uptake and translocation of other mineral elements, including phosphorus (P), Fe, and magnesium (Mg) [14–16]. As a result, Mn toxicity leads to the appearance of toxicity symptoms, including chlorosis in young leaves, necrotic dark spots on mature leaves, and crinkled leaves, ultimately inhibiting plant growth. Symptoms of Mn toxicity vary widely among plant species and varieties. For example, chlorosis and necrosis have been reported in leaves of common bean (*Phaseolus vulgaris*) [17], clover (*Trifolium repens*) [18], ryegrass (*Lolium perenne*) [19], and stylo (*Stylosanthes guianensis*) [20]. Brown spots surrounded by irregular areas of chlorotic tissues are observed in cowpea (*Vigna unguiculata*) [21], soybean (*Glycine max*) [22], and barley (*Hordeum vulgare*) [23]. The diverse expressions of Mn toxicity probably indicate different Mn-tolerant capabilities among plant species and cultivars. For example, among different legumes, *Medicago sativa*, *Trifolium fragifevum*, *Leucaena leucocephala*, and *Medicago tvuncatula* are considered the most sensitive to Mn toxicity, while *Centrosemapubescens*, *Lotononis bainesii*, Townsville stylo (*Stylosanthes humilis*), and *Desmodium mcinatum* are the most tolerant plant species [24].

Over the last few decades, there have been major advances in elucidating the mechanisms underlying plant tolerance to Mn toxicity at multiple levels, from physiological changes to biochemical responses (Figure 1). For example, activation of the antioxidant system, including the free radical-mitigating antioxidant enzymes and nonenzymatic components, is thought to be vital for plants alleviating excess Mn-induced oxidative stress [25]. The important roles of the regulation of Mn uptake, translocation, and distribution have been implicated in many plants' responses to Mn toxicity, such as rice (*Oryza sativa*) [26,27], Arabidopsis (*Arabidopsis thaliana*) [28], and Caribbean stylo (*Stylosanthes hamata*) [29]. Furthermore, plants can sequester Mn into subcellular compartments, such as vacuoles, the endoplasmic reticulum (ER), Golgi apparatuses, and cell walls, to withstand the toxic effects of high Mn stress [30,31]. In addition, free Mn ions can be chelated with protein-based, organic, and inorganic compounds to form inactive Mn complexes, combating the deleterious effects of Mn toxicity [18–20].

To date, a variety of genes and proteins have been shown to be involved in the responses to Mn toxicity of plants, such as orange (*Citrus sinensis*) [32], common bean [33], tomato (*Solanum lycopersicum*) [34], stylo [20,35], cowpea [21,36], soybean [22], rice, and barley [23]. Many of the identified genes have been functionally integrated into specific pathways, illuminating the molecular processes of the plant response to Mn toxicity. Furthermore, the functions of numerous genes involved in Mn detoxification through regulation of Mn uptake, distribution, and accumulation have been well characterized in plants [29,37–39]. Therefore, the purpose of this review is mainly to focus on Mn as a toxic transition metal to plants and the mechanisms of plant tolerance to Mn stress. This review will discuss the current understanding of plant genes involved in Mn uptake, distribution, and accumulation, which contribute to Mn detoxification. Furthermore, we also highlight the candidate genes that can potentially be used for breeding crop varieties tolerant to Mn toxicity via genetic improvement approaches.

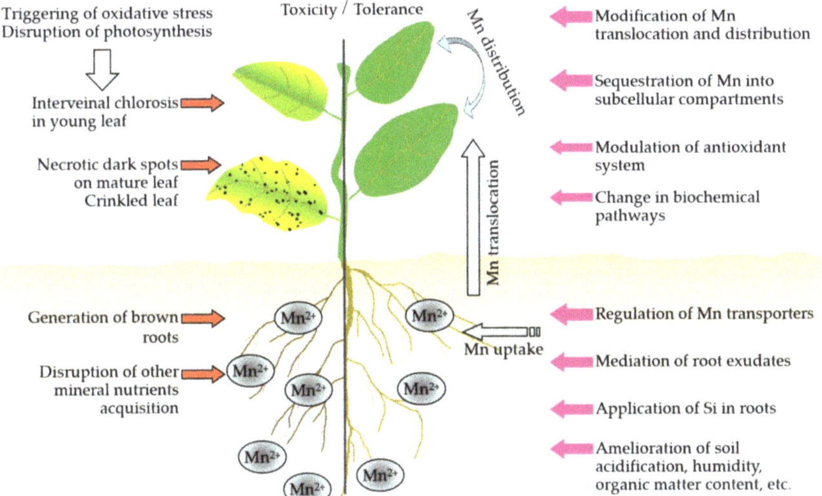

Figure 1. Schematic representation of Mn toxicity and strategies for increasing Mn tolerance in plants. Mn toxicity can trigger oxidative stress and disrupt photosynthesis, which may result in the generation of interveinal chlorosis in young leaves, necrotic dark spots on mature leaves, and crinkled leaf. Furthermore, Mn toxicity can lead to the formation of brown roots and prevent the uptake and translocation of other mineral elements. In plants, Mn tolerance strategies include modification of Mn translocation and distribution, sequestration of Mn into subcellular compartments, modulation of the antioxidant system, changes in biochemical pathways, and regulation of Mn transporters. In addition, the mediation of root exudates, the application of Si in roots, and the amelioration of soil acidification, humidity, and organic matter content also contribute to increase plant Mn tolerance. Red arrows indicate the toxic effects of excess Mn to plants. Purple arrows represent the adaptive strategies of plants to Mn toxicity.

2. Activation of the Antioxidant System

As a toxic metal, excess Mn can generate reactive oxygen species (ROS) and trigger oxidative stress in plants, causing lipid peroxidation and damaging photosynthetic pigments and proteins if ROS are not well scavenged [25,35]. One of the adaptive changes that alleviates the toxic effects of high Mn in plants involves the activation of the antioxidant system via antioxidant enzymes, such as superoxide dismutase (SOD), peroxidase (POD), catalase (CAT), ascorbate peroxidase (APX), and glutathione reductase (GR), and nonenzymatic antioxidant components, including ascorbate (AsA) and glutathione (GSH) [35,40]. Increases in the activities of antioxidant enzymes under Mn toxicity are generally associated with enhanced Mn tolerance in common bean [41], cucumber (*Cucumis sativus*) [42], and perennial ryegrass [40]. In perennial ryegrass, for example, the Mn-tolerant ryegrass cultivar Kingston exhibits higher SOD activity than the Mn-sensitive ryegrass cultivar Nui—a higher expression of the *Fe–SOD* gene is observed in Kingston compared to that in Nui [40]. Thus, the induced *Fe–SOD* expression in Kingston is likely to contribute to its high Mn-toxicity tolerance. Additional studies in cowpea have shown that both the activities of H_2O_2-producing and H_2O_2-consuming PODs are enhanced by Mn toxicity in the leaf apoplast [21]. Furthermore, proteomic analysis indicated that the protein accumulation of PODs in the leaf apoplast is increased by high Mn [21]. Similar results have been implicated in citrus and stylo, in which the expression of *POD* genes is enhanced when plants are subjected to Mn toxicity [35,43]. Therefore, it is probable that SOD and POD represent two key proteins in the plant defense against oxidative damage caused by Mn toxicity. However, considering the damage caused by Mn toxicity, ROS-scavenging systems, through regulation of the antioxidant system, seem to be insufficient to alleviate oxidative stress, which might be a general response of plants to Mn toxicity.

3. Regulation of Mn Uptake

Although Mn is required in relatively small amounts, the Mn content accumulated in most plants is approximately 30–500 mg per kilogram dry weight, which is higher than their normal growth requirements [8,14,44]. Therefore, it is reasonable to propose that there are some key transporter genes responsible for Mn acquisition in response to high Mn stress (Figure 2). Studying the mechanisms of plant Mn transport can greatly increase our understanding of how plants acquire and transport Mn under variable environmental Mn levels.

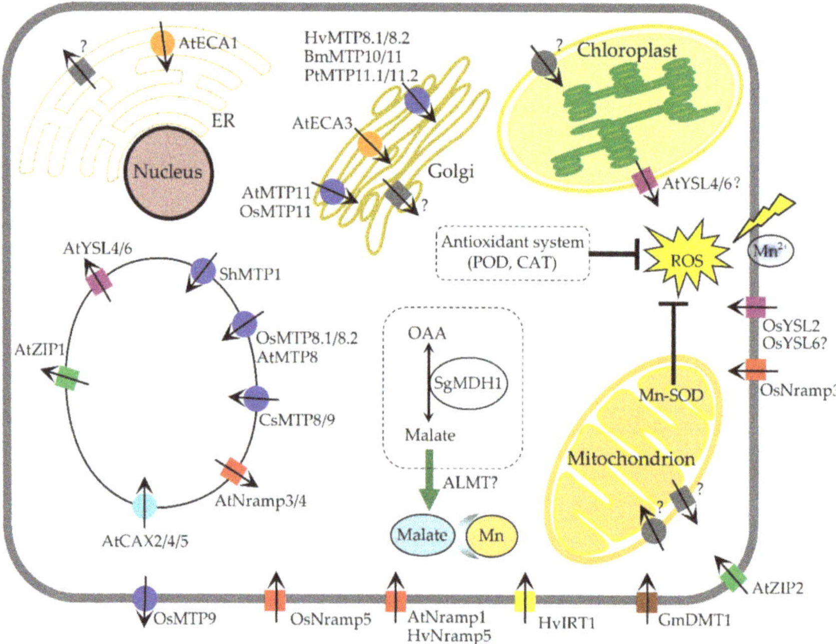

Figure 2. Summary of genes affecting Mn transport and tolerance in plants. Squares: Import into the cytosol; circles: Export out of the cytosol; blue: MTP family; green: ZTP family; red: Nramp family; purple: YSL family; yellow: IRT family; orange: ECA family; cyan: CAX family; brown: DMT family; gray: unknown. ER: Endoplasmic reticulum; Nramp: Natural resistance-associated macrophage protein; MTP: Metal tolerance protein; DMT: Divalent metal transporter; ZIP/IRT: Zinc-regulated transporter/iron-regulated transporter-like proteins; YSL: Yellow stripe-like protein; CAX: Cation exchanger; ECAs: ER-type calcium ATPases; MDH: Malate dehydrogenase; ALMT: Aluminum-activated malate transporter; OAA: Oxaloacetate; ROS: Reactive oxygen species; SOD: Superoxide dismutase; POD: Peroxidase; CAT: Catalase. At: *Arabidopsis thaliana*; Os: *Oryza sativa*; Gm: *Glycine max*; Hv: *Hordeum vulgare*; Mt: *Medicago tvuncatula*; Cs: *Cucumis sativus*; Sh: *Stylosanthes hamata*; Sg: *Stylosanthes guianensis*; Le: *Lycopersicon esculentum*; Bm: *Beta vulgaris* subspecies maritima; Pt: *Populus trichocarpa*. Question marks behind some genes mean that the exact roles of these genes or their localizations remain to be further clarified.

The major transporters responsible for Mn acquisition in plants are members of the natural resistance-associated macrophage protein (Nramp) family, which have so far been functionally characterized in many plants, for example, AtNramp1 from Arabidopsis, OsNramp5 from rice, and HvNramp5 from barley [26,45]. In Arabidopsis, AtNramp1, belonging to the Nramp family, is the major high-affinity Mn transporter involved in Mn uptake. AtNramp1 is localized to the plasma membrane. The transcripts of *AtNramp1* are mainly detected in roots, where their levels are ten times greater than in shoots. Furthermore, *AtNramp1* transcripts are increased by Mn deficiency in the roots [46]. AtNramp1 can complement the phenotype of a yeast mutant, *smf1*, which is defective in Mn uptake when grown in medium containing the divalent cation chelator EGTA [47]. Furthermore, when cultivated in a medium lacking Mn, the T-DNA insertion mutant Atnramp1-1 produces less biomass than wild-type Arabidopsis. The growth inhibition of the mutant can be attributed to less Mn accumulation compared to the wild-type plants under Mn-deficient conditions [46].

In rice, Mn uptake is mediated by OsNramp5, a homolog of AtNramp1 [48]. In contrast to Arabidopsis, OsNramp5 is constitutively expressed in roots, and its expression is enhanced by Fe and zinc (Zn) deficiency but does not respond to different Mn levels in roots [48]. As OsNramp5 can complement the growth of yeast strains defective in Mn and Fe transport, OsNramp5 is implicated in Mn and Fe transport [49]. As OsNramp5 is polarly located at the distal side of both the exodermis and endodermis of mature roots, OsNramp5 is likely to act as an influx transporter and acquire Mn from the soil to the exodermal cells as well as from the apoplastic solution to endodermal cells [48]. Knockout of *OsNramp5* resulted in a decreased concentration of Mn and Fe but not Zn in the shoots, suggesting that OsNramp5 is able to transport Fe in addition to Mn. However, the growth of *OsNramp5* knockout lines is unaffected when the Fe concentration in the external solution is decreased, and the Fe concentrations in the shoots and roots are similar to those of the wild type under Fe deficiency. Thus, the authors conclude that the uptake of Fe required for growth is mediated by other transporters, and OsNramp5 is responsible for additional Fe uptake [48]. A similar key role has been assigned to metal tolerance protein 9 (OsMTP9), the other type of transporter belonging to the cation diffusion facilitator (CDF) family that participates in Mn uptake and translocation in rice roots [50]. *OsMTP9* shows higher expression in roots, but its expression is not influenced by external Mn levels [50]. Tissue- and cell-specific localization analysis revealed that OsMTP9 is localized to the proximal sides of both the exodermis and endodermis of mature root zones, which is opposite to the sites of of OsNramp5 localization in rice roots. Further evidence shows that OsMTP9 acts as an efflux transporter and is responsible for Mn translocation to the root stele [50]. Therefore, the different polar localizations of OsNramp5 and OsMTP9 facilitate Mn uptake from the soil solution to the stele in rice.

Similar results have also been found for HvNramp5, which is localized to the plasma membranes of the epidermal cells of the root tips in the outer root cell layers of barley [51]. There is evidence that HvNramp5 displays transport activity for both Mn and cadmium (Cd) when expressed in yeast cells, and disruption of *HvNramp5* results in growth reduction in barley under low Mn supply [51]. Therefore, HvNramp5 is a transporter required for Mn uptake in barley. In addition, GmDMT1 (divalent metal transporter 1), a nodule-enhanced transporter belonging to the Nramp family in soybean, has also been found to transport Mn in addition to Fe when expressed in yeast [52], although further investigation is needed to understand the physiological roles of GmDMT1 in Mn acquisition in soybean. In addition, members of the zinc-regulated transporter/iron-regulated transporter-like proteins (ZRT/IRT) family were found to have the ability to transport Mn, such as HvIRT1 from barley [53].

Considering the particular importance of the transporter genes controlling Mn uptake in plants, it is reasonable to propose that increased Mn detoxification can be achieved through decreased Mn accumulation from decreasing excess Mn uptake and root-to-shoot Mn translocation, by downregulating transporter genes specific for Mn uptake under high Mn stress. Therefore, manipulation of these transporter genes is an alternative strategy to facilitate the plant response to varying Mn levels through regulation of Mn acquisition.

4. Regulation of Mn Translocation and Distribution

After Mn is taken up by roots, most Mn is translocated from roots to shoots and further delivered to various tissues for growth requirements. Thus, it is important to understand the long-distance and whole-plant translocation of Mn in plants in response to different Mn levels, from limited to excessive. In Arabidopsis, two ZIP members, AtZIP1 and AtZIP2, are implicated in Mn translocation from roots to shoots [54]. Both AtZIP1 and AtZIP2 are mainly expressed in the root stele and do not respond to external Mn levels at the transcriptional level. AtZIP1 and AtZIP2 localize to the tonoplast and plasma membrane, respectively. It is probable that AtZIP1 functions in the remobilization of Mn from vacuoles to the cytoplasm in root stellar cells, while AtZIP2 plays a role in Mn movement to the root vasculature for further translocation to the shoots [54]. The loss-of-function mutants of the *AtZIP1* gene in Arabidopsis show severe sensitivity to Mn deficiency. However, the T-DNA *AtZIP2* knockout lines display more tolerance to Mn toxicity than the wild type [54]. Furthermore, Mn concentration in the roots of *AtZIP2* knockout lines is much higher than that in wild-type plants, but no significant differences in shoot Mn concentrations are observed between knockout lines and wild-type plants [54]. Considering that AtZIP2 has high root expression in the stele, AtZIP2 is likely to play a role in Mn transport into the root vasculature, which ultimately helps to provide Mn to the xylem parenchyma, where other transporters such as the heavy metal ATPase, AtHMA2/4, may mediate xylem loading of Mn to the shoot in the transpiration stream as proposed by the authors [54].

OsYSL2, belonging to the yellow stripe-like family, has been characterized to function in long-distance Mn transport and distribution in rice [55]. *OsYSL2* is mainly expressed in leaves, flowers, and developing seeds [55,56]. Electrophysiological measurements using *Xenopus laevis* oocytes show that OsYSL2 is involved in the transportation of Mn–nicotianamine (NA) in addition to Fe–NA complexes [56]. The phloem and seed localization of OsYSL2 suggests that OsYSL2 transports Mn–NA and Fe–NA complexes via the phloem and then loads these complexes into the grain [56]. Overexpression of *OsYSL2* leads to increases in Mn accumulation in the grain [55], suggesting that OsYSL2 is involved in the translocation of Mn into the grain. In addition, some evidence suggests that Mn complexes may be delivered by other transporters, such as *AtOPT3* (a putative oligopeptide transporter) and *AtYSLs* from Arabidopsis [57–59], and *ZmYS1* from maize [60], but the exact roles of these genes remain to be clarified.

Additional studies have shown that rice OsNramp3 is a plasma membrane-localized influx transporter for the distribution of Mn, but not Fe and Cd [37]. *OsNramp3* displays higher expression in the nodes and is not affected by external Mn at the transcriptional level. It is noteworthy that the OsNramp3 protein is rapidly degraded within a few hours when plants are exposed to high Mn stress [37]. OsNramp3 is proposed to function with the following patterns: Under Mn deficiency, OsNramp3 preferentially transports Mn to young leaves and panicles via intervascular transfer, but in contrast, under excess Mn conditions, due to rapid OsNramp3 protein degradation, Mn is delivered to old tissues, protecting developing tissues from the toxic effects of excess Mn [37]. Therefore, the authors suggest that OsNramp3 functions as a node-based switch for Mn distribution, which turns the protein on or off in response to variable environmental Mn levels. These findings above provide a major advancement in the understanding of Mn distribution in plants through the regulation of transporters at the post-translational level.

5. Intracellular Mn Detoxification in Subcellular Compartments

As the amount of Mn accumulated in most plants usually exceeds their normal growth requirements, plants must cope with excess Mn via internal detoxification. In this regard, one of the key strategies for plant tolerance to Mn toxicity is the compartmentalization of Mn into subcellular compartments [14]. Therefore, transporters that localize to the endomembrane compartments are suggested to be critical for intracellular Mn detoxification in plant cells.

The vacuole, an organelle that comprises approximately 90% of the total cell volume, is the dominant sink for various toxic compounds, including Mn [61]. Some transporters belonging to the

CDF family act as proton antiporters for efflux metals (e.g., Zn, Fe, Mn, and Cd) out of the cytoplasm or into subcellular compartments (e.g., vacuoles) [62]. ShMTP1, the first functionally characterized CDF for Mn transport into the vacuoles, was isolated from Caribbean stylo, a tropical legume with superior Mn tolerance [29,35,45]. Evidence shows that ShMTP1 is localized to the tonoplast, and overexpression of *ShMTP1* confers Mn tolerance in yeast cells and Arabidopsis via sequestration of Mn into the vacuoles [29]. In addition to ShMTP1, other CDF members, such as OsMTP8.1, also participate in delivering Mn to vacuoles for Mn sequestration [63]. The transcript of *OsMTP8.1* is mainly detected in shoots and is enhanced by high Mn levels. OsMTP8.1 is expressed in all cells of leaf blades and is also localized to the tonoplast. In rice, knockout of *OsMTP8.1* results in the generation of symptoms of Mn toxicity when plants are exposed to high Mn toxicity [63]. However, OsMTP8.1 is not a unique CDF in mediating Mn transport into vacuoles in rice. OsMTP8.2, a homolog of OsMTP8.1, is also involved in Mn sequestration, and loss of function of *OsMTP8.2* results in severe growth inhibition of both shoots and roots of the *osmtp8.1* mutant in the presence of high Mn [64]. Therefore, it is probable that OsMTP8.2 mediates Mn tolerance together with OsMTP8.1 by sequestering Mn into vacuoles. To date, a set of *MTP* homolog genes have been characterized with similar functions in sequestering Mn into vacuoles, such as *AtMTP8* from Arabidopsis [37], *CsMTP8/9* from cucumber [65,66], and *CsMTP8* from the tea plant (*Camellia sinensis*) [67]. The conserved function of MTPs among different plant species fully supports the dominant roles of MTPs in Mn detoxification.

Another major transporter for intracellular Mn sequestration into vacuoles is a member of the cation exchanger (CAX) family with metal/H^+ antiport activity. In Arabidopsis, the role of AtCAX2 in Mn transport was confirmed by its ability to confer tolerance to Mn toxicity when its expression was heterologous in *pmc1vcx1cnb*, a Mn-sensitive yeast mutant. A three-amino acid Mn-binding region (Cys–Ala–Phe) in AtCAX2 was subsequently found to be critical for Mn-transport activity [68–70]. Further analysis showed that overexpression of *AtCAX2* in tobacco (*Nicotiana tabacum*) increases the resistance to Mn toxicity via mediating the sequestration of Mn into the vacuoles [68]. In addition to AtCAX2, AtCAX4 and AtCAX5, which localize to the vacuolar membrane, also display Mn^{2+}/H^+ antiport activity [71,72]. The transcripts of both *AtCAX4* and *AtCAX5* in roots are increased under conditions of high Mn [71–73]. Phenotypic analysis shows that transgenic tobacco overexpressing *AtCAX4* displays tolerance to Mn toxicity, while *AtCAX5* can rescue the growth of Mn-sensitive yeast, suggesting their roles in conferring Mn tolerance [72,74]. Arabidopsis mutants, including *cax1*, *cax2*, *cax3*, *cax1/cax2*, and *cax2/cax3*, have been generated and analyzed for their growth performances under excess Mn levels. Among these mutants, *cax2* and *cax2/cax3* displayed severe sensitivity to high Mn stress [75].

An alternate mechanism of intracellular-Mn tolerance in plants is the sequestration of Mn into the Golgi apparatus or endoplasmic reticulum (ER) [15]. AtMTP11 is suggested to be involved in this process in Arabidopsis. AtMTP11 can rescue the growth of yeast mutant *pmr1*, which is defective in a Ca^{2+}/Mn^{2+}–ATPase, in the presence of excess Mn. Arabidopsis mutants impaired in *AtMTP11* are sensitive to high Mn levels, whereas plants overexpressing *AtMTP11* are more tolerant to Mn toxicity [76]. In contrast to ShMTP1, OsMTP8.1, and OsMTP8.2 mentioned above, AtMTP11 is localized to a punctate endomembrane compartment probably in the trans-Golgi, but not to the plasma membrane and vacuole. Therefore, a secretory pathway involving vesicular trafficking and exocytosis mediated by AtMTP11 is believed to help increase Mn tolerance in Arabidopsis [28]. Similar functions of other MTPs in sequestering Mn into the Golgi apparatus have been reported for OsMTP11 from rice [27], HvMTP8.1 and HvMTP8.2 from barley [77], PtMTP11.1 and PtMTP11.2 from poplar (*Populus trichocarpa*) [28], as well as BmMTP10 and BmMTP11 from beets (*Beta vulgaris*) [78].

It has been well demonstrated that ER-type calcium ATPases (ECAs), belonging to the Ca^{2+}–ATPase subfamily, can use energy from ATP hydrolysis to catalyze the translocation of cations across membranes [79,80]. There are four predicted ECAs in Arabidopsis (AtECA1–4) and three in rice (OsECA1–3) [79]. In Arabidopsis, AtECA1 and AtECA3 are localized to the ER and Golgi compartments, respectively [81–83]. The expression of AtECA1 and AtECA3 was found in all major organs of

Arabidopsis, especially in the roots [81,83]. Both AtECA1 and AtECA3 are able to rescue the growth of yeast under high Mn stress [81,84]. Furthermore, under excess-Mn conditions, the Arabidopsis *ateca1-1* mutants display inhibited root growth, and the growth of the *ateca1-1* mutant is rescued by overexpression of *AtECA1* [81]. Similarly, the root growth of the *ateca3* mutant is impaired by excess Mn, confirming that AtECA3 is also necessary for Mn detoxification in Arabidopsis [84]. Therefore, AtECA1 and AtECA3 are the two key components required for delivering Mn into the ER and Golgi compartments for Mn tolerance. In addition, the YSL family is also implicated in the sequestration of Mn into endomembrane compartments. AtYSL4 and AtYSL6 are reported to be localized to vacuole membranes and internal membranes resembling the ER in Arabidopsis. Significant decreases in fresh weight have been observed in single mutants and double mutants of *AtYSL4* and *AtYSL6* compared to wild-type Arabidopsis grown in high Mn for 21 d [59]. The authors suggest a role for AtYSL4 and AtYSL6 in the sequestration or efflux of this metal into intracellular compartments [59]. However, future characterization of YSL as well as ECAs in other crop species is needed to confirm their exact roles in Mn detoxification via sequestration of Mn into intracellular compartments.

OsYSL6 is reported to transport Mn from the apoplast to the symplast, which is required for the detoxification of excess Mn in rice [84]. Although the expression of *OsYSL6* does not respond to either deficiency or toxicity of Mn, ectopic expression of *OsYSL6* in the yeast mutant indicates transport activity for the Mn–NA complex. Furthermore, knockout of *OsYSL6* in rice increases Mn accumulation in the leaf apoplast but not in the symplast under high Mn stress, resulting in the development of necrosis in the old leaves, a symptom of Mn toxicity [84]. As divalent Mn accumulated in the apoplast can potentially be oxidized to trivalent Mn, which further oxidizes proteins and lipids, causing deleterious effects of Mn toxicity [21], OsYSL6 is likely to alleviate excess Mn toxicity via the transport of Mn from the apoplast to the symplast in rice.

Most of the Mn transporter genes mentioned above display no or only slight responses to varying Mn levels, which may partially explain why plants accumulate large amounts of Mn that far exceed their growth requirements. Therefore, it is of great importance to investigate the regulatory mechanisms of the plant response to external Mn in the future.

6. Si Application Alleviates Mn Toxicity

Another strategy for increasing Mn tolerance can be achieved by the application of silicon (Si) to the roots of plants such as rice [85], cowpea [86,87], and cucumber [88]. The mechanisms for Si-alleviated Mn toxicity include decreasing the Mn accumulation in shoots, promoting Mn oxidation in roots and increasing the cell wall-binding capacity for Mn [88–90]. A recent study showed that supplementation with Si successfully decreases the Mn concentration in the shoots but increases Mn in the roots of rice under high Mn stress, alleviating Mn toxicity [90]. However, Si application cannot alleviate Mn toxicity in the rice *lsi1* mutant, which is defective in Si uptake. OsLsi1 is a Si transporter that transports Si from the external solution to the root cells in rice [91]. Interestingly, the expression of *OsNramp5* is decreased by long-term exposure to Si in the wild type but not in the *lsi1* mutant. The authors suggest that the Si-alleviated Mn toxicity in rice can be attributed to inhibition of root-to-shoot translocation of Mn and decreased Mn uptake by downregulation of Mn transporters, such as OsNramp5 and OsMTP9 [90]. Therefore, OsLsi1 might participate in Mn detoxification through regulation of Si uptake, which deserves further clarification.

7. Organic Acid Mediates Mn Detoxification

Mn can be chelated with protein-based, organic, and inorganic compounds to form Mn complexes, thus decreasing Mn uptake and/or Mn phytotoxicity. Regulation of organic acid metabolism is an important strategy in Mn detoxification. Intracellular Mn in cowpea, *Gossia bidwillii*, and *Phytolacca acinosa* is found to be chelated in complexes with internal citrate, malate, and oxalate, respectively [92–94]. The complexation of Mn by organic acids in the apoplast is proposed to decrease Mn phytotoxicity in cowpea [87]. Increases in internal malate concentrations are observed in leaves and roots of the Mn-tolerant stylo genotype Fine-stem under high Mn stress, and are closely linked to its Mn tolerance capabilities [20]. Accordingly, Mn might be chelated by malate to form Mn–malate complexes, ultimately conferring Mn tolerance in stylo. Subsequent analysis shows that malate synthesis in stylo could be attributed to a Mn-enhanced malate dehydrogenase (SgMDH1), which catalyzes the reversible conversion of oxaloacetate to malate. Due to successful increases in resistance to Mn toxicity in both yeast cells and Arabidopsis, SgMDH1 is hypothesized to be involved in Mn detoxification through mediated malate synthesis [20].

On the other hand, increases in organic acid exudation from roots in response to Mn toxicity are found in stylo, clover, and ryegrass [18–20]. Increased root exudates of oxalate and citrate in Mn-tolerant ryegrass cultivars have been implicated in increasing Mn tolerance by decreasing Mn uptake from the rhizosphere [19]. Similar results are also reported in stylo, where increased malate exudation from roots helps to confer Mn tolerance, and exogenous malate application to the growth medium increases the resistance of the Mn-sensitive stylo genotype to the toxic effects of Mn [20]. Interestingly, the expression of an aluminum-activated malate transporter (*SgALMT1*) is enhanced by high Mn stress in the Mn-tolerant stylo genotype [20], which likely functions in mediating malate efflux from roots, as observed in aluminum detoxification [95]. Therefore, it is reasonable to hypothesize that coordinated regulation of malate synthesis and exudation by *SgMDH1* combined with *SgALMT1* might facilitate the tolerance of stylo to Mn toxicity.

8. Other Aspects

In recent years, the development of biotechnologies, such as RNA-seq and proteomics, has provided favorable platforms to reveal complex responses of plants to biotic and abiotic stresses [35,96,97]. Many differentially expressed genes and proteins have been previously identified in plants' responses to Mn toxicity. For example, various Mn-responsive genes have been isolated from leaves of citrus using cDNA–AFLP technology, and the identified genes can be classified into different functional categories, such as biological regulation and signal transduction (e.g., protein phosphatase 2a and Myb family transcription factor), carbohydrate and energy metabolism (e.g., ATP synthase subunit alpha and UDP-glycosyltransferases), nucleic acid metabolism (e.g., DNA polymerase phi subunit and histone H4), protein metabolism (e.g., ribosomal proteins, eukaryotic initiation factors, and glutathione S-transferase Tau2), cell wall metabolism (e.g., cell wall-associated hydrolase and glycoside hydrolase family 28 protein), stress responses (e.g., *CAT*, *POD42*, and monodehydroascorbate reductase), and cell transport (e.g., ABC transporter family protein) [32]. In addition, a set of Mn-regulated proteins were identified in the Mn-tolerant stylo genotype through proteomic analysis. These proteins are mainly involved in defense responses, photosynthesis, carbon fixation, metabolism, cell wall modulation, and signaling [35]. Further analysis shows that some of the identified proteins related to the phenylpropanoid pathway, including phenylalanine ammonia-lyase (PAL), chalcone synthase (CHS), chalcone–flavonone isomerase family protein (CFI), and isoflavone reductase (IFR), are regulated by external Mn in stylo [35]. As secondary metabolites, such as phenolics, flavonoids, phenylalanine, and callose, have been reported to be regulated by excess Mn in plants [12,16,98], the regulation of the phenylpropanoid pathway seems to facilitate plants' adaptations to Mn toxicity. Furthermore, combined with the physiological and proteomic analysis, the molecular responses involved in stylo adaptation to Mn toxicity are suggested to include enhancing defense responses and phenylpropanoid pathways, adjusting photosynthesis and metabolic processes, and modulating protein synthesis and turnover [35]. Despite the advances in the identification of various genes and proteins responding to Mn toxicity, there remains a scarcity of work designed to investigate how these genes are involved in plant tolerance to Mn toxicity, and future work is needed in these areas.

9. Future Perspectives

Of the mineral nutrients essential for plant growth, Mn can cause phytotoxicity at excess levels, especially in acid soils. Even with the examination of the physiological and molecular mechanisms and characterization of genes controlling Mn tolerance over the last few decades, relatively little is known about the molecular mechanisms regulating Mn homeostasis and detoxification in plants, which are critical to allow plants to adjust their Mn requirements and to avoid toxicity. Furthermore, as many genes responsible for Mn transport and distribution are not or are only slightly responsive to external Mn, future work is required to elucidate the possible regulatory mechanisms, such as transcriptional regulatory networks and post-translational protein modifications (e.g., phosphorylation, ubiquitination, and glycosylation), by which these components facilitate plant adaptions to changing Mn levels.

Although some genes have been implicated in Mn detoxification via ectopic expression in model yeast cells or Arabidopsis, the exact roles of these genes need to be determined at both the cellular and whole-plant levels, considering molecular and physiological aspects in planta. Aside from the model plants Arabidopsis and rice, candidate genes in other crop species should be identified to clarify their roles in Mn acquisition and detoxification, which might be more complicated depending on the physiological, biochemical, and molecular responses in different crops. Once identified, these genes can potentially be used to breed crop varieties with high Mn acquisition efficiency under Mn deficiency in alkaline soils, or with increased Mn tolerance under Mn toxicity in acid soils. Additionally, in some hyperaccumulator plants that can store high levels of toxic metals without displaying obvious toxicity, excess Mn has been shown to accumulate in the non-photosynthetic tissues for detoxification [99,100]. However, the mechanisms underlying Mn hyperaccumulation and the responses of hyperaccumulators to Mn remain poorly understood. Candidate genes responsible for Mn detoxification in Mn-hyperaccumulator plants have yet to be reported. These are some of the future directions that should be taken into account, as these resources can be exploited to develop genetically engineered plants used for Mn phytoremediation.

To date, most of the studies conducted to investigate gene functions in Mn detoxification have mainly focused on Mn transport, distribution, or homeostasis. Genes associated with other pathways, such as biological regulation and signal transduction, photosynthesis, carbohydrate and energy metabolism, and secondary metabolism, which can potentially influence Mn tolerance mechanisms, have received little attention. Future efforts to investigate these areas are of great importance for increasing our understanding of how plants detoxify Mn.

10. Conclusions

Although Mn is an essential element for plants, excess Mn can cause phytotoxicity, inhibiting plant growth. This review shows that increasing plant Mn tolerance can be achieved by coordination of Mn absorption, translocation, and distribution, as well as by complex regulations of physiological changes and biochemical responses. This review highlights that Mn detoxification is regulated by a variety of genes and proteins associated with specific pathways, such as Mn transport and homeostasis, which can potentially be used to breed crop varieties with high Mn tolerance. This review also provides some of the future areas that could be taken into account in terms of gaining a better understanding of how plants tolerate Mn toxicity.

Author Contributions: Z.C. and G.L. designed the review; J.L., Y.J., R.D., and P.L. accessed the information; R.H., Z.W., and Z.C. revised the manuscript; J.L., Y.J., R.D., and X.L. designed the charts; J.L., Y.J., and Z.C. wrote the paper. All authors read and approved the final manuscript.**Funding:** This work was supported by the National Natural Science Foundation of China (31801951, 31861143013), the Young Talents Academic Innovation Project of Hainan Association for Science and Technology (QCXM201715), the Key Research and Development Program of Hainan (ZDYF2018048), the Central Public-interest Scientific Institution Basal Research Fund for CATAS (1630032017086), the Modern Agro-Industry Technology Research System (CARS-34), the Construction of World First Class Discipline of Hainan University (No.RZZX201905), and the Integrated Demonstration of Key Techniques for the Industrial Development of Featured Crops in Rocky Desertification Areas of Yunnan–Guangxi–Guizhou Provinces.

Acknowledgments: We sincerely thank American Journal Experts (www.aje.com) for their help with English language editing.

Conflicts of Interest: The authors declare no conflict of interest.

References

1. Geszvain, K.; Butterfield, C.; Davis, R.E.; Madison, A.S.; Lee, S.W.; Parker, D.L.; Soldatova, A.; Spiro, T.G.; Luther, G.W.; Tebo, B.M. The molecular biogeochemistry of manganese (II) oxidation. *Biochem. Soc. Trans.* **2012**, *40*, 1244–1248. [CrossRef] [PubMed]
2. Hernandez-Soriano, M.C.; Degryse, F.; Lombi, E.; Smolders, E. Manganese toxicity in barley is controlled by solution manganese and soil manganese speciation. *Soil Sci. Soc. Am. J.* **2012**, *76*, 399–407. [CrossRef]
3. Sparrow, L.A.; Uren, N.C. Manganese oxidation and reduction in soils: Effects of temperature, water potential, pH and their interactions. *Soil Res.* **2014**, *52*, 483–494. [CrossRef]
4. Watmough, S.A.; Eimers, M.C.; Dillon, P.J. Manganese cycling in central Ontario forests: Response to soil acidification. *Appl. Geochem.* **2007**, *22*, 1241–1247. [CrossRef]
5. Niu, L.L.; Yang, F.X.; Xu, C.; Yang, H.Y.; Liu, W.P. Status of metal accumulation in farmland soils across China: From distribution to risk assessment. *Environ. Pollut.* **2013**, *176*, 55–62. [CrossRef]
6. Antibachi, D.; Kelepertzis, E.; Kelepertsis, A. Heavy metals in agricultural soils of the Mouriki-Thiva area (Central Greece) and environmental impact implications. *Soil Sediment Contam.* **2012**, *21*, 434–450. [CrossRef]
7. Reisenauer, H.M. Determination of plant-available soil manganese. In *Manganese in Soils and Plants. Developments in Plant and Soil Sciences*; Graham, R.D., Hannam, R.J., Uren, N.C., Eds.; Springer: Dordrecht, The Netherlands, 1988; Volume 33, pp. 87–98.
8. Marschner, P. *Marschner's Mineral Nutrition of Higher Plants*, 3rd ed.; Academic Press: Boston, MA, USA, 2012; pp. 483–651.
9. Goussias, C.; Boussac, A.; Rutherford, A.W. Photosystem II and photosynthetic oxidation of water: An overview. *Philos. Trans. R. Soc. Lond. B Biol. Sci.* **2002**, *357*, 1369–1381. [CrossRef]
10. Nickelsen, J.; Rengstl, B. Photosystem II assembly: From cyanobacteria to plants. *Annu. Rev. Plant Biol.* **2013**, *64*, 609–635. [CrossRef]
11. Broadley, M.R.; Brown, P.H.; Cakmak, I.; Rengel, Z.; Zhao, F. Function of nutrients: Micronutrients. In *Marschner's Mineral Nutrition of Higher Plants*; Marschner, P., Ed.; Academic Press: Boston, MA, USA, 2012; pp. 191–248.
12. Lidon, F.C.; Barreiro, M.G.; Ramalho, J.C. Manganese accumulation in rice: Implications for photosynthetic functioning. *J. Plant Physiol.* **2004**, *161*, 1235–1244. [CrossRef]
13. Madhumita, J.M.; Sharma, A. Manganese in cell metabolism of higher plants. *Bot. Rev.* **1991**, *57*, 117–149.
14. Millaleo, R.; Reyes-Díaz, M.; Ivanov, A.G.; Mora, M.L.; Alberdi, M. Manganese as essential and toxic element for plants: Transport, accumulation and resistance mechanism. *J. Soil Sci. Plant Nutr.* **2010**, *10*, 470–481. [CrossRef]
15. Ducic, T.; Polle, A. Transport and detoxification of manganese and copper in plants. *Braz. J. Plant Physiol.* **2005**, *17*, 103–112. [CrossRef]
16. Lei, Y.B.; Korpelainen, H.; Li, C. Physiological and biochemical responses to high Mn concentrations in two contrasting *Populus cathayana* populations. *Chemosphere* **2007**, *68*, 686–694. [CrossRef] [PubMed]
17. Fernando, D.R.; Lynch, J.P. Manganese phytotoxicity: New light on an old problem. *Ann. Bot.* **2015**, *116*, 313–319. [CrossRef]
18. Rosas, A.; Rengel, Z.; de la Luz Mora, M. Manganese supply and pH influence growth, carboxylate exudation and peroxidase activity of ryegrass and white clover. *J. Plant Nutr.* **2007**, *30*, 253–270. [CrossRef]
19. De la Luz Mora, M.; Rosas, A.; Ribera, A.; Rengel, Z. Differential tolerance to Mn toxicity in perennial ryegrass genotypes: Involvement of antioxidative enzymes and root exudation of carboxylates. *Plant Soil* **2009**, *320*, 79–89. [CrossRef]
20. Chen, Z.; Sun, L.; Liu, P.; Liu, G.; Tian, J.; Liao, H. Malate synthesis and secretion mediated by a manganese-enhanced malate dehydrogenase confers superior manganese tolerance in *Stylosanthes guianensis*. *Plant Physiol.* **2015**, *167*, 176–188. [CrossRef]
21. Fecht-Christoffers, M.M.; Braun, H.P.; Lemaitre-Guillier, C.; VanDorsselaer, A.; Horst, W.J. Effect of manganese toxicity on the proteome of the leaf apoplast in cowpea. *Plant Physiol.* **2003**, *133*, 1935–1946. [CrossRef]
22. Chen, Z.; Yan, W.; Sun, L.; Tian, J.; Liao, H. Proteomic analysis reveals growth inhibition of soybean roots by manganese toxicity is associated with alteration of cell wall structure and lignification. *J. Proteom.* **2016**, *30*, 151–160. [CrossRef]

23. Führs, H.; Behrens, C.; Gallien, S.; Heintz, D.; Dorsselaer, A.V.; Braun, H.P.; Horst, W.J. Physiological and proteomic characterization of manganese sensitivity and tolerance in rice (*Oryza sativa*) in comparison with barley (*Hordeum vulgare*). *Ann. Bot.* **2010**, *105*, 1129–1140. [CrossRef]
24. Andrew, C.S.; Hegarty, M.P. Comparative responses to manganese excess of eight tropical and four temperate pasture legume species. *Aust. J. Agric. Res.* **1969**, *20*, 687–696. [CrossRef]
25. Sheng, H.; Zeng, J.; Liu, Y.; Wang, X.; Wang, Y.; Kang, H.; Fan, X.; Sha, L.; Zhang, H.; Zhou, Y. Sulfur mediated alleviation of Mn toxicity in polish wheat relates to regulating Mn allocation and improving antioxidant system. *Front. Plant Sci.* **2016**, *7*, 1382. [CrossRef] [PubMed]
26. Shao, J.F.; Yamaji, N.; Shen, R.F.; Ma, J.F. Key to Mn homeostasis in plants: Regulation of Mn transporters. *Trends Plant Sci.* **2016**, *22*, 215. [CrossRef] [PubMed]
27. Tsunemitsu, Y.; Genga, M.; Okada, T.; Yamaji, N.; Ma, J.F.; Miyazaki, A.; Kato, S.; Iwasaki, K.; Ueno, D. A member of cation diffusion facilitator family, MTP11, is required for manganese tolerance and high fertility in rice. *Planta* **2018**, *248*, 231–241. [CrossRef] [PubMed]
28. Peiter, E.; Montanini, B.; Gobert, A.; Pedas, P.; Husted, S.; Maathuis, F.J.M.; Blaudez, D.; Chalot, M.; Sanders, D. A secretory pathway-localized cation diffusion facilitator confers plant manganese tolerance. *Proc. Natl. Acad. Sci. USA* **2007**, *104*, 8532–8537. [CrossRef] [PubMed]
29. Delhaize, E.; Kataoka, T.; Hebb, D.M.; White, R.G.; Ryan, P.R. Genes encoding proteins of the cation diffusion facilitator family that confer manganese tolerance. *Plant Cell* **2003**, *15*, 1131–1142. [CrossRef]
30. Williams, L.E.; Pittman, J.K. Dissecting pathways involved in manganese homeostasis and stress in higher plants cells. In *Cell Biology of Metals and Nutrients. Plant Cell Monographs*; Hell, R., Mendel, R.R., Eds.; Springer: Berlin, Germany, 2010; Volume 17, pp. 95–117.
31. Yang, S.; Yi, K.; Chang, M.M.; Ling, G.Z.; Zhao, Z.K.; Li, X.F. Sequestration of Mn into the cell wall contributes to Mn tolerance in sugarcane (*Saccharum officinarum* L.). *Plant Soil* **2019**, *436*, 475–487. [CrossRef]
32. You, X.; Yang, L.; Lu, Y.; Li, H.; Zhang, S.; Chen, L. Proteomic changes of *Citrus* roots in response to long-term manganese toxicity. *Trees* **2014**, *28*, 1383–1399. [CrossRef]
33. Valdés-López, O.; Yang, S.S.; Aparicio-Fabre, R.; Graham, P.H.; Reyes, J.L.; Vance, C.P.; Hernández, G. MicroRNA expression profile in common bean (*Phaseolus vulgaris*) under nutrient deficiency stresses and manganese toxicity. *New Phytol.* **2010**, *187*, 805–818. [CrossRef]
34. Ceballos-Laita, L.; Gutierrez-Carbonell, E.; Imai, H.; Abadía, A.; Uemura, M.; Abadía, J.; López-Millán, A.F. Effects of manganese toxicity on the protein profile of tomato (*Solanum lycopersicum*) roots as revealed by two complementary proteomic approaches, two-dimensional electrophoresis and shotgun analysis. *J. Proteom.* **2018**, *185*, 51–63. [CrossRef]
35. Liu, P.; Huang, R.; Hu, X.; Jia, Y.; Li, J.; Luo, J.; Liu, Q.; Luo, L.; Liu, G.; Chen, Z. Physiological responses and proteomic changes reveal insights into *Stylosanthes* response to manganese toxicity. *BMC Plant Biol.* **2019**, *19*, 212. [CrossRef] [PubMed]
36. Führs, H.; Hartwig, M.; Molina, L.E.; Heintz, D.; Van, D.A.; Braun, H.P.; Horst, W.J. Early manganese-toxicity response in *Vigna unguiculata* L.—A proteomic and transcriptomic study. *Proteomics* **2008**, *8*, 149–159. [CrossRef] [PubMed]
37. Yamaji, N.; Sasaki, A.; Xia, J.X.; Yokosho, K.; Ma, J.F. A node-based switch for preferential distribution of manganese in rice. *Nat. Commun.* **2013**, *4*, 2442. [CrossRef] [PubMed]
38. Eroglu, S.; Meier, B.; Wirén, N.V.; Peiter, E. The vacuolar manganese transporter MTP8 determines tolerance to iron deficidency-induced chlorosis in Arabidopsis. *Plant Physiol.* **2016**, *170*, 1030–1045. [CrossRef]
39. Tsunemitsu, Y.; Yamaji, N.; Ma, J.F.; Kato, S.I.; Iwasaki, K.; Ueno, D. Rice reduces Mn uptake in response to Mn stress. *Plant Signal. Behav.* **2018**, *13*, e1422466. [CrossRef]
40. Ribera-Fonseca, A.; Inostroza-Blancheteau, C.; Cartes, P.; Rengel, Z.; Mora, M.L. Early induction of *Fe-SOD* gene expression is involved in tolerance to Mn toxicity in perennial ryegrass. *Plant Physiol. Bioch.* **2013**, *73*, 77–82. [CrossRef]
41. González, A.; Steffen, K.L.; Lynch, J.P. Light and excess manganese implications for oxidative stress in common bean. *Plant Physiol.* **1998**, *118*, 493–504. [CrossRef]
42. Shi, Q.H.; Zhu, Z.J.; Li, J.; Qian, Q.Q. Combined Effects of excess Mn and low pH on oxidative stress and antioxidant enzymes in cucumber roots. *Agric. Sci. China* **2006**, *5*, 767–772. [CrossRef]

43. Zhou, C.P.; Qi, Y.P.; You, X.; Yang, L.T.; Guo, P.; Ye, X.; Zhou, X.X.; Ke, F.J.; Chen, L.S. Leaf cDNA-AFLP analysis of two *Citrus* species differing in manganese tolerance in response to long-term manganese-toxicity. *BMC Genom.* **2013**, *14*, 621. [CrossRef]
44. Clarkson, D.T. The uptake and translocation of manganese by plant roots. In *Manganese in Soils and Plants. Developments in Plant and Soil Sciences*; Graham, R.D., Hannam, R.J., Uren, N.C., Eds.; Springer: Dordrecht, The Netherlands, 1988; Volume 33, pp. 101–111.
45. Socha, A.L.; Guerinot, M.L. Mn-euvering manganese: The role of transporter gene family members in manganese uptake and mobilization in plants. *Front. Plant Sci.* **2014**, *5*, 106. [CrossRef]
46. Cailliatte, R.; Schikora, A.; Briat, J.F.; Mari, S.; Curie, C. High-affinity manganese uptake by the metal transporter NRAMP1 is essential for *Arabidopsis* growth in low manganese conditions. *Plant Cell* **2010**, *22*, 904–917. [CrossRef] [PubMed]
47. Thomine, S.; Wang, R.; Ward, J.M.; Crawford, N.M.; Schroeder, J.I. Cadmium and iron transport by members of a plant metal transporter family in Arabidopsis with homology to *Nramp* genes. *Proc. Natl. Acad. Sci. USA* **2000**, *97*, 4991–4996. [CrossRef] [PubMed]
48. Sasaki, A.; Yamaji, N.; Yokosho, K.; Ma, J.F. Nramp5 is a major transporter responsible for manganese and cadmium uptake in rice. *Plant Cell* **2012**, *24*, 2155–2167. [CrossRef] [PubMed]
49. Ishimaru, Y.; Takahashi, R.; Bashir, K.; Shimo, H.; Senoura, T.; Sugimoto, K.; Ono, K.; Yano, M.; Ishikawa, S.; Arao, T.; et al. Characterizing the role of rice NRAMP5 in manganese, iron and cadmium transport. *Sci. Rep.* **2012**, *2*, 286. [CrossRef]
50. Ueno, D.; Sasaki, A.; Yamaji, N.; Fujii, Y.; Takemoto, Y.; Moriyama, S.; Che, J.; Moriyama, Y.; Lwasaki, K.; Ma, J.F. A polarly localized transporter for efficient manganese uptake in rice. *Nat. Plants* **2015**, *1*, 15170. [CrossRef]
51. Wu, D.; Yamaji, N.; Yamane, M.; Kashino, F.M.; Sato, K.; Ma, J.F. The HvNramp5 transporter mediates uptake of cadmium and manganese, but not iron. *Plant Physiol.* **2016**, *172*, 1899–1910. [CrossRef]
52. Kaiser, B.N.; Moreau, S.; Castelli, J.; Lambert, A.; Bogliolo, S.; Puppo, A.; Day, D.A. The soybean NRAMP homologue, GmDMT1, is a symbiotic divalent metal transporter capable of ferrous iron transport. *Plant J.* **2003**, *35*, 295–304. [CrossRef]
53. Pedas, P.; Ytting, C.K.; Fuglsang, A.T.; Jahn, T.P.; Schjoerring, J.K.; Husted, S. Manganese efficiency in barley: Identification and characterization of the metal ion transporter HvIRT1. *Plant Physiol.* **2008**, *148*, 455–466. [CrossRef]
54. Milner, M.J.; Seamon, J.; Craft, E.; Kochian, L.V. Transport properties of members of the ZIP family in plants and their role in Zn and Mn homeostasis. *J. Exp. Bot.* **2013**, *64*, 369–381. [CrossRef]
55. Ishimaru, Y.; Masuda, H.; Bashir, K.; Inoue, H.; Tsukamoto, T.; Takahashi, M.; Nakanishi, H.; Aoki, N.; Hirose, T.; Ohsugi, R.; et al. Rice metal–nicotianamine transporter, OsYSL2, is required for the long-distance transport of iron and manganese. *Plant J.* **2010**, *62*, 379–390. [CrossRef]
56. Koike, S.; Inoue, H.; Mizuno, D.; Takahashi, M.; Nakanishi, H.; Mori, S.; Nishizawa, N.K. OsYSL2 is a rice metal-nicotianamine transporter that is regulated by iron and expressed in the phloem. *Plant J.* **2004**, *39*, 415–424. [CrossRef] [PubMed]
57. Stacey, M.G.; Koh, S.; Becker, J.; Stacey, G. AtOPT3, a member of the oligopeptide transporter family, is essential for embryo development in Arabidopsis. *Plant Cell* **2002**, *14*, 2799–2811. [CrossRef] [PubMed]
58. Wintz, H.; Fox, T.; Wu, Y.Y.; Feng, V.; Chen, W.; Chang, H.S.; Zhu, T.; Vulpe, C. Expression profiles of *Arabidopsis thaliana* in mineral deficiencies reveal novel transporters involved in metal homeostasis. *J. Biol. Chem.* **2003**, *278*, 47644–47653. [CrossRef] [PubMed]
59. Conte, S.S.; Chu, H.H.; Rodriguez, D.C.; Punshon, T.; Vasques, K.A.; Salt, D.E. *Arabidopsis thaliana* Yellow Stripe1-Like4 and Yellow Stripe1-Like6 localize to internal cellular membranes and are involved in metal ion homeostasis. *Front. Plant Sci.* **2013**, *4*, 283. [CrossRef] [PubMed]
60. Schaaf, G.; Ludewig, U.; Erenoglu, B.E.; Mori, S.; Kitahara, T.; von Wirén, N. ZmYS1 functions as a proton-coupled symporter for phytosiderophore- and nicotianamine-chelated metals. *J. Biol. Chem.* **2004**, *279*, 9091–9096. [CrossRef]
61. Pittman, J.K. Managing the manganese: Molecular mechanisms of manganese transport and homeostasis. *New Phytol.* **2005**, *167*, 733–742. [CrossRef]
62. Gustin, J.L.; Zanis, M.J.; Salt, D.E. Structure and evolution of the plant cation diffusion facilitator family of ion transporters. *BMC Evol. Biol.* **2011**, *11*, 76. [CrossRef]

63. Chen, Z.; Fujii, Y.; Yamaji, N.; Masuda, S.; Takemoto, Y.; Kamiya, T.; Yusuyin, Y.; Iwasaki, K.; Kato, S.I.; Maeshima, M.; et al. Mn tolerance in rice is mediated by MTP8.1, a member of the cation diffusion facilitator family. *J. Exp. Bot.* **2013**, *64*, 4375–4387. [CrossRef]
64. Takemoto, Y.; Tsunemitsu, Y.; Fujii, K.M.; Mitani, U.N.; Yamaji, N.; Ma, J.F.; Kato, S.I.; Iwasaki, K.; Ueno, D. The tonoplast-localized transporter MTP8.2 contributes to manganese detoxification in the shoots and roots of *Oryza sativa* L. *Plant Cell Physiol.* **2017**, *58*, 1573–1582. [CrossRef]
65. Migocka, M.; Papierniak, A.; Dziubińska, E.M.; Poździk, P.; Posyniak, E.; Garbiec, A.; Filleur, S. Cucumber metal transport protein MTP8 confers increased tolerance to manganese when expressed in yeast and *Arabidopsis thaliana*. *J. Exp. Bot.* **2014**, *65*, 5367–5384. [CrossRef]
66. Migocka, M.; Papierniak, A.; Kosieradzka, A.; Posyniak, E.; Maciaszczyk-Dziubinska, E.; Biskup, R.; Garbiec, A.; Marchewka, T. Cucumber metal tolerance protein CsMTP9 is a plasma membrane H^+-coupled antiporter involved in the Mn^{2+} and Cd^{2+} efflux from root cells. *Plant J.* **2015**, *84*, 1045–1058. [CrossRef] [PubMed]
67. Li, Q.; Li, Y.; Wu, X.; Zhou, L.; Zhu, X.; Fang, W. Metal transport protein 8 in *Camellia sinensis* confers superior manganese tolerance when expressed in yeast and *Arabidopsis thaliana*. *Sci. Rep.* **2017**, *7*, 39915. [CrossRef] [PubMed]
68. Hirschi, K.D.; Korenkov, V.D.; Wilganowski, N.L.; Wagner, G.J. Expression of Arabidopsis *CAX2* in tobacco. Altered metal accumulation and increased manganese tolerance. *Plant Physiol.* **2000**, *124*, 125–133. [CrossRef] [PubMed]
69. Shigaki, T.; Pittman, J.K.; Hirschi, K.D. Manganese specificity determinants in the Arabidopsis metal/H^+ antiporter CAX2. *J. Biol. Chem.* **2003**, *278*, 6610–6617. [CrossRef] [PubMed]
70. Pittman, J.K.; Shigaki, T.; Marshall, J.L.; Morris, J.L.; Cheng, N.H.; Hirschi, K.D. Functional and regulatory analysis of the *Arabidopsis thaliana* CAX2 cation transporter. *Plant Mol. Biol.* **2004**, *56*, 959–971. [CrossRef] [PubMed]
71. Cheng, N.H.; Pittman, J.K.; Shigaki, T.; Hirschi, K.D. Characterization of CAX4, an Arabidopsis H^+/cation antiporter. *Plant Physiol.* **2002**, *128*, 1245–1254. [CrossRef] [PubMed]
72. Edmond, C.; Shigaki, T.; Ewert, S.; Nelson, M.D.; Connorton, J.M.; Chalova, V.; Noordally, Z.; Pittman, J.K. Comparative analysis of CAX2-like cation transporters indicates functional and regulatory diversity. *Biochem. J.* **2009**, *418*, 145–154. [CrossRef]
73. Mei, H.; Cheng, N.H.; Zhao, J.; Park, S.; Escareno, R.A.; Pittman, J.K.; Hirschi, K.D. Root development under metal stress in *Arabidopsis thaliana* requires the H^+/cation antiporter CAX4. *New Phytol.* **2009**, *183*, 95–105. [CrossRef]
74. Korenkov, V.; Hirschi, K.; Crutchfield, J.D.; Wagner, G.J. Enhancing tonoplast Cd/H antiport activity increases Cd, Zn, and Mn tolerance, and impacts root/shoot Cd partitioning in *Nicotiana tabacum* L. *Planta* **2007**, *226*, 1379–1387. [CrossRef]
75. Connorton, J.M.; Webster, R.E.; Cheng, N.; Pittman, J.K. Knockout of multiple Arabidopsis cation/H^+ exchangers suggests isoform-specific roles in metal stress response, germination and seed mineral nutrition. *PLoS ONE* **2012**, *7*, e47455. [CrossRef]
76. Delhaize, E.; Gruber, B.D.; Pittman, J.K.; White, R.G.; Leung, H.; Miao, Y.; Jiang, L.; Ryan, P.R.; Richardson, A.E. A role for the *AtMTP11* gene of Arabidopsis in manganese transport and tolerance. *Plant J.* **2007**, *51*, 198–210. [CrossRef] [PubMed]
77. Pedas, P.; Stokholm, M.S.; Hegelund, J.N.; Ladegård, A.H.; Schjoerring, J.K.; Husted, S. Golgi localized barley MTP8 proteins facilitate Mn transport. *PLoS ONE* **2014**, *9*, e113759. [CrossRef] [PubMed]
78. Erbasol, I.; Bozdag, G.O.; Koc, A.; Pedas, P.; Karakaya, H.C. Characterization of two genes encoding metal tolerance proteins from *Beta vulgaris* subspecies maritima that confers manganese tolerance in yeast. *Biometals* **2013**, *26*, 795–804. [CrossRef] [PubMed]
79. Baxter, I.; Tchieu, J.; Sussman, M.R.; Boutry, M.; Palmgren, M.G.; Gribskov, M.; Harper, J.F.; Axelsen, K.B. Genomic comparison of P-type ATPase ion pumps in Arabidopsis and rice. *Plant Physiol.* **2003**, *132*, 618–628. [CrossRef]
80. Huda, K.M.; Banu, M.S.; Tuteja, R.; Tuteja, N. Global calcium trans ducer P-type Ca^{2+}-ATPases open new avenues for agriculture by regulating stress signalling. *J. Exp. Bot.* **2013**, *64*, 3099–3109. [CrossRef] [PubMed]

81. Wu, Z.; Liang, F.; Hong, B.; Young, J.C.; Sussman, M.R.; Harper, J.F.; Sze, H. An endoplasmic reticulum-bound Ca^{2+}/Mn^{2+} pump, ECA1, supports plant growth and confers tolerance to Mn^{2+} stress. *Plant Physiol.* **2002**, *130*, 128–137. [CrossRef] [PubMed]
82. Li, X.; Chanroj, S.; Wu, Z.; Romanowsky, S.M.; Harper, J.F.; Sze, H. A distinct endosomal Ca^{2+}/Mn^{2+} pump affects root growth through the secretory process. *Plant Physiol.* **2008**, *147*, 1675–1689. [CrossRef]
83. Mills, R.F.; Doherty, M.L.; Lopez-Marques, R.L.; Weimar, T.; Dupree, P.; Palmgren, M.G.; Pittman, J.K.; Williams, L.E. ECA3, a Golgi-localized P2A-type ATPase, plays a crucial role in manganese nutrition in Arabidopsis. *Plant Physiol.* **2008**, *146*, 116–128. [CrossRef]
84. Sasaki, A.; Yamaji, N.; Xia, J.; Ma, J.F. OsYSL6 is involved in the detoxification of excess manganese in rice. *Plant Physiol.* **2011**, *157*, 1832–1840. [CrossRef]
85. Ma, J.F.; Yamaji, N. Silicon uptake and accumulation in higher plants. *Trends Plant Sci.* **2006**, *11*, 392–397. [CrossRef]
86. Führs, H.; Götze, S.; Specht, A.; Erban, A.; Gallien, S.; Heintz, D.; Dorsselaer, A.V.; Kopka, J.; Braun, H.P.; Horst, W.J. Characterization of leaf apoplastic peroxidases and metabolites in *Vigna unguiculata* in response to toxic manganese supply and silicon. *J. Exp. Bot.* **2009**, *60*, 1663–1678. [CrossRef] [PubMed]
87. Horst, W.J.; Fecht, M.; Naumann, A.; Wissemeier, A.H.; Maier, P. Physiology of manganese toxicity and tolerance in *Vigna unguiculata* (L.) Walp. *J. Plant Nutr. Soil Sci.* **1999**, *162*, 263–274. [CrossRef]
88. Rogalla, H.; Romheld, V. Role of leaf apoplast in silicon-mediated manganese tolerance of *Cucumis sativus* L. *Plant Cell Environ.* **2002**, *25*, 549–555. [CrossRef]
89. Dragišić Maksimović, J.; Mojović, M.; Maksimović, V.; Römheld, V.; Nikolic, M. Silicon ameliorates manganese toxicity in cucumber by decreasing hydroxyl radical accumulation in the leaf apoplast. *J. Exp. Bot.* **2012**, *63*, 2411–2420. [CrossRef] [PubMed]
90. Che, J.; Yamaji, N.; Shao, J.F.; Ma, J.F.; Shen, R.F. Silicon decreases both uptake and root-to-shoot translocation of manganese in rice. *J. Exp. Bot.* **2016**, *67*, 1535–1544. [CrossRef]
91. Ma, J.F.; Tamai, K.; Yamaji, N.; Mitani, N.; Konishi, S.; Katsuhara, M.; Ishiguro, M.; Murata, Y.; Yano, M. A silicon transporter in rice. *Nature* **2006**, *440*, 688–691. [CrossRef] [PubMed]
92. Xu, X.; Shi, J.; Chen, X.; Chen, Y.; Hu, T. Chemical forms of manganese in the leaves of manganese hyperaccumulator *Phytolacca acinosa* Roxb.(Phytolaccaceae). *Plant Soil* **2009**, *318*, 197–204. [CrossRef]
93. Fernando, D.R.; Mizuno, T.; Woodrow, I.E.; Baker, A.J.; Collins, R.N. Characterization of foliar manganese (Mn) in Mn (hyper)accumulators using X-ray absorption spectroscopy. *New Phytol.* **2010**, *188*, 1014–1027. [CrossRef]
94. Kopittke, P.M.; Lombi, E.; McKenna, B.A.; Wang, P.; Donner, E.; Webb, R.I.; Blamey, F.P.; de Jonge, M.D.; Paterson, D.; Howard, D.L.; et al. Distribution and speciation of Mn in hydrated roots of cowpea at levels inhibiting root growth. *Physiol. Plant.* **2013**, *147*, 453–464. [CrossRef]
95. Sasaki, T.; Yamamoto, Y.; Ezaki, B.; Katsuhara, M.; Ahn, S.J.; Ryan, P.R.; Delhaize, E.; Matsumoto, H. A wheat gene encoding an aluminum-activated malate transporter. *Plant J.* **2004**, *37*, 645–653. [CrossRef]
96. Liang, C.; Tian, J.; Liao, H. Proteomics dissection of plant responses tomineral nutrient deficiency. *Proteomics* **2013**, *13*, 624–636. [CrossRef] [PubMed]
97. Stark, R.; Grzelak, M.; Hadfield, J. RNA sequencing: The teenage years. *Nat. Rev. Genet.* **2019**, *1*, 1–26. [CrossRef] [PubMed]
98. Fecht-Christoffers, M.M.; Fuhrs, H.; Braun, H.P.; Horst, W.J. The role of hydrogen peroxide-producing and hydrogen peroxide-consuming peroxidases in the leaf apoplast of cowpea in manganese tolerance. *Plant Physiol.* **2006**, *140*, 1451–1463. [CrossRef] [PubMed]
99. Kramer, U. Metal hyperaccumulation in plants. *Ann. Rev. Plant Biol.* **2010**, *61*, 517–534. [CrossRef] [PubMed]
100. Fernando, D.R.; Marshall, A.; Baker, A.J.; Mizuno, T. Microbeam methodologies as powerful tools in manganese hyperaccumlation research: Present status and future directions. *Front. Plant Sci.* **2013**, *4*, 219. [CrossRef]

© 2019 by the authors. Licensee MDPI, Basel, Switzerland. This article is an open access article distributed under the terms and conditions of the Creative Commons Attribution (CC BY) license (http://creativecommons.org/licenses/by/4.0/).

International Journal of
Molecular Sciences

Article

Combined Effect of Cadmium and Lead on Durum Wheat

Alessio Aprile [1], Erika Sabella [1,*], Enrico Francia [2], Justyna Milc [2], Domenico Ronga [2], Nicola Pecchioni [2], Erika Ferrari [3], Andrea Luvisi [1], Marzia Vergine [1] and Luigi De Bellis [1]

1 Department of Biological and Environmental Sciences and Technologies (DiSTeBA), Salento University, Via Prov. le Lecce-Monteroni, 73100 Lecce, Italy; alessio.aprile@unisalento.it (A.A.); andrea.luvisi@unisalento.it (A.L.); marzia.vergine@unisalento.it (M.V.); luigi.debellis@unisalento.it (L.D.B.)

2 Department of Life Science, University of Modena and Reggio Emilia, via Amendola 2, 42122 Reggio Emilia, Italy; enrico.francia@unimore.it (E.F.); justynaanna.milc@unimore.it (J.M.); domenico.ronga@unimore.it (D.R.); nicola.pecchioni@unimore.it (N.P.)

3 Department of Chemical and Geological Sciences, University of Modena and Reggio Emilia, via G. Campi 103, 41125 Modena, Italy; erika.ferrari@unimore.it

* Correspondence: erika.sabella@unisalento.it

Received: 22 October 2019; Accepted: 22 November 2019; Published: 24 November 2019

Abstract: Cadmium (Cd) and lead (Pb) are two toxic heavy metals (HMs) whose presence in soil is generally low. However, industrial and agricultural activities in recent years have significantly raised their levels, causing progressive accumulations in plant edible tissues, and stimulating research in this field. Studies on toxic metals are commonly focused on a single metal, but toxic metals occur simultaneously. The understanding of the mechanisms of interaction between HMs during uptake is important to design agronomic or genetic strategies to limit contamination of crops. To study the single and combined effect of Cd and Pb on durum wheat, a hydroponic experiment was established to examine the accumulation of the two HMs. Moreover, the molecular mechanisms activated in the roots were investigated paying attention to transcription factors (bHLH family), heavy metal transporters and genes involved in the biosynthesis of metal chelators (nicotianamine and mugineic acid). Cd and Pb are accumulated following different molecular strategies by durum wheat plants, even if the two metals interact with each other influencing their respective uptake and translocation. Finally, we demonstrated that some genes (*bHLH 29, YSL2, ZIF1, ZIFL1, ZIFL2, NAS2* and *NAAT*) were induced in the durum wheat roots only in response to Cd.

Keywords: cadmium; lead; nicotianamine; mugineic acid; heavy metal; toxic metal; durum wheat

---

1. Introduction

The distribution of heavy metals (HMs) in soils is variable from one place to another. In some regions, the HM natural background is higher than in other ones, but industrialization and human activities had strongly, in last decades, modified the concentration of many HMs worldwide [1]. A concern with HMs is that they can easily enter the food chain through consumption of vegetables and plant parts. Indeed, the toxicity is due mainly to chronic exposure by eating HM-contaminated foods. Moreover, since low levels of HMs in soil generally do not affect plant growth and development (no visible symptoms), HMs could endanger human health [2] if adequate counteractions are not implemented. Soil metal contamination usually occurs with a combination of different metals. Cadmium (Cd) and lead (Pb) are considered environmental hazards, as they are toxic for humans and other living organisms [3,4] and the Codex Alimentarius (CDX 193-1995, Amended 2019) has set a maximum level of 0.2 mg kg^{-1} for both Cd and Pb in wheat [5].

Cd is an element of group II B in the periodic table and its atomic number is 48, while Pb belongs to group IV A and its atomic number is 82. Cd and Pb can form complexes with other compounds. In particular, Cd could form complexes with ammonia, amines, halide ions and cyanide [6]. In soils, Pb makes complexes with inorganic constituents (e.g., HCO_3^-, CO_3^{2-}, SO_4^{2-} and Cl^-), or may occur as organic ligands [7].

Although, Cd and Pb are not essential elements, plants are able to adsorb these metals from the soil and store them on different edible organs [8,9]. For these reasons, the control of HM accumulation in plant edible organs is a key point to preserve human health.

Plants respond to HM toxicity activating several physiological and molecular mechanisms. Such responses include immobilization, exclusion, chelation and compartmentalization of the metal ions, and the expression of common stress-related genes such as those involved in ethylene pathway and genes coding for stress proteins [10]. The current approach in risk-evaluation of HMs accumulation in plants is almost always based on the effects of single contaminants [11], but the combined effect of HMs is still low investigated.

Durum wheat is a staple food used to produce pasta (mainly in Europe and America), couscous and freekeh (Africa), bulgur (Asia) and bread (South Italy); minor crop if compared to bread wheat, it is mainly cultivated in Europe (Italy and France), America (Canada, USA, Mexico and Argentina), North Africa and Asia (Ukraine, Russia, Kazakhstan Indi and, China). These regions have different climate conditions and soil types with variable levels of toxic metals. In recent years, many authors have reported information about metal accumulation in durum wheat tissues, such as Cd [12,13], copper [14], arsenic [15], nickel [16], Pb, copper and chromium [17]. However, little is known about the combined effect of these metals on plant development, metal compartmentalization in plant tissues and molecular responses. Recently Shafiq et al. [18] reported how the expression of Heavy Metal ATPase 2 and ATP-Binding Cassette and promoter methylation could have a central role in Cd, Pb and zinc accumulation.

The aim of this work was to investigate the uptake and translocation of Cd and Pb, their interaction in root and leaf tissues and the molecular mechanisms activated by the aboveground presence of one or both metals during the growth of durum wheat plants. We employed two near-isogenic lines (NILs) [19] with an opposite behavior concerning Cd accumulation in leaves, low Cd (L-Cd NIL) and high Cd (H-Cd NIL), respectively. Since Cd accumulation is strongly affected by specific genomic regions [20], we should observe similar accumulation trends in these plants if the same genome regions are involved in Pb accumulation. Moreover, we added to the experiment design two commercial cultivars already characterized for their accumulation and responses to Cd [13] and, with this experiment, we evaluated how the contemporaneous presence of the Cd and Pb can affect each other.

2. Results

2.1. Levels of Cd and Pb in Root and Leaf of Wheat Plants

Samples were collected 42 days after germination, at the tillering stage, from plants grown in hydroponic solutions with the addition of Cd or Pb or with both metals at the concentrations of 0.5 and 2.0 μM, respectively. Figures 1 and 2 show the concentration of the two HMs in roots and leaves of the wheat plants; in roots, Cd and Pb concentrations range from 10 to 40 μg/g dry weight. The presence of a higher level of Cd and Pb in roots is evident (Figures 1 and 2): approximately, the concentration in roots is ten-times higher than in leaves. The Cd concentration in leaves expressed as μg/g dry weight is also twice as high as that of Pb and about four times if expressed as molarity because of the different atomic weight of the two metals.

L-Cd NIL, the near isogenic line characterized by the ability to accumulate a low level of Cd in leaves, collected, as expected, a low level of Cd in leaves compared to H-Cd NIL and to Svevo (Figure 1), whereas it showed the presence of a high level of Pb in leaves in comparison with all other genotypes (Figure 2), suggesting the presence of different molecular mechanism for the transport of the two metals. In L-Cd NIL the co-presence of Pb reduced the Cd accumulation in L-Cd NIL leaves (Figure 1).

Despite H-Cd NIL theoretically shares about the 95% of the genome with L-Cd NIL, it accumulates a high level of Cd in leaves [17], as confirmed in Figure 1. The co-presence of Pb did not alter significantly the concentration of Cd, which remained at comparable levels in both roots and leaves. Instead, L-Cd NIL and H-Cd NIL showed a similar behavior concerning the uptake and translocation of Pb and the co-presence of Cd altered considerably the Pb accumulation only in L-Cd NIL leaves (Figure 2).

The accumulation of Cd in Creso was influenced by the co-presence of Pb in hydroponic solution: when Creso was treated with both metals, the Cd concentrations in roots and leaves were slightly lower indicating a negative effect of Pb on Cd uptake/translocation.

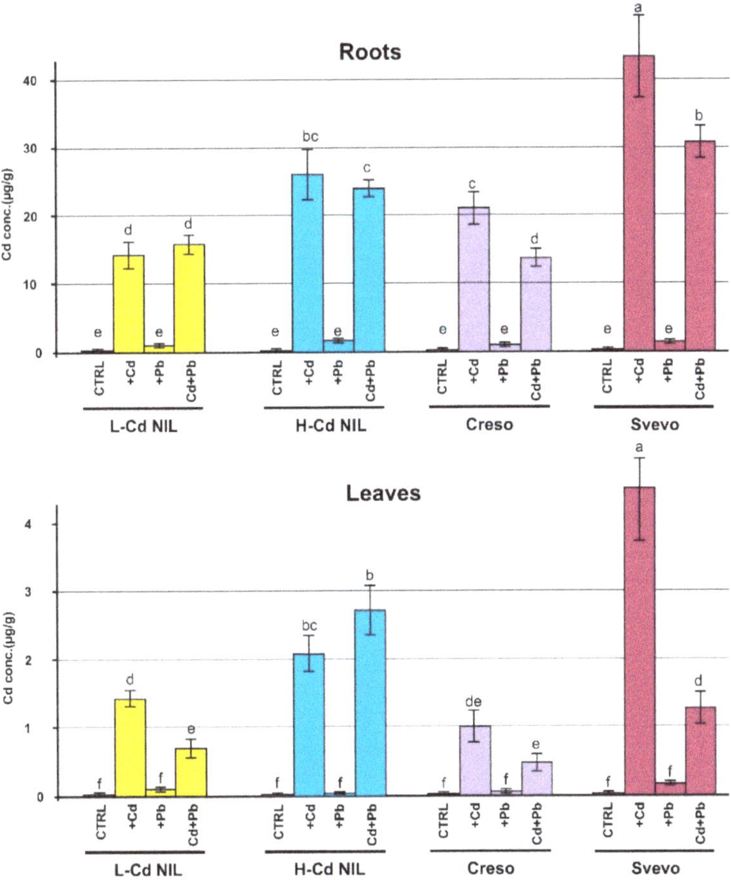

Figure 1. Cadmium (Cd) concentration in roots and leaves of low cadmium near-isogenic line (L-Cd NIL), high cadmium near-isogenic line (H-Cd NIL), Creso and Svevo. Cd concentrations in durum wheat genotypes grown in standard hydroponic solution in the presence of Cd 0.5 µM, lead (Pb) 2.0 µM or in the presence of both heavy metals (HMs) (Cd 0.5 µM plus Pb 2.0 µM). Roots and leaves were collected 42 days after germination (at the tillering stage). Cd concentration was quantified by inductively coupled plasma mass spectrometer (ICP-MS). Statistical analysis was performed through ANOVA (p-value < 0.05, n = 3) followed by Tukey-HSD post hoc test. Different letters correspond to statistically different means.

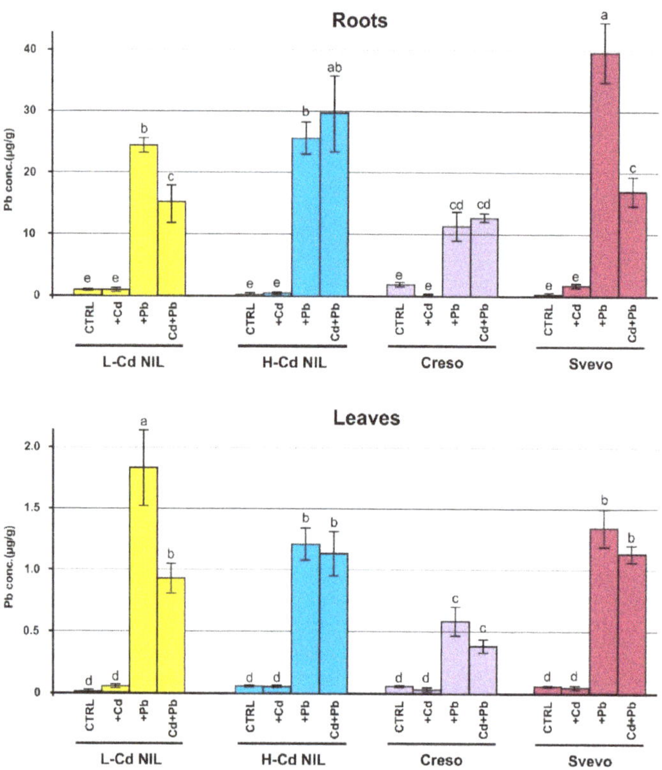

Figure 2. Pb concentration in roots and leaves of L-Cd NIL, H-Cd NIL, Creso and Svevo. Pb concentrations in durum wheat genotypes grown in standard hydroponic solution in the presence of Cd 0.5 µM, Pb 2.0 µM or in the presence of both HMs (Cd 0.5 µM plus Pb 2.0 µM). Roots and leaves were collected 42 days after germination (at the tillering stage). Pb concentration was quantified by ICP-MS. Statistical analysis was performed through ANOVA (p-value < 0.05, n = 3) followed by Tukey-HSD post hoc test. Different letters correspond to statistically different means.

Svevo, compared to the other genotypes, resulted more sensitive to the occurrence at the same time of two metals in the hydroponic solution. Indeed, in Svevo the addition of a second metal strongly reduced the accumulation of the other one. This behavior was observed in roots (both Cd and Pb) and in leaves (only Cd was significantly reduced after the application of Pb).

An important parameter to study HMs uptake and translocation in plants is the translocation factor (Figure 3); it is the ratio of the metal concentration in other plant tissues in relation to roots [3]. Figure 3 indicates that the translocation factors from roots to leaves for Cd and Pb were clearly lower than 1.0 (ranging around 0.1 or less as it results from the values on the axes) indicating a robust limitation in Cd and Pb transport in durum wheat and an immobilization in the root cells; this strategy is widely used by plants to protect the photosynthetic tissues from damages caused by the HMs [21]. Anyway, translocation factor values showed that in the genotypes L-Cd NIL and Creso, Cd and Pb were translocated from root to shoot with a similar ratio; conversely, in Svevo and H-Cd NIL, Cd was translocated from root to shoot more efficiently than Pb compared with the other genotypes (Figure 3a). This is an expected result since Svevo and H-Cd NIL are well-known genotypes with high grain-Cd accumulation [13,17]. During the combined treatment with Cd and Pb, the genotype H-Cd NIL kept higher translocation factor values for Cd (Figure 3b), while in Svevo, the translocation factor of Pb became significantly higher than the translocation factors of Cd (Figure 3b). This is explainable since in Svevo the HMs combined treatment affected Cd accumulation more at leaves level than in roots (Figure 1) causing a decrease in translocation factor values. In contrast, the combined treatment impacted Pb accumulation at root level more significantly than in leaves (Figure 2), determining an increase in translocation factors values.

Such data may represent a first evidence of the existence of different response mechanisms to Cd and Pb.

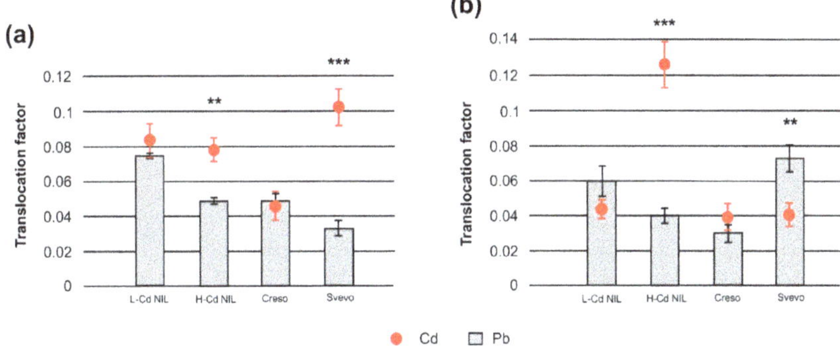

Figure 3. Translocation factor of Cd and Pb from root to shoot of wheat plants, (**a**) after the single treatments with Cd and Pb and (**b**) after the combined treatment with the two HMs. The significant differences, between the single heavy metal treatments, were highlighted according to Student's *t*-test (* $p < 0.05$; ** $p < 0.01$; *** $p < 0.001$).

2.2. Gene Expression in Response to Cd and Pb

The differences in Cd and Pb accumulation observed in the durum wheat genotypes may be induced by differential expression of gene categories involved in metal ion response and transport. The role of these genes has already been investigated in tissues of plants treated with Cd [22]. To study the molecular response to Cd, Pb and their combined effect, we considered the level of expression of genes already known to be involved in HMs responses. We focused the attention on transcription factors (bHLH and WRKY family), eight metal ion transporters and genes coding for the enzyme responsible of the synthesis of nicotianamine and mugineic acid, that are two typical metal chelators of graminaceous plants [23]. The expression levels of these genes were analyzed both in roots and leaves, but we observed no expression or no differential expression in durum wheat leaves, suggesting a tissue-specific regulation/expression. Below, only the root transcription data were reported.

2.2.1. Expression of the Transcription Factors Basic Helix-Loop-Helix (bHLH) and WRKY33

As shown in Figure 4, the expression of *bHLH29/FIT* and *bHLH38/ORG2* was clearly up-regulated in the roots of the four analyzed genotypes when treated with Cd and with Cd plus Pb; on the contrary, *bHLH47/PYE* (Figure 4) was only up-regulated in the genotype H-Cd NIL both after a treatment with Cd or Pb and following the combined stress determined by Cd plus Pb. *WRKY33*, member of the WRKY transcription factors family, did not show a significant modulation in roots of the four genotypes during treatments with the HMs (Figure 4).

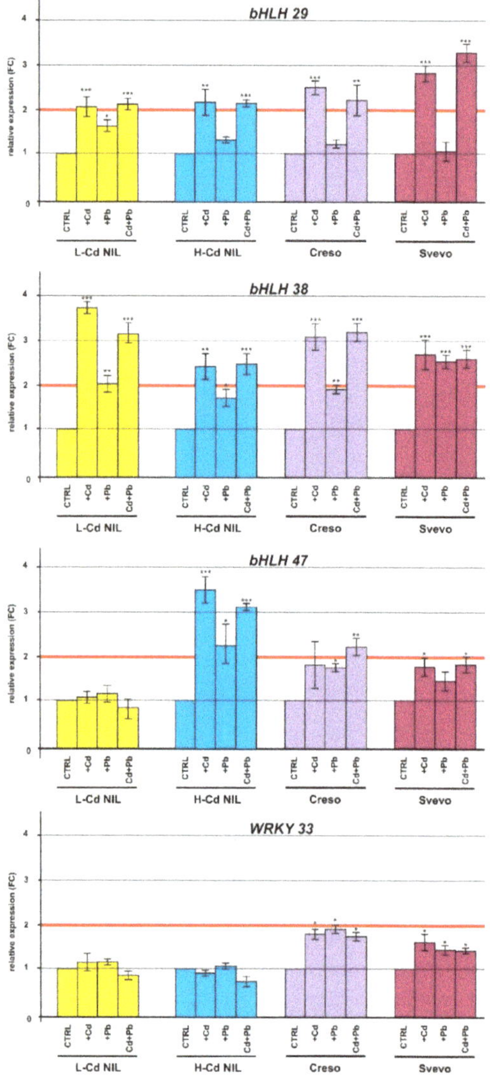

Figure 4. Relative expression as fold change (FC) of the *bHLH29/FIT*, *bHLH38/ORG2*, *bHLH47/PYE* and *WRKY33* genes in root tissues of the durum wheat genotypes (L-Cd NIL, H-Cd NIL, Creso and Svevo) grown in the presence of Cd 0.5 μM, Pb 2.0 μM or in the presence of both HMs (Cd 0.5 μM plus Pb 2.0 μM). Error bars indicate standard deviation of the mean of three technical replicates resulting from a bulk of three biological replicates. ANOVA results were reported basing on their statistical significance. * $p < 0.05$, ** $p < 0.01$, *** $p < 0.001$.

2.2.2. Expression of HMs Transporters

Plants treated with HMs showed no significant change in transcripts of the genes coding for the HM transporters ZIP4 and ZTP29 in root tissues (Figure 5). *YSL1* was slightly up-regulated in the roots of the L-Cd NIL and Creso when treated with Cd or with Cd plus Pb (Figure 5); conversely, expression level of *YSL2* was strongly up-regulated in roots of L-Cd NIL and Creso when treated with Cd. The presence of both Cd and Pb resulted in a significant increase of the transcript levels in roots of L-Cd NIL; in roots of the cultivar Creso, the combined stress induced a slight upregulation (Figure 5). A slight up-regulation was observed also in H-Cd NIL and Svevo roots treated with Cd and in the combined treatment Cd plus Pb (Figure 5). The vacuolar zinc transporter genes *ZIF* and *ZIF*-like genes (*ZIFL1* and *ZIFL2*) were strongly up-regulated in the roots of the four durum wheat genotypes treated with Cd and Cd plus Pb (Figure 6). The transcript levels were significantly higher in L-Cd NIL and Creso (Figure 6). Finally, the plasma membrane-localized transporter *HMA5* was significantly up-regulated in roots when treated both with one HM, or with the combined Cd plus Pb (Figure 6).

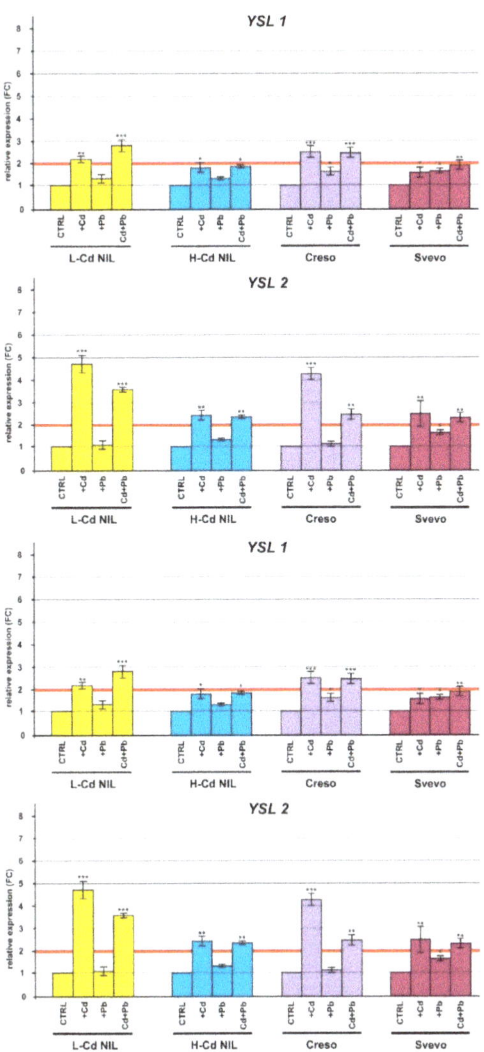

Figure 5. Relative expression (FC) of the *ZIP4*, *ZTP29*, *YSL1* and *YSL2* genes in root tissues of the durum wheat genotypes (L-Cd NIL, H-Cd NIL, Creso and Svevo) grown in the presence of Cd 0.5 μM, Pb 2.0 μM or in the presence of both HMs (Cd 0.5 μM plus Pb 2.0 μM). Error bars indicate standard deviation of the mean of three technical replicates derived from a bulk of three biological replicates. ANOVA results were reported basing on their statistical significance. * $p < 0.05$, ** $p < 0.01$, *** $p < 0.001$.

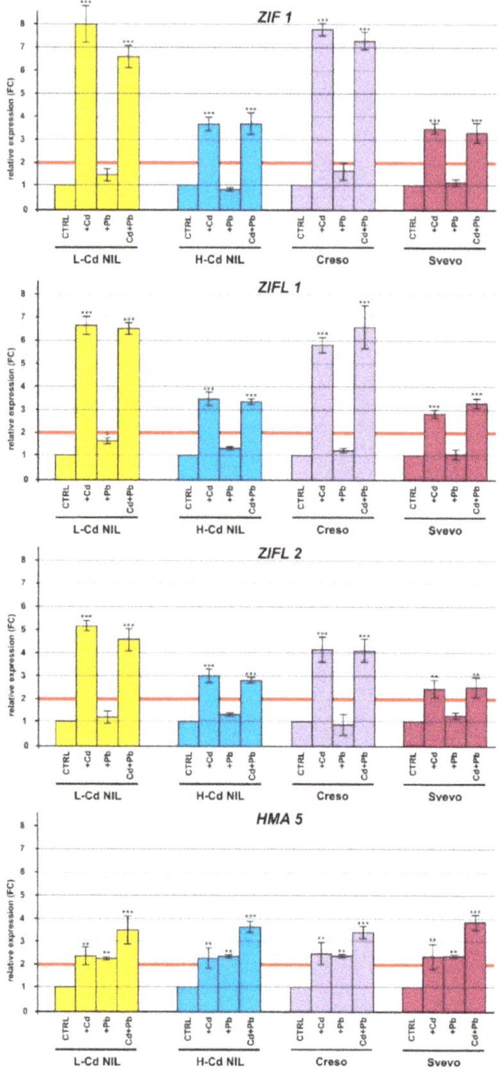

Figure 6. Relative expression (FC) of the *ZIF1*, *ZIFL1*, *ZIFL2* and *HMA5* genes in root tissues of the durum wheat genotypes (L-Cd NIL, H-Cd NIL, Creso and Svevo) grown in the presence of Cd 0.5 μM, Pb 2.0 μM or in the presence of both HMs (Cd 0.5 μM plus Pb 2.0 μM). Error bars indicate standard deviation of the mean of three technical replicates derived from a bulk of three biological replicates. ANOVA results were reported basing on their statistical significance. * $p < 0.05$, ** $p < 0.01$, *** $p < 0.001$.

2.2.3. Expression of the Nicotianamine Synthase Genes (NAS) and Nicotianamine Aminotransferase (NAAT)

In the current study, the transcripts levels of the nicotianamine synthase genes (*NAS2*, *NAS3* and *NAS4*) confirmed their up-regulation in roots of plants treated with Cd (Figure 7) as observed by other authors [19]. Interestingly, a significant increase of the transcript levels for the genes *NAS2*, *NAS3* and *NAS4* were also found in roots of plants treated with Cd plus Pb (Figure 7). Moreover, the gene *NAS3* resulted modulated in response to Pb treatment too, with a minor induction in roots (Figure 7). The transamination of nicotianamine produces mugineic acid and the enzyme that catalyzes the synthesis is called nicotianamine aminotransferase (NAAT); the relative gene is expressed at highest levels in roots of L-Cd NIL and Creso grown in the presence of Cd and Cd plus Pb; a significant induction was also found in H-Cd NIL and Svevo exposed to Cd and Cd plus Pb (Figure 7).

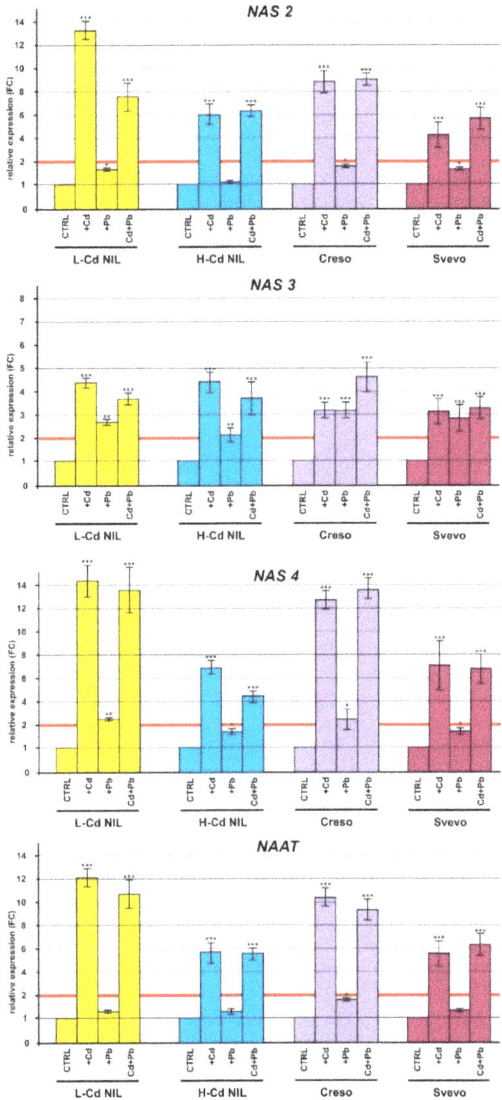

Figure 7. Relative expression (FC) of the nicotianamine synthase genes (*NAS2*, *NAS3*, *NAS4*) and nicotianamine aminotransferase gene (*NAAT*) in root tissues of the durum wheat genotypes (L-Cd, H-Cd, Creso and Svevo) grown in the presence of Cd 0.5 μM, Pb 2.0 μM or in the presence of both HMs (Cd 0.5 μM plus Pb 2.0 μM). Error bars indicate standard deviation of the mean of three technical replicates derived from a bulk of three biological replicates. ANOVA results were reported basing on their statistical significance. * $p < 0.05$, ** $p < 0.01$, *** $p < 0.001$.

3. Discussion

The data support the hypothesis that the effects produced by combinations of HMs could have a different impact on both accumulation and gene expression in comparison to the individual effects of each metal. The analyzed durum wheat genotypes showed different distribution of Cd and Pb in roots and leaves when treated with the two HMs in comparison with a treatment with only one metal. In general, the four genotypes showed a reduced accumulation of both Cd and Pb in the combined treatment (Figures 1 and 2). Several studies described that the presence of one metal influenced the uptake of another metal [24,25]. Xie et al. [25] investigated the effects of combined HMs toxicity on two rice genotypes differing in Cd accumulation; they found that the application of Pb, Cd, chromium (Cr) and copper (Cu) significantly affected grain Cd accumulations. In rice, Zeng et al. [26] also reported a significant Cd and Pb reduction in grains when exposed to both metals. Other studies were carried out on cucumber (*Cucumis sativus* that could retain greater amount of metals in the roots due to its root morphology) [27] to assess HMs toxicity in soils contaminated by Cu, Cd and Pb separately and in combinations; according to their results, bioaccumulation of one metal was influenced by the presence of other metals and, in general, the HMs accumulation patterns reflected antagonistic and/or synergistic plant's responses. In the binary combination of Cd and Pb, they found a synergistic response with a reduction of toxicity effect [28]. Cd/Pb synergisms have been previously reported by Zaray et al. [29]. On *Lemna minor*, a metal pollution sensitive plant, the combined toxicity of Pb and Cd was found to be less effective when compared to the toxicity of the individual treatment [30].

The data about Cd and Pb accumulation in leaves highlighted different behaviors among genotypes in relation to the two toxic metals. Svevo and H-Cd NIL are good accumulators of both Cd and Pb in leaves. On the contrary Creso accumulates lower level of Cd and Pb if compared to the other genotypes. The L-Cd NIL has a contrasting behavior: accumulates low level of Cd and high level of Pb. It is interesting to note that the expression levels of the tested genes in Svevo and H-Cd NIL are generally less up-regulated or not-regulated at all, suggesting that these genes are involved in some molecular mechanisms to stuck Cd and Pb at root level (e.g., genes involved in vacuole compartmentalization of toxic metals).

These differences (both in accumulation and distribution of HMs) could be, in part, due to differential expression of genes involved in HM uptake, cellular sequestration and translocation from root to shoot. Since among the HMs, Cd is accumulated in the grain of durum wheat to levels exceeding the Codex Alimentarius Commission standards [5] Cd uptake, cellular sequestration and translocation have been thoroughly studied [31,32]; some of the genes, with a key role in these physiological steps in durum wheat plants during Cd treatment [22], were investigated during the combined exposure to Cd and Pb.

Cd can enter root cells as Cd^{2+} through ZIP (zinc regulated transporter/iron regulated transporter-like protein) transporters or as Cd-chelates through YSL (Yellow-Stripe 1-Like) proteins [33]. Some investigations have proven that the ZIP family transporters participate in Cd absorption and accumulation in plants [34]; in our work, the selected gene *ZIP4* was not significantly regulated (Figure 5) in response to Cd treatment. Yamaguchi et al. [35] also found that genes coding for ZIP transporters did not change their expression patterns in *Solanum torvum* roots during treatment with low Cd concentrations and they postulated that the absence of changes in these metal transporters may explain why the mild Cd exposure did not induce serious competitive inhibitory effects on metal ion homeostasis in roots. The same effect due to the exposure to a low HM concentration could explain the unchanged expression of the gene *ZTP29* (Figure 5). This gene, coding for a zinc transporter with homology to the *Arabidopsis ZTP29*, is localized into the endoplasmic reticulum and it is thought to play a role in the unfolded protein response [36], Liu et al. [37] found that Cd up-regulated *ZTP29* in roots of *Cosmos bipinnatus* Cav. When they are under 40 μmol/L Cd stress (80 times more concentrated than the concentration used in this work).

Instead, the genes coding for the YSL (Yellow-Stripe 1-Like) proteins responsible for the transport of the Cd-chelates resulted up-regulated in Cd and Cd plus Pb treatments with higher FC values in the genotypes L-Cd NIL and Creso in which also *YSL2* was vigorously activated (Figure 5). This upregulation in response to Cd treatment is in accordance with RNA sequencing data reported in Aprile et al. [22]. The YSL family of transporters represents a candidate for the transport of nicotianamine (NA)–metal chelates across plant cell membranes [38]; several members of the YSL family are localized to the plasma membrane and function as transporters of metals that are bound to the metal chelator nicotianamine or the related set of mugineic acid family chelators known as phytosiderophores [39]. Other YSL members are localized to the vacuole membranes and to the internal membranes and may play a role in detoxification by HM sequestration in the vacuole [40]; this compartmentalization mechanism could contribute with the characteristic trait of low Cd-accumulation in durum wheat grain [13]. The two genotypes (L-Cd NIL and Creso) with lower Cd and Pb accumulation in leaves, had also the higher level of expression of the gene *YSL2*, suggesting a possible regulatory role in Cd and Pb compartmentalization in roots. The treatment with Pb did not affect the expression level of the *YSL* transporters while the Cd-Pb combined treatment affects the *YSL2* expression in the L-Cd NIL and Creso by reducing the amount of mRNA in comparison with the expression level in the single Cd treatment; the exposure to the combined HMs can activate simultaneously several nonspecific defense systems and it is reported that the interactions among HMs affect their uptake and accumulation in plants [25,26,31,41]. Other NA vacuolar transporter genes are the *ZIF* and *ZIF*-like genes, their expression levels were up-regulated by Cd and Cd plus Pb treatments in all the analyzed genotypes while the Pb treatment did not change expression pattern (Figure 6); so it is reasonable to assume that the Cd-nicotinamine chelates could enter into the vacuoles through the ZIF and ZIFL.

HMs are loaded from the symplast into the xylem by heavy metal P_{1B}-ATPases, known as heavy metal ATPases (HMAs) that play an important role in metal transport in plants [20,33].

In our experiment, the gene coding for the transporter HMA5 was upregulated both in Cd and Pb single treatment and the combined treatment increased the amount of mRNA if compared with the HM single treatment as if there was a cumulative effect (Figure 6). Functional studies on the HMAs have shown that these transporters can be divided into two subgroups based on their metal-substrate specificity: a copper (Cu)/silver (Ag) group and a zinc (Zn)/cobalt (Co)/Cd/Pb group [42]; this indication is in accordance with our data since *HMA5* was induced both by Cd and Pb. Besides genes coding for HMs transporters, previous studies have characterized several transcription factors (TFs) involved in Cd response: ERF, WRKY and bHLH TF families [43]. Obtained gene expression patterns supported the involvement of these genes in the response to Cd stress since *bHLH29/FIT*, *bHLH38/ORG2* and *bHLH47/PYE* were significantly induced only in response to Cd treatment (Figure 4). A significant up-regulation in response to Pb treatment was recorded exclusively in the genotype Svevo for the gene *bHLH38/ORG2* (Figure 4). For the TF *WRKY33* not significant induction was observed (Figure 4). Since many transcription factors are transiently regulated by stresses/treatments, the long-term exposure to Cd and Pb and the sample collection at 42 days after germination were not suitable to observe a transcription variation. In *Arabidopsis thaliana* treated with Cd, real-time PCR analyses demonstrated that the *WRKY13* transcript was rapidly induced by Cd stress, reaching the maximum level after 1 h of Cd treatment and gradually decreasing, thereafter [44]. On the other hand, Long et al. [45] described how the co-overexpression of *FIT* and *ORG2* enhanced the expression of nicotianamine synthase 1 (*NAS1*) and *NAS2*, resulting in the accumulation of nicotianamine, a crucial chelator for Fe transportation and homeostasis.

Figure 7 showed a strong induction for the genes *NAS2*, *NAS3* and *NAS4* in response to Cd treatment while no gene expression modulation was recorded in response to Pb treatment. These data suggested that the molecular mechanisms regulated by the bHLH TFs are well conserved between *Arabidopsis* and durum wheat and also gave evidence of a detoxification mechanism specific for Cd stress since Pb treatment did not activate the synthesis of nicotianamine. According to Pal and Ray [46], phytochelatins synthesis is influenced by the metal ion treatment and a number of phytochelatins variants have been found among plant species. Nicotianamine (NA) is a non-proteogenic amino acid chelator having more than one binding centers, which confer high affinity for Fe, but also for other metals such as Zn, Cu, Mn, Ni and Cd [46]. Another chelator, involved in iron uptake in graminaceous plant species, is the mugineic acid [47]. In the mucigenic acid biosynthetic pathway, nicotianamine aminotransferase (NAAT) is implicated in the formation of 2′-deoxymugineic acid (DMA) from nicotianamine [48]. With our expression data, we described that *NAAT* was highly expressed in root tissues of all the analyzed genotypes treated with Cd (Figure 7) indicating a critical role for the mucigenic acid in response to Cd.

4. Materials and Methods

4.1. Genetic Materials

To identify the effects of the combined stress of Cd and Pb in durum wheat (*Triticum turgidum* L. subsp. *Durum*), a pot experiment was conducted in a growth chamber. Two low grain Cd accumulation (L-Cd NIL and Creso) and two high grain Cd accumulation (H-Cd NIL and Svevo) were analyzed. Creso and Svevo accession numbers are, respectively, K-53049 and RICP-01C0107074, and all pedigree information is browsable at CIMMYT database (http://www.wheatpedigree.net). L-Cd NIL and H-Cd NIL are two near-isogenic lines (NILs) of durum wheat (*Triticum turgidum* L. subsp. *durum*) that differ in grain Cd accumulation TL 8982-H (H-Cd NIL) and TL 8982_L (L-Cd NIL) [17].

4.2. Experimental Design

Cd and Pb management to the four genotypes was set up by hydroponic system. After external sterilization, seeds were germinated in Petri dishes with moist filter paper, in the dark at 8 °C. After germination (6–7 days), plantlets were located into 0.4 L plastic pots (7 cm × 7 cm × 8 cm) filled with perlite, moistened with tap water, and immediately transferred to the 10 L polyethylene tanks of the hydroponic system, as described by Harris and Taylor [49] with little modifications. In each pot three seedlings were lodged in and for each treatment, three different pots were considered for three biological replicates. The positions of the pots in the growth chamber were completely randomized and changed weekly with a new randomization. Plants were grown in the growth chamber under long days, 16 h light/8 h night, 21 °C/16 °C. The hydroponic solution was given with systematic pauses, irrigating for 15 min every 2 h during the all day; while during the night no fertigation. In this way, the perlite substrate was constantly dampened with hydroponic solution, avoiding stagnation. The nutrient solution was prepared using reverse osmosis (RO) water (<30 μS cm^{-1}) and contained: 0.3 mM NH_4NO_3, 0.25 mM KNO_3, 0.1 mM K_2SO_4, 50 μM KCl, 1.0 mM $Ca(NO_3)_2$, 0.3 mM $Mg(NO_3)_2$, 100 μM $Fe(NO_3)_3$, 1.0 μM $MnSO_4$, 10.0 μM H_3BO_3, 10.0 μM $ZnSO_4$, 0.2 μM Na_2MoO_4, 2.0 μM $CuSO_4$, 2.0 μM $Cu(NO_3)_2$, 0.1 mM K_2HPO_4, 138.6 μM *N*-(2-hydroxyethyl) ethylenediaminetriacetic acid (HEDTA), and 2 mM 2-(*N*-Morpholino) ethanesulfonic acid hydrate. After preparation of solution, pH was adjusted by 1.42 mM KOH.

The pH of the nutrient solution and its electrical conductivity (EC) were constantly monitored every 2 days, EC was used to estimate water depletion by keeping EC in the main tank between 550 and 600 μS cm^{-1}, and nitric acid (1% *v*/*v*) was used to adjust pH between 5.5 and 6.0, when needed.

In cultivated fields the presence of Cd and Pb is usually not toxic for the crops. In this experiment plants were treated by adding to the nutrient solution 0.5 μM $CdCl_2$, or 2.0 μM $Pb(NO_3)_2$, or even both in the case of the double metal stress. These concentrations do not cause a significant toxic effect on root and leaf biomass as reported respectively by Harris and Taylor [49] and Sun et al. [50], while Cd 0.5 μM has been employed in previous works of the University of Salento [12,22].

The control plants were cultivated without Cd or Pb in the same hydroponic solution. Hydroponic solution was constantly aerated. One plant for each pot (three for each treatment) was sampled 42 days after germination, at tillering stage (roots and leaves). Roots were easily extracted from perlite substrate and washed manually to remove the perlite beads adherent to roots, and possible excess of hydroponic solution. Leaf samples were washed immediately on harvest in RO water for 30 s, while root samples were triple rinsed (RO water, 1 min; 1 mM $CaCl_2$, 5 min; RO water, 1 min), and blotted dry. Samples for quantitative RT-PCR were frozen in liquid nitrogen and then stored at −80 °C.

4.3. Inductively Coupled Plasma Mass Spectrometry (ICP-MS) Analysis

Measurements of Cd and Pb uptake in roots were performed by using X SeriesII inductively coupled plasma mass spectrometer (ICP-MS; Thermo Fisher Scientific, Waltham, MA, USA) equipped with Peltier cooled (3 °C) spray chamber. Samples were introduced by the autosampler CETAC ASX 520 into the nebulizer, and the positively charged ions were then produced by a high-temperature, inductively coupled plasma. The ions passed through a sampling cone interface into a high-performance quadrupole mass spectrometer that is computer-controlled to carry out multi-element analysis. Data were analyzed by PlasmaLab software. The instrument was tuned daily with an ICP-MS tuning solution. Yttrium in HNO_3 4% (100 ppb) was used as internal standard. Cd and Pb standards ranging from 0.2 to 100 ppb were freshly prepared before each analysis and used to build calibration curve. Each sample was analyzed at least in three independent measurements and each experiment comprised three repetitions. Results are given as mean value ± standard deviation.

Each sample was mineralized by a microwave-assisted procedure performed with an ultraWAWE microwawe digestion system (Milestone Inc., Shelton, CT, USA), as recently reported by Durante et al. [51]. The sample (~300 mg) was accurately weighed in the microwave-quartz vessels before adding 2 mL of ultrapure HNO_3 65% *w/w* and 4 mL of Milli-Q water. At the end of the digestion process, an almost colorless, pale yellow sample was obtained. The solution was then diluted up to 10 g with Milli-Q water in test tubes. Between each mineralization cycle, a washing cycle was carried out.

Ultrapure HNO_3 65% *w/w* was obtained from analytical grade nitric acid (Carlo Erba, Milan, Italy) after sub-boiling distillation performed with a sub-boiler SAVILLEX DST 1000 (Savillex Corp. Eden Prairie, MN, USA) apparatus.

4.4. Total RNA Isolation, cDNA Synthesis and qPCR Analysis of Gene Expression

To evaluate the response of durum wheat plants to Cd and Pb treatments, we carried out a transcriptomic analysis in a small, but well-defined, group of genes [22]. Total RNA was extracted from root and leaf tissues using TRIZOL reagent according to the method published by Marè et al. [52]. To assess RNA quality and quantity, several dilutions of each sample were analyzed using the Agilent RNA 6000 nano Kit and Agilent Bioanalyzer 2100. cDNA synthesis was performed using TaqMan® Reverse Transcription Reagents (Applied Biosystems, Foster City, CA, USA). qPCR was performed with the Power SYBR Green RT-PCR Master mix (Applied Biosystems, Foster City, CA, USA) according to the manufacturer's instructions. To calculate the relative expression levels between a reference sample and the related treatments, the fold change (FC) formula was used:

$$FC = 2^{\wedge-\Delta\Delta CT},$$

where $\Delta\Delta CT = (CT_{targetgene} - CT_{referencegene})_{treatedsample} - (CT_{targetgene} - CT_{referencegene})_{controlsample}$.

Sequences related to genes coding for transcription factors (*bHLH29*, *bHLH38*, *bHLH47* and *WRKY33*), membrane transporters (*HMA5*, *YSL1*, *YSL2*, *ZIF1*, *ZIFL1*, *ZIFL2*, *ZIP4* and *ZTP29*) and genes involved in the metal chelator pathway (*NAS2*, *NAS3*, *NAS4* and *NAAT*) were downloaded from the site https://www.ebi.ac.uk/arrayexpress, at European Bioinformatics Institute (EMBL-EBI). The accession code is E-MTAB-7266. Then, the sequences were compared to the NCBI database using the BLAST algorithm and the most similar sequences were used to design the relative real-time PCR primers (Primer Express™ Software v3.0, Applied Biosystems, Foster City, CA, USA) Supplementary Table S1. TheNADH ubiquinone reductase gene was used as the reference gene to normalize the expression levels of the target genes.

4.5. Statistical Analyses

Means of quantitative data related to Cd and Pb concentrations were determined for each tissue (root and leaf) and were subjected to two-way ANOVA analysis (genotype X treatment), followed by Tukey-HSD (honestly significant difference) post hoc test ($p < 0.05$). Translocation factor data were subjected to two-way ANOVA (genotype X treatment). A *t*-test was employed to find statistical differences between Cd and Pb translocation factors on each genotype. A one-way ANOVA and *t*-test were applied to expression gene data.

Analyses were achieved using R version 3.5.3.

5. Conclusions

The mechanisms activated by plants to tolerate the presence of HMs were studied by exposing plants to not only a single HM but to a combination of Cd and Pb. Uptake and translocation strategies are not regulated by the same genes as suggested by the strong differences observed among genotypes in response to the two toxic HMs. In fact, the level of Cd in durum wheat roots and leaves is influenced by the co-presence of Pb and vice versa even if the phenomenon has different extent among genotypes.

Furthermore, nicotianamine and mucigenic acid seem to play a key role in response to Cd stress, probably by chelating the metal and avoiding its translocation to the plant shoots. The combined stress with Cd and Pb did not affect this mechanism, which appeared to be specific for Cd.

Supplementary Materials: Supplementary Materials can be found at http://www.mdpi.com/1422-0067/20/23/5891/s1.

Author Contributions: A.A., E.S., N.P. and L.D.B. have planned the experimental design. E.F. (Enrico Francia), J.M. and D.R. have grown plants in hydroponic conditions and collected samples for further investigations. E.F. (Erika Ferrari) has carried out the chemical analysis on Cd and Pb concentration in tissues. M.V. and A.L. processed samples for mRNA extraction. E.S. and A.A. have run the real-time PCR expression analysis and the relative data analysis. A.A., E.S. and L.D.B. wrote the manuscript and N.P., E.F. (Enrico Francia), A.A. and L.D.B. reviewed and edited the final version.

Funding: This work was economically supported by MIUR (Italian Minister of University and Research) project "Sviluppo tecnologico e innovazione per la sostenibilità e competitività della cerealicoltura meridionale MIUR-UE (PON01_01145/1-ISCOCEM)" and by Regione Puglia, "FutureInResearch" project "Frumento duro "Cappelli": valorizzazione delle componenti genetiche alla base della tolleranza allo stress idrico", code 2I19HY5. These funding bodies had no role in the design of this study, during its execution, analyses, interpretation of the data, or decision to submit results.

Acknowledgments: We thank Curtis Pozniak and John Clarke for the kind sharing of the near isogenic lines TL 8982-H (H-Cd NIL) and TL 8982-L (L-Cd NIL). We are thankful to 'Centro Interdipartimentale Grandi Strumenti (CIGS)' of the University of Modena and Reggio Emilia that supplied ICP-MS, and to Prof. Andrea Marchetti for his precious advice and FKV autoclave.

Conflicts of Interest: The authors declare no conflict of interest.

References

1. Romic, M.; Romic, D. Heavy metals distribution in agricultural topsoils in urban area. *Environ. Geol.* **2003**, *43*, 795–805. [CrossRef]
2. Jaishankar, M.; Tseten, T.; Anbalagan, N.; Mathew, B.B.; Beeregowda, K.N. Toxicity, mechanism and health effects of some heavy metals. *Interdiscip. Toxicol.* **2014**, *7*, 60–72. [CrossRef] [PubMed]
3. Wang, G.; Su, M.; Chen, Y.; Lin, F.; Luo, D.; Gao, S. Transfer characteristics of cadmium and lead from soil to the edible parts of six vegetable species in southeastern China. *Environ. Pollut.* **2006**, *144*, 127–135. [CrossRef] [PubMed]
4. Ghaderi, A.A.; Abduli, M.A.; Karbassi, A.R.; Nasrabadi, T.; Khajeh, M. Evaluating the Effects of Fertilizers on Bioavailable Metallic Pollution of soils, Case study of Sistan farms, Iran. *Int. J. Environ. Res.* **2012**, *6*, 565–570.
5. Codex Alimentarius. General Standard for Contaminants and Toxins in Food and Feed. CXS 193-1995 (Amended 2019). Available online: http://www.fao.org/fao-who-codexalimentarius/sh-proxy/en/?lnk=1&url=https%253A%252F%252Fworkspace.fao.org%252Fsites%252Fcodex%252FStandards%252FCXS%2B193-1995%252FCXS_193e.pdf (accessed on 12 November 2019).

6. Hasan, S.A.; Fariduddin, Q.; Ali, B.; Hayat, S.; Ahmad, A. Cadmium: Toxicity and tolerance in plants. *J. Environ. Biol.* **2009**, *30*, 165–174.
7. Pourrut, B.; Shahid, M.; Dumat, C.; Winterton, P.; Pinelli, E. Lead Uptake, Toxicity, and Detoxification in Plants. *Rev. Environ. Contam. Toxicol.* **2011**, *213*, 113–136.
8. Florijn, P.J.; Van Beusichem, M.L. Uptake and distribution of cadmium in maze inbred lines. *Plant Soil* **1993**, *150*, 25–32. [CrossRef]
9. Li, Y.M.; Channey, L.R.; Scheiter, A.A. Genotypic variation in kernel cadmium concentration in sunflower germplasm under varying soil conditions. *Crop Sci.* **1995**, *35*, 137–141. [CrossRef]
10. DalCorso, G.; Fasani, E.; Manara, A.; Visioli, G.; Furini, A. Heavy Metal Pollutions: State of the Art and Innovation in Phytoremediation. *Int. J. Mol. Sci.* **2019**, *20*, 3412. [CrossRef]
11. Backhaus, T.; Faust, M. Predictive environmental risk assessment of chemical mixtures: A conceptual framework. *Environ. Sci. Technol.* **2012**, *4*, 2564–2573. [CrossRef]
12. Vergine, M.; Aprile, A.; Sabella, E.; Genga, A.; Siciliano, M.; Rampino, P.; Lenucci, M.S.; Luvisi, A.; De Bellis, L. Cadmium Concentration in Grains of Durum Wheat (*Triticum turgidum* L. subsp. *durum*). *J. Agr. Food Chem.* **2017**, *65*, 6240–6246. [CrossRef] [PubMed]
13. Arduini, I.; Masoni, A.; Mariotti, M.; Pampana, S.; Ercoli, L. Cadmium uptake and translocation in durum wheat varieties differing in grain-Cd accumulation. *Plant Soil Environ.* **2014**, *60*, 43–49. [CrossRef]
14. Bravin, M.L.; Le Merrer, B.; Denaix, L.; Schneider, A.; Hinsinger, P. Copper uptake kinetics in hydroponically-grown durum wheat (*Triticum turgidum durum* L.) as compared with soil's ability to supply copper. *Plant Soil* **2010**, *331*, 91–104. [CrossRef]
15. Shi, G.L.; Li, D.J.; Wang, Y.F.; Liu, C.H.; Hu, Z.B.; Lou, L.Q.; Rengel, Z.; Cai, Q.S. Accumulation and distribution of arsenic and cadmium in winter wheat (*Triticum aestivum* L.) at different developmental stages. *Sci. Total Environ.* **2019**, *667*, 532–539. [CrossRef] [PubMed]
16. Dalir, N.; Khoshgoftarmanesh, A.H. Symplastic and apoplastic uptake and root to shoot translocation of nickel in wheat as affected by exogenous amino acids. *J. Plant Physiol.* **2014**, *171*, 531–536. [CrossRef]
17. Frost, H.L.; Ketchum, L.H.J. Trace metal concentration in durum wheat from application of sewage sludge and commercial fertilizer. *Adv. Environ. Res.* **2000**, *4*, 347–355. [CrossRef]
18. Shafiq, S.; Zeb, Q.; Ali, A.; Sajjad, Y.; Nazir, R.; Widemann, E.; Liu, L. Lead, Cadmium and Zinc Phytotoxicity Alter DNA Methylation Levels to Confer Heavy Metal Tolerance in Wheat. *Int. J. Mol. Sci.* **2019**, *20*, 4676. [CrossRef]
19. Clarke, J.M.; Leisle, D.; DePauw, R.M.; Thiessen, L.L. The registration of genetic stocks: Registration of five pairs of near-isogenic lines for cadmium concentration. *Crop Sci.* **1995**, *37*, 297. [CrossRef]
20. Maccaferri, M.; Harris, N.S.; Cattivelli, L.; Twardziok, S.O.; Pasam, R.K.; Gundlach, H.; Spannagl, M.; Ormanbekova, D.; Lux, T.; Prade, V.M.; et al. Durum wheat genome highlights past domestication signatures and future improvement targets. *Nat. Genet.* **2019**, *51*, 885–895. [CrossRef]
21. Lakhdar, A.; Iannelli, M.A.; Debez, A.; Massacci, A.; Jedidi, N.; Abdelly, C. Effect of municipal solid waste compost and sewage sludge use on wheat (*Triticum durum*): Growth, heavy metal accumulation, and antioxidant activity. *J. Sci. Food Agric.* **2010**, *90*, 965–971. [CrossRef]
22. Aprile, A.; Sabella, E.; Vergine, M.; Genga, A.; Siciliano, M.; Nutricati, E.; Rampino, P.; De Pascali, M.; Luvisi, A.; Miceli, A.; et al. Activation of a gene network in durum wheat roots exposed to cadmium. *BMC Plant Biol.* **2018**, *18*, 238. [CrossRef] [PubMed]
23. Haydon, M.J.; Cobbett, C.S. Transporters of ligands for essential metal ions in plants. *New Phytol.* **2007**, *174*, 499–506. [CrossRef] [PubMed]
24. Peralta-Videaa, J.R.; Gardea-Torresdeya, J.L.; Gomezc, E.; Tiemanna, K.J.; Parsonsa, J.G.; Carrillod, G. Effect of mixed cadmium, copper, nickel and zinc at different pHs upon alfalfa growth and heavy metal uptake. *Environ. Pollut.* **2002**, *119*, 291–301. [CrossRef]
25. Xie, L.; Hao, P.; Cheng, Y.; Ahmed, I.M.; Cao, F. Effect of combined application of lead, cadmium, chromium and copper on grain, leaf and stem heavy metal contents at different growth stages in rice. *Ecotoxicol. Environ. Saf.* **2018**, *162*, 71–76. [CrossRef]
26. Zeng, F.; Mao, Y.; Cheng, W.; Wu, F.; Zhang, G. Genotypic and environmental variation in chromium, cadmium and lead concentrations in rice. *Environ. Pollut.* **2008**, *153*, 309–314. [CrossRef]
27. An, Y.J. Soil ecotoxicity assessment using cadmium sensitive plants. *Environ. Pollut.* **2004**, *127*, 21–26. [CrossRef]

28. An, Y.J.; Kimb, Y.M.; Kwonb, T.I.; Jeong, S.-W. Combined effect of copper, cadmium, and lead upon *Cucumis sativus* growth and bioaccumulation. *Sci. Total Environ.* **2004**, *326*, 85–93. [CrossRef]
29. Zaray, G.; Phuong, D.D.T.; Varga, I.; Kantor, T.; Cseh, E.; Fodor, F. Influences of lead contamination and complexing agents on the metal uptake of cucumber. *Microchem. J.* **1995**, *51*, 207–213. [CrossRef]
30. Mohan, B.S.; Hosetti, B.B. Potential phytotoxicity of lead and cadmium to *Lemna minor* grown in sewage stabilization ponds. *Environ. Pollut.* **1997**, *98*, 233–238. [CrossRef]
31. Hart, J.J.; Welch, R.M.; Norvell, W.A.; Kochian, L.V. Transport interactions between cadmium and zinc in roots of bread and durum wheat seedlings. *Physiol. Plant.* **2002**, *116*, 73–78. [CrossRef]
32. Salsman, E.; Kumar, A.; AbuHammad, W.; Abbasabadi, A.O.; Dobrydina, M.; Chao, S.; Li, X.; Manthey, F.A.; Elias, E.M. Development and validation of molecular markers for grain cadmium in durum wheat. *Mol. Breed.* **2018**, *38*, 28. [CrossRef]
33. Lux, A.; Martinka, M.; Vaculík, M.; White, P.J. Root responses to cadmium in the rhizosphere: A review. *J. Exp. Bot.* **2011**, *62*, 21–37. [CrossRef] [PubMed]
34. Wu, H.; Wang, J.; Li, B.; Ou, Y.; Jiang, W.; Liu, D.; Zou, J. Uptake and Accumulation of Cadmium and Relative Gene Expression in Roots of Cd-resistant *Salix matsudana* Koidz. *Pol. J. Environ. Stud.* **2016**, *25*, 2717–2723. [CrossRef]
35. Yamaguchi, H.; Fukuoka, H.; Arao, T.; Ohyama, A.; Nunome, T.; Miyatake, K.; Negoro, S. Gene expression analysis in cadmium-stressed roots of a low cadmium-accumulating solanaceous plant, *Solanum torvum*. *J. Exp. Bot.* **2010**, *61*, 423–437. [CrossRef]
36. Wang, M.; Xu, Q.; Yu, J.; Yuan, M. The putative Arabidopsis zinc transporter ZTP29 is involved in the response to salt stress. *Plant Mol. Biol.* **2010**, *73*, 467–479. [CrossRef]
37. Liu, Y.; Yu, X.; Feng, Y.; Zhang, C.; Wang, C.; Zeng, J.; Huang, Z.; Kang, H.; Fan, X.; Sha, L.; et al. Physiological and transcriptome response to cadmium in cosmos (*Cosmos bipinnatus* Cav.) seedlings. *Sci. Rep.* **2017**, *7*, 14691. [CrossRef]
38. Curie, C.; Cassin, G.; Couch, D.; Divol, F.; Higuchi, K.; Jean, M.L.; Misson, J.; Schikora, A.; Czernic, P.; Mari, S. Metal movement within the plant: Contribution of nicotianamine and yellow stripe 1-like transporters. *Ann. Bot.* **2009**, *103*, 1–11. [CrossRef]
39. Chu, H.-H.; Chiecko, J.; Punshon, T.; Lanzirotti, A.; Lahner, B.; Salt, D.E.; Walker, E.L. Successful Reproduction Requires the Function of Arabidopsis YELLOW STRIPE-LIKE1 and YELLOW STRIPE-LIKE3 Metal-Nicotianamine Transporters in Both Vegetative and Reproductive Structure. *Plant Physiol.* **2010**, *154*, 197–210. [CrossRef]
40. Conte, S.S.; Chu, H.H.; Chan Rodriguez, D.; Punshon, T.; Vasques, K.A.; Salt, D.E.; Walker, E.L. *Arabidopsis thaliana* Yellow Stripe1- Like4 and Yellow Stripe1-Like6 localize to internal cellular membranes and are involved in metal ion homeostasis. *Front. Plant Sci.* **2013**, *4*, 283. [CrossRef]
41. John, R.; Ahmad, P.; Gadgil, K.; Sharma, S. Effect of cadmium and lead on growth, biochemical parameters and uptake in *Lemna polyrrhiza* L. *Plant Soil Environ.* **2008**, *54*, 262–270. [CrossRef]
42. Williams, L.E.; Mills, R.F. P(1B-)ATPases—An ancient family of transition metal pumps with diverse functions in plants. *Trends Plant Sci.* **2005**, *10*, 491–502. [CrossRef]
43. DalCorso, G.; Farinati, S.; Furini, A. Regulatory networks of cadmium stress in plants. *Plant Signal. Behav.* **2010**, *5*, 663–667. [CrossRef]
44. Sheng, Y.; Yan, X.; Huang, Y.; Han, Y.; Zhang, C.; Ren, Y.; Fan, T.; Xiao, F.; Liu, Y.; Cao, S. The WRKY transcription factor, WRKY13, activates PDR8 expression to positively regulate cadmium tolerance in Arabidopsis. *Plant Cell Environ.* **2019**, *42*, 891–903. [CrossRef] [PubMed]
45. Long, T.A.; Tsukagoshi, H.; Busch, W.; Lahner, B.; Salt, D.E.; Benfey, P.N. The bHLH transcription factor POPEYE regulates response to Iron deficiency in Arabidopsis roots. *Plant Cell* **2010**, *22*, 2219–2236. [CrossRef]
46. Pal, R.; Rai, J.P.N. Phytochelatins: Peptides Involved in Heavy Metal Detoxification. *Appl. Biochem. Biotechnol.* **2010**, *160*, 945–963. [CrossRef] [PubMed]
47. Álvarez-Fernández, A.; Díaz-Benito, P.; Abadía, A.; López-Millán, A.-F.; Abadía, J. Metal species involved in long distance metal transport in plants. *Front. Plant Sci.* **2014**, *5*, 105. [CrossRef]
48. Beasley, J.T.; Bonneau, J.P.; Johnson, A.A.T. Characterisation of the nicotianamine aminotransferase and deoxymugineic acid synthase genes essential to Strategy II iron uptake in bread wheat (*Triticum aestivum* L.). *PLoS ONE* **2017**, *12*, e0177061. [CrossRef]

49. Harris, N.S.; Taylor, G.J. Cadmium uptake and partitioning in durum wheat during grain filling. *BMC Plant Biol.* **2013**, *13*, 103. [CrossRef]
50. Sun, Q.; Wang, X.-R.; Ding, S.-M.; Yuan, X.-F. Effects of Interaction Between Cadmium and Plumbum on Phytochelatins and Glutathione Production in Wheat (*Triticum aestivum* L.). *J. Integr. Plant Biol.* **2015**, *47*, 435–442. [CrossRef]
51. Durante, C.; Bertacchini, L.; Cocchi, M.; Manzini, D.; Marchetti, A.; Rossi, M.C.; Sighinolfi, S.; Tassi, L. Development of 87Sr/86Sr maps as targeted strategy to support wine quality. *Food Chem.* **2018**, *255*, 139–146. [CrossRef]
52. Marè, C.; Aprile, A.; Roncaglia, E.; Tocci, E.; Corino, L.G.; De Bellis, L.; Cattivelli, L. Rootstock and soil induce transcriptome modulation of phenylpropanoid pathway in grape leaves. *J. Plant Interact.* **2014**, *8*, 334–349. [CrossRef]

© 2019 by the authors. Licensee MDPI, Basel, Switzerland. This article is an open access article distributed under the terms and conditions of the Creative Commons Attribution (CC BY) license (http://creativecommons.org/licenses/by/4.0/).

International Journal of
Molecular Sciences

Article

Lead, Cadmium and Zinc Phytotoxicity Alter DNA Methylation Levels to Confer Heavy Metal Tolerance in Wheat

Sarfraz Shafiq [1,2,*], Qudsia Zeb [3], Asim Ali [2], Yasar Sajjad [4], Rashid Nazir [2], Emilie Widemann [5] and Liangyu Liu [3,*]

1. Department of Anatomy and Cell Biology, University of Western Ontario, 1151 Richmond St, London, ON N6A5B8, Canada
2. Department of Environmental Sciences, COMSATS University Islamabad, Abbottabad campus, Pakhtunkhwa 22060, Pakistan; asim21pk@hotmail.com (A.A.); rashidnazir@ciit.net.pk (R.N.)
3. College of Life Sciences, Capital Normal University, Beijing 100084, China; zeb_qudsia@yahoo.com
4. Department of Biotechnology, COMSATS University Islamabad, Abbottabad campus, Pakhtunkhwa 22060, Pakistan; yasarsajjad@ciit.net.pk
5. Department of Biology, University of Western Ontario, 1151 Richmond St, London, ON N6A5B8, Canada; ewidema4@uwo.ca

* Correspondence: sshafiq2@uwo.ca (S.S.); liangyu.liu@cnu.edu.cn (L.L.)

Received: 17 August 2019; Accepted: 17 September 2019; Published: 20 September 2019

Abstract: Being a staple food, wheat (*Triticum aestivum*) nutritionally fulfills all requirements of human health and also serves as a significant link in the food chain for the ingestion of pollutants by humans and animals. Therefore, the presence of the heavy metals such as lead (Pb) and cadmium (Cd) in soil is not only responsible for the reduction of wheat crop yield but also the potential threat for human and animal health. However, the link between DNA methylation and heavy metal stress tolerance in wheat has not been investigated yet. In this study, eight high yielding wheat varieties were screened based on their phenotype in response to Pb stress. Out of these, Pirsabak 2004 and Fakhar-e-sarhad were identified as Pb resistant and sensitive varieties, respectively. In addition, Pirsabak 2004 and Fakhar-e-sarhad varieties were also found resistant and sensitive to Cd and Zinc (Zn) stress, respectively. Antioxidant activity was decreased in Fakhar-e-sarhad compared with control in response to Pb/Cd/Zn stresses, but Fakhar-e-sarhad and Pirsabak 2004 accumulated similar levels of Pb, Cd and Zn in their roots. The expression of *Heavy Metal ATPase 2 (TaHMA2)* and *ATP-Binding Cassette (TaABCC2/3/4)* metal detoxification transporters are significantly upregulated in Pirsabak 2004 compared with Fakhar-e-sarhad and non-treated controls in response to Pb, Cd and Zn metal stresses. Consistent with upregulation of metal detoxification transporters, CG DNA hypomethylation was also found at the promoter region of these transporters in Pirsabak 2004 compared with Fakhar-e-sarhad and non-treated control, which indicates that DNA methylation regulates the expression of metal detoxification transporters to confer resistance against metal toxicity in wheat. This study recommends the farmers to cultivate Pirsabak 2004 variety in metal contaminated soils and also highlights that DNA methylation is associated with metal stress tolerance in wheat.

Keywords: DNA methylation; ABCC transporters; HMA2; wheat; metal stress tolerance

---

1. Introduction

Plants encounter many environmental stresses during their life cycles and have consequently developed the ability to combat those variations that adversely affect growth, development and reproduction. Among them, heavy metal stress affects the fitness, survival and yield of crop plants during the course of their development by impairing the molecular, biochemical and physiological processes [1]. The heavy metals, lead (Pb) and cadmium (Cd) are highly toxic trace pollutants for humans, animals and plants [2,3]. Pb exposure to plants results in impaired root growth and germination, alterations in membrane permeability, water regime, hormonal status, disarrays in mineral nutrition, decrease in photosynthesis, transpiration, DNA synthesis and increased generation of reactive oxygen species (ROS) [2,4]. Similar to Pb, Cd toxicity is also associated with impaired plant growth, development, metabolism, enzyme activities, etc. [5,6]. In contrast to these toxic metals, essential metals like zinc (Zn) and iron (Fe) are needed for plants during their development in order to perform their vital physiological and biochemical functions [7,8]. Therefore, Zn deficiency is associated with impaired plant growth, yield and grain quality [9]. However, excess of Zn may also cause toxicity and affect the plant physiology [9]. Therefore, uptake, storage and utilization of these heavy metals are tightly controlled in plants to maintain their concentration in different cellular compartments.

To cope with heavy metals, plants have evolved either avoidance of uptake or tolerant mechanisms, including detoxification. The detoxification of heavy metals is associated with the exclusion of heavy metals from the cells, phytochelatin synthesis, sequestration of heavy metals into the vacuoles, binding to glutathione and amino acids [10]. The multidrug resistance-associated proteins (MRPs) belong to a subclass of ATP-binding cassette (ABC) transporters and are involved in heavy metal detoxifications, vacuolar sequestration of metabolites, pathogen response and plant development in *Arabidopsis* (*Arabidopsis thaliana*) [11–14]. Similar to *Arabidopsis*, ABCC-MRP from yeast and human have been reported to play a role in metal detoxifications [15,16], suggesting a conserved mechanism of ABCC-MRP transporters among different organisms. Wheat (*Triticum aestivum*) contains 18 *ABCC-MRP* genes [17] and *TaABCC3* has earlier been reported to play a role in grain development and resistance against secreted mycotoxin from *Fusarium* [18]. Another *ABCC-MRP* partial gene has been suggested to play a role in xenobiotic detoxification in wheat [19], indicating the important function of TaABCCs in wheat plant resistance.

In parallel to ABCC-MRP, plants have also evolved another system to prevent a cytotoxic concentration by effluxing the metals from the cytosol to the apoplast through the action of heavy metal ATPases (HMAs), also known as PIB type-ATPases. HMAs have been reported to play a role in heavy metal tolerance in *Arabidopsis* [20], and are well-conserved proteins among different organisms [21]. HMAs are associated with the transport of Zn, Cd, cobalt (Co), Pb, copper (Cu) and silver (Ag) [22,23]. The heavy metal ATPase2, TaHMA2, is a plasma membrane transporter from wheat that was suggested to export the Zn/Cd toward the apoplast [24]. Yeast expressing wheat *TaHMA2* was found resistant to Zn/Cd and the over expression of *TaHMA2* conferred a mild resistance against Zn and Cd in *Arabidopsis*, suggesting the important function of TaHMA2 in metal tolerance in wheat.

Chromatin landscape becomes dynamic in response to environmental and developmental cues, thus modulates the DNA accessibility to regulate gene expression and thus controls the various cellular and physiological processes [25]. DNA methylation is involved in various biological processes including flowering time, imprinting, flower and leaf morphogenesis, fertility through gene silencing [26,27]. Different DNA methyltransferases are involved in DNA cytosine methylation of three different sequence contexts, i.e., CG, CHG and CHH [28]. In *Arabidopsis*, the DNA methylation of CGs is maintained by methyltransferase 1 (MET1), a homolog of mammalian DNA-methyltransferase 1 (Dnmt1) [29], while plant specific chromomethylase 3 (CMT3) is required for the maintenance of CHGs [30]. Furthermore, all the methylation contexts, especially CHH methylation, are maintained by the *de novo* DNA methyltransferase Domains rearranged methyltransferase 2 (DRM2), the homolog of mammalian DNA-methyltransferase 3 (Dnmt3) [31]. In wheat, five putative DNA methyltransferases have been identified and classified into different categories based on their similarity to *Arabidopsis* or mammals [32]. DNA hypomethylation or hypermethylation can happen in response to various stresses, and thus regulate gene expression and subsequent plant physiology [27,33,34]. DNA methylation is also affected in response to Cd, arsenic (As) and nickel (Ni) in human and mouse [35,36]. Similarly in plants, DNA methylation is altered in white clover (*Trifolium repens* L.), industrial hemp (*Cannabis sativa* L.) plants, oil seed rape (*Brassica napus*) and radish (*Raphanus sativus* L.) in response to metal stress [37–39].

In *Posidonia oceanica*, Cd treatment induces the DNA hypermethylation and heterochromatinization [40]. In rice (*Oryza sativa*), DNA methylation levels were altered in response to Cd [41], and the progenies of rice plants that have been exposed to metal stress in their life cycle exhibited altered DNA methylation levels [42], indicating that DNA methylation plays an important role in plant response to metal stress. However, the link between DNA methylation and metal stress tolerance in crop plants, especially wheat, and the underlying epigenetic mechanism have not been investigated yet.

Here, we first screened several wheat-cultivated varieties against Pb toxicity. The Pirsabak 2004 and Fakhar-e-sarhad varieties that presented highest Pb resistance and sensitivity, respectively, were further characterized for their resistance against Cd and Zn toxicity. The resistance level was evaluated by measuring antioxidant activities and the accumulation of Pb, Cd and Zn in their roots. We hypothesized that the variation of resistance could be due to a different efficiency of metal detoxification such as subcellular sequestration or transportation. Therefore, we evaluated the expression level of the root-expressed *TaABCCs* and *TaHMA2* transporters in the presence of Pb, Cd or Zn in the resistant Pirsabak 2004 and sensitive Fakhar-e-sarhad varieties. To explore the underlying epigenetic mechanisms that regulate gene expression of transporters, we investigated the expression of DNA methyltransferases and quantified the DNA methylation levels at the promoter of the selected *TaABCCs* and *TaHMA2* metal transporters.

2. Results

2.1. Genetic Diversity of Wheat Varieties Against Pb Toxicity

In order to investigate the metal toxicity mechanism in wheat, the high yielding wheat varieties were selected and screened against Pb toxicity to narrow down the genetic potential of each variety. The selected varieties were first screened for their germination capability in the presence of Pb (Figure 1). The tested varieties showed wide genetic diversity regarding Pb toxicity. Compared to the control, no effect on the germination rate was observed for Attahabib and Punjab 85 at 0.5 mM and 1 mM concentration of $Pb(NO_3)_2$, but a slight decrease in germination was observed at 2 mM. A higher sensitivity was observed for Fakhar-e-sarhad, Khyber 87, Janbaz and Pak 81, which showed a dose-dependent decrease in the germination rate. On the contrary, $Pb(NO_3)_2$ had no effect on the germination rate of Pirsabak 2004 at the tested concentrations. To further validate the genetic diversity observed in these varieties, we scored their primary root length and epicotyl length (Figure 1). Pb toxicity affected the primary root and epicotyl length in a dose-dependent manner in all the tested varieties, except Pirsabak 2004 whose growth was unaltered. Fakhar-e-sarhad showed a maximum decrease in the root length at 1 mM $Pb(NO_3)_2$ compared to control and no stronger effect was observed at 2 mM $Pb(NO_3)_2$. Similar to the root growth, the epicotyl growth was also severely impaired in Fakhar-e-sarhad at 1 mM and 2 mM $Pb(NO_3)_2$ compared to other varieties.

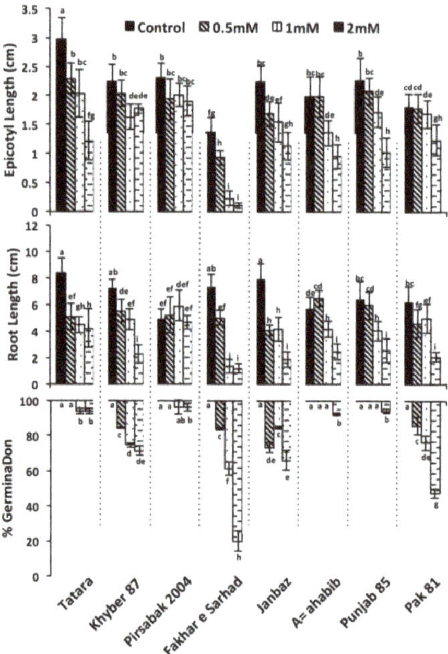

Figure 1. Screening of high-yielding wheat varieties based on their phenotypic characteristics against Pb toxicity. Seeds were grown on Murashige and Skoog (MS) media for 6 days under different concentrations of $Pb(NO_3)_2$. The results shown are the average of three biological replicates. Different letters indicate significant difference by a least significant difference (LSD) test ($p \leq 0.05$). Error bars represent SD.

2.2. Pirsabak 2004 and Fakhar-e-sarhad Sensitivity to Pb, Cd and Zn

We further evaluated the resistant Pirsabak 2004 and sensitive Fakhar-e-sarhad varieties response to Pb treatment in hydroponics (Figure 2). In the hydroponic culture, 0.5 mM Pb was found toxic to seedlings, therefore, the heavy metal treatment was done by adding 100 μM of $Pb(NO_3)_2$. Fakhar-e-sarhad showed a decrease in root length in response to Pb toxicity in hydroponics, while the root length of Pirsabak 2004 did not change compared with control. This further confirms our result that Pirsabak 2004 is resistant to Pb toxicity. We hypothesized that Pb resistant varieties could also be resistant to other divalent ions, such as Cd and Zn. We therefore investigated the response of Pirsabak 2004 and Fakhar-e-sarhad varieties in hydroponic experiment supplemented with Cd or Zn (Figure 2). The results showed that the root length in Pirsabak 2004 did not change in response to Cd and Zn stresses, whereas the root length of Fakhar-e-sarhad decreased compared with control. This indicates that Pirsabak 2004 is also resistant to Cd and Zn stresses in hydroponic culture.

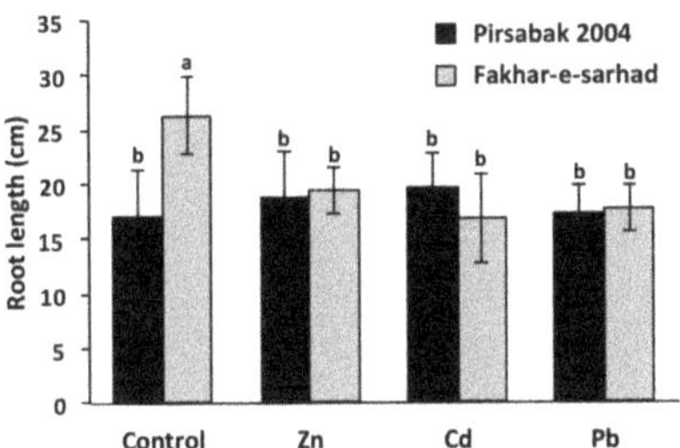

Figure 2. Pirsabak 2004 and Fakhar-e-sarhad varieties response to Cd and Zn stresses in hydroponic culture. The wheat seedlings were grown in hydroponic culture with 100 μM of $Pb(NO_3)_2$, $ZnSO_4$ or $CdCl_2$ and the root length was measured after two weeks of treatment. The results shown are the average of three biological replicates. Different letters indicate a significant difference by an LSD test ($p \leq 0.05$). Error bars represent SD.

2.3. Antioxidant Activity in Pirsabak 2004 and Fakhar-e-sarhad in Response to Metal Stress

The evaluation of a wheat response to metal toxicity can be achieved through the measurement of antioxidant activity of the superoxide dismutase (SOD), peroxidase (POD) or catalase (CAT). Pirsabak 2004 showed increased levels of SOD, POD and CAT activities in response to all the tested metals compared with control (Figure 3). Although Fakhar-e-sarhad showed slightly higher levels of SOD, POD and CAT than control, their levels were still significantly lower than Pirsabak 2004, indicating that the antioxidant activity is decreased in Fakhar-e-sarhad compared with Pirsabak 2004.

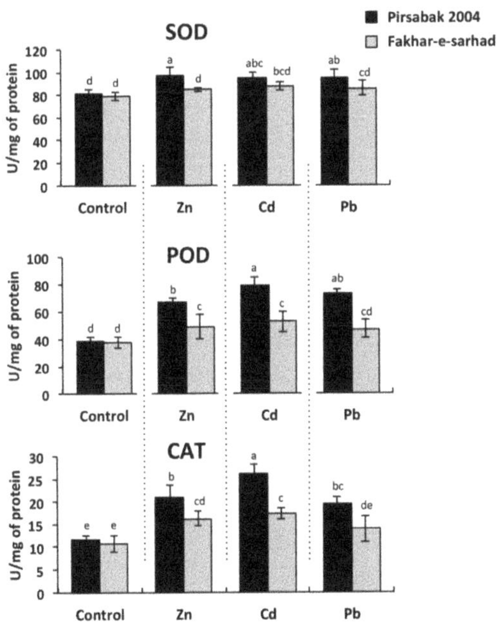

Figure 3. Superoxide dismutase (SOD), peroxidase (POD) or catalase (CAT) anti-oxidant levels in Pirsabak 2004 and Fakhar-e-sarhad varieties in response to 100 μM of $Pb(NO_3)_2$, $ZnSO_4$ and $CdCl_2$ in the hydroponic culture. The results shown are the average of three biological replicates. Different letters indicate a significant difference by an LSD test ($p \leq 0.05$). Error bars represent SD.

2.4. Accumulation of Pb, Cd and Zn in Pirsabak 2004 and Fakhar-e-Sarhad

We next investigated the accumulation of Pb, Cd and Zn in the roots of Fakhar-e-sarhad and Pirsabak 2004 varieties (Figure 4). Pirsabak 2004 and Fakhar-e-sarhad showed a similar amount of Pb, Cd and Zn in their roots. Moreover, Pb accumulation was higher than Cd and Zn in both varieties, which indicates that plants prefer to accumulate Pb compared with Zn and Cd. Pirsabak 2004 and Fakhar-e-sarhad showed non-significant amounts of Pb, Cd and Zn in shoots, indicating that both varieties do not differ in metal accumulation, and accumulated metals were mainly confined in roots. Together, these results indicate that the difference in toxicity was not due to a difference in metal accumulation.

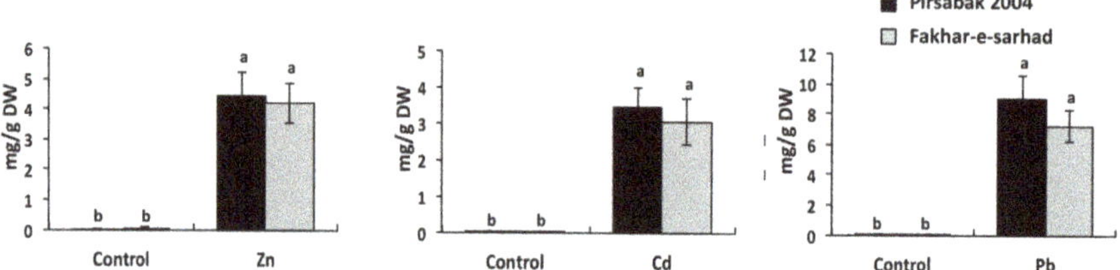

Figure 4. Pb/Cd/Zn accumulation in the roots of Pirsabak 2004 and Fakhar-e-sarhad varieties. The plants were grown in hydroponic culture with 100 μM of $Pb(NO_3)_2$, $ZnSO_4$ or $CdCl_2$, and metal accumulation was investigated after two weeks of treatment. The results shown are the average of three biological replicates. Different letters indicate a significant difference by an LSD test ($p \leq 0.05$). Error bars represent SD.

2.5. Expression of Root Expressed TaABCCs and TaHMA2 transporters in Response to Pb, Cd and Zn Metal Stresses

Since the resistant Pirsabak 2004 and sensitive Fakhar-e-sarhad varieties showed similar levels of metal accumulation in roots (Figure 4), we hypothesized that perhaps the resistant variety has transported the metals more efficiently in vacuoles compared to the sensitive variety. We quantified the expression of root expressed *TaABCC* transporters in both varieties (Figure 5). In general, Pb, Cd and Zn influenced the expression of several root expressed *TaABCC* genes, and all the tested genes responded differentially to each metal. The expression of *TaABCC2*, *TaABCC3* and *TaABCC4* was induced in both varieties (Pirsabak 2004 and Fakhar-e-sarhad) upon Pb treatment compared to their controls (Figure 5A), but Pirsabak 2004 showed a higher level of *TaABCC3* and *TaABCC4* transcripts than Fakhar-e-sarhad. Furthermore, *TaABCC9* and *TaABCC12* expression was down regulated in Pirsabak 2004 in response to Pb. However, the expression level of *TaABCC14* was largely unaltered by Pb treatment in both varieties. In response to Cd treatment (Figure 5B), the expression of *TaABCC2* and *TaABCC4* was induced only in Pirsabak 2004, whereas the expression of *TaABCC3* and *TaABCC4* was down regulated in Fakhar-e-sarhad compared to control. However, the expression of *TaABCC2* was not changed in Fakhar-e-sarhad compared to control in response to Cd treatment. Furthermore, the expression of *TaABCC9*, *TaABCC11*, *TaABCC12* and *TaABCC14* was down regulated in both varieties compared to the control in response to Cd treatment, but their expression levels were higher in Fakhar-e-sarhad than that of Pirsabak 2004. In response to Zn (Figure 5C), the expression of *TaABCC3*, *TaABCC4* and *TaABCC11* was increased in both varieties compared with control. However, *TaABCC3* and *TaABCC4* expression was much more induced in Pirsabak 2004 compared to Fakhar-e-sarhad, whereas the *TaABCC11* expression was more induced in Fakhar-e-sarhad compared with Pirsabak 2004. The expression of *TaABCC2* was increased in Pirsabak 2004 compared with control, whereas, the Fakhar-e-sarhad showed lower expression of *TaABCC2*. Furthermore, the expression of *TaABCC9* and *TaABCC12* was only induced in Fakhar-e-sarhad compared with control in response to Zn treatment. In addition, the expression of *TaABCC14* was down regulated in both varieties compared with control in response to Zn treatment. In brief, these results suggest that *TaABCC2*, *TaABCC3* and *TaABCC4* could contribute to Pb, Cd and Zn metal stress tolerance.

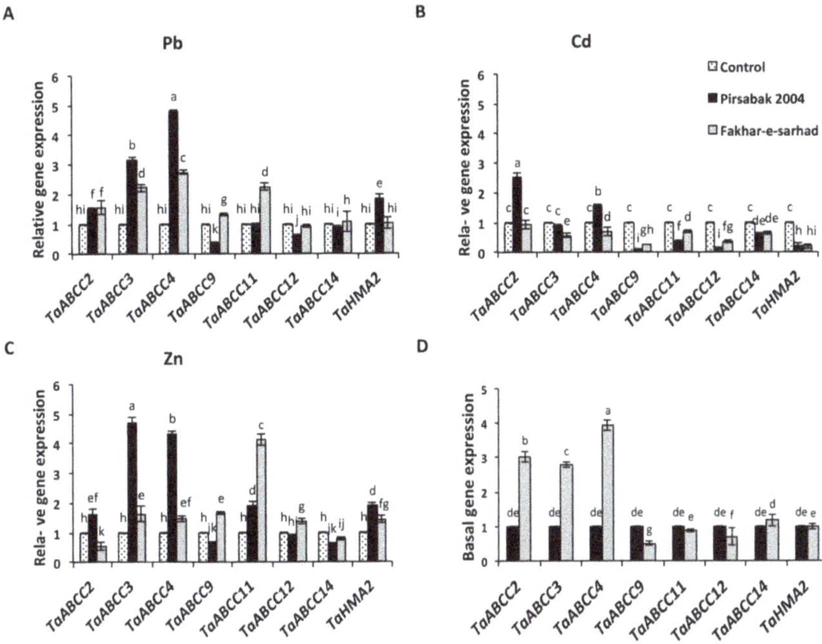

Figure 5. The expression of *TaABCCs/TaHMA2* transporters in response to Pb (**A**), Cd (**B**), Zn metals (**C**) and basal transcript levels (**D**) in the roots of Pirsabak 2004 and Fakhar-e-sarhad varieties. *18SrRNA* was used as an internal control. The results shown are the average of three biological replicates. Different letters indicate a significant difference by an LSD test ($p \leq 0.05$). Error bars represent SD.

We also quantified the expression of *TaHMA2* in response to different metals in both varieties (Figure 5A–C). The *TaHMA2* expression in both varieties was increased in response to Zn compared with control. However, Pirsabak 2004 presented more induced expression compared with Fakhar-e-sarhad in response to Zn treatment (Figure 5C). In addition, *TaHMA2* expression was also increased in Pirsabak 2004 compared with control in response to Pb treatment, but did not change in Fakhar-e-sarhad (Figure 5A). The expression of *TaHMA2* was decreased in both varieties in response to Cd (Figure 5B). Thus, the expression of *TaHMA2* is specifically regulated in response to particular metals.

Since Pirsabak 2004 and Fakhar-e-sarhad showed different root lengths without any treatment, we wondered if this difference of phenotype could be to a differential basal expression of some genes due to their genetic background (Figure 5D). To our expectation, we found that the expression of *TaABCC2*, *TaABCC3* and *TaABCC4* was higher in Fakhar-e-sarhad compared with Pirsabak 2004 without any treatment, whereas, the expression of *TaABCC9* and *TaABCC12* was down regulated in Fakhar-e-sarhad compared with Pirsabak 2004, indicating that both varieties have different basal gene expression. However, the expression of *TaABCC11*, *TaABCC14* and *TaHMA2* did not show any remarkable difference between the two varieties.

2.6. DNA Methyltransferase Expression in Response to Pb, Cd and Zn Metal Stresses

We next investigated the expression of DNA methyltransferases, and our results showed their differential expression in response to Pb, Cd and Zn metal treatment (Figure 6). In response to Pb (Figure 6A), the expression of *TaMET1* was decreased in both Pirsabak 2004 and Fakhar-e-sarhad compared with control, whereas, the expression of *TaMET2a*, *TaMET2b* and *TaMET3* was increased in both varieties compared with control. However, the levels of *TaMET2a*, *TaMET2b* and *TaMET3* were different in Pirsabak 2004 and Fakhar-e-sarhad. The expression of *TaCMT1* was higher in Pirsabak 2004 compared with control in response to Pb treatment, whereas its expression was found lower in Fakhar-e-sarhad compared with control. In response to Cd treatment (Figure 6B), the expression of *TaMET2b* and *TaCMT1* was increased in Pirsabak 2004 and Fakhar-e-sarhad compared with control, but their expression was higher in Pirsabak 2004 compared with Fakhar-e-sarhad. On the contrary, the expression of *TaMET3* was strongly decreased in both Pirsabak 2004 and Fakhar-e-sarhad compared with control in response to Cd treatment. Similarly, the expression of *TaMET1* and *TaMET2a* was down regulated in Fakhar-e-sarhad compared with control in response to Cd treatment. In Pirsabak 2004, the *TaMET1* expression was also down regulated compared with control, but the levels in Pirsabak 2004 were higher than that of Fakhar-e-sarhad in response to Cd. However, the expression of *TaMET2a* did not change in Pirsabak 2004 in response to Cd treatment.

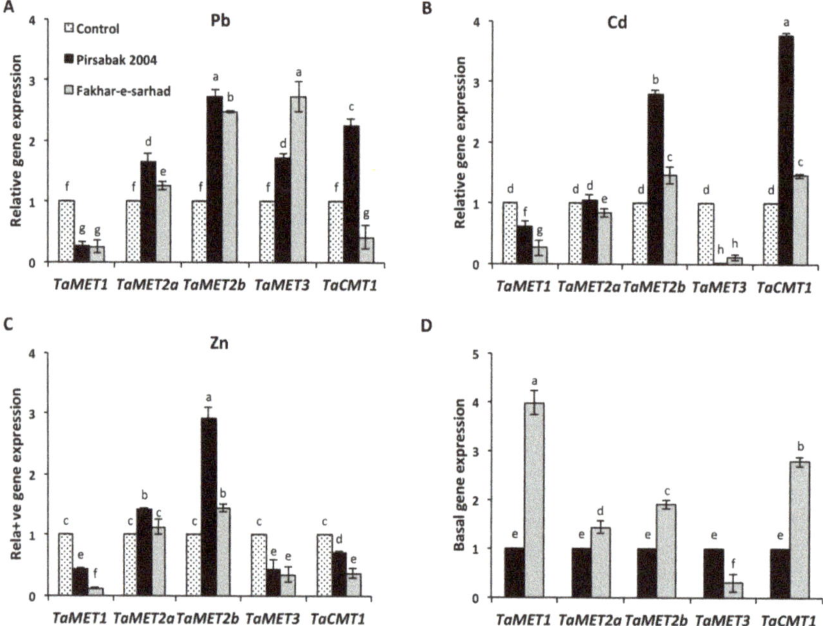

Figure 6. The expression of DNA methyltransferases in response to Pb (**A**), Cd (**B**), Zn metals (**C**), and their basal expression (**D**) in the roots of Pirsabak 2004 and Fakhar-e-sarhad varieties. *18SrRNA* was used as an internal control. The data presented are the average of three biological replicates. Different letters indicate a significant difference by an LSD test ($p \leq 0.05$). Error bars represent SD.

In response to Zn treatment (Figure 6C), the expression of *TaMET2b* was increased in both varieties compared with control, but Pirsabak 2004 levels were higher than that of Fakhar-e-sarhad. In contrast to these, *TaMET1*, *TaMET3* and *TaCMT1* expression was down regulated in response to Zn treatment in Pirsabak 2004 and Fakhar-e-sarhad. However, *TaMET1* and *TaCMT1* expression was higher in Pirsabak 2004 compared with Fakhar-e-sarhad in response to Zn treatment, indicating that the expression of a particular DNA methyltransferase is regulated depending on the metal stress and genetic background.

As we found that the expression of some *TaABCC* transporters was genetically different between these varieties without treatment (Figure 5D), we also compared the basal expression of methyltransferases in both varieties (Figure 6D). We found that the basal expression of *TaMET1*,

TaMET2a, *TaMET2b* and *TaCMT1* was higher in Fakhar-e-sarhad compared with Pirsabak 2004, while the expression of *TaMET3* was lower in Fakhar-e-sarhad compared with Pirsabak 2004, which indicate that both varieties could have different DNA methylation levels and/or sites, thus could explain the difference in basal expression of some *TaABCC* transporters in Fakhar-e-sarhad and Pirsabak 2004.

2.7. DNA Hypomethylation of Pirsabak 2004 in Response to Pb, Cd and Zn Metal Stresses

We next quantified the DNA methylation levels at the promoter of *TaABCCs* and *TaHMA2* transporters in response to Pb, Cd and Zn treatments. DNA hypomethylation was observed at the promoters of the tested transporters in Pirsabak 2004 compared with control in response to Pb, Cd and Zn (Figure 7A–C, Figure S1). In response to Pb (Figure 7A), CG DNA methylation levels were reduced at the promoters of *TaABCC2*, *TaABCC3*, *TaABCC4*, *TaABCC9*, *TaABCC12* and *TaHMA2* in Pirsabak 2004 compared to the control. In Fakhar-e-sarhad, the DNA methylation levels were slightly higher at the promoter of *TaABCC2* and *TaABCC3* compared with control in response to Pb treatment. However, the CG DNA methylation levels did not change at *TaABCC4*, *TaABCC9*, *TaABCC12* and *TaHMA2* in Fakhar-e-sarhad compared with the control in response to Pb treatment. Similar to Pb, in response to Cd and Zn (Figure 7B,C), CG DNA methylations were also reduced at the promoters of all the tested *TaABCCs* and *TaHMA2* transporters in Pirsabak 2004 compared with control. Moreover, CG DNA methylation levels were decreased in response to Cd in Fakhar-e-sarhad, while the levels were generally increased in response to Zn compared with control. In general, DNA methylation levels in response to Pb, Cd and Zn were lower in Pirsabak 2004 compared with Fakhar-e-sarhad at the promoter of tested *TaABCCs* and *TaHMA2* transporters, thus probably explains the increase of *TaABCC2*, *TaABCC3*, *TaABCC4*, and *TaHMA2* expressions in Pirsabak 2004 in response to metal stress.

We also investigated whether the different basal expression of *TaABCCs* in both varieties is due to the different levels of DNA methylation on these genes. Basal DNA methylation levels were decreased at the promoter of *TaABCC2*, *TaABCC3*, *TaABCC4*, *TaABCC9*, *TaABCC12* and *TaHMA2* in Fakhar-e-sarhad compared with Pirsabak 2004 (Figure S1 and Figure 7D). Among the tested genes, only *TaABCC3* showed the CHH/CHG DNA methylation at his promoter. Fakhar-e-sarhad showed lower CHH/CHG DNA methylation compared with Pirsabak 2004 at the promoter of *TaABCC3* (Figure S2).

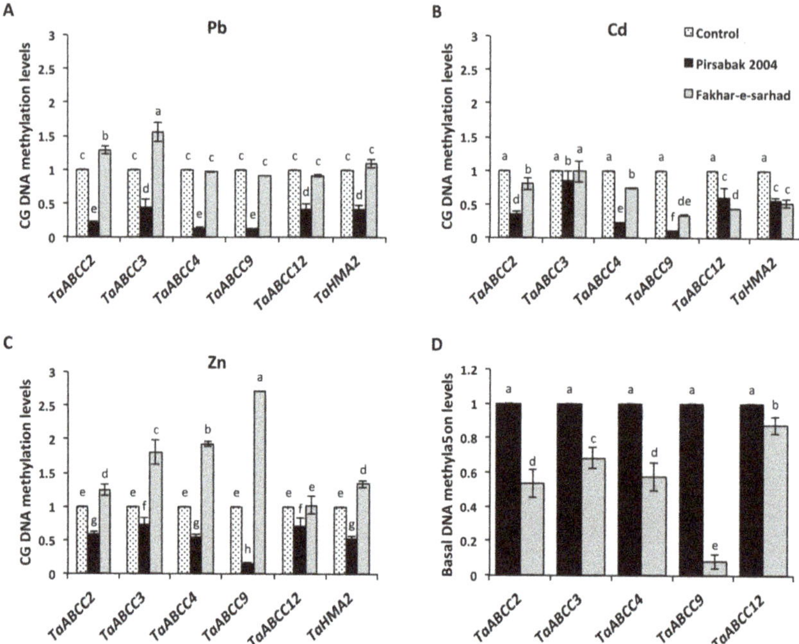

Figure 7. CG DNA methylation levels at the promoter of *TaABCC* transporters in response to Pb (**A**), Cd (**B**) and Zn metals (**C**) and the basal DNA methylation levels (**D**) in the roots of Pirsabak 2004 and Fakhar-e-sarhad varieties. DNA was digested with *AciI* and *hpaII* for CG DNA methylation. Equal amount of digested and undigested DNA were used as template for qPCR, % to non-digested DNA was calculated and relative to control is presented. The control of each variety was set to 1, therefore, presented only once in the graph. Basal DNA methylation represents the levels of DNA methylation of each variety in control conditions. The data presented are the average of three biological replicates. Different letters indicate a significant difference by an LSD test ($p \leq 0.05$). Error bars represent SD.

3. Discussion

Heavy metal toxicity for the environment, plants and human life has become a major global problem. Heavy metals do not easily degrade or volatilize, which leads to their accumulation in the soil over years. Among these heavy metals, Pb and Cd are the most harmful because they can enter into the food chain through the soil, thus imposing a serious threat not only to plants but also to humans and livestock [2,4,6]. In this particular scenario, screening high yielding plant varieties against metal toxicity and their adoption in plant breeding programs is essential. Therefore, we first screened eight high yielding wheat varieties for their phenotypic sensitivity to Pb toxicity at different doses. We found that Pb resistant Pirsabak 2004 and sensitive Fakhar-e-sarhad wheat varieties are also resistant or sensitive to Cd and Zn. Furthermore, Pb, Cd and Zn metal stresses induce DNA hypomethylation at the promoter of some selected *TaABCC* and *TaHMA2* metal detoxification transporters in Pirsabak 2004, which is correlated with their increased gene expression and metal resistant phenotype.

In order to reduce the Pb and Cd concentration in the soils, a lot of efforts have been made in previous years, including the use of hyper-accumulating plants [43]. However, due to low biomass, long remediation time and narrow biological adaptability, the usage of hyper-accumulating plants could not meet the demands of large-scale applications. Therefore, in parallel to hyper-accumulating plants, evaluating the genetic potential of crop plants against heavy metal toxicity could be a valuable choice. In this study we chose high yielding wheat varieties and evaluated their response to Pb toxicity. Seed germination and seedling growth are some of the most important and earlier physiological processes that are affected in wheat plants in response to metal stress [44]. Thus, the ability of a seed to germinate and the increase in seedling growth in the presence of metal stress would indicate the level of tolerance to metal stress. Our results showed that germination percentage, epicotyl length and root length were largely unaffected in Pirsabak 2004 in response to Pb stress, while these phenotypes were the most severely affected in Fakhar-e-sarhad among all the tested varieties (Figure 1). Furthermore, the root length of Pirsabak 2004 was also found not affected in response to Cd and Zn stresses, while Fakhar-e-sarhad root length was significantly affected (Figure 2). These results indicate that Pirsabak 2004 and Fakhar-e-sarhad were the most resistant and sensitive varieties, respectively. Since Pb, Cd and Zn affected the germination percentage and seedling growth in Fakhar-e-sarhad, we expected the low crop yield of Fakhar-e-sarhad in metal contaminated soils. While on the contrary, we expected the better performance of Pirsabak 2004 in metal contaminated soils. Thus we recommend farmers to cultivate Pirsabak 2004 in metal contaminated soils to ensure the better crop yield compared with all the tested varieties.

Plants exposed to heavy metals generate reactive oxygen species (ROS) such as O_2^- and OH^-, which cause oxidative damage to the cellular structure and functions [45]. Therefore, plants have developed a complex antioxidant response, including the production of antioxidant enzymes, such as SOD, POD and CAT. SOD catalyzes the conversion of O_2^- into molecular O_2 and H_2O_2, and CAT and/or POD further detoxify the H_2O_2 [46,47]. This indicates that the levels of antioxidant activities would indicate the ability of the plant to cope with the metal stress by limiting the impact of ROS. The levels of SOD, POD and CAT were significantly increased in response to metal stress in Pirsabak 2004 and Fakhar-e-sarhad compared with control, but the levels of SOD, POD and CAT in Pirsabak 2004 were significantly higher than that of Fakhar-e-sarhad (Figure 3). This indicates that the antioxidant activities are decreased in Fakhar-e-sarhad in response to Pb, Cd and Zn, which may contribute to its sensitive phenotype.

The genetic diversity of plants has been extensively studied based on morphological and biochemical evaluation in the pre-genomic era, while DNA (or molecular) markers were studied in the post-genomic era [48]. Besides genetic variation, epigenetic modifications can create epialleles that can be inherited independently and epigenetic variations evolve more quickly [27,49]. Therefore, epigenetic variations could be used in plant breeding programs [27]. Our data showed the DNA hypomethylation at the promoter of *TaABCC* genes in Fakhar-e-sarhad compared with Pirsabak 2004 (Figure 7D) in the control samples, indicating the epigenetic variations between Pirsabak 2004 and Fakhar-e-sarhad. In addition, Pirsabak 2004 and Fakhar-e-sarhad also differed in basal transcriptional responses (Figures 5D and 6D) and root length in the control samples (Figure 1). Especially, the expression of *TaABCC2, TaABCC3* and *TaABCC4* was higher in Fakhar-e-sarhad compared with Pirsabak 2004, which is consistent with DNA hypomethylation at their promoters in Fakhar-e-sarhad. Together, our results indicate the genetic and epigenetic diversity of Pirsabak 2004 and Fakhar-e-sarhad. However, further studies are required to explore their epigenetic diversity.

DNA methylation events in response to a metal exposure have also been reported in *Vicia faba*, rape seedlings, and *Arabidopsis* [38,50,51]. *Trifolium repens* L. and *Cannabis sativa* L. plants have already been reported to have different basal DNA methylation levels in their roots [37]. Moreover, the Cd and Ni metal treatments induce DNA hypomethylation in both *Trifolium repens* L. and *Cannabis sativa* L. DNA methylation changes in response to Cd stress depends on the plants, e.g., in *Brassica napus*, *Trifolium repens* L. and *Cannabis sativa* L., the Cd induces the DNA hypomethylation [37,38,50], while in *Vicia faba*, Cd induces the DNA hypermethylation. Our results showed that Pirsabak 2004 presents CG DNA hypomethylation in response to Pb, Cd and Zn metal stresses at the promoter of *TaABCC2, TaABCC3, TaABCC4* and *TaHMA2* transporters (Figure 7), while Fakhar-e-sarhad showed hypermethylation considering their basal DNA methylation level, and increased DNA methylation levels compared with control, especially in the case of Zn. These observations suggest that DNA methylation plays an important role in the resistance mechanism of metal stress in Pirsabak 2004.

Plants have evolved a system to prevent a cytotoxic concentration by effluxing the metals from the cytosol to the apoplast through the action of heavy metal ATPases (HMAs), also known as PIB-ATPases. The heavy metal ATPase2, TaHMA2, is a plasma membrane located transporter from wheat that was suggested to export the Zn/Cd toward the apoplast [24]. A yeast expressing wheat *TaHMA2* was found resistant to Zn/Cd and furthermore, the over expression of *TaHMA2* conferred a mild resistance against Zn and Cd in *Arabidopsis*, indicating the important function of TaHMA2 in metal tolerance in wheat. Interestingly, our results showed that the expression of *TaHMA2* was higher in Pirsabak 2004 compared with Fakhar-e-sarhad (Figure 5A–C) in response to Pb, Cd and Zn, which is consistent with DNA hypomethylation in Pirsabak 2004 compared with Fakhar-e-sarhad (Figure 7, Figure S1). This suggests that the increase in *TaHMA2* expression in response to metal stress likely contributes to the resistance in Pirsabak 2004. In parallel to HMAs, ABCC transporters have been reported to enhance resistance against metal stress in plants as well as in yeast by vacuole sequestration. Yeast Cadmium Factor 1 (YCF1), an ABCC transporter, has been reported to play an important role in metal tolerance in yeast [52,53]. Over-expression of *YCF1* in *Arabidopsis*, poplar and *Brassica* enhances the tolerance to Cd and Pb [54–56], suggesting the important function of ABCC transporters in metal detoxification. Furthermore, *Arabidopsis AtABCC1/AtABCC2* genes also play a role in conferring a resistance to Cd and mercury (Hg) stresses by vacuole sequestration [14]. Our results also showed the DNA hypomethylation (Figure 7, Figure S1) and increased expression of *TaABCC2*, *TaABCC3* and *TaABCC4* in Pirsabak 2004 compared with Fakhar-e-sarhad in response to Pb, Cd and Zn (Figure 5A–C). Notably, Pirsabak 2004 and Fakhar-e-sarhad accumulate similar amounts of Pb, Cd and Zn in their roots, which indicate that the resistance of Pirsabak 2004 is not due to less accumulation of toxic metals in the roots, but is likely due to the detoxification mechanism of plants. In this scenario, the enhanced activity of TaABCC2, TaABCC3, TaABCC4 and TaHMA2 transporters in Pirsabak 2004 likely contributes to the metal resistance of Pirsabak 2004. Therefore, we proposed that upon the exposure to Pb, Cd and Zn stresses, DNA hypomethylation occurred at the promoters of *TaABCCs* and *TaHMA2* in Pirsabak 2004, which will eventually lead to an increase in their transcription. The increased activity of TaABCCs may efficiently sequestrate accumulated Pb, Cd and Zn into the vacuole and in the meanwhile increased TaHMA2 activity may send the toxic metals back to the apoplast. The resulting metal concentration is not toxic to the cells, and meanwhile, the increased activity of SOD, POD and CAT scavenges the impact of ROS generated from metal toxicity. Together, these processes may lead to confer the resistance phenotype of Pirsabak 2004 (Figure 8). Furthermore, in response to Pb, Cd and Zn, TaABCCs mediated vacuole sequestration of toxic metals as well as export back of toxic metals to apoplast through the activity of TaHMA2 may not be sufficient in Fakhar-e-sarhad. The resulting metal concentration becomes higher in cells, and the antioxidant response may not be fully able to overcome the metal toxicity. Thus, relatively lower detoxification efficiency mediated by TaABCCs and TaHMA2, and decreased antioxidant activity in Fakhar-e-sarhad compared with Pirsabak 2004 may explain its sensitive phenotype. However, more functional studies of TaABCCs are required to validate this model in wheat.

In summary, our results demonstrated that the DNA methylation difference between resistant Pirsabak 2004 and sensitive Fakhar-e-sarhad varieties in response to Pb, Cd and Zn contributes to the metal tolerance through the regulation of the expression of metal detoxification transporters. This study highlights that the DNA methylation is an important parameter to confer heavy metal resistance in Pirsabak 2004. This study also recommends the cultivation of Pirsabak 2004 in metal contaminated soils.

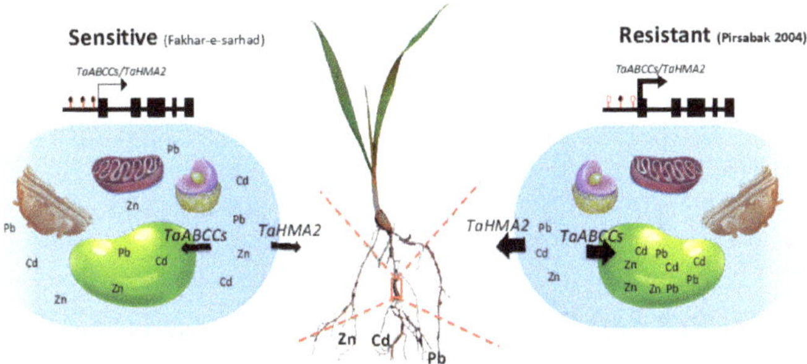

Figure 8. Proposed model for Pirsabak 2004 resistant and Fakhar-e-sarhad sensitive phenotypes. In response to metal stress, DNA methylation levels differently changed at the promoter of *TaABCCs* and *TaHMA2* transporters, and consequently changed their expression levels in Pirsabak 2004 and Fakhar-e-sarhad. Thus, vacuole sequestration of toxic metals through TaABCCs and export back to apoplast through TaHMA2 activity maintain the metal homeostasis and confer the resistant phenotype. Black filled circles represent DNA methylation, while open circles represent hypomethylation. The intensity of the arrow represents the expression level of genes.

4. Materials and Methods

4.1. Plant Material

The selected high yielding varieties were collected from the seed stock of COMSATS University Islamabad, Abbottabad campus, Pakistan. The names of the selected varieties are Tatara, Khyber 87, Pirsabak 2004, Fakhar-e-sarhad, Janbaz, Attahabib, Punjab 85 and PAK 81. The pedigree detail of these varieties is given in Table S2.

4.2. Sowing and Growth Conditions for MS Media

The seeds were sterilized by dipping in 0.1% $HgCl_2$ for 15–20 min, and then washed with double distilled water. Then the seeds were washed again in 70% ethanol for 10 min followed by four washes with distilled water. No $Pb(NO_3)_2$ as a control or different doses of $Pb(NO_3)_2$ i.e., 0.5 mM, 1 mM and 2 mM were added to Murashige and Skoog (MS) medium [57], supplemented with agar. Sterilized seeds of the selected varieties were sown on MS media in a growth chamber having 25 °C temperature and 16/8 h light/dark conditions. The data of morphological traits, root length, epicotyl length and germination percentage were scored after six days of sowing. The root and epicotyl lengths were measured from fifteen seedlings per replicate and the germination percentage was calculated from hundred seeds per replicate. All the experiments were performed in three biological replicates for control and treatments. The length of the primary root and the epicotyl length were measured by Image J, http://rsbweb.nih.gov/ij/.

4.3. Sowing and Growth Conditions for Hydroponics

The seeds were placed in a tray containing moist filter paper and placed in a growth chamber for 48 h. After germination, healthy seedlings (five seedlings per pot) having the same root length were wrapped in a foam layer and fixed in plastic cups, which were inserted into the plastic pots containing hydroponic solutions. The hydroponic solutions were composed of 0.2 mM KH_2PO_4, 1.0 mM K_2SO_4, 2.0 mM $Ca(NO_3)_2$, 2.0 mM $CaCl_2$, 0.5 mM $MgSO_4$ and 0.2 mM $FeSO_4.7H_2O$ as a source of macronutrients while 5.0 µM H_3BO_3, 2.0 µM $MnSO_4$, 0.5 µM $ZnSO_4$, 0.3 µM $CuSO_4$ and 0.01 µM $(NH_4)_2Mo_7O24$ as micronutrients [58]. The nutrient solution was replaced twice in a week with freshly prepared solution of same strength. The heavy metal treatments were applied by adding 100 µM of $CdCl_2$, $Pb(NO_3)_2$ and $ZnSO_4$ in a hydroponic solution, while the control contained only hydroponic solution. All the experiments were performed in three biological replicates for control and treatments.

4.4. Atomic Absorption Analysis

After two weeks of the application of $Pb(NO_3)_2$, $ZnSO_4$ and $CdCl_2$ in the hydroponic culture medium plants were harvested. Root length was measured with measuring tape. In order to measure the uptake of Pb, Cd and Zn metals in root, dried roots were crushed and dried at 37 °C. The dried sample was ashed at 550 °C for 4–5 h in the furnace and allowed to cool down. Samples were digested for 3–4 h by adding 2 mL of 4 M HNO_3. After 4 h, 8 mL of distilled water was added to make the final volume of 10 mL. Finally; the digested diluted plant material was filtered by using filter paper and analyzed for Pb, Cd and Zn on atomic absorption spectrophotometer (model, AAnalyst 700, PerkinElmer Inc., Shelton, CT, USA).

4.5. Extraction and Measurement of Antioxidant Enzymes

Leaf samples were placed in liquid nitrogen immediately after their harvesting and stored at −80 °C until their analysis. The frozen leaf samples (0.5 g) were homogenized in 2.5 mL of 100 mM freshly prepared potassium phosphate buffer of pH 7 supplemented with 0.1 mM EDTA. Then the samples were centrifuged at 15,000 × g for 10 min at 4 °C and a supernatant was collected in eppendorf tubes, which were used for the analysis of antioxidants. Catalase (CAT), peroxidase (POD) and superoxide dismutase (SOD) activities were determined as described in [59]. The CAT activity was determined by monitoring the decomposition of H_2O_2 at 240 nm through spectrophotometer (U2020 IRMECO, Germany), while the activity of POD was measured by using guaiacol as substrate. The reaction mixture contained 0.1 M phosphate buffet of pH 7, 1% guaiacol, 0.4 M H_2O_2 and enzyme extract. Change in absorbance per unit time was measured at 470 nm. SOD activity was measured by photoreduction of nitroblue tetrazolium (NBT). Reaction mixture comprised 50 mM phosphate buffer of pH 7.8, 0.1 mM EDTA, 20 mM L-methionine, 750 μM NBT, 20 μM riboflavin and enzyme extract. The mixture was exposed to light for 15 min and absorbance was measured at 560 nm. The protein content in the leaves was measured by following the Bradford method [60]. The enzyme activity was expressed as U/mg of protein.

4.6. Gene Expression Analysis

Total RNA was extracted by using Trizol (Invitrogen, Waltham, MA, USA,) according to the manufacturer's instructions from roots after 48 h of treatment with Pb, Cd and Zn. After DNase I treatment, reverse transcription was performed with Superscript III (Invitrogen) using the gene-specific primers. RT-qPCR was performed with the gene-specific primers using SYBR green Master Mix (Roche, Indianapolis, IN, USA) as described in [61]. The expression was corrected by using *18SrRNA* as an internal reference gene. The relative gene expression presented in Figures 5 and 6 corresponds to the fold change of expression that was calculated by normalizing the expression of metal treated sample to the expression in the respective control sample of each variety. In order to clarify the fold change in response to a particular metal, the control of each variety was set to 1. The expression in control samples is referred as basal expression in the text. To compare the basal gene expression in Pirsabak 2004 and Fakhar-e-sarhad without any treatment, the expression levels of Pirsabak 2004 were set to 1 for each gene and then the expression in Fakhar-e-sarhad was normalized to the expression level of Pirsabak 2004. Primer sequences for gene expression are listed in Table S1.

4.7. DNA Methylation Chop-Quantitative PCR (Chop-qPCR)

Chop-PCR was performed as previously described by [62]. Briefly, genomic DNA was extracted with the CTAB method from wheat roots after 48 h of Pb, Cd and Zn treatments. Then the DNA was digested with *AciI* (R0551S New England Biolabs, Ipswich, MA, USA) and *hpaII* (R0171S New England Biolabs, Ipswich, MA, USA) for CG DNA methylation, and with *AluI* (New England Biolabs, Ipswich, MA, USA USA) and *haeIII* (R0108S New England Biolabs, Ipswich, MA, USA) for CHH/CHG methylation. Equal amount of digested and undigested DNA were used as template for qPCR, and normalized to undigested DNA. Basal DNA methylation represents the levels of DNA methylation of each variety in control conditions. To compare the basal gene DNA methylation levels in Pirsabak 2004 and Fakhar-e-sarhad, the DNA methylation levels of Pirsabak 2004 were set to 1 for each gene and then the DNA methylation levels in Fakhar-e-sarhad were normalized to Pirsabak 2004. For total DNA methylation levels, the *McrBC* enzyme (M0272S New England Biolabs, Ipswich, MA, USA USA) was used that specially digested methylated DNA, therefore, bands represent the non-DNA methylation levels. Chop-qPCR primers are listed in the Table S1.

4.8. Statistical Analysis

Experiments were conducted in a completely randomized design (CRD) with three replicates. The Shapiro–Wilk normality test was performed to test the normal distribution of data and the homogeneity of variance was tested by using the Levene's test. After that analysis of variance (ANOVA) was performed followed by the least significant difference (LSD) test at p value ≤ 0.05 for each parameter. Statistical analyses were performed by using the Statistical Analysis System (SAS) software (SAS Institute Inc., Kerry, NC, USA) and the Statistical Package for Social Sciences (SPSS) software (version 11.0- SPSS Inc., Chicago, IL, USA).

Supplementary Materials: Supplementary materials can be found at http://www.mdpi.com/1422-0067/20/19/4676/s1.

Author Contributions: S.S., Q.Z. and A.A. performed the experiments; S.S., Y.S., R.N., E.W., L.L. analyze the data; S.S., E.W. and L.L. conceived the idea and S.S. wrote the paper.

Funding: This work was supported by grant from the National Natural Science Foundation of China (No.31571258 and No.31800224), NRPU.20-3657/R&D/HEC/14/704, and PD-IPFP/HRD/HEC/2013/1129 from Higher Education Commission of Pakistan.

Conflicts of Interest: The authors declare no conflict of interest. The funders had no role in the design of the study; in the collection, analyses, or interpretation of data; in the writing of the manuscript, or in the decision to publish the results.

References

1. Babst-Kostecka, A.A.; Waldmann, P.; Frérot, H.; Vollenweider, P. Plant adaptation to metal polluted environments—Physiological, morphological, and evolutionary insights from Biscutella laevigata. *Environ. Exp. Bot.* **2016**, *127*, 1–13. [CrossRef]
2. Sengar, R.S.; Gautam, M.; Sengar, R.S.; Garg, S.K.; Sengar, K.; Chaudhary, R. Lead stress effects on physiobiochemical activities of higher plants. *Rev. Env. Contam. Toxicol.* **2008**, *196*, 73–93.
3. Hartwig, A. Cadmium and cancer. *Met. Ions. Life Sci.* **2013**, *11*, 491–507. [CrossRef] [PubMed]
4. Pourrut, B.; Shahid, M.; Dumat, C.; Winterton, P.; Pinelli, E. Lead Uptake, Toxicity, and Detoxification in Plants. In *Reviews of Environmental Contamination and Toxicology*; Whitacre, D.M., Ed.; Springer: New York, NY, USA, 2011; Volume 213, pp. 113–136.
5. Gill, S.S.; Tuteja, N. Cadmium stress tolerance in crop plants. *Plant. Signal. Behav.* **2011**, *6*, 215–222. [CrossRef] [PubMed]
6. Andresen, E.; Küpper, H. Cadmium Toxicity in Plants. In *Cadmium: From Toxicity to Essentiality*; Sigel, A., Sigel, H., Sigel, R.K.O., Eds.; Springer: Dordrecht, The Netherlands, 2013; pp. 395–413.
7. Kobayashi, T.; Nishizawa, N.K. Iron Uptake, Translocation, and Regulation in Higher Plants. *Annu. Rev. Plant. Biol.* **2012**, *63*, 131–152. [CrossRef] [PubMed]
8. Sadeghzadeh, B.; Rengel, Z. Zinc in Soils and Crop Nutrition. In *The Molecular and Physiological Basis of Nutrient Use Efficiency in Crops*; Wiley-Blackwell: Brussels, Belgium, 2011; pp. 335–375.
9. Broadley, M.R.; White, P.J.; Hammond, J.P.; Zelko, I.; Lux, A. Zinc in plants. *New Phytol.* **2007**, *173*, 677–702. [CrossRef] [PubMed]
10. Kushwaha, A.; Rani, R.; Kumar, S.; Gautam, A. Heavy metal detoxification and tolerance mechanisms in plants: Implications for phytoremediation. *Environ. Rev.* **2015**, *24*, 39–51. [CrossRef]

11. Ji, H.; Peng, Y.; Meckes, N.; Allen, S.; Stewart, C.N., Jr.; Traw, M.B. ATP-dependent binding cassette transporter G family member 16 increases plant tolerance to abscisic acid and assists in basal resistance against Pseudomonas syringae DC3000. *Plant. Physiol.* **2014**, *166*, 879–888. [CrossRef] [PubMed]
12. Liu, G.; Sanchez-Fernandez, R.; Li, Z.S.; Rea, P.A. Enhanced multispecificity of arabidopsis vacuolar multidrug resistance-associated protein-type ATP-binding cassette transporter, AtMRP2. *J. Biol. Chem.* **2001**, *276*, 8648–8656. [CrossRef] [PubMed]
13. Nagy, R.; Grob, H.; Weder, B.; Green, P.; Klein, M.; Frelet-Barrand, A.; Schjoerring, J.K.; Brearley, C.; Martinoia, E. The Arabidopsis ATP-binding cassette protein AtMRP5/AtABCC5 is a high affinity inositol hexakisphosphate transporter involved in guard cell signaling and phytate storage. *J. Biol. Chem.* **2009**, *284*, 33614–33622. [CrossRef]
14. Park, J.; Song, W.-Y.; Ko, D.; Eom, Y.; Hansen, T.H.; Schiller, M.; Lee, T.G.; Martinoia, E.; Lee, Y. The phytochelatin transporters AtABCC1 and AtABCC2 mediate tolerance to cadmium and mercury. *Plant. J.* **2012**, *69*, 278–288. [CrossRef] [PubMed]
15. Li, Z.-S.; Szczypka, M.; Lu, Y.-P.; Thiele, D.J.; Rea, P.A. The Yeast Cadmium Factor Protein (YCF1) Is a Vacuolar Glutathione S-Conjugate Pump. *J. Biol. Chem.* **1996**, *271*, 6509–6517. [CrossRef] [PubMed]
16. Sharma, K.G.; Mason, D.L.; Liu, G.; Rea, P.A.; Bachhawat, A.K.; Michaelis, S. Localization, Regulation, and Substrate Transport Properties of Bpt1p, a Saccharomyces cerevisiae MRP-Type ABC Transporter. *Eukaryot. Cell* **2002**, *1*, 391–400. [CrossRef] [PubMed]
17. Bhati, K.K.; Sharma, S.; Aggarwal, S.; Kaur, M.; Shukla, V.; Kaur, J.; Mantri, S.; Pandey, A.K. Genome-wide identification and expression characterization of ABCC-MRP transporters in hexaploid wheat. *Front. Plant. Sci.* **2015**, *6*, 488. [CrossRef] [PubMed]
18. Walter, S.; Kahla, A.; Arunachalam, C.; Perochon, A.; Khan, M.R.; Scofield, S.R.; Doohan, F.M. A wheat ABC transporter contributes to both grain formation and mycotoxin tolerance. *J. Exp. Bot.* **2015**, *66*, 2583–2593. [CrossRef] [PubMed]
19. Theodoulou, F.L.; Clark, I.M.; He, X.L.; Pallett, K.E.; Cole, D.J.; Hallahan, D.L. Co-induction of glutathione-S-transferases and multidrug resistance associated protein by xenobiotics in wheat. *Pest. Manag. Sci.* **2003**, *59*, 202–214. [CrossRef]
20. Morel, M.; Crouzet, J.; Gravot, A.; Auroy, P.; Leonhardt, N.; Vavasseur, A.; Richaud, P. AtHMA3, a P1B-ATPase allowing Cd/Zn/Co/Pb vacuolar storage in Arabidopsis. *Plant Physiol.* **2009**, *149*, 894–904. [CrossRef]
21. Williams, L.E.; Mills, R.F. P(1B)-ATPases–an ancient family of transition metal pumps with diverse functions in plants. *Trends Plant Sci.* **2005**, *10*, 491–502. [CrossRef]
22. Arguello, J.M. Identification of ion-selectivity determinants in heavy-metal transport P1B-type ATPases. *J. Membr. Biol.* **2003**, *195*, 93–108. [CrossRef]
23. Axelsen, K.B.; Palmgren, M.G. Evolution of Substrate Specificities in the P-Type ATPase Superfamily. *J. Mol. Evol.* **1998**, *46*, 84–101. [CrossRef]
24. Tan, J.; Wang, J.; Chai, T.; Zhang, Y.; Feng, S.; Li, Y.; Zhao, H.; Liu, H.; Chai, X. Functional analyses of TaHMA2, a P1B-type ATPase in wheat. *Plant Biotechnol. J.* **2013**, *11*, 420–431. [CrossRef] [PubMed]
25. Berr, A.; Shafiq, S.; Shen, W.H. Histone modifications in transcriptional activation during plant development. *Biochim. Biophys. Acta* **2011**, *1809*, 567–576. [CrossRef]
26. Zhang, M.; Kimatu, J.N.; Xu, K.; Liu, B. DNA cytosine methylation in plant development. *J. Genet. Genom.* **2010**, *37*, 1–12. [CrossRef]
27. Shafiq, S.; Khan, A.R. Plant Epigenetics and Crop Improvement. In *PlantOmics: The Omics of Plant Science*; Barh, D., Khan, M.S., Davies, E., Eds.; Springer: New Delhi, India, 2015; pp. 157–179.
28. Law, J.A.; Jacobsen, S.E. Establishing, maintaining and modifying DNA methylation patterns in plants and animals. *Nat. Rev. Genet.* **2010**, *11*, 204–220. [CrossRef] [PubMed]
29. Kankel, M.W.; Ramsey, D.E.; Stokes, T.L.; Flowers, S.K.; Haag, J.R.; Jeddeloh, J.A.; Riddle, N.C.; Verbsky, M.L.; Richards, E.J. Arabidopsis MET1 cytosine methyltransferase mutants. *Genetics* **2003**, *163*, 1109. [PubMed]
30. Lindroth, A.M.; Cao, X.; Jackson, J.P.; Zilberman, D.; McCallum, C.M.; Henikoff, S.; Jacobsen, S.E. Requirement of CHROMOMETHYLASE3 for maintenance of CpXpG methylation. *Science* **2001**, *292*, 2077. [CrossRef] [PubMed]
31. Cao, X.; Jacobsen, S.E. Role of the arabidopsis DRM methyltransferases in de novo DNA methylation and gene silencing. *Curr. Biol.* **2002**, *12*, 1138–1144. [CrossRef]

32. Dai, Y.; Ni, Z.; Dai, J.; Zhao, T.; Sun, Q. Isolation and expression analysis of genes encoding DNA methyltransferase in wheat (*Triticum aestivum* L.). *Biochim. Biophys. Acta* **2005**, *1729*, 118–125. [CrossRef] [PubMed]
33. Peng, H.; Zhang, J. Plant genomic DNA methylation in response to stresses: Potential applications and challenges in plant breeding. *Prog. Nat. Sci.* **2009**, *19*, 1037–1045. [CrossRef]
34. Dowen, H.R.; Pelizzola, M.; Schmitz, R.; Lister, R.; Dowen, J.; Nery, J.; Dixon, J.; R Ecker, J. Widespread dynamic DNA methylation in response to biotic stress. *P. Nat. Acad. Sci. USA* **2012**, *109*, E2183–E2191. [CrossRef] [PubMed]
35. Vilahur, N.; Vahter, M.; Broberg, K. The Epigenetic Effects of Prenatal Cadmium Exposure. *Curr. Environ. Health. Rep.* **2015**, *2*, 195–203. [CrossRef] [PubMed]
36. Arita, A.; Costa, M. Epigenetics in metal carcinogenesis: Nickel, Arsenic, Chromium and Cadmium. *Metallomics* **2009**, *1*, 222–228. [CrossRef] [PubMed]
37. Aina, R.; Sgorbati, S.; Santagostino, A.; Labra, M.; Ghiani, A.; Citterio, S. Specific hypomethylation of DNA is induced by heavy metals in white clover and industrial hemp. *Physiol. Plant* **2004**, *121*, 472–480. [CrossRef]
38. Filek, M.; Keskinen, R.; Hartikainen, H.; Szarejko, I.; Janiak, A.; Miszalski, Z.; Golda, A. The protective role of selenium in rape seedlings subjected to cadmium stress. *J. Plant Physiol.* **2008**, *165*, 833–844. [CrossRef] [PubMed]
39. Yang, J.L.; Liu, L.W.; Gong, Y.Q.; Huang, D.Q.; Wang, F.; He, L.L. Analysis of genomic DNA methylation level in radish under cadmium stress by methylation-sensitive amplified polymorphism technique. *J. Plant Physiol. Mol. Biol.* **2007**, *33*, 219–226.
40. Greco, M.; Chiappetta, A.; Bruno, L.; Bitonti, M.B. In Posidonia oceanica cadmium induces changes in DNA methylation and chromatin patterning. *J. Exp. Bot.* **2012**, *63*, 695–709. [CrossRef] [PubMed]
41. Feng, S.J.; Liu, X.S.; Tao, H.; Tan, S.K.; Chu, S.S.; Oono, Y.; Zhang, X.D.; Chen, J.; Yang, Z.M. Variation of DNA methylation patterns associated with gene expression in rice (Oryza sativa) exposed to cadmium. *Plant Cell Environ.* **2016**, *39*, 2629–2649. [CrossRef] [PubMed]
42. Ou, X.; Zhang, Y.; Xu, C.; Lin, X.; Zang, Q.; Zhuang, T.; Jiang, L.; von Wettstein, D.; Liu, B. Transgenerational inheritance of modified DNA methylation patterns and enhanced tolerance induced by heavy metal stress in rice (Oryza sativa L.). *PLoS ONE* **2012**, *7*, e41143. [CrossRef] [PubMed]
43. Rascio, N.; Navari-Izzo, F. Heavy metal hyperaccumulating plants: How and why do they do it? And what makes them so interesting? *Plant Sci.* **2011**, *180*, 169–181. [CrossRef]
44. Wang, H.; Zhong, G.; Shi, G.; Pan, F. Toxicity of Cu, Pb, and Zn on Seed Germination and Young Seedlings of Wheat (*Triticum aestivum* L.). In *International Conference on Computer and Computing Technologies in Agriculture*; Springer: Heidelberg/Berlin, Germany, 2010; pp. 231–240.
45. Apel, K.; Hirt, H. Reactive oxygen species: Metabolism, oxidative stress, and signal transduction. *Annu. Rev. Plant. Biol.* **2004**, *55*, 373–399. [CrossRef]
46. Elstner, E.F. Oxygen Activation and Oxygen Toxicity. *Annu. Rev. Plant. Physiol.* **1982**, *33*, 73–96. [CrossRef]
47. Bowler, C.; Montagu, M.V.; Inze, D. Superoxide Dismutase and Stress Tolerance. *Annu. Rev. Plant Physiol. Plant. Mol. Biol.* **1992**, *43*, 83–116. [CrossRef]
48. Govindaraj, M.; Vetriventhan, M.; Srinivasan, M. Importance of Genetic Diversity Assessment in Crop Plants and Its Recent Advances: An Overview of Its Analytical Perspectives. *Genet. Res. Int.* **2015**, *2015*, 14. [CrossRef] [PubMed]
49. Mirouze, M.; Vitte, C. Transposable elements, a treasure trove to decipher epigenetic variation: Insights from Arabidopsis and crop epigenomes. *J. Exp. Bot.* **2014**, *65*, 2801–2812. [CrossRef] [PubMed]
50. Taspinar, M.S.; Agar, G.; Alpsoy, L.; Yildirim, N.; Bozari, S.; Sevsay, S. The protective role of zinc and calcium in Vicia faba seedlings subjected to cadmium stress. *Toxicol. Ind. Health* **2010**, *27*, 73–80. [CrossRef] [PubMed]
51. Li, Z.; Chen, X.; Li, S.; Wang, Z. Effect of nickel chloride on Arabidopsis genomic DNA and methylation of 18S rDNA. *Electron. J. Biotechnol.* **2014**, *283*, 51–57. [CrossRef]
52. Szczypka, M.S.; Wemmie, J.A.; Moye-Rowley, W.S.; Thiele, D.J. A yeast metal resistance protein similar to human cystic fibrosis transmembrane conductance regulator (CFTR) and multidrug resistance-associated protein. *J. Biol. Chem.* **1994**, *269*, 22853–22857. [PubMed]
53. Ghosh, M.; Shen, J.; Rosen, B.P. Pathways of As(III) detoxification in Saccharomyces cerevisiae. *Proc. Natl. Acad. Sci. USA* **1999**, *96*, 5001–5006. [CrossRef]

54. Song, W.-Y.; Ju Sohn, E.; Martinoia, E.; Jik Lee, Y.; Yang, Y.-Y.; Jasinski, M.; Forestier, C.; Hwang, I.; Lee, Y. Engineering tolerance and accumulation of lead and cadmium in transgenic plants. *Nat. Biotech.* **2003**, *21*, 914–919. [CrossRef]
55. Shim, D.; Kim, S.; Choi, Y.I.; Song, W.Y.; Park, J.; Youk, E.S.; Jeong, S.C.; Martinoia, E.; Noh, E.W.; Lee, Y. Transgenic poplar trees expressing yeast cadmium factor 1 exhibit the characteristics necessary for the phytoremediation of mine tailing soil. *Chemosphere* **2013**, *90*, 1478–1486. [CrossRef]
56. Bhuiyan, M.S.U.; Min, S.R.; Jeong, W.J.; Sultana, S.; Choi, K.S.; Song, W.Y.; Lee, Y.; Lim, Y.P.; Liu, J.R. Overexpression of a yeast cadmium factor 1 (YCF1) enhances heavy metal tolerance and accumulation in Brassica juncea. *Plant. Cell Tiss. Org. Cult.* **2011**, *105*, 85–91. [CrossRef]
57. Murashige, T.; Skoog, F. A Revised Medium for Rapid Growth and Bio Assays with Tobacco Tissue Cultures. *Physiol. Plant* **1962**, *15*, 473–497. [CrossRef]
58. Shahzad, M.; Witzel, K.; Zörb, C.; Mühling, K.H. Growth-Related Changes in Subcellular Ion Patterns in Maize Leaves (*Zea mays* L.) under Salt Stress. *J. Agron. Crop. Sci.* **2012**, *198*, 46–56. [CrossRef]
59. Csiszár, J.; Lantos, E.; Tari, I.; Emilian, M.; Wodala, B.; Vashegyi, A.; Horvath, F.; Pécsváradi, A.; Szabo, M.; Bartha-Dima, B.; et al. Antioxidant enzyme activities in Allium species and their cultivars under water stress. *Plant. Soil Environ.* **2007**, *53*, 517–523. [CrossRef]
60. Bradford, M.M. A rapid and sensitive method for the quantitation of microgram quantities of protein utilizing the principle of protein-dye binding. *Anal. Biochem.* **1976**, *72*, 248–254. [CrossRef]
61. Shafiq, S.; Berr, A.F.; Shen, W.H. Combinatorial functions of diverse histone methylations in Arabidopsis thaliana flowering time regulation. *New Phytol.* **2014**, *201*, 312–322. [CrossRef] [PubMed]
62. Zhang, H.; Tang, K.; Wang, B.; Duan, C.-G.; Lang, Z.; Zhu, J.-K. Protocol: A beginner's guide to the analysis of RNA-directed DNA methylation in plants. *Plant Methods* **2014**, *10*, 18. [CrossRef] [PubMed]

© 2019 by the authors. Licensee MDPI, Basel, Switzerland. This article is an open access article distributed under the terms and conditions of the Creative Commons Attribution (CC BY) license (http://creativecommons.org/licenses/by/4.0/).

International Journal of
Molecular Sciences

Article

MAPK Pathway under Chronic Copper Excess in Green Macroalgae (Chlorophyta): Influence on Metal Exclusion/Extrusion Mechanisms and Photosynthesis

Paula S. M. Celis-Plá [1,†], Fernanda Rodríguez-Rojas [1,†], Lorena Méndez [1], Fabiola Moenne [1], Pamela T. Muñoz [1,2,3], M. Gabriela Lobos [4], Patricia Díaz [4], José Luis Sánchez-Lizaso [5], Murray T. Brown [6], Alejandra Moenne [7] and Claudio A. Sáez [1,8,*]

1. Laboratory of Aquatic Environmental Research, Centro de Estudios Avanzados, Universidad de Playa Ancha, Viña del Mar 2520000, Chile; paulacelispla@upla.cl (P.S.M.C.-P.); fernanda.rodriguez@upla.cl (F.R.-R.); lorena.mendez@alumnos.uv.cl (L.M.); fabiola.moenne@upla.cl (F.M.); pamela.munoz@upla.cl (P.T.M.)
2. Doctorado Interdisciplinario en Ciencias Ambientales, Facultad de Ciencias Naturales y Exactas, Universidad de Playa Ancha, Valparaíso 2340000, Chile
3. Doctorado en Ciencias del Mar y Biología Aplicada, Departamento de Ciencias del Mar y Biología Aplicada, Universidad de Alicante, 03080 Alicante, Spain
4. Laboratory of Environmental and Analytical Chemistry, Instituto de Química y Bioquímica, Facultad de Ciencias, Universidad de Valparaíso, Valparaíso 234000, Chile; gabriela.lobos@uv.cl (M.G.L.); patriciaediazg@gmail.com (P.D.)
5. Departamento de Ciencias del Mar y Biología Aplicada, Universidad de Alicante, 03080 Alicante, Spain; jl.sanchez@ua.es
6. School of Biological and Marine Sciences, University of Plymouth, Plymouth PL4 8AA, UK; mt.brown@plymouth.ac.uk
7. Laboratory of Marine Biotechnology, Facultad de Química y Biología, Universidad de Santiago de Chile, Santiago 9170020, Chile; alejandra.moenne@usach.cl
8. HUB-AMBIENTAL UPLA, Universidad de Playa Ancha, Valparaíso 2340000, Chile

* Correspondence: claudio.saez@upla.cl

† These authors contributed equally to this work.

Received: 8 July 2019; Accepted: 9 September 2019; Published: 13 September 2019

Abstract: There is currently no information regarding the role that whole mitogen activated protein kinase (MAPK) pathways play in counteracting environmental stress in photosynthetic organisms. To address this gap, we exposed *Ulva compressa* to chronic levels of copper (10 μM) specific inhibitors of Extracellular Signal Regulated Kinases (ERK), c-Jun *N*-terminal Kinases (JNK), and Cytokinin Specific Binding Protein (p38) MAPKs alone or in combination. Intracellular copper accumulation and photosynthetic activity (in vivo chlorophyll *a* fluorescence) were measured after 6 h, 24 h, 48 h, and 6 days of exposure. By day 6, when one (except JNK) or more of the MAPK pathways were inhibited under copper stress, there was a decrease in copper accumulation compared with algae exposed to copper alone. When at least two MAPKs were blocked, there was a decrease in photosynthetic activity expressed in lower productivity (ETR_{max}), efficiency (α_{ETR}), and saturation of irradiance (EkETR), accompanied by higher non-photochemical quenching (NPQ_{max}), compared to both the control and copper-only treatments. In terms of accumulation, once the MAPK pathways were partially or completely blocked under copper, there was crosstalk between these and other signaling mechanisms to enhance metal extrusion/exclusion from cells. Crosstalk occurred among MAPK pathways to maintain photosynthesis homeostasis, demonstrating the importance of the signaling pathways for physiological performance. This study is complemented by a parallel/complementary article Rodríguez-Rojas et al. on the role of MAPKs in copper-detoxification.

Keywords: in vivo chlorophyll *a* florescence; physiology; mitogen activated protein kinases; *Ulva compressa*; metal accumulation

1. Introduction

In eukaryotes, including mammals, plants, and algae, stimuli such as reactive oxygen species (ROS), growth factors, cytokines, and metal excess can activate mitogen activated protein kinases (MAPKs). These enzymes participate in an important cellular signaling pathway which regulates metabolic and physiological processes, including transcription, mitosis, and cell differentiation, in response to biotic and environmental stressors [1–3]. MAPKs are serine/threonine kinases that are activated by a sequenced phosphorylation cascade involving MAPK kinases (MAPKK), which are induced upstream by MAPKK kinases (MAPKKK). There are three main MAPK pathways, which correspond to those leading to MAPK Extracellular Signal Regulated Kinases (ERK), c-Jun *N*-terminal Kinases (JNK), and Cytokinin Specific Binding Protein (p38) [2,3]. The exact mechanism by which ROS activate MAPK pathways is still not well understood, but it has been established that the accumulation of ROS, such as superoxide anions, hydrogen peroxide and hydroxyl radicals, oxidizes thioredoxins (TRX) and binding apoptosis stimulating kinase 1 (ASK1), leading to the dissociation of TRX from ASK1; the latter has been shown to trigger at least the JNK and p38 MAPK pathways [4]. In addition, cysteine residues present in the receptors of cytokines and growth factors are oxidized by ROS leading to their activation, as evidenced by antioxidants inhibiting EGF-1-induced ROS production and activation of the ERK pathway [5]. Furthermore, hydrogen peroxide (H_2O_2) directly induces growth factor receptors and MAPK pathways in different cell types [6]. A potential alternative mechanism is the direct ROS-oxidation and degradation of phosphatases that inhibit MAPK pathways [7]. Another factor known to activate MAPKs is excess cellular metal concentrations, although it is generally agreed that the mechanisms by which metals trigger the pathway relate to metal-mediated overproduction of ROS [8]. For example, copper can induce ROS by substituting iron in a Fenton-like reaction, disrupting electron transport chains in chloroplasts and mitochondria, and facilitating excess energy transfer to oxygen [9,10]. The signaling role of ROS is well-known; however, beyond certain threshold levels there can be an imbalance in reactive oxygen metabolism resulting in oxidative stress and, in severe cases, damage [9].

The biological implications of copper excess are well-documented in photoautotrophs from terrestrial and aquatic ecosystems [11–14]. Due to domestic and industrial activities, copper is recognized as one of the most pernicious pollutants in the marine environment [15]. As well as their role as primary producers, marine macroalgae or seaweeds are regarded as significant habitat-forming organisms, and the detrimental effects of copper pollution under both laboratory and field assessments have been explored in some detail [12,16–18]. The main recognized biological strategies in macroalgae to counteract copper (and other metals) excess are related to cellular exclusion/extrusion mechanisms, detoxification through the syntheses of metal chelators, and activation of antioxidant defences. Most of these processes were observed to be transcriptionally regulated [9,18,19]. Despite this, there is a paucity of information on signal transduction pathways involved in metal tolerance. Records in other photoautotrophs indicate that the main pathways underlying metal tolerance are associated with ROS, nitric oxide (NO), MAPKs, hormones, and calcium signaling [8]. Of these, the best described for macroalgae are the calcium, NO, and ROS signaling mechanisms, using the copper-tolerant green macroalga *Ulva compressa* as a model species [9]. Under chronic copper exposure of 10 μM, *U. compressa* internalizes calcium from the extracellular media mediated by transient dependent potential (TRP) and voltage-gated calcium channels (VDCC), processes that induce calcium release from the endoplasmic reticulum (ER). This, in turn, induces calcium crosstalk with ROS and NO, the activation of calmodulins (CaMs) and calcium-dependent protein kinases (CDPKs), and the subsequent upregulation of enzymes involved in ROS detoxification [20–24]. Recent findings also suggest that the MAPK ERK pathway may play a role in the regulation of antioxidant enzymes and in the synthesis of metallothioneins (MTs), which are metal chelators [20].

However, the dynamics and overall contributions of the ERK and MAPK pathways in regulating metal-tolerance mechanisms in macroalgae remain unclear.

Once intracellular copper concentrations reach toxic levels, biological impairment manifests itself in physiological processes, with photosynthetic activity being an important example [25]. A reliable, non-invasive technique for the measurement of photosynthesis is pulse amplitude modulation (PAM) chlorophyll *a* fluorescence associated with photosystem II (PSII), which is used as an indicator of eco-physiological performance [26,27]. Information on various functional components of the photosynthetic apparatus is acquired, including the maximal quantum yield of PSII (F_v/F_m), which implies the state of photoinhibition [28], the maximal electron transport rate (ETR_{max}), the photosynthetic efficiency (α_{ETR}), and the saturation of irradiance (EkETR), which estimate photosynthetic capacity, efficiency, and productivity, and non-photochemical quenching (NPQ), which is considered to be an indicator of photoprotection that is accomplished through xanthophyll cycle-mediated energy dissipation [12,29]. Several published studies on the effects of high copper levels on photosynthetic activity in *Ulva* spp. report a decrease in F_v/F_m and rETR [30] and high NPQ [31]. These patterns are indicative of copper-induced overstimulation of electron transport chains, triggering decreased photosynthetic activity, and the activation of energy dissipation mechanisms to avoid damage in PSII. As with other photoautotrophs, no records have been published regarding the involvement of the MAPK signaling pathway in photosynthesis in macroalgae.

In this investigation, we studied the roles of the three main pathways within the MAPK signaling cascade on the physiological processes associated with copper accumulation and photosynthetic activity in green macroalgae, using *U. compressa* as a model organism. Specific inhibitors of the ERK, JNK, and p38 pathways were applied alone or in combination under copper excess. The implications for metal extrusion/exclusion mechanisms and photosynthetic activity, as determined from measurements of F_v/F_m, ETR_{max}, α_{ETR}, Ek_{ETR}, and NPQ_{max}, were evaluated. In a parallel study by Rodríguez-Rojas et al. [1], we extended this research to assess the role of the MAPK pathway in the toxicity/detoxification mechanisms of *U. compressa* under copper excess.

2. Results

2.1. Copper Accumulation

Copper accumulation was significantly higher than in controls when *U. compressa* was exposed to copper, although the patterns differed between treatments in the presence of MAPK inhibitors and at different times ($p < 0.01$, Figure 1, Table S1). Average intracellular concentrations were around 81% of the total accumulated copper (intracellular plus extracellular).

Copper accumulation at 6 h was highest in treatments containing copper and ERK inhibitor (T3, T6, T7, and T9) and JNK (T4) or p38 (T5) inhibitors. Moreover, seaweeds exposed to all treatments with copper and inhibitors (T3–T9) had higher intracellular copper levels that those exposed to only copper (Figure 1A). In contrast, at 24 h, the copper content was always lower in treatments containing copper and inhibitors compared with only copper (T3–T9). Among treatments with copper and inhibitors, the lowest levels of accumulation were detected when JNKi, p38i, and ERK were applied together (T9) (Figure 1B). Patterns at 48 h were similar to those observed at 6 h, although the most accumulation was recorded in treatments with copper and p38 inhibitor (T5, T7, T8, and T9) and with either ERK (T3) or JNK inhibitors (T4) (Figure 1C). For all treatments, the accumulation of copper was highest after exposure for 6 d (Figure 1A–D). Following the oscillatory trends found throughout the experiment, except for copper with JNK inhibitor (T4), at day 6, the metal accumulation levels in treatments combining copper and inhibitors (T3–T9) were significantly lower than with copper alone (T2) (Figure 1D). Intracellular copper concentrations in treatments with copper and inhibitors were as follows: ERK and p38 (T7) < p38 (T5) < ERK < p38 and JNK (T9) (Figure 1D).

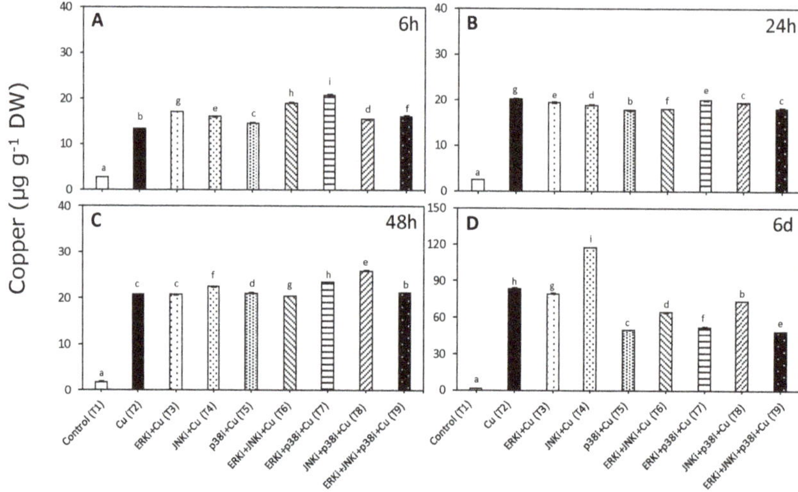

Figure 1. Intracellular copper accumulation in *Ulva compressa* under copper and/or exposure to mitogen activated protein kinase (MAPK) inhibitors. Treatments: T1—control only with seawater; T2—copper only exposure as 10 µM $CuSO_4$ (Cu); T3—copper + 5 µM MAPK Extracellular Signal Regulated Kinase (ERK) inhibitor PD98059 (Cu + ERKi); T4—copper + 5 µM MAPK c-Jun *N*-terminal Kinase (JNK) inhibitor SP600125 (Cu + JNKi); T5—copper + MAPK Cytokinin Specific Binding Protein (p38) inhibitor SB203580 (Cu + p38i); T6—Cu + ERKi + JNKi; T7—Cu + ERKi + p38i; T8—Cu + p38i + JNKi; and T9—Cu + ERKi + JNKi + p38i. Samples were analyzed after (**A**) 6 h, (**B**) 24 h, (**C**) 48 h, and (**D**) 6 days. Different letters within each graph represent significant differences at the 95% confidence interval ($p < 0.05$). Plots are mean ± SE ($n = 3$).

2.2. Photosynthetic Activity According to In Vivo Chlorophyll a Florescence

While patterns were identified for many different parameters associated with photosynthetic activity, the most significant were those associated with the electron transport rate (ETR) and non-photochemical quenching (NPQ). In the case of the maximum quantum yield (F_v/F_m), there were no significant differences between treatments ($p < 0.05$, Figure 2, Table S2).

There was a significant interaction between time and treatment factors for the photosynthetic efficiency (α_{ETR}) data ($p < 0.05$, Table S2). Similar trends were observed in α_{ETR} at different experimental time periods. For most, there were significant differences in the α_{ETR} between controls (T1) and exposure to 10 µM copper (T2), although at 48 h, the α_{ETR} was significantly lower in the copper treatment than controls (Figure 3). The copper/MAPK inhibitor-treated material at all measurement periods tended to have lower α_{ETR} levels in all combinations than copper alone, although these differences were only significant at 6 h and 6 d (except for T9 after 6 d) (Figure 3A,B). There were no clear trends between treatments with combinations of inhibitors under copper excess at any measurement period, nor between treatments with single MAPK inhibitors and copper (Figure 3A–D).

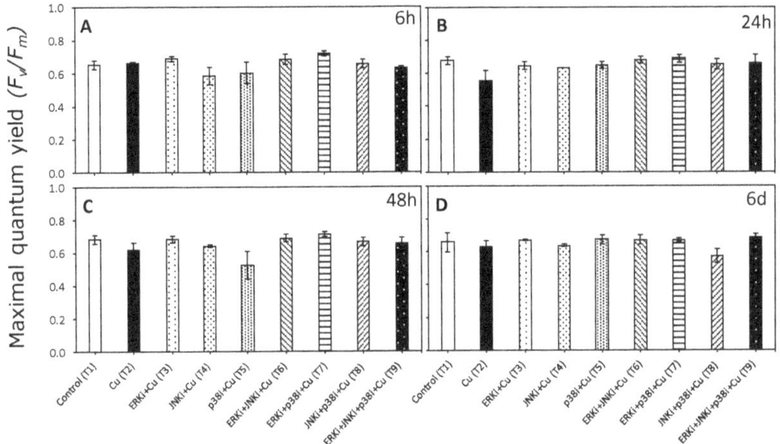

Figure 2. Maximum quantum yield of photosystem II (PSII) (F_v/F_m) in *U. compressa* under copper and/or exposure to MAPK inhibitors. Treatments: T1—control only with seawater; T2—copper only exposure as 10 μM $CuSO_4$ (Cu); T3—copper + 5 μM MAPK ERK inhibitor PD98059 (Cu + ERKi); T4—copper + 5 μM MAPK JNK inhibitor SP600125 (Cu + JNKi); T5—copper + MAPK p38 inhibitor SB203580 (Cu + p38i); T6—Cu + ERKi + JNKi; T7—Cu + ERKi + p38i; T8—Cu + p38i + JNKi; and T9—Cu + ERKi + JNKi + p38i. Samples were analyzed after (**A**) 6 h, (**B**) 24 h, (**C**) 48 h, and (**D**) 6 d of treatment. No significant differences were detected in any experimental times at the 95% confidence interval ($p > 0.05$). Plots are mean ± SE ($n = 3$).

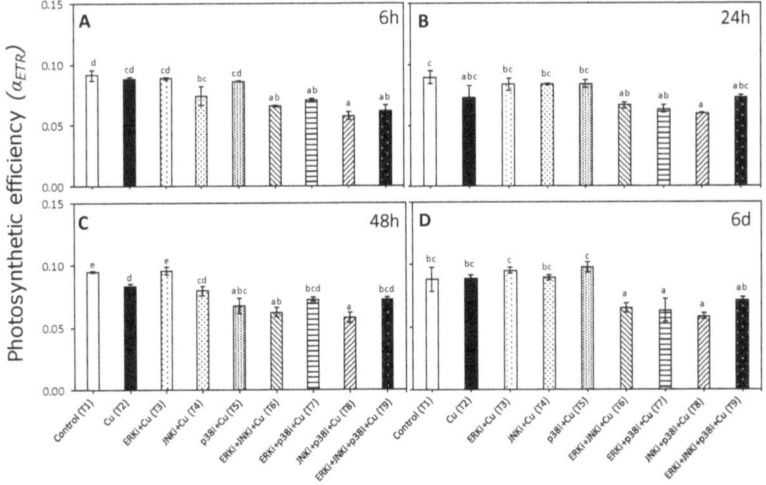

Figure 3. Photosynthetic efficiency (αETR) in *U. compressa* under copper and/or exposure to MAPK inhibitors. Treatments: T1—control only with seawater; T2—copper only exposure as 10 μM $CuSO_4$ (Cu); T3—copper + 5 μM MAPK ERK inhibitor PD98059 (Cu + ERKi); T4—copper + 5 μM MAPK JNK inhibitor SP600125 (Cu + JNKi); T5—copper + MAPK p38 inhibitor SB203580 (Cu + p38i); T6—Cu + ERKi + JNKi; T7—Cu + ERKi + p38i; T8—Cu + p38i + JNKi; and T9—Cu + ERKi + JNKi + p38i. Samples were analyzed after (**A**) 6 h, (**B**) 24 h, (**C**) 48 h, and (**D**) 6 d of treatment. Different letters within each graph represent significant differences at the 95% confidence interval ($p < 0.05$). Plots are mean ± SE ($n = 3$).

Patterns in ETR_{max} and Ek_{ETR} were similar to those observed for α_{ETR} with a significant interaction between treatments and times ($p < 0.05$, Table S2). There were no significant differences in ETR_{max} and Ek_{ETR} between control (T1) and copper exposure (T2) at any time period, except at 24 h (Figures 4B and 5B, respectively). No clear trends were apparent between treatments combining MAPK inhibitors with copper (T6–T9) for any time period, although values of ETR_{max} and Ek_{ETR} were always lower than when exposed only to copper (T2) (Figures 4 and 5).

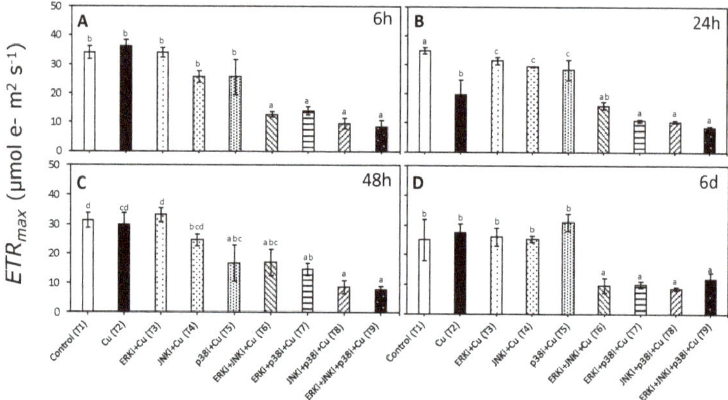

Figure 4. Maximal electron transport rate (ETR_{max}) in *U. compressa* under copper and/or exposure to MAPK inhibitors. Treatments: T1—control only with seawater; T2—copper only exposure as 10 µM $CuSO_4$ (Cu); T3—copper + 5 µM MAPK ERK inhibitor PD98059 (Cu + ERKi); T4—copper + 5 µM MAPK JNK inhibitor SP600125 (Cu + JNKi); T5—copper + MAPK p38 inhibitor SB203580 (Cu + p38i); T6—Cu + ERKi + JNKi; T7—Cu + ERKi + p38i; T8—Cu + p38i + JNKi; and T9)—Cu + ERKi + JNKi + p38i. Samples were analyzed after (**A**) 6 h, (**B**) 24 h, (**C**) 48 h, and (**D**) 6 d of treatment. Different letters within each graph represent significant differences at the 95% confidence interval ($p < 0.05$). Plots are mean ± SE ($n = 3$).

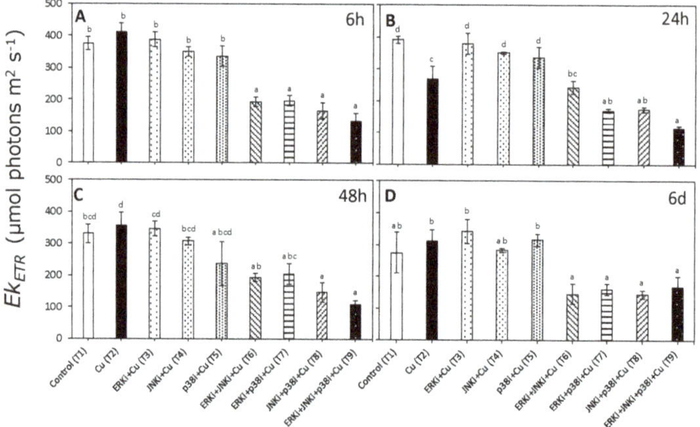

Figure 5. Saturation of the irradiance of ETR (EkETR) in *U. compressa* under copper and/or exposure to MAPK inhibitors. Treatments: T1—control only with seawater; T2—copper only exposure as 10 µM $CuSO_4$ (Cu); T3—copper + 5 µM MAPK ERK inhibitor PD98059 (Cu + ERKi); T4—copper + 5 µM MAPK JNK inhibitor SP600125 (Cu + JNKi); T5—copper + MAPK p38 inhibitor SB203580 (Cu + p38i); T6—Cu + ERKi + JNKi; T7—Cu + ERKi + p38i; T8—Cu + p38i + JNKi; and T9)—Cu + ERKi + JNKi + p38i. Samples were analyzed after (**A**) 6 h, (**B**) 24 h, (**C**) 48 h, and (**D**) 6 d of treatment. Different letters within each graph represent significant differences at the 95% confidence interval ($p < 0.05$). Plots are mean ± SE ($n = 3$).

There were significant differences ($p < 0.05$) in non-photochemical quenching (NPQ_{max}) between certain treatments but only at day 6 (Figure 6 Table S2). Values of, NPQ_{max} were significantly lower in copper (T2) than controls, and treatments with copper combined with ERK inhibitor (T6, T7 and T9) (Figure 6D).

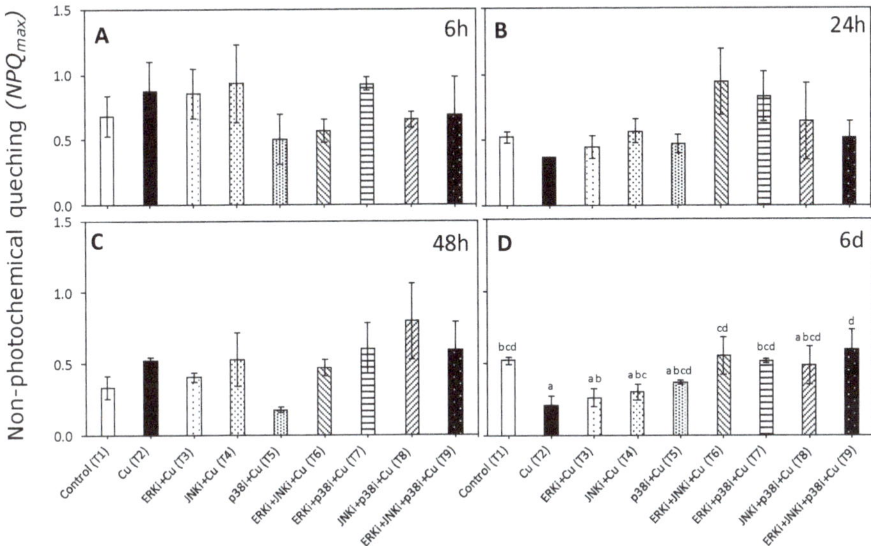

Figure 6. Maximal non-photochemical quenching (NPQ_{max}) in *U. compressa* under copper and/or exposure to MAPK inhibitors. Treatments: T1—control only with seawater; T2—copper only exposure as 10 μM $CuSO_4$ (Cu); T3—copper + 5 μM MAPK Extracellular Signal Regulated Kinase (ERK) inhibitor PD98059 (Cu + ERKi); T4—copper + 5 μM MAPK c-Jun *N*-terminal Kinase (JNK) inhibitor SP600125 (Cu + JNKi); T5—copper + MAPK Cytokinin Specific Binding Protein (p38) inhibitor SB203580 (Cu + p38i); T6—Cu + ERKi + JNKi; T7—Cu + ERKi + p38i; T8—Cu + p38i + JNKi; and T9—Cu + ERKi + JNKi + p38i. Samples were analyzed after (**A**) 6 h, (**B**) 24 h, (**C**) 48 h, and (**D**) 6 d of treatment. Different letters within each graph represent significant differences at the 95% confidence interval ($p < 0.05$). No significant differences were detected at experimental times of 6, 24, and 48 h ($p > 0.05$). Plots are represented as mean ± SE ($n = 3$).

2.3. Principal Component Ordination (PCO) Analysis

The PCO showed a positive correlation of the first axis (54.8% of total variation) with ETR_{max}, Ek_{ETR}, α_{ETR} and, to a lesser extent, F_v/F_m. Conversely, NPQ_{max} and intracellular copper accumulation were negatively correlated according to Spearman's correlation (Figure 7). Time had a large effect on these factors (Figure 7a), in particular on NPQ_{max} and intracellular copper when the treatment used contained copper and two or more MAPK inhibitors (Figure 7b). In both graphical representations, the combination of the first two axes explained 80.2% of the total variation (Figure 7a,b).

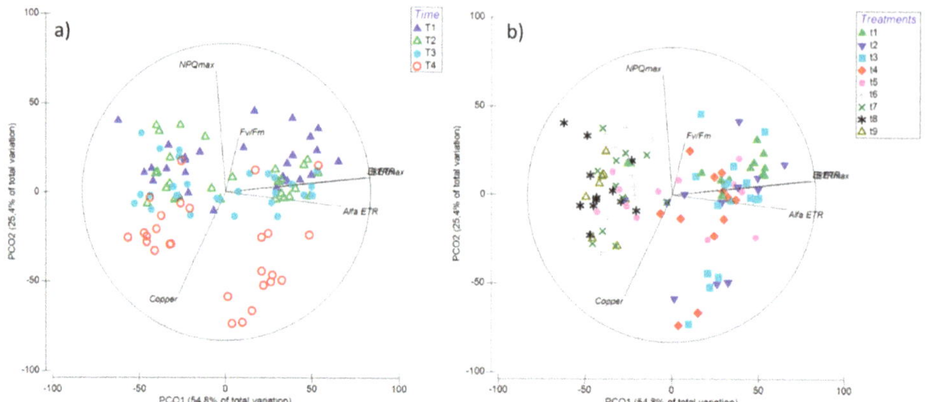

Figure 7. Principal Component Ordination (PCO) analysis diagrams expressed in relation to (**a**) time: T1—6 h; T2—24 h; T3—48 h; and T4—6 d; and (**b**) treatments: t1—control only with seawater; t2—copper only exposure as 10 μM $CuSO_4$ (Cu); t3—copper + 5 μM MAPK ERK inhibitor PD98059 (Cu + ERKi); t4—copper + 5 μM MAPK JNK inhibitor SP600125 (Cu + JNKi); t5—copper + MAPK p38 inhibitor SB203580 (Cu + p38i); t6—Cu + ERKi + JNKi; t7—Cu + ERKi + p38i; t8—Cu + p38i + JNKi; and t9—Cu + ERKi + JNKi + p38i. Vector overlays (Spearman's rank correlation) indicate the relationship between the PCO axes and the physiological variables, namely, intracellular copper accumulation (copper), photoinhibition (F_v/F_m), productivity (ETR_{max}), efficiency (α_{ETR}), and saturation of irradiance (EkETR), accompanied by higher non-photochemical quenching (NPQ_{max}).

3. Discussion

In this investigation, *Ulva compressa* was exposed to an excess copper concentration of 10 μM. Moreover, under this stress condition, the three main MAPK pathways ending in ERK, JNK, and/or p38 were blocked in order to evaluate the importance of the whole signaling pathway on metal exclusion/extrusion mechanisms and photosynthetic performance in green macroalgae under copper excess. Despite, to our knowledge, metal uptake and accumulation having important influences on the extent of toxic responses and effects, there is no published information regarding the potential roles of the MAPK pathways in these processes in eukaryotes. This is relevant given that these processes are, at least partly, transcriptionally regulated phenomena [8]. For example, it has been recorded that *Arabidopsis thaliana* mutants overexpressing the cadmium extrusion transporter AtPDR8 have a greater tolerance to cadmium excess [32]. Also, the expression of the copper-transporting ATPase HMA5 from the metallophyte *Silene vulgaris* confers increased copper tolerance in *A. thaliana* [33]. Moreover, a whole transcriptome study revealed that *Chlamydomonas reinhardtii* exposed to high silver levels overexpressed eight genes encoding hydroxyproline-rich glycoprotein components, which are constituents of the cell wall, and glycoside hydrolase, which aids in remodelling of the cell wall [34]. These findings are important, since metal-extrusion/exclusion mechanisms in macroalgae are known to involve chelation to cell wall components and energy-dependent actions of membrane transporters [9]. Although the role of the MAPK signaling pathway in activating the metal-extrusion/exclusion mechanisms in algae has not been explored, results from studies with plants provide insight into its potential participation. For instance, it was observed that several transcription factors regulating metal tolerance responses, such as C2H2-type zinc finger transcription factor (ZAT), basic region leucine zipper (bZIP), and myeloblastosis (MYB), were controlled upstream by MAPKs [8,35,36]. Wang et al. [37] observed that the MYB transcription factor OsARM1(ARSENITE-RESPONSIVE MYB1) bound to the promoters of OsLsi1, OsLsi2, and OsLsi6 that encoded and regulated the expression of key arsenic transporters. Interestingly, when *U. compressa* was exposed to copper and at least one (except JNK) MAPK pathway inhibitor, intracellular copper accumulation decreased markedly compared to when exposed to copper alone. Evidence from other photoautotrophs suggested that partial or complete inhibition of the MAPK pathways overstimulates other signaling pathways and enhances copper exclusion/extrusion. This indicated crosstalk between, for example, calcium, NO, and ROS signaling, as previously described in *U. compressa* under copper excess [9,23]. Even though we did not find other similar investigations on photoautotrophs, support for this view came from a study on human bone-marrow stroma cells (Doan et al. 2012). Inhibition of the JNK or ERK pathways induced an increase in calcium content and deposition of these cells, overstimulating bone differentiation [38]. In contrast, when blocking the p38 pathway, there were lower concentrations and deposition of calcium, followed by decreased rates of bone differentiation [38]. Given that crosstalk between calcium and MAPK signaling pathways has been described for plants [39,40], it is possible that impairment of one, with the exception of JNK, or more pathways within the MAPKs may overstimulate other signaling mechanisms, enhancing exclusion/extrusion mechanisms and decreasing total intracellular accumulation under copper excess in *U. compressa*. Furthermore, our findings suggested that only blocking JNK results in increased intracellular copper accumulation, while blocking p38 (alone or in combination) gave rise to the lowest cellular concentrations. These results suggested that JNK was the least relevant and p38 the most pertinent MAPK pathway for copper exclusion/extrusion in *U. compressa*; these findings warrant further detailed investigation.

Although reports of involvement of the MAPK pathways in photosynthesis are scarce, a connection was described for *Glycine max* (soya bean). By blocking MEKK1 (JNK pathway), a dramatic downregulation of genes was observed in the primary metabolism, including those related to photosynthesis [41]. In contrast, in the microalga *Vischeria helvetica* (Ochrophyta), synthesis of carotenoids increased when p38 and ERK (but not JNK) MAPK pathways were inhibited [42]. In other studies, no effects were found; for example, in *A. thaliana* mutants deprived of MPK4 (upstream JNK), there were no significant differences with wildtypes in terms of F_v/F_m, rETR, and NPQ [43], and in the microalga *Asterochloris erici* (chlorophyte), when the p38 and JNK pathways were inhibited, no changes in F_v/F_m were observed, even under hyperosmotic conditions [44]. Despite these mixed results, the MAPK pathways do appear to play a role in photosynthesis, although the overall contribution of the ERK, JNK, and p38 MAPK pathways in regulating photosynthetic activity in photoautotrophs is not clear and their involvement under environmental stress, such as metal pollution, is comparatively unknown.

Negative controls exposing *U. compressa* to MAPK pathway inhibitors without added copper showed no patterns in in vivo chlorophyll *a* fluorescence parameters associated with PSII compared to algae grown in seawater (see Section 4). Therefore, the observed differences in photosynthetic activity under copper excess compared with controls and those subjected to copper and inhibitors, especially when more than one MAPK was blocked, were indeed a consequence of exposure to the metal and inhibition of MAPK pathways. Since no changes were recorded for F_v/F_m, our findings suggested that no transient damage to the antenna complex was induced under different treatments [12,29]. In contrast, during most of the measurement periods, α_{ETR}, ETR_{max}, and EK_{ETR} under copper in combination with two or more blocked MAPK pathways were always lower than controls and when only exposed to copper. This trend was especially evident after 6 d of exposure, and behaviour was similar for algae under copper and mixed MAPK inhibitors. Rapid light curves (RLCs), from which α_{ETR}, ETR_{max}, and EK_{ETR} were derived, indicated that the algae exposed to a combination of blocked MAPK pathways subject to copper excess displayed diminished photosynthetic efficiency, maximal electron transport rate, and saturation of irradiance of ETR, respectively [12,26,29]. The latter is coincidental, since it is known that copper excess inhibits photosynthesis by reducing the chlorophyll content as a consequence of copper-induced iron deficiency [25]. Moreover, it is well recognized that copper stress disrupts electron transport chains in PSII, with the excess energy being transferred to oxygen, for example, via the triplet state of chlorophyll, thereby inducing overproduction of singlet oxygen ($^1O_2$), a highly oxidizing ROS [9,43]. Excess energy in chloroplasts is mainly counteracted by the synthesis of carotenoids, among which xanthophylls are the most important [43,45,46]. Thermal dissipation can be measured as non-photochemical PSII fluorescence quenching (NPQ), which is triggered by the trans-thylakoidal proton gradient (ΔpH) and zeaxanthin (ZEA) synthesis through the xanthophyll cycle [29]. Our results from day 6 revealed that when two or more MAPK pathways were blocked under copper (except JNK and p38), NPQ_{max} values were higher than for algae subjected only to copper, reaching control values. Although there is limited related information, it was recently discovered in *A. thaliana* that upstream activation of p38, MPK3, and MPK6 induces downregulation of different genes associated with photosynthesis, among which are those associated with light reactions, plastid organization, responses to light stimulus, electron transport chains, metabolism of porphyrin-associated compounds, tetrapyrrole metabolic processes, light harvesting, and pigment biosynthesis [47]. Our results suggested that, in *U. compressa*, blocking two pathways or the whole MAPK pathway under copper excess caused a decrease in photosynthesis and excess energy transfer, which may have been counteracted by increased energy dissipation through heat. Furthermore, our data supported the idea that there was crosstalk between ERK, JNK, and p38 pathways under copper stress, and that photosynthetic activity could be maintained even when one pathway within the MAPK group was impaired. However, if more than one pathway was blocked then photosynthetic performance was compromised, demonstrating the importance of the whole MAPK pathway to ensure balanced photosynthetic activity in *U. compressa* under environmental stress.

Considering the whole dataset presented in this investigation through PCO analysis, the differences in the various measured parameters were mainly induced after 6 days of exposure, by which time the highest correlations were represented by α_{ETR}, ETR_{max}, and EK_{ETR}. Furthermore, most of the significant differences were induced under copper when blocking two or more MAPK pathways.

In this investigation, we identified that the whole MAPK pathway plays an important role in the physiological responses associated with metal accumulation and photosynthetic activity in *Ulva compressa* under copper stress. In particular, blocking one or more of the MAPK pathways, with the exception of JNK, decreased intracellular copper accumulation, which may have been a consequence of enhanced metal exclusion/extrusion mechanisms through crosstalk and activation of other cellular signaling pathways (e.g., calcium, NO, ROS). Moreover, by inhibiting most of the MAPK pathways (at least two out of ERK, JNK, and p38), photosynthetic activity was impaired, supporting the idea of potential crosstalk between the different studied MAPK pathways. To confirm most of these hypotheses, a whole transcriptome representation of the blocked MAPK pathways under copper stress in *U. compressa* is currently being explored in our ongoing investigations. To complement this article, in a parallel manuscript [1], we explore the roles of the ERK, JNK, and p38 MAPK pathways in the copper-detoxification mechanisms associated with antioxidant metabolism and synthesis of metal chelators in *U. compressa*.

4. Materials and Methods

4.1. Sample Collection and Culture Environment Conditions

All samples of *Ulva compressa* L. (Chlorophyta, Ulvaceae) were collected from Cachagua beach, Valparaíso, Chile (32°34′59″S; 71°26′16″W), a location with no history of metal pollution. Adult individuals were sampled from the intertidal zone, placed in plastic bags with seawater, and transported to the laboratory in a cooler. Algal material was successively washed with filtered seawater (2 μm), and stored inside plastic containers in a culture chamber with constant aeration at 16 °C, with a photoperiod cycle of 12:12 (day/night); daylight conditions corresponded to 160 $\mu mol \cdot m^2 \cdot s^{-1}$ photosynthetic active radiation (PAR). *Ulva compressa* was acclimated for 48 h prior to the start of the experiment.

4.2. Copper and MAPK Inhibitor Treatments

Following the acclimation period, 5 g fresh weight (FW) samples of macroalgae were placed in 300 mL plastic containers with 200 mL of 2 μm filtered and autoclaved seawater. A chronic copper concentration of 10 μM was chosen, based on previously obtained physiological viability measurements [48]. There was a total of nine treatments, with three replicates each: T1—control with only seawater; T2—chronic copper exposure at 10 μM $CuSO_4$ (Sigma-Aldrich, St. Louis, MO, USA) (Cu); T3—Cu + 5 μM MAPK ERK inhibitor PD98059 (Tocris Bioscience, St. Louis, MO, USA)) (Cu + ERKi); T4—Cu + 5 μM MAPK JNK inhibitor SP600125 (Tocris Bioscience) (Cu + JNKi); T5—Cu + MAPK p38 inhibitor SB203580 (Tocris Bioscience, St. Louis, MO, USA)) (Cu + p38i); T6—Cu + ERKi + JNKi; T7—Cu + ERKi + p38i; T8—Cu + p38i + JNKi; and T9—Cu + ERKi + JNKi + p38i. Negative controls were included, comprising seawater and ERKi, JNKi, or p38; there were no significant differences in photosynthetic activity between these treatments (see Supplementary Figures) nor in gene expression (see Supplementary Figure in Rodríguez-Rojas et al.) [1]. The inhibitors were initially developed to target human ERK, JNK, and p38 MAPKs, although they were subsequently used successfully in studies on invertebrates [49,50], vertebrates [51,52], plants [53–55], and algae [56,57]. Moreover, ERK, JNK, and p38 mammalian-like MAPKs were identified in macroalgae species, including the ulvophytes *Ulva rigida* and *Chaetomorpha aerea* [58,59]. ERK-encoding genes showed high degrees of homology between eukaryotes, among which included the microalga *Dunaliella viridis*, the plant *Arapidopsis thaliana*, the amoeba *Dictyostelium discoideum*, the protozoan *Plasmodium falciparum*, the mammal *Mus musculus*, and humans [60]; thus, it was not unexpected that human-developed inhibitors were also effective in a wide range of eukaryotes due to the highly conserved nature of MAPKs.

Exposure to all treatments took 6 d, with measurements of copper accumulation and photosynthetic activity taken at 6 h, 24 h, 48 h, and 6 d. Treatment solutions were changed every 48 h to avoid nutrient, copper, and inhibitor limitations.

4.3. Intracellular Copper Accumulation

At each measurement point, 1 g fresh weight (FW) of macroalgae was collected and washed twice for 15 min with 10 mM ethylene diamine tetra acetic acid (EDTA), as described in Roncarati et al. [19]; the latter allowed for the extraction of metal bound to cell walls and intracellular spaces and provided a measure of only intracellular copper accumulation. Macroalgae were then dried until a constant weight in an oven at 60 °C. Samples were digested with modifications to the protocol of US EPA 3052. Initially, samples were pre-digested overnight at room temperature in teflon vials with 6 mL of 65% nitric acid (Merck®) and 2 mL of 30% hydrogen peroxide (Merck®), then placed in a microwave oven (Milestone®, model Ethos Easy). Samples were digested according to the following cycle: 10 min at 200 °C (1800 W power) and then 20 min at 200 °C (1800 W power). Once the digestion process was completed, samples were cooled at room temperature for 30 min. Digests were then brought to 25 mL with ultrapure (Milli Q; Adrona®, model CB-1903, Riga, Letonia) water and stored in the dark at room temperature for analyses. Copper was measured using inductively coupled plasma optical emission spectrometry (ICP-OES; Perkin-Elmer®, model ICP Optima 2000 DV, Wellesley, MA, USA). To ensure precision and accuracy of the results, the methodology was also applied to certified reference material (CRM) of sea lettuce (*U. lactuca*; BCR-279). There was significant agreement between the measured and certified levels, with values ranging between 92% and 105% ($p > 0.05$).

4.4. Photosynthesis and Energy Dissipation through In Vivo Chlorophyll a Fluorescence

In vivo chlorophyll *a* fluorescence associated with photosystem II (PSII) was determined using a portable fluorometer Junior PAM fluorometer with WinControl-3.2 software (Walz GmbH, Effeltrich, Germany). Samples of *U. compressa* were collected at each measurement point (6 h, 24 h, 48 h, and 6 days) and placed in 10 mL incubation chambers to obtain rapid light curves (RLC) for each treatment (three RLCs per treatment). RLCs represented the saturation characteristics of PSII electron transport and overall photosynthetic performance [26,29].

4.5. Maximum Quantum Yield of PSII (Fv/Fm).

In order to conduct RLCs, F_o (basal fluorescence from fully oxidized reaction centers of PSII) and F_m (maximal fluorescence from fully reduced PSII reaction center), were measured after 15 min in darkness. The Maximal quantum yield (F_v/F_m), was derived from these parameters, with F_v being the difference between F_m and Fo [28].

Electron transport rate (ETR) is an indicator of productivity and photosynthetic capacity [26,29]. ETRs were determined after 20 s of the exposure period under twelve incremental irradiances of blue light (E1 = 25, E2 = 45, E3 = 66, E4 = 90, E5 = 125, E6 = 190, E7 = 285, E8 = 420, E9 = 625, E10 = 845, E11 = 1150, and E12 = 1500 μmol $m^2 \cdot s^{-1}$).

The ETR was calculated according to Schreiber et al. [28]:

$$ETR\ (\mu\text{mol e}^- \cdot \text{m}^{-2} \cdot \text{s}^{-1}) = \Delta F/F_m' \times E \times A \times FII \quad (1)$$

where $\Delta F/F_m'$ is the effective quantum yield, $\Delta F = (F_m' - F_t)$, F_t is the intrinsic fluorescence of alga incubated in light, and F_m' is the maximal fluorescence reached after a saturation pulse of the alga incubated in light, E is the incident PAR irradiance expressed in μmol photons $m^{-2}\ s^{-1}$, A is the thallus absorbance as a fraction of incident irradiance that is absorbed by the alga, and FII is the fraction of chlorophyll related to PSII (400–700 nm), being 0.5 in green macroalgae [26].

The photosynthetic parameters, i.e., the *maximal electron transport rate* (ETR_{max}, estimate of maximal photosynthetic capacity) and the *photosynthetic efficiency* (α_{ETR}, estimate of photosynthetic efficiency in the RLC), were obtained from the tangential model function reported by Eilers and Peeters [61]. The irradiance at which ETR (Ek_{ETR}) was saturated was calculated from the intercept between ETR_{max} and α_{ETR}.

Non-photochemical quenching (NPQ), as an indicator of energy dissipation and photoprotection, was also obtained from the RLC. The NPQ was calculated according to Schreiber et al. [28] as:

$$NPQ = (F_m/F_m')/F_m' \quad (2)$$

Maximal non-photochemical quenching (NPQ_{max}) and the initial slope of NPQ versus irradiance curves (α_{NPQ}) were obtained from the tangential model function of NPQ vs irradiance function.

4.6. Statistical Analyses

Effects on physiological variables and intracellular copper content were analyzed using ANOVA with the software SPSS v.21 (IBM, Armonk, NY, USA). This test was performed per time treatment (one-way) as a fixed factor for all variables (mean ± SE, n = 3). Each time period (6 h, 24 h, 48 h, and 6 days) included a treatment factor with nine levels, T1, T2, T3, T4, T5, T6, T7, T8, and T9, with a level of probability applied in the statistical analyses at $p < 0.05$. Homogeneity and homoscedasticity were tested using Cochran tests and by visual inspection of the residuals. In the case of significant ANOVAs, a posteriori Student Newman Keuls tests (SNK) were performed to evaluate the differences between groups.

A multi-dimensional scaling (MDS) or Principal Coordinates Ordination (PCO) analysis [62] was performed to detect patterns among parameters on the basis of Euclidean distance using PERMANOVA + PRIMER6 package. PCO analyses were conducted for F_v/F_m, α_{ETR}, ETR_{max}, Ek_{ETR}, NPQ_{max} and intracellular copper accumulation for each treatment and time period. This procedure was an equivalent ordination to a principal component analysis (PCA), calculating the percentage variation explained by each of the axes in the multidimensional scale. The overlay of the vectors onto the PCO was performed using Spearman's correlation [63].

Supplementary Materials: Supplementary materials can be found at http://www.mdpi.com/1422-0067/20/18/4547/s1.

Author Contributions: Conceptualization, P.S.M.C.-P., F.R.R., A.M., and C.A.S.; methodology and investigation, P.S.M.C.-P., F.R.R., L.M., F.M., P.T.M., G.L., and P.D.; formal analysis and writing original draft preparation, P.S.M.C.-P., F.R.R., and C.A.S.; writing review and editing, J.L.S.-L., M.T.B., A.M., P.S.M.C.-P., F.R.R., and C.A.S.; supervision, C.A.S.; funding acquisition, C.A.S.

Funding: Project FONDECYT NO. 11160369 granted to C.A. Sáez.

Conflicts of Interest: The authors declare no conflict of interest.

Abbreviations

MDPI	Multidisciplinary Digital Publishing Institute
MAPK	mitogen activated protein kinases
ERK	Extracellular Signal Regulated Kinases
JNK	c-Jun *N*-terminal Kinases
p38	Cytokinin Specific Binding Protein
ROS	reactive oxygen species
F_v/F_m	Maximal quantum yield of photosystem II
ETR_{max}	maximal electron transport rate of photosynthesis
EkETR	Saturation of the irradiance of electron transport rate
α_{ETR}	photosynthetic efficiency
NPQ_{max}	maximal non-photochemical quenching

References

1. Rodríguez-Rojas, F.; Celis-Pla, P.S.M.; Méndez, L.; Moenne, F.; Muñoz, P.; Lobos, M.; Díaz, P.; Brown, M.; Moenne, A.; Sáez, C.A. MAPK pathway under chronic copper excess in green macroalgae (Chlorophyta): Involvement in the regulation of detoxification mechanisms. *Int. J. Mol. Sci.* **2019**, in press.
2. Song, Y.; Cheong, Y.K.; Kim, N.H.; Chung, H.T.; Kang, D.G.; Pae, H.O. Mitogen-activated protein kinases and reactive oxygen species: How can ROS activate MAPK pathways? *J. Signal Transduct.* **2011**, *2011*, 792639.
3. Sinha, A.K.; Jaggi, M.; Raghuram, B.; Tuteja, N. Mitogen-activated protein kinase signaling in plants under abiotic stress. *Plant Signal. Behav.* **2011**, *6*, 196–203. [CrossRef] [PubMed]
4. Matsuzawa, A.; Ichijo, H. Redox control of cell fate by MAP kinase: Physiological roles of ASK1-MAP kinase pathway in stress signaling. *Biochim. Et Biophys. Acta* **2008**, *11*, 16. [CrossRef] [PubMed]
5. Meng, D.; Shi, X.; Jiang, B.H.; Fang, J. Insulin-like growth factor-I (IGF-I) induces epidermal growth factor receptor transactivation and cell proliferation through reactive oxygen species. *Free Radic. Biol. Med.* **2007**, *42*, 1651–1660. [CrossRef] [PubMed]
6. Guyton, K.Z.; Liu, Y.; Gorospe, M.; Xu, Q.; Holbrook, N.J. Activation of mitogen-activated protein kinase by H_2O_2. Role in cell survival following oxidant injury. *J. Biol. Chem.* **1996**, *271*, 4138–4142. [CrossRef]
7. Kamata, H.; Honda, S.; Maeda, S.; Chang, L.; Hirata, H.; Karin, M. Reactive oxygen species promote TNFalpha-induced death and sustained JNK activation by inhibiting MAP kinase phosphatases. *Cell* **2005**, *120*, 649–661. [CrossRef]
8. Jalmi, S.K.; Bhagat, P.K.; Verma, D.; Noryang, S.; Tayyeba, S.; Singh, K.; Sharma, D.; Sinha, A.K. Traversing the Links between Heavy Metal Stress and Plant Signaling. *Front. Plant Sci.* **2018**, *9*, 21. [CrossRef]
9. Moenne, A.; González, A.; Sáez, C.A. Mechanisms of metal tolerance in marine macroalgae, with emphasis on copper tolerance in Chlorophyta and Rhodophyta. *Aquat. Toxicol.* **2016**, *176*, 30–37. [CrossRef]
10. Kungolos, A.; Emmanouil, C.; Tsiridis, V.; Tsiropoulos, N. Evaluation of toxic and interactive toxic effects of three agrochemicals and copper using a battery of microbiotests. *Sci. Total Environ.* **2009**, *407*, 4610–4615. [CrossRef]
11. Ferreira, J.G.; Andersen, J.H.; Borja, A.; Bricker, S.B.; Camp, J.; da Silva, M.C.; Garces, E.; Heiskanen, A.S.; Humborg, C.; Ignatiades, L.; et al. Overview of eutrophication indicators to assess environmental status within the European Marine Strategy Framework Directive. *Estuar. Coast. Shelf Sci.* **2011**, *93*, 117–131. [CrossRef]
12. Celis-Plá, P.S.M.; Brown, M.T.; Santillán-Sarmiento, A.; Korbee, N.; Sáez, C.A.; Figueroa, F.L. Ecophysiological and metabolic responses to interactive exposure to nutrients and copper excess in the brown macroalga *Cystoseira tamariscifolia*. *Mar. Pollut. Bull.* **2018**, *128*, 214–222. [CrossRef] [PubMed]
13. Sáez, C.A.; González, A.; Contreras, R.A.; Moody, A.J.; Moenne, A.; Brown, M.T. A novel field transplantation technique reveals intra-specific metal-induced oxidative responses in strains of *Ectocarpus siliculosus* with different pollution histories. *Environ. Pollut.* **2015**, *199*, 130–138. [CrossRef] [PubMed]

14. Sáez, C.A.; Pérez-Matus, A.; Lobos, M.G.; Oliva, D.; Vásquez, J.A.; Bravo, M. Environmental assessment in a shallow subtidal rocky habitat: Approach coupling chemical and ecological tools. *Chem. Ecol.* **2012**, *28*, 1–15. [CrossRef]
15. Sheppard, C. Introduction to World Seas: An Environmental Evaluation. In *World Seas: An Environmental Evaluation (Second Edition)*; Academic Press: New York, NY, USA, 2018.
16. Ratkevicius, N.; Correa, J.A.; Moenne, A. Copper accumulation, synthesis of ascorbate and activation of ascorbate peroxidase in *Enteromorpha compressa* (L.) Grev. (Chlorophyta) from heavy metal-enriched environments in northern Chile. *Plant Cell Environ.* **2003**, *26*, 1599–1608. [CrossRef]
17. Sáez, C.A.; Lobos, M.G.; Macaya, E.; Oliva, D.; Quiroz, W.; Brown, M.T. Variation in patterns of metal accumulation in thallus parts of *Lessonia trabeculata* (Laminariales; Phaeophyceae): Implications for biomonitoring. *PLoS ONE* **2012**, *7*, e50170. [CrossRef] [PubMed]
18. Sáez, C.A.; Roncarati, F.; Moenne, A.; Moody, J.A.; Brown, M.T. Copper-induced intra-specific oxidative damage and antioxidant responses in strains of the brown alga *Ectocarpus siliculosus* with different pollution histories. *Aquat. Toxicol.* **2015**, *159*, 81–89. [CrossRef]
19. Roncarati, F.; Sáez, C.A.; Greco, M.; Gledhill, M.; Bitonti, M.B.; Brown, M.T. Response differences between *Ectocarpus siliculosus* populations to copper stress involve cellular exclusion and induction of the phytochelatin biosynthetic pathway. *Aquat. Toxicol.* **2015**, *159*, 167–175. [CrossRef]
20. Laporte, D.; Valdés, N.; González, A.; Sáez, C.A.; Zúñiga, A.; Navarrete, A.; Meneses, C.; Moenne, A. Copper-induced overexpression of genes encoding antioxidant system enzymes and metallothioneins involve the activation of CaMs, CDPKs and MEK1/2 in the marine alga *Ulva compressa*. *Aquat. Toxicol.* **2016**, *177*, 433–440. [CrossRef]
21. Gómez, M.; González, A.; Sáez, C.A.; Moenne, A. Copper-induced membrane depolarizations involve the induction of mosaic TRP channels, which activate VDCC leading to calcium increases in *Ulva compressa*. *Front. Plant Sci.* **2016**, *7*, 754. [CrossRef]
22. González, A.; Cabrera Mde, L.; Mellado, M.; Cabello, S.; Márquez, S.; Morales, B.; Moenne, A. Copper-induced intracellular calcium release requires extracellular calcium entry and activation of L-type voltage-dependent calcium channels in *Ulva compressa*. *Plant Signal. Behav.* **2012**, *7*, 728–732. [CrossRef] [PubMed]
23. González, A.; Cabrera Mde, L.; Henriquez, M.J.; Contreras, R.A.; Morales, B.; Moenne, A. Cross talk among calcium, hydrogen peroxide, and nitric oxide and activation of gene expression involving calmodulins and calcium-dependent protein kinases in *Ulva compressa* exposed to copper excess. *Plant Physiol.* **2012**, *158*, 1451–1462. [CrossRef] [PubMed]
24. Gómez, M.; González, A.; Sáez, C.A.; Morales, B.; Moenne, A. Copper-induced activation of TRP channels promotes extracellular calcium entry and activation of CaMs and CDPKs leading to copper entry and membrane depolarization in *Ulva compressa*. *Front. Plant Sci.* **2015**, *6*, 182. [PubMed]
25. Yruela, I. Copper in plants: Acquisition, transport and interactions. *Funct. Plant Biol.* **2009**, *36*, 409–430. [CrossRef]
26. Figueroa, F.L.; Conde-Álvarez, R.; Gómez, I. Relations between electron transport rates determined by pulse amplitude modulated chlorophyll fluorescence and oxygen evolution in macroalgae under different light conditions. *Photosynth. Res.* **2003**, *75*, 259–275. [CrossRef] [PubMed]
27. Celis-Pla, P.S.M.; Korbee, N.; Gomez-Garreta, A.; Figueroa, F.L. Seasonal photoacclimation patterns in the intertidal macroalga *Cystoseira tamariscifolia* (Ochrophyta). *Sci. Mar.* **2014**, *78*, 377–388. [CrossRef]
28. Schreiber, U.; Bilger, W.; Neubauer, C. Chlorophyll fluorescence as a nonintrusive indicator for rapid assessment of in vivo photosynthesis. In *Ecophysiology of Photosynthesis*; Schulze, E.-D., Caldwell, M.M., Eds.; Springer: Berlin/Heidelberg, Germany, 1995; pp. 49–70.
29. Celis-Plá, P.S.M.; Bouzon, Z.L.; Hall-Spencer, J.M.; Schmidt, E.C.; Korbee, N.; Figueroa, F.L. Seasonal biochemical and photophysiological responses in the intertidal macroalga *Cystoseira tamariscifolia* (Ochrophyta). *Mar. Environ. Res.* **2016**, *115*, 89–97. [CrossRef]
30. Kumar, K.S.; Han, Y.-S.; Choo, K.-S.; Kong, J.-A.; Han, T. Chlorophyll fluorescence based copper toxicity assessment of two algal species. *Toxicol. Environ. Health Sci.* **2009**, *1*, 17–23. [CrossRef]
31. Gao, G.; Liu, Y.; Li, X.; Feng, Z.; Xu, Z.; Wu, H.; Xu, J. Expected CO_2-induced ocean acidification modulates copper toxicity in the green tide alga *Ulva prolifera*. *Environ. Exp. Bot.* **2017**, *135*, 63–72. [CrossRef]
32. Kim, D.-Y.; Bovet, L.; Maeshima, M.; Martinoia, E.; Lee, Y. The ABC transporter AtPDR8 is a cadmium extrusion pump conferring heavy metal resistance. *Plant J.* **2007**, *50*, 207–218. [CrossRef]

33. Li, Y.; Iqbal, M.; Zhang, Q.; Spelt, C.; Bliek, M.; Hakvoort, H.W.J.; Quattrocchio, F.M.; Koes, R.; Schat, H. Two *Silene vulgaris* copper transporters residing in different cellular compartments confer copper hypertolerance by distinct mechanisms when expressed in *Arabidopsis thaliana*. *New Phytol.* **2017**, *215*, 1102–1114. [CrossRef] [PubMed]
34. Simon, D.F.; Domingos, R.F.; Hauser, C.; Hutchins, C.M.; Zerges, W.; Wilkinson, K.J. Transcriptome sequencing (RNA-seq) analysis of the effects of metal nanoparticle exposure on the transcriptome of *Chlamydomonas reinhardtii*. *Appl. Environ. Microbiol.* **2013**, *79*, 4774–4785. [CrossRef] [PubMed]
35. Roelofs, D.; Aarts, M.G.M.; Schat, H.; Van Straalen, N.M. Functional ecological genomics to demonstrate general and specific responses to abiotic stress. *Funct. Ecol.* **2008**, *22*, 8–18. [CrossRef]
36. Li, S.; Wang, W.; Gao, J.; Yin, K.; Wang, R.; Wang, C.; Petersen, M.; Mundy, J.; Qiu, J.-L. MYB75 Phosphorylation by MPK4 Is Required for Light-Induced Anthocyanin Accumulation in *Arabidopsis*. *Plant Cell* **2016**, *28*, 2866–2883. [CrossRef] [PubMed]
37. Wang, X.; Wang, C.; Sheng, H.; Wang, Y.; Zeng, J.; Kang, H.; Fan, X.; Sha, L.; Zhang, H.; Zhou, Y. Transcriptome-wide identification and expression analyses of ABC transporters in dwarf polish wheat under metal stresses. *Biol. Plant.* **2017**, *61*, 293–304. [CrossRef]
38. Doan, T.K.P.; Park, K.S.; Kim, H.K.; Park, D.S.; Kim, J.H.; Yoon, T.R. Inhibition of JNK and ERK pathways by SP600125- and U0126-enhanced osteogenic differentiation of bone marrow stromal cells. *Tissue Eng. Regen. Med.* **2012**, *9*, 283–294. [CrossRef]
39. Ludwig, A.A.; Saitoh, H.; Felix, G.; Freymark, G.; Miersch, O.; Wasternack, C.; Boller, T.; Jones, J.D.G.; Romeis, T. Ethylene-mediated cross-talk between calcium-dependent protein kinase and MAPK signaling controls stress responses in plants. *Proc. Natl. Acad. Sci. United States Am.* **2005**, *102*, 10736–10741. [CrossRef] [PubMed]
40. Mehlmer, N.; Wurzinger, B.; Stael, S.; Hofmann-Rodrigues, D.; Csaszar, E.; Pfister, B.; Bayer, R.; Teige, M. The Ca^{2+}-dependent protein kinase CPK3 is required for MAPK-independent salt-stress acclimation in *Arabidopsis*. *Plant J.* **2010**, *63*, 484–498. [CrossRef]
41. Xu, H.-Y.; Zhang, C.; Li, Z.-C.; Wang, Z.-R.; Jiang, X.-X.; Shi, Y.-F.; Tian, S.-N.; Braun, E.; Mei, Y.; Qiu, W.-L.; et al. The MAPK Kinase Kinase GmMEKK1 regulates cell death and defense responses. *Plant Physiol.* **2018**, *178*, 907–922. [CrossRef]
42. Aburai, N.; Abe, K. Metabolic switching: Synergistic induction of carotenogenesis in the aerial microalga, *Vischeria helvetica*, under environmental stress conditions by inhibitors of fatty acid biosynthesis. *Biotechnol. Lett.* **2015**, *37*, 1073–1080. [CrossRef]
43. Gawroński, P.; Witoń, D.; Vashutina, K.; Bederska, M.; Betliński, B.; Rusaczonek, A.; Karpiński, S. Mitogen-activated protein kinase 4 is a salicylic acid-independent regulator of growth but not of photosynthesis in *Arabidopsis*. *Mol. Plant* **2014**, *7*, 1151–1166. [CrossRef] [PubMed]
44. Gasulla, F.; Barreno, E.; Parages, M.L.; Cámara, J.; Jiménez, C.; Dörmann, P.; Bartels, D. The role of phospholipase D and MAPK signaling cascades in the adaption of lichen microalgae to desiccation: Changes in membrane lipids and phosphoproteome. *Plant Cell Physiol.* **2016**, *57*, 1908–1920. [CrossRef] [PubMed]
45. Silvan, J.M.; Reguero, M.; de Pascual-Teresa, S. A protective effect of anthocyanins and xanthophylls on UVB-induced damage in retinal pigment epithelial cells. *Food Funct.* **2016**, *7*, 1067–1076. [CrossRef] [PubMed]
46. Sytar, O.; Kumar, A.; Latowski, D.; Kuczynska, P.; Strzalka, K.; Prasad, M.N.V. Heavy metal-induced oxidative damage, defense reactions, and detoxification mechanisms in plants. *Acta Physiol. Plant.* **2013**, *35*, 985–999. [CrossRef]
47. Su, J.; Yang, L.; Zhu, Q.; Wu, H.; He, Y.; Liu, Y.; Xu, J.; Jiang, D.; Zhang, S. Active photosynthetic inhibition mediated by MPK3/MPK6 is critical to effector-triggered immunity. *PLOS Biol.* **2018**, *16*, e2004122. [CrossRef] [PubMed]
48. González, A.; Vera, J.; Castro, J.; Dennett, G.; Mellado, M.; Morales, B.; Correa, J.A.; Moenne, A. Co-occurring increases of calcium and organellar reactive oxygen species determine differential activation of antioxidant and defense enzymes in *Ulva compressa* (Chlorophyta) exposed to copper excess. *Plant Cell Environ.* **2010**, *33*, 1627–1640. [CrossRef] [PubMed]
49. Franchi, N.; Schiavon, F.; Betti, M.; Canesi, L.; Ballarin, L. Insight on signal transduction pathways involved in phagocytosis in the colonial ascidian *Botryllus schlosseri*. *J. Invertebr. Pathol.* **2013**, *112*, 260–266. [CrossRef] [PubMed]

50. Taylor, E.; Heyland, A. Thyroid hormones accelerate initiation of skeletogenesis via mapk (ERK1/2) in larval sea urchins (*Strongylocentrotus purpuratus*). *Front. Endocrinol.* **2018**, *9*, 16. [CrossRef]
51. Zhao, H.Y.; Liu, W.; Wang, Y.; Dai, N.N.; Gu, J.H.; Yuan, Y.; Liu, X.Z.; Bian, J.C.; Liu, Z.P. Cadmium induces apoptosis in primary rat osteoblasts through caspase and mitogen-activated protein kinase pathways. *J. Vet. Sci.* **2015**, *16*, 297–306. [CrossRef]
52. Zhu, D.L.; Wang, K.L.; Chen, P.L.; Li, Y. The c-Jun N-terminal kinases (JNK)/mitogen-activated Protein Kinase (MAPK) is responsible for the protection of tanshinol (Danshensu) upon H_2O_2-induced L-6 rat myoblast cell injury. *Acta Sci. Vet.* **2014**, *42*, 8.
53. Khokon, M.A.R.; Salam, M.A.; Jammes, F.; Ye, W.; Hossain, M.A.; Uraji, M.; Nakamura, Y.; Mori, I.C.; Kwak, J.M.; Murata, Y. Two guard cell mitogen-activated protein kinases, MPK9 and MPK12, function in methyl jasmonate-induced stomatal closure in *Arabidopsis thaliana*. *Plant Biol.* **2015**, *17*, 946–952. [CrossRef]
54. Livanos, P.; Galatis, B.; Gaitanaki, C.; Apostolakos, P. Phosphorylation of a p38-like MAPK is involved in sensing cellular redox state and drives atypical tubulin polymer assembly in angiosperms. *Plant Cell Environ.* **2014**, *37*, 1130–1143. [CrossRef]
55. de Oliveira, E.A.G.; Romeiro, N.C.; Ribeiro, E.D.; Santa-Catarina, C.; Oliveira, A.E.A.; Silveira, V.; de Souza, G.A.; Venancio, T.M.; Cruz, M.A.L. Structural and functional characterization of the protein kinase Mps1 in *Arabidopsis thaliana*. *PLoS ONE* **2012**, *7*, 9. [CrossRef]
56. Charneco, G.O.; Parages, M.L.; Camarena-Gomez, T.; Jimenez, C. Phosphorylation of MAP Kinases crucially controls the response to environmental stress in *Dunaliella viridis*. *Environ. Exp. Bot.* **2018**, *156*, 203–213. [CrossRef]
57. Jimenez, C.; Berl, T.; Rivard, C.J.; Edelstein, C.L.; Capasso, J.M. Phosphorylation of MAP kinase-like proteins mediate the response of the halotolerant alga *Dunaliella viridis* to hypertonic shock. *Biochim. Et Biophys. Acta-Mol. Cell Res.* **2004**, *1644*, 61–69. [CrossRef]
58. Parages, M.L.; Capasso, J.M.; Niell, F.X.; Jiménez, C. Responses of cyclic phosphorylation of MAPK-like proteins in intertidal macroalgae after environmental stress. *J. Plant Physiol.* **2014**, *171*, 276–284. [CrossRef]
59. Parages, M.L.; Figueroa, F.L.; Conde-Alvarez, R.M.; Jimenez, C. Phosphorylation of MAPK-like proteins in three intertidal macroalgae under stress conditions. *Aquat. Biol.* **2014**, *22*, 213–226. [CrossRef]
60. Jimenez, C.; Cossio, B.R.; Rivard, C.J.; Berl, T.; Capasso, J.M. Cell division in the unicellular microalga *Dunaliella viridis* depends on phosphorylation of extracellular signal-regulated kinases (ERKs). *J. Exp. Bot.* **2007**, *58*, 1001–1011. [CrossRef]
61. Eilers, P.H.C.; Peeters, J.C.H. A model for the relationship between light intensity and the rate of photosynthesis in phytoplankton. *Ecol. Model.* **1988**, *42*, 199–215. [CrossRef]
62. GOWER, J.C. Some distance properties of latent root and vector methods used in multivariate analysis. *Biometrika* **1966**, *53*, 325–338. [CrossRef]
63. Anderson, M.; Gorley, R.N.; Clarke, K. *PERMANOVA+ for Primer: Guide to Software and Statistical Methods*; PRIMER-e: Plymouth, UK, 2008; p. 218.

© 2019 by the authors. Licensee MDPI, Basel, Switzerland. This article is an open access article distributed under the terms and conditions of the Creative Commons Attribution (CC BY) license (http://creativecommons.org/licenses/by/4.0/).

International Journal of
Molecular Sciences

Article

MAPK Pathway under Chronic Copper Excess in Green Macroalgae (Chlorophyta): Involvement in the Regulation of Detoxification Mechanisms

Fernanda Rodríguez-Rojas [1,2,†], Paula S. M. Celis-Plá [1,2,†], Lorena Méndez [1], Fabiola Moenne [1,2], Pamela T. Muñoz [1,3,4], M. Gabriela Lobos [5], Patricia Díaz [5], José Luis Sánchez-Lizaso [6], Murray T. Brown [7], Alejandra Moenne [8] and Claudio A. Sáez [1,2,*]

1 Laboratory of Aquatic Environmental Research, Centro de Estudios Avanzados, Universidad de Playa Ancha, Viña del Mar 2520000, Chile
2 HUB-AMBIENTAL UPLA, Universidad de Playa Ancha, Valparaíso 2340000, Chile
3 Doctorado Interdisciplinario en Ciencias Ambientales, Facultad de Ciencias Naturales y Exactas, Universidad de Playa Ancha, Valparaíso 2340000, Chile
4 Doctorado en Ciencias del Mar y Biología Aplicada, Departamento de Ciencias del Mar y Biología Aplicada, Universidad de Alicante, 03080 Alicante, Spain
5 Laboratory of Environmental and Analytical Chemistry, Instituto de Química y Bioquímica, Facultad de Ciencias, Universidad de Valparaíso, Valparaíso 234000, Chile
6 Departamento de Ciencias del Mar y Biología Aplicada, Universidad de Alicante, 03080 Alicante, Spain
7 School of Biological and Marine Sciences, University of Plymouth, PL4 8AA Plymouth, UK
8 Laboratory of Marine Biotechnology, Facultad de Química y Biología, Universidad de Santiago de Chile, Santiago 9170020, Chile
* Correspondence: claudio.saez@upla.cl
† These authors contributed equally to this work.

Received: 18 July 2019; Accepted: 1 September 2019; Published: 13 September 2019

Abstract: Following the physiological complementary/parallel Celis-Plá et al., by inhibiting extracellular signal regulated kinases (ERK), c-Jun N-terminal kinases (JNK), and cytokinin specific binding protein (p38), we assessed the role of the mitogen-activated protein kinases (MAPK) pathway in detoxification responses mediated by chronic copper (10 μM) in *U. compressa*. Parameters were taken at 6, 24, and 48 h, and 6 days (d). H_2O_2 and lipid peroxidation under copper and inhibition of ERK, JNK, or p38 alone increased but recovered by the sixth day. By blocking two or more MAPKs under copper, H_2O_2 and lipid peroxidation decayed even below controls. Inhibition of more than one MAPK (at 6 d) caused a decrease in total glutathione (reduced glutathione (GSH) + oxidised glutathione (GSSG)) and ascorbate (reduced ascorbate (ASC) + dehydroascorbate (DHA)), although in the latter it did not occur when the whole MAPK was blocked. Catalase (*CAT*), superoxide dismutase (*SOD*), thioredoxin (*TRX*) ascorbate peroxidase (*APX*), dehydroascorbate reductase (*DHAR*), and glutathione synthase (*GS*), were downregulated when blocking more than one MAPK pathway. When one MAPK pathway was blocked under copper, a recovery and even enhancement of detoxification mechanisms was observed, likely due to crosstalk within the MAPKs and/or other signalling processes. In contrast, when more than one MAPK pathway were blocked under copper, impairment of detoxification defences occurred, demonstrating that MAPKs were key signalling mechanisms for detoxification in macroalgae.

Keywords: mitogen-activated protein kinases; *Ulva compressa*; antioxidant; oxidative stress; metal chelator

---

1. Introduction

It has been recognised that besides metal exclusion/extrusion mechanisms, the main strategies by which macroalgae counteract metal excess in polluted environments are detoxification of bioavailable metals through the syntheses of chelating compounds and the inactivation of metal-mediated excess of reactive oxygen species (ROS) by the production and activities of antioxidants and antioxidant enzymes [1,2]. In green macroalgae (Ulvophyceae), metal chelation is importantly controlled by cysteine-rich compounds, among which the most important are the gene encoded metallothioneins and the enzymatically synthesised glutathione (GSH) and its poly-oligomers phytochelatins (PCs) [2,3]. GSH, in a recycling process with ascorbate (ASC), termed the Foyer–Halliwell–Asada cycle, is recognised as central to antioxidant metabolism in photoautotrophs, algae, and plants, and is one of the main mechanisms by which these organisms tolerate metal-mediated oxidative stress [2,4,5]. The process is controlled by complementary de novo synthesis and recycling of GSH and ASC. Briefly, GSH is synthesised by subsequent enzymatically mediated reactions controlled by γ-glutamate-cysteine ligase (γ-GCL) and glutathione synthase (GS), and in the case of ASC by L-galatose dehydrogenase (L-GDH) and l-galatonolactone dehydrogenase (L-GLDH) [4]. Moreover, the enzyme monodehydroascorbate reductase (MDHAR) uses GSH as a substrate to reduce dehydroascorbate (DHA) to ASC, also producing glutathione disulphide (GSSG). Then, the enzyme glutathione reductase (GR) reduces back GSSG to GSH using nicotinamide adenine dinucleotide phosphate (NADPH) [4]. Other enzymatic complexes also participate in the antioxidant metabolism, directly or indirectly linked with the GSH-ASC cycle. For instance, superoxide radicals ($O_2\cdot^-$) are dismutated to hydrogen peroxide (H_2O_2), a lower oxidative-power ROS, by the enzyme superoxide dismutase (SOD) [6]. H_2O_2 is then reduced to water by either ascorbate peroxidase (APX) using ASC as electron donor, or by the enzyme catalase (CAT) [7]. Another important group of enzymes within the antioxidant metabolism of photoautotrophs is thioredoxins (TRXs), which principally reduce oxidised cysteine residues and split disulfide bonds [8].

Information on the role of antioxidant metabolism on copper-stressed green macroalgae is mostly derived from studies using the model *Ulva compressa*. For example, *U. compressa* from copper polluted sites, compared with non-impacted areas, of northern Chile displayed high levels of lipid peroxidation, enhanced activity of APX, increased ascorbate (as DHA), and consumption of glutathione pools [9]. Likewise, when exposed to chronic copper levels of 10 µM for up to 10 days (d) in the laboratory, *U. compressa* displayed increased levels of lipid peroxidation, accompanied by greater content of ASC, GSH, and PCs, higher expression of *γ-GCL*, *GS*, *L-GDH*, *L-GLDH*, *DHAR*, and *GR*, and increased transcripts and activities of the enzymes APX, GR, and SOD [10–13]. Similar induction of copper- and ROS-detoxification mechanisms have been reported in other *Ulva* species, including *U. lactuca* [14] and *U. fasciata* [15]. Together, these results indicate that the activation of antioxidant defences under chronic chronic copper excess is a transcriptionally regulated process in green macroalgae [2].

Although the main mechanisms by which green macroalgae withstand chronic chronic copper excess are now fairly well described, the cellular signalling pathways by which these organisms activate such metabolic machinery is poorly understood, and this is also true for exposure to environmental stress more generally. To date, the signalling pathways that have been observed to play a role in copper tolerance in green macroalgae relate to ROS, nitric oxide (NO), and calcium (Ca^{+2}) signalling. The potential participation of the mitogen-activated protein kinases (MAPKs) has been proposed, but not fully confirmed [2,12]. MAPKs are serine/threonine kinases following a cascade of activations from MAPK kinase kinase (MAPKKK), MAPK kinase (MAPKK) to MAPK, thus finally inducing gene expression [1,16,17]. There are three main sub-pathways within the MAPKs, these ending in the MAPK ERK (extracellular signal regulated kinases), JNK (c-Jun *N*-terminal kinases), and p38 (cytokinin specific binding protein) [1,16,17]. While a full description of the mechanistic contribution of these MAPK pathways to tolerance machinery under environmental stress in photoautotrophs has not been elucidated, their presence and induction have been observed in both micro and macroalgae [18–21].

Here, we report on a holistic investigation to disclose for the first time in photoautotrophs the role of different pathways within the MAPKs to withstand environmental stress, in this case caused by chronic copper excess in green macroalgae. The green seaweed *U. compressa* under 10 μM of copper was exposed to specific inhibitors of MAPK pathways, ERK, JNK, and/or p38, with measurements taken at different time points throughout the experiment (6 h, 24 h, 48 h, and 6 d). In a complementary investigation, we demonstrated that under chronic copper excess when two or more MAPK pathways were inhibited, there was a general trend of decreasing intracellular copper concentrations when compared to exposure only to copper [1]. Moreover, the lowest levels of photosynthetic performance, as determined from measurements of photosynthetic efficiency (α_{ETR}), productivity (ETR_{max}), and saturation of the irradiance (Ek_{ETR}), were under copper in the presence of two or more MAPK inhibitors, [1]. Coincidentally, on the sixth d non-photochemical quenching (NPQ_{max}) increased in treatments with copper and a mixture of MAPK inhibitors, confirming that the perturbed electron transport in PSII was being countered by the dissipation of heat [1]. These data suggest that by blocking most of the MAPK pathways, an alternative cellular signalling pathway(s) may be over-stimulated, thus enhancing mechanisms of copper exclusion/extrusion. Moreover, results indicate that the MAPK pathways may have a central role in photosynthetic performance, even under the presence of environmental perturbation (in this case, chronic copper excess) [1]. Here, we extend this study to investigate the participation of MAPK pathways in the reactive oxygen metabolism (ROM) of *U. compressa* under chronic copper excess. More specifically, we evaluated parameters of oxidative stress and damage, concentrations of glutathione (GSH and GSSG), and ascorbate (ASC and DHA), as well as the expression of enzymes involved in antioxidant metabolism (*APX*, *CAT*, *SOD*, *GS*, *DHAR*, and *TRX*).

2. Results

2.1. Hydrogen Peroxide Levels

The levels of H_2O_2 were measured as an indicator of ROS imbalance (Figure 1). At 6 h, all treatments compared with control samples induced a significantly higher accumulation of H_2O_2, especially treatments with combinations of inhibitors that included p38i (Figure 1A). Treatments that had significantly higher levels than those of T2 (copper only) were: Cu + ERKi + JNKi (T6) and Cu + ERKi + JNKi + p38i (T9). Similar patterns were observed at 24 h, although the Cu + ERKi + p38i (T7) and Cu + ERKi + JNKi + p38i (T9) treatments were the ones with significantly higher levels of H_2O_2 (Figure 1B). Inverted patterns could be detected at 48 h and 6 d, where concentrations of H_2O_2 in all treatments containing copper and inhibitors were either the same as or significantly lower than control and copper-only treatments (Figure 1C,D).

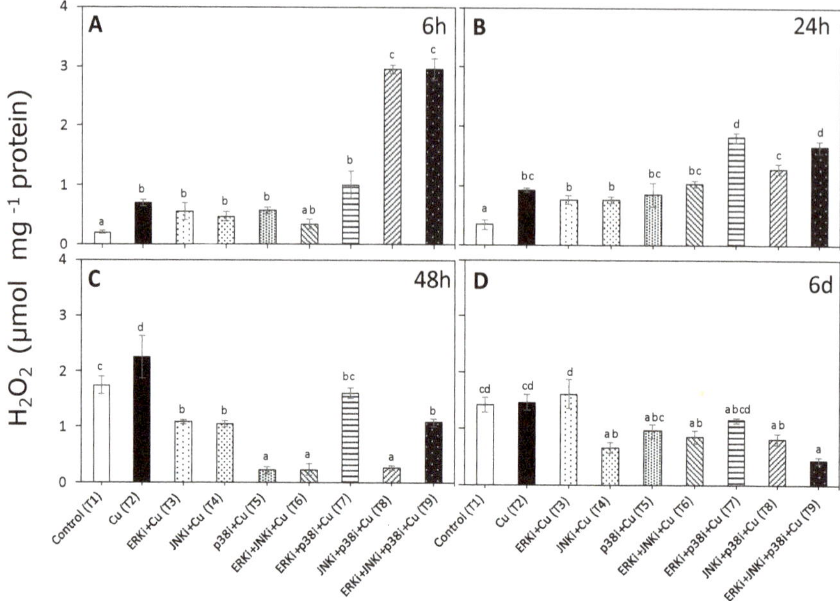

Figure 1. Hydroxide peroxide (H_2O_2) levels in *U. compressa* under copper and/or exposure to mitogen-activated protein kinases (MAPK) inhibitors. Treatments consisted in: T1) control only with seawater; T2) solely copper exposure as 10 μM of $CuSO_4$ (Cu); T3) copper + 5 μM MAPK extracellular signal regulated kinases (ERK) inhibitor PD98059 (Cu + ERKi); T4) copper + 5 μM MAPK c-Jun *N*-terminal kinases (JNK) inhibitor SP600125 (Cu + JNKi); T5) copper + MAPK cytokinin specific binding protein (p38) inhibitor SB203580 (Cu + p38i); T6) Cu + ERKi + JNKi; T7) Cu + ERKi + p38i; T8) Cu + p38i + JNKi; and T9) Cu + ERKi + JNKi + p38i. Samples were analysed after 6 h (**A**), 24 h (**B**), 48 h (**C**), and 6 d (**D**) treatments. Different letters represent significant difference at 95% confidence interval ($p < 0.05$). Plots are represented as mean ± SE ($n = 3$).

2.2. Lipid Peroxidation

Exposure to copper did not induce significantly higher levels of TBARS than controls at 6 and 24 h, nor did the majority of the treatments containing inhibitors, however copper treatments with p38 (T5) and all three inhibitors (T9) had significantly higher levels of TBARS (Figure 2A,B). At 48 h, in treatments with copper without and with inhibitors (except with ERKi alone), there were significant decreases in TBARS compared to the controls, although there were no significant differences between these treatments (Figure 2C). At day 6, no differences could be detected between the controls and treatments with only copper and those containing copper and ERKi (T3), JNKi (T4), and ERKi + p38i (T7). In all other treatments with copper plus inhibitors, there were significantly lower levels of TBARS than controls and under copper only, but with no differences between them (Figure 2D).

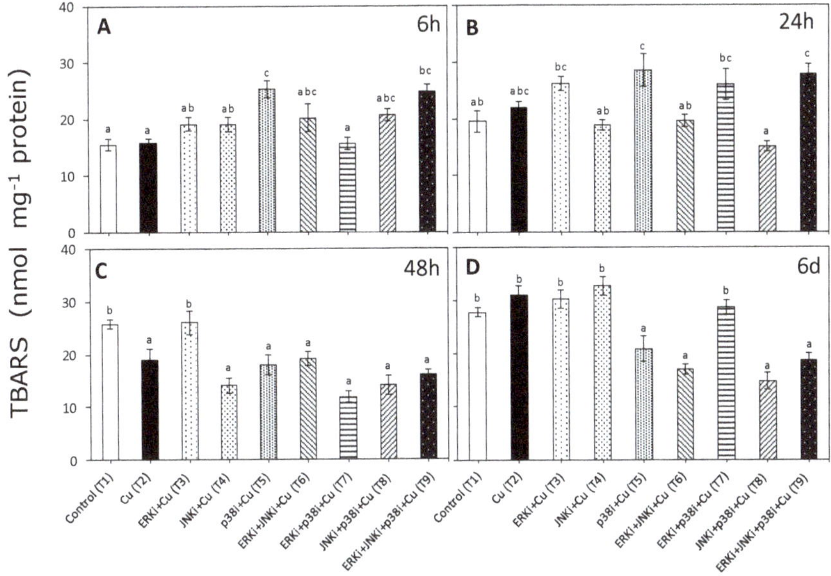

Figure 2. Thiobarbituric acid reactive substances (TBARS) accumulation in *U. compressa* under copper and/or exposure to MAPK inhibitors. Treatments consisted in: T1) control only with seawater; T2) solely copper exposure as 10 μM of $CuSO_4$ (Cu); T3) copper + 5 μM MAPK ERK inhibitor PD98059 (Cu + ERKi); T4) copper + 5 μM MAPK JNK inhibitor SP600125 (Cu + JNKi); T5) copper + MAPK p38 inhibitor SB203580 (Cu + p38i); T6) Cu + ERKi + JNKi; T7) Cu + ERKi + p38i; T8) Cu + p38i + JNKi; and T9) Cu + ERKi + JNKi + p38i. Samples were analyzed after 6 h (**A**), 24 h (**B**), 48 h (**C**) and 6 d (**D**) treatments. Different letters represent significant difference at 95% confidence interval ($p < 0.05$). Plots are represented as mean ± SE ($n = 3$).

2.3. Levels of Reduced and Oxidised Glutathione

At 6 h, there was a significant decrease of total glutathione and GSH in all treatments compared to controls (Figure 3A). Moreover, in most treatments with copper (T2) and copper plus inhibitors (T3-T9) there were the same or significantly lower levels of GSSG than controls; the exceptions were copper with single MAPK inhibitor treatments (T3, T4 and T5) in which significantly higher GSSG concentrations were measured (Figure 3A). It is important to highlight that controls presented over 4-fold levels of GSH in comparison to GSSG, whereas the majority of the other treatments had higher GSSG than GSH (Figure 3A). At 24 h, similar concentrations within each treatment were recorded for GSH and GSSG; in addition, all treatments containing only copper (T2) and copper with inhibitors (T3-T9) displayed significantly lower levels of GSH and GSSG than controls (Figure 3B). Only in the case of treatment Cu + ERKi + p38i (T7), were concentrations of GSSG higher than in copper alone (T2) (Figure 3B). As at 6 h, at 48 h the levels of GSH were 4 times higher than GSSG in controls (Figure 3C). In treatments with only copper, copper with either ERKi or JNKi, and copper with all the blocked MAPKs, GSSG concentrations were higher than those of GSH (Figure 3C). In most treatments containing copper, concentrations of GSH were significantly lower than controls; the exception was treatment under copper and either ERKi (T3) or JNKi (T4) (Figure 3C). There were no clear patterns in GSSG across treatments, although in treatments with copper and either p38 or all three inhibitors, concentrations of GSSG were significantly higher than controls and copper only treatments (Figure 3C). By the 6th d of the experiment, GSH concentrations in the copper only treatment were significantly higher than controls, and levels in the controls were significantly higher than when inhibitors were present; there was no clear pattern amongst these latter treatments (Figure 3D). Similarly for significantly higher concentrations than in controls were measured in the copper only and copper plus ERKi, both having the largest pool of total glutathione (Figure 3D).

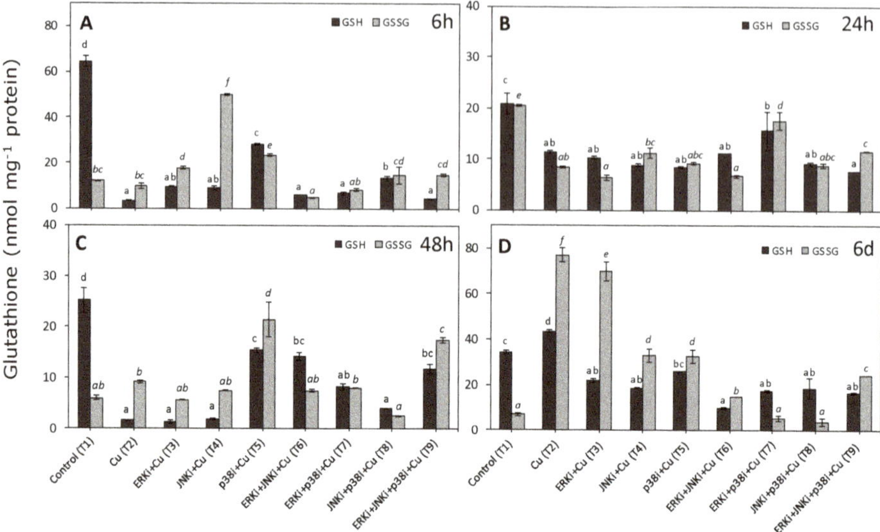

Figure 3. Total reduced (GSH) and oxidised glutathione (GSSG) content in *U. compressa* under copper and/or exposure to MAPK inhibitors. Treatments consisted in: T1) control only with seawater; T2) Solely copper exposure as 10 µM $CuSO_4$ (Cu); T3) copper + 5 µM MAPK ERK inhibitor PD98059 (Cu + ERKi); T4) copper + 5 µM MAPK JNK inhibitor SP600125 (Cu + JNKi); T5) copper + MAPK p38 inhibitor SB203580 (Cu + p38i); T6) Cu + ERKi + JNKi; T7) Cu + ERKi + p38i; T8) Cu + p38i + JNKi; and T9) Cu + ERKi + JNKi + p38i. Samples were analysed after 6 h (**A**), 24 h (**B**), 48 h (**C**), and 6 d (**D**) treatments. Different letters represent significant difference at 95% confidence interval ($p < 0.05$). Plots are represented as mean ± SE ($n = 3$).

2.4. Concentrations of Reduced and Oxidised Ascorbate

The lowest concentrations of all ASC, DHA, and total ascorbate pool were measured at 6 h (Figure 4A). ASC levels were higher than in controls in treatments with only copper, Cu + ERKi (T3), and Cu + ERKi + JNKi + p38i (T9) (Figure 4A). The remainder of the treatments always had significantly lower levels of ASC compared to the controls (Figure 4A). There was a similar trend for DHA, although levels were only higher than in controls for treatment-only copper (T2) and copper with all combined inhibitors (T9) (Figure 4A). With the exception of T9, higher ASC concentrations were significantly higher than those of DHA (Figure 4A). At 24 h, the total ascorbate pool, mainly due to DHA, increased considerably compared to 6 h in all treatments containing only copper and copper with inhibitors (Figure 4B). Excluding the treatment with copper and all inhibitors combined (T9), ASC levels were always significantly lower in treatments containing copper than the controls (Figure 4B). In contrast, treatments containing copper had higher DHA concentrations than the controls, the exceptions being those with ERKi (T3, T6, T7, and T9) (Figure 4B). At 48 h, the general trend was for higher ASC than DHA levels, except for treatments with only copper and with copper and ERKi (Figure 4C). Treatments combining copper with two inhibited MAPK pathways had significant lower levels of ASC than the controls, whereas in treatments with only copper, single inhibitor exposure, ERKi (T3) or p38i (T5), and all three inhibitors (T9) there were no significant differences with controls, however in only Cu + JNKi (T4) was ASC significantly higher than the controls (Figure 4C). Concentrations of DHA were significantly higher under exposure to copper alone (T2), copper and ERKi (T3), and copper and JNKi (T4), compared to the controls (T1) (Figure 4C). The lowest levels of DHA were measured in all treatments having combinations of inhibitors (T6-T9) and in Cu + p38i (T5), all being significantly lower than controls (Figure 4C). The highest levels of total ascorbate were measured on d 6, with concentrations of 60 and 70 nmol mg^{-1} protein in treatments with only copper (T2) and copper plus all three inhibitors (T9), respectively (Figure 4D). Treatments containing copper generally had higher DHA than ASC levels, with the exceptions of copper and ERKi (T3) or JNKi (T4) (Figure 4D). Finally, in ascending order, the highest concentrations of DHA were observed at solely copper and subject to copper combined with all inhibitors (Figure 4D).

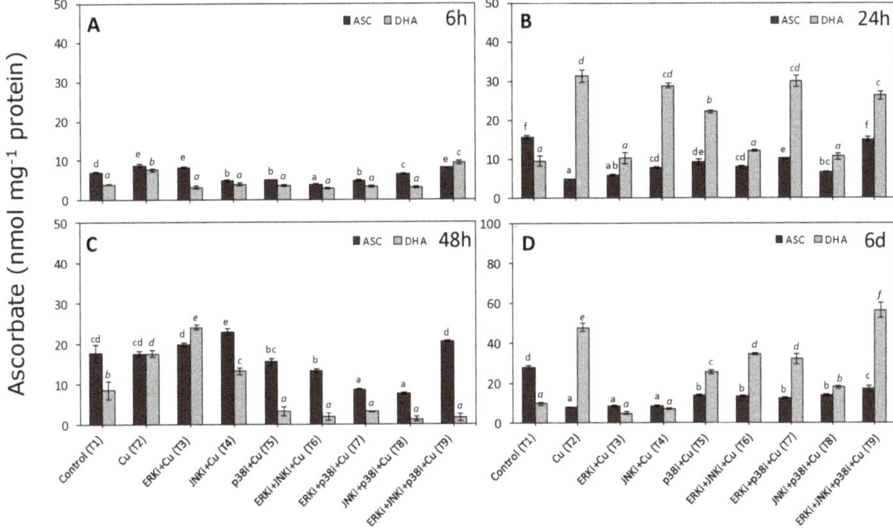

Figure 4. Total reduced ascorbate (ASC) and dehydroascorbate (DHA) concentrations in *U. compressa* under copper and/or exposure to MAPK inhibitors. Treatments consisted in: T1) control only with seawater; T2) solely copper exposure as 10 μM of $CuSO_4$ (Cu); T3) copper + 5 μM MAPK ERK inhibitor PD98059 (Cu + ERKi); T4) copper + 5 μM MAPK JNK inhibitor SP600125 (Cu + JNKi); T5) copper + MAPK p38 inhibitor SB203580 (Cu + p38i); T6) Cu + ERKi + JNKi; T7) Cu + ERKi + p38i; T8) Cu + p38i + JNKi; and T9) Cu + ERKi + JNKi + p38i. Samples were analysed after 6 h (**A**), 24 h (**B**), 48 h (**C**), and 6 d (**D**) treatments. Different letters represent significant difference at 95% confidence interval ($p < 0.05$). Plots are represented as mean ± SE ($n = 3$).

2.5. Gene Expression Profiles

Similar behaviour could be described in the expression of the studied genes. Under copper alone (T2), all genes, at most time points, were upregulated. When the alga was subjected to copper and single inhibitor exposure (T3-T5), the general trend was also upregulation at most time points with an oscillatory pattern of decrease at 24 and 48 h and then increase at 6 d, although there were certain exceptions to this, as for *CAT* and *DHAR* (Figure 5A,C, respectively). Conversely, when two or more MAPK pathways were blocked, the general trend was downregulation, even though at most time periods the differences were not significant (Figures 5 and 6). There were several exceptions to this general trend for example, at 48 h and 6 d *CAT, SOD*, and *TRX* were upregulated in treatments containing copper and p38i with either ERKi (T7) or JNKi (T8) (Figure 5A,B, and Figure 6A, respectively).

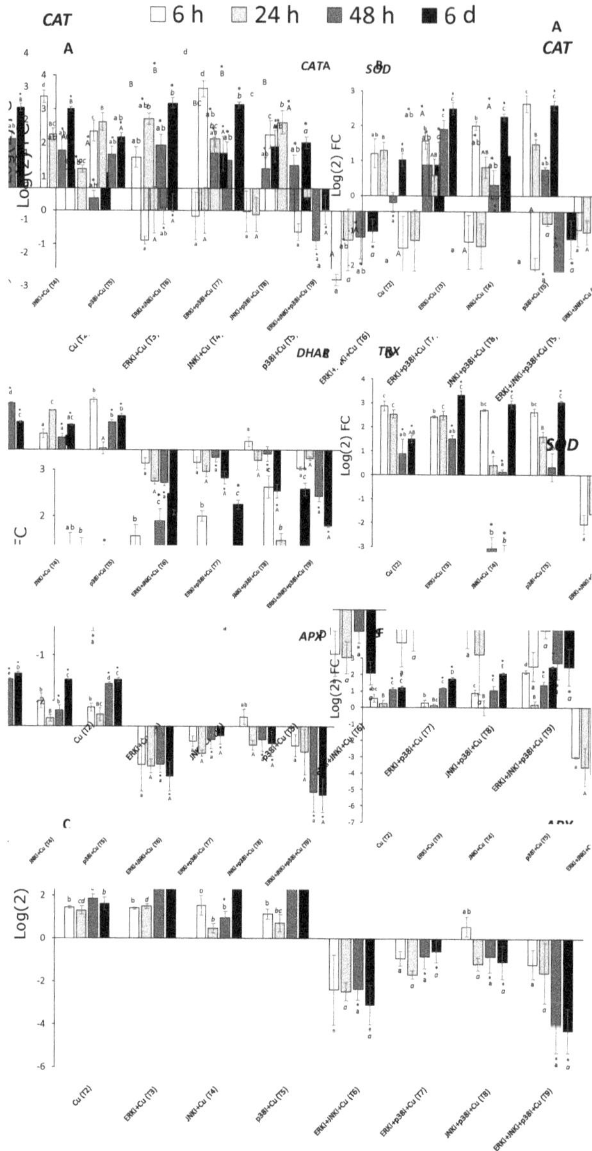

Figure 5. Relative expression of the genes catalase (*CAT*; **A**), superoxide dismutase (*SOD*; **B**), dehydroascorbate reductase (*DHAR*; **C**) in *U. compressa* under copper and/or exposure to MAPK inhibitors. Expression was relative to the levels of transcripts in *U. compressa* under control conditions. Treatments consisted in: T1) solely copper exposure as 10 μM of $CuSO_4$ (Cu); T2) copper + 5 μM MAPK ERK inhibitor PD98059 (Cu + ERKi); T3) copper + 5 μM MAPK JNK inhibitor SP600125 (Cu + JNKi); T4) copper + MAPK p38 inhibitor SB203580 (Cu + p38i); T5) Cu + ERKi + JNKi; T6) Cu + ERKi + p38i; T7) Cu + p38i + JNKi; and T8) Cu + ERKi + JNKi + p38i. Samples were analysed after 6 h (A), 24 h (B), 48 h (C), and 6 d (D) treatments. Different letters, in italics and/or with asterisk represent significant difference at 95% confidence interval ($p < 0.05$) within each experimental time. Plots are represented as mean ± SE ($n = 3$).

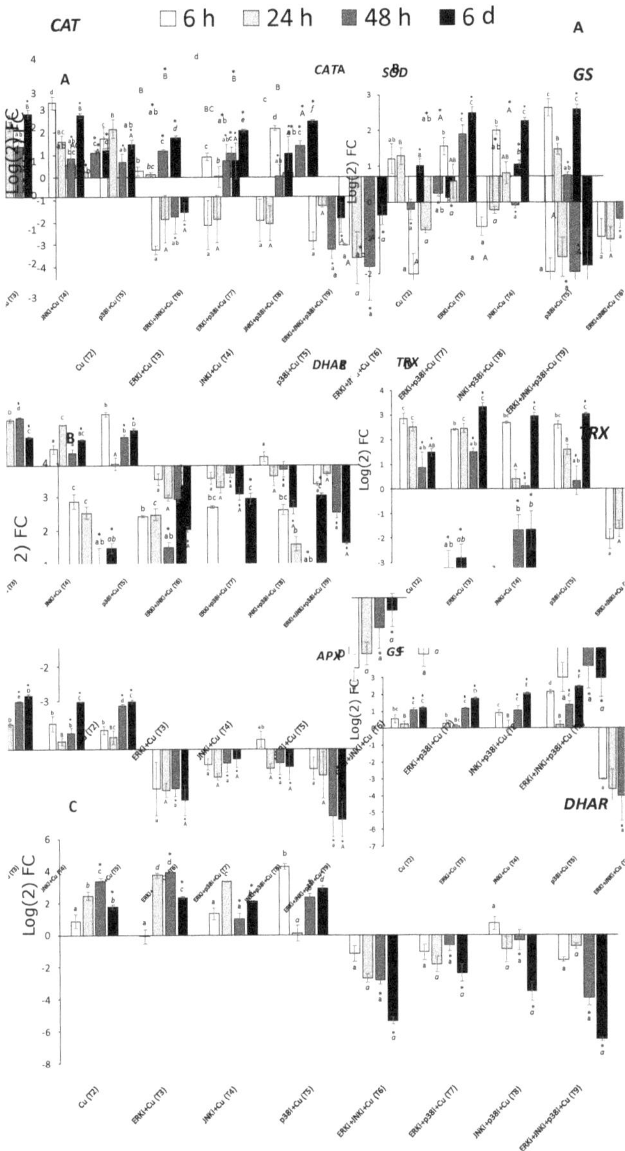

Figure 6. Relative expression of the genes glutathione synthase (*GS*; **A**), thioredoxin (*TRX*; **B**), and dehydroascorbate reductase (*DHAR*; **C**) in *U. compressa* under copper and/or exposure to MAPK inhibitors. Expression was relative to the levels of transcripts in *U. compressa* under control conditions. Treatments consisted in: T1) solely copper exposure as 10 μM of $CuSO_4$ (Cu); T2) copper + 5 μM MAPK ERK inhibitor PD98059 (Cu + ERKi); T3) copper + 5 μM MAPK JNK inhibitor SP600125 (Cu + JNKi); T4) copper + MAPK p38 inhibitor SB203580 (Cu + p38i); T5) Cu + ERKi + JNKi; T6) Cu + ERKi + p38i; T7) Cu + p38i + JNKi; and T8) Cu + ERKi + JNKi + p38i. Samples were analysed after 6 h (A), 24 h (B), 48 h (C), and 6 d (D) treatments. Different letters, in italics and/or with asterisk represent significant difference at 95% confidence interval ($p < 0.05$) within each experimental time. Plots are represented as mean ± SE ($n = 3$).

2.6. Principal Coordinate Ordination (PCO) Analysis

Considering all whole data presented in this article, evidence of clustering of the measured parameters in relation to time period and treatment can be discerned from the PCO (Figure 7). In terms of time, there was an association after the 6th d in the expression of all studied genes, in agreement with the general pattern of increase in transcripts towards the end of the experiment (Figure 7a). Another grouping between GSH, GSSG, DHA, and thiobarbituric acid reactive substances (TBARS) is evident, albeit with no clear trend between time periods, that is attributable to the pattern of decrease in treatments containing copper with two or more MAPK inhibitors (T6-T9) (Figure 7a). The final cluster is between ASC and H_2O_2, which can be ascribed to the oscillatory pattern of increase and decrease in treatments containing copper and mixed inhibitors (T6-T9) between time periods (Figure 7a). Furthermore, certain clusters can be detected within different treatments, particularly in those with copper and two or more MAPK inhibitors (T6-T9), which is in agreement, at least in part, with the marked general patterns in gene expression, and the parameters of TBARS, glutathione, and ascorbate (Figure 7b).

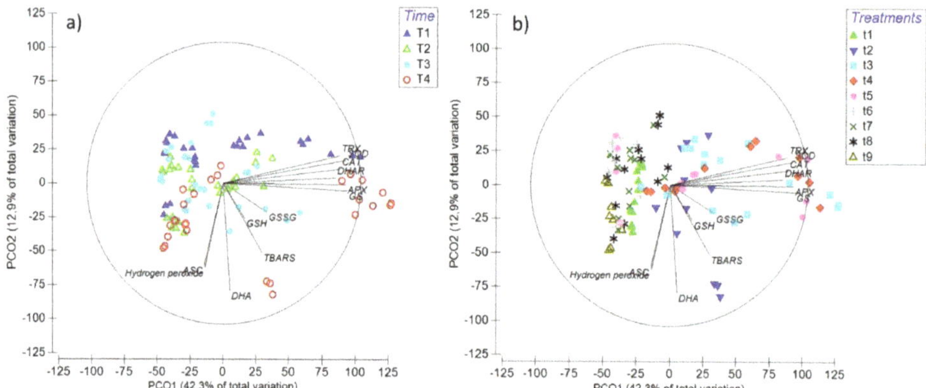

Figure 7. Principal components ordination (PCO) analysis diagrams in relation with time (**a**): T1: 6 h, T2: 24 h, T3: 48 h, and T4: 6 d; and treatments (**b**): t1: Control only with seawater, t2: Solely copper exposure as 10 µM of $CuSO_4$ (Cu), t3: copper + 5 µM MAPK ERK inhibitor PD98059 (Cu + ERKi), t4: Copper + 5 µM MAPK JNK inhibitor SP600125 (Cu + JNKi), t5: Copper + MAPK p38 inhibitor SB203580 (Cu + p38i), t6: Cu + ERKi + JNKi, t7: Cu + ERKi + p38i, t8: Cu + p38i + JNKi, and t9: Cu + ERKi + JNKi + p38i. Vectors overlay (Sperman rank correlation) indicate the relationship between the PCO axes and the parameters H_2O_2, TBARS (thiobarbituric acid reactive substance), GSH (reduced glutathione), GSSG (oxidised glutathione), ASC (reduced ascorbate), DHA (dehydroascorbate), and relative gene expression catalase (*CAT*), superoxide dismutase (*SOD*), dehydroascorbate reductase (*DHAR*), thioredoxin (*TRX*), ascorbate peroxidase (*APX*), and glutathione synthase (*GS*).

Taking into account the results presented in this article and those of the parallel-complementary paper Celis-Plá et al. [1], a second PCO was prepared combining the whole dataset from of both articles (Figure 8). Clear trends are apparent in the grouping between gene expression and photosynthetic parameters associated with electron transport rates (ETR_{max}, α_{ETR} and EkETR), which have in common the marked decrease under copper and mixed inhibitor treatments (T6-T9) (Figure 8). The GSH, GSSH, DHA and TBARS cluster was maintained but with the addition of intracellular copper accumulation [1]; this obey to, not as marked as in gene expression and ETR associated parameters, a decrease in concentrations upon exposure to copper and mixed MAPK inhibitors (T6-T9) (Figure 8). The grouping between ASC and H_2O_2 is also maintained but with the addition of the photosynthetic parameters of photoinhibition, Fv/Fm and photoprotection (NPQ_{max}); the latter was the most isolated as at 6 d the trend was a general increase in copper treatments containing combined inhibitors (T6-T9) compared to that with solely copper (T2) (Figure 8). Upon time, clustering was mostly observed at d 6, attributed to the numeric dominance of ETR related parameters, levels of transcripts, GSH, GSSG, DHA and TBARS (Figure 8a). Grouping was repeated mostly within treatments considering copper and two or more combined MAPK inhibitors (T6-T9) (Figure 8b).

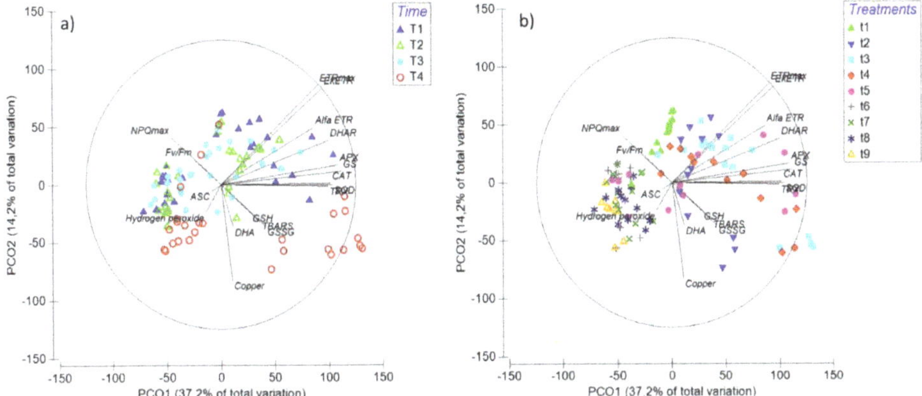

Figure 8. Principal Components Ordination (PCO) analysis diagrams considering the data covered in this article and in the parallel/complementary Celis-Plá et al. [1]. PCOs are related with time (**a**): T1: 6 h, T2: 24 h, T3: 48 h, and T4: 6 d; and treatments (**b**): t1: control only with seawater, t2: solely copper exposure as 10 µM $CuSO_4$ (Cu), t3: copper + 5 µM MAPK ERK inhibitor PD98059 (Cu + ERKi), t4: copper + 5 µM MAPK JNK inhibitor SP600125 (Cu + JNKi), T5: copper + MAPK p38 inhibitor SB203580 (Cu + p38i), t6: Cu + ERKi + JNKi, t7: Cu + ERKi + p38i, t8: Cu + p38i + JNKi, and t9: Cu + ERKi + JNKi + p38i. Vectors overlay (Sperman rank correlation) indicate the relationship between the PCO axes and the parameters H_2O_2, TBARS (thiobarbituric acid reactive substance), GSH (reduced glutathione), GSSG (oxidised glutathione), ASC (reduced ascorbate), DHA (dehydroascorbate), and relative gene expression catalase (*CAT*), superoxide dismutase (*SOD*), dehydroascorbate reductase (*DHAR*), thioredoxin (*TRX*), ascorbate peroxidase (*APX*), and glutathione synthase (*GS*). Furthermore, the physiological variables intracellular copper accumulation (copper photoinhibition (*Fv*/*Fm*) productivity (ETRmax), efficiency (αETR) and saturation of irradiance (EkETR), accompanied by higher non-photochemical quenching (NPQmax) [1].

3. Discussion

In this investigation, we exposed the green macroalga *Ulva compressa* to chronic copper excess (10 µM), and the relevance of the three main MAPK pathways, those ending in ERK, JNK, and p38, were studied upon their role within detoxification mechanisms associated with metal chelation, but mostly, with antioxidant responses. We observed that up to 24 h, levels of ROS H_2O_2 increased markedly compared to the controls under the treatment with only copper, and that in treatments blocking p38 under single or combined exposure (T5, T7-T9) with other inhibitors concentrations were the same or even higher than under copper alone (T2). That pattern at 48 h was inversed, and by the sixth day of the experiment, treatments with copper and solely p38i (T5) and combined inhibitors, excepting ERKi + p38i (T7), displayed even lower levels of H_2O_2 than under copper-only treatments and controls. Interestingly, almost the same pattern was observed for lipid peroxidation. For antioxidants, most treatments with copper displayed higher levels of GSSG and DHA than their reduced forms GSH and ASC, respectively, although this was more evident for ascorbate. Towards the end of the experiment, while the greatest concentrations of total glutathione were observed with copper alone (T2), the lowest levels were measured under copper and mixed MAPK inhibitor treatments (T6-T9). In contrast, the ascorbate pool was highest in the treatment with copper and all three MAPK pathways blocked, followed by copper alone (mostly due to DHA). Clear general patterns could be observed among the expression of the genes *CAT, SOD, DHAR, TRX, APX*, and *GS*. Upregulation, sometimes higher than under copper alone, was detected in treatments with copper and ERKi (T3), JNKi (T4), or p38i (T5), and mostly downregulation when the alga was subjected to copper and more than one inhibited MAPK pathway (T6-T9).

It is well recognised that chronic copper excess can induce oxidative stress and, in severe cases, cause damage in photoautotrophs [2,4]. Although chronic copper excess can produce oxidative stress in *U. compressa* under field conditions [9,22], this has also been described for other green macroalgae species subject to laboratory-controlled experiments. For example, it has been detected that *Ulva fasciata* exposed to copper concentrations of over 20 μM for 4 d displayed increased lipid peroxidation [15]. Notwithstanding the latter result, in this investigation no differences in lipid peroxidation were detected between controls and exposure to copper alone over the experimental period. While at most time periods levels of H_2O_2 were higher under copper alone compared to controls, the data suggest that these concentrations of H_2O_2 by themselves, or in combination with other ROS, were insufficient to cause oxidative damage, and ROS concentrations remained within homeostatic-tolerable levels. A different pattern was observed when blocking certain MAPK pathways, which started at 48 h but became more marked by 6 d. Inhibiting p38 (T5), ERK + JNK (T6), JNK + p38 (T8), and ERK + JNK + p38 (T9) under chronic copper excess, coincidentally decreased both H_2O_2 and lipid peroxidation relative to controls (T1) and under copper-only treatment (T2). The information demonstrates that under short-term copper exposure (especially at 6 h), the inhibition of more than one MAPK pathway causes increased oxidative stress and damage. However, over longer periods of copper exposure (mainly at 6 d), the patterns observed during short-term exposure become generally inverted, with resilient (adaptation response) lower levels of oxidative stress and damage in certain single and combined MAPK inhibitor treatments.

To counteract oxidative stress, the main mechanism recognised in photoautotrophs is associated with the glutathione-ascorbate, or Foyer-Halliwell-Asada, pathway [4,23]. As in previous investigations [13], *U. compressa* displayed a higher total glutathione pool under chronic copper excess in relation to controls, although at late exposures (6 d). At 48 h and beyond, patterns of higher GSSG upon GSH were observed in copper-exposed macroalgae, which is coincident with an induction of ROS over-production and GSH oxidation, indeed, a decrease in GSH/GSSG ratios have been recorded in several photoautotrophs under chronic copper excess and is fairly recognised as a good indicator of an oxidative stress condition [2,4,5,23,24]. In the presence of copper and MAPK inhibitors, most consistent trends were observed at 6 d. In most treatments with copper and one or more MAPK blocked pathways, with the exceptions of ERKi + p38i (T7) and JNKi + p38i (T8), GSH levels were always lower than GSSG. In spite of the latter, the total glutathione pool was always lower in treatments containing MAPK inhibitors (T3–T9) compared to that with solely copper (T2), and especially in those inhibiting more than one MAPK pathway (T6–T9). Similar tendencies were detected for ascorbate, particularly after 6 d. In most of the copper exposure treatments, ASC levels were higher than DHA, with the exception of treatments blocking ERK (T3) or JNK (T4). Total ascorbate was generally lower under copper and suppression of MAPK pathways, drawing certain exceptions (see results). Chronic copper excess also induced upregulation of *CAT, SOD, DHAR, TRX, APX*, and *GS* at most time periods and generally increasing upon exposure time. A similar behaviour in the expression of these genes was detected when chronic copper excess was applied and ERK (T3), JNK (T4), or p38 (T5) were blocked, with a recovery process throughout the experimental period (more details in results). In contrast, when more than one MAPK pathway was inhibited (T6-T9), at most of the experimental times the studied genes displayed downregulation. An interesting feature of our investigation lies in the fact that both GS expression and GSH levels in *U. compressa* displayed similar behaviour among treatments and time. Previous investigations under high copper in *U. compressa* have shown that the increased induction of GS and GSH is accompanied by a parallel increment in the synthesis and polymerisation of GSH oligomers PCs [13]. Thus, this information suggests that MAPKs may also play an important role in the production of metal-chelating thiols, not only GSH but also PCs. Although records about the regulation of the glutathione-ascorbate cycle through MAPK signalling are scarce in eukaryotes, and especially in photoautotrophs, Karuppanapandian and Kim [25] observed that the plant *Brassica juncea* under 100 μM of cobalt, activated 44 kDa and 46 kDa MAPKs, which was accompanied by an increases in GSH, GSSG, ASC, and DHA. Furthermore, Shan and Sun [26] by blocking ERK (with PD98059 and U0126), observed that jasmonic acid-mediated production of nitric oxide activated ERK which, in turn, induced increased activities of APX, GR, DHAR, MDHAR, γ-GCL, and L-GLDH, which also enhanced syntheses of ASC and GSH. Moreover, Qi et al. [27] detected that by inhibiting ERK (PD98059) in the plant *Cucumis sativus* under nitrate excess induced decreased activities of SOD, CAT, APX, and peroxidases (POD), being responsible, at least in part, for the observed greater lipid peroxidation and H_2O_2 content. To the extent of our knowledge, there are no published studies describing the role of the three main MAPK pathways, ERK, JNK, and p38 alone or complemented in the antioxidant metabolism of photoautotrophs under environmental stress. In our investigation, we demonstrated that the whole MAPK pathway has an important role in the antioxidant metabolism of macroalgae. The data shows that under single and combined inhibitors of the three main MAPK pathways, there is a progressive affection of glutathione production. Although all three MAPK pathways show to have an influence in GSH production, JNK and p38 appear to be more relevant than ERK. Furthermore, when more than one MAPK pathway is affected, total glutathione is almost impaired. Similar effects can be observed for ascorbate, however, this was observed at late copper exposure, and the main pathways regarding its production were related to the ERK, JNK, and JNK + p38 MAPK pathways. In relation to the expression of genes involved in antioxidant metabolism and the synthesis of metal chelators, when one of the three pathways is affected independently, in time it appears to enhance the upregulation mediated by chronic copper excess. Crosstalk is a common phenomenon among signalling mechanisms and has been observed within MAPK pathways as well as with other signalling processes as those related to phytohormones, calcium, NO, ROS, and among others [28–31]. In this context, it is possible that to counteract the inability of certain MAPK pathways, others MAPKs and/or more cellular signalling mechanisms could be triggered and elicit an intensified transcriptomic response in *U. compressa*. Although no similar investigations have been conducted in photoautotrophs, ERK inhibition in human melanoma cells have demonstrated to be counteracted through the activation of alternative signalling pathways, such as those mediated by phosphoinositide 3-kinases (PI3K) and Protein kinase B (Akt) [32–36]. Furthermore, inhibition of ERK induces an upregulation of the transcription factor Forkhead box D3 (*FOXD3*), involved in the development of embryo stem cells, mediated by the activation of PI3K and receptor tyrosine-protein kinase (ErbB3) [37,38]. Considering that in our investigation there was mostly gene downregulation when more than one MAPK pathway was blocked under chronic copper excess, the information suggests that the gene over-expression in *U. compressa* is promoted mainly by a crosstalk within the whole MAPK signalling pathway. On the other hand, it could be hypothesised that gene repression under copper and combined MAPK inhibitors is caused by the low levels of intracellular copper detected in these treatments in Celis-Plá et al. [1]. Indeed, it has been observed that the expression of the enzymes SOD, CAT, and APX had a positive correlation with levels of intracellular copper in the brown macroalga *Ectocarpus siliculosus* [5,39]. However, this explanation appears unlikely considering that the decline of antioxidant defences in treatments with copper and combined MAPK inhibitors most of the time reached levels even below control conditions. Thus, MAPK signalling demonstrates to be an important mechanism for the activation of copper-detoxification defences in *U. compressa*.

Multivariate analyses confirmed that the data contained in this article was importantly influenced by time, but mostly by treatments. Furthermore, considering the results of the complementary/parallel Celis-Plá et al. [1], the important agreement between physiological, biochemical, and transcriptomic responses in MAPK-intervened *U. compressa* can be highlighted. In this regard, general tendencies demonstrate that the inactivation of single MAPK pathways ending in ERK, JNK, and p38 subject to high copper can be progressively counteracted by complementary biological strategies, which appear to be related to crosstalk within the MAPKs and/or with other signalling pathways. Despite the latter, when more than one of these MAPK pathways is disabled under chronic copper excess, biological impairment start to manifest, which is traduced in diminished photosynthetic activity [1], ROS-overproduction, increased oxidative damage, and affected production and regulation of antioxidant mechanisms. These investigations provide the first complete evidence on the importance of the MAPK pathway to counteract environmental stress in photoautotrophs, and further confirm that copper-detoxification mechanisms are importantly founded on transcriptionally regulated metabolic and physiological processes in green macroalgae. Certainly, additional research on the whole transcriptome of *U. compressa* subject to modified MAPK activation/inactivation will ascertain on open questions regarding green macroalgae biological strategies to withstand stress, in this case, induced by chronic copper excess.

4. Methodology

4.1. Samples Collection and Culture Conditions

Ulva compressa (Chlorophyta, Ulvaceae) was sampled from Cachagua beach, Valparaíso Region, Chile (32°34′59 S; 71°26′16 W). Algae material was washed with 2 µm of filtered seawater and stored in plastic containers with seawater under constant aeration at 16 °C. The photoperiod cycle was 12:12 (day/night) with daylight conditions of 120 µmol $m^2 \cdot s^{-1}$ of photosynthetic active radiation (PAR). Algae were acclimated for 48 h before experiments.

4.2. Copper and MAPK Inhibitors Treatments

Following the acclimation period, 5 g of fresh weight (FW) *U. compressa* were separated in different plastic containers with 200 mL of filtered (2 µm) and autoclaved seawater. Nine different treatments, with three replicates each, were conducted as: T1) control with just seawater; T2) only copper exposure as 10 µM of CuSO4 (Sigma-Aldrich, St. Louis, MO, USA) (Cu); T3) copper + 5 µM MAPK ERK inhibitor PD98059 (Tocris Bioscience, St. Louis, MO, USA) (Cu + ERKi); T4) copper + 5 µM MAPK JNK inhibitor SP600125 (Tocris Bioscience, St. Louis, MO, USA) (Cu + JNKi); T5) copper + MAPK p38 inhibitor SB203580 (Tocris Bioscience) (Cu + p38i); T6) Cu + ERKi + JNKi; T7) Cu + ERKi + p38i; T8) Cu + p38i + JNKi; and T9) Cu + ERKi + JNKi + p38i. All treatments were performed under the same environmental conditions mentioned in Section 2.1. Negative controls to assess the effectiveness of the ERK, JNK, and p38 inhibitors were conducted by measuring the expression of *SOD* and *CAT* (protocols in Sections 2.5 and 2.6) in the absence of copper (Figure S1) and these were complemented by negative controls contained as Supplementary Materials in Celis-Pla et al. [1]. Even though the inhibitors used were first developed to target human MAPKs, this and prior investigations demonstrate their successful application to other eukaryote models given the conserved nature of MAPKs (further details) in [1].

During the experiments, sub-samples were taken from each treatment at 6 h, 24 h, 48 h, and 6 d, immediately frozen in liquid nitrogen and stored at −80 °C for biochemical and molecular analyses.

4.3. Hydrogen Peroxide (H_2O_2) Quantification

Concentrations of H_2O_2 were quantified using a modification of the colorimetric method by which H_2O_2 is induced by its interaction with potassium iodide (KI), as described by Junglee et al. [40]. Briefly, 100 mg of liquid nitrogen-ground sample was lysed and acidified with 100 µL of 0.1%TCA, 100 µL of 10 mM potassium phosphate buffer pH 7, 100 µL of FARB lysis buffer (Favorgen, Biotech Corp, Ping-Tung, Taiwan), and 500 µL of 1M KI. Subsequently, samples were vortexed for 10 min at room temperature in darkness and centrifuged at 12,000× *g* for 15 min at 4°C. Then, 300 µL of recovered supernatant were placed in a 96-well plate and absorbance was measure at 350 nm in a microplate reader SPECTROStar Nano (BMG LABTECH, Offenburg, Germany). A standard curve using commercial H_2O_2 (Merck, Darmstadt, Germany) was constructed for assay calibration. Data values were normalised to protein concentrations, as determined by the Bradford method [41].

4.4. Determination of Lipid Peroxidation

As a proxy for lipid peroxidation, the levels of thiobarbituric acid reactive substance (TBARS) were measured according to Sáez et al. [24]. A total of 100 mg of grounded tissue were lysed with 300 μL of 0.1 % TCA, adding four glass beads (2 mm), and vortexed for 5 min at room temperature. After vortexing, samples were centrifuged at 17,800 g for 15 min at 4°C, and 200 μL of supernatant were mixed with 200 μL of 0.5 % TBA, and incubated at 95 °C for 45 min. Then, samples were cooled to room temperature and the absorbance of 200 μL of the mixture was measured at 532 nm in a microplate reader. Commercial malondialdehyde (MDA; Sigma-Aldrich, St. Louis, MO, USA) was used to construct a standard curve. Data were normalised to protein concentration, as determined by the Bradford method [41].

4.5. Oxidised and Reduced Glutathione

Reduced (GSH) and oxidised (GSSG) glutathione levels were determined using the enzymatic recycling method [42] as described for brown macroalgae in Sáez et al. [24], but optimised for *U. compressa*. Briefly, for total glutathione (GSH + GSSG), 100 mg of liquid nitrogen-grinded sample was lysed with 500 μL of 0.1M HCl and vortexed for 10 min at room temperature. Supernatant was recovered by centrifugation at 7400× *g* for 15 min at 4 °C, and neutralised with 500 μL of 500 mM sodium phosphate buffer pH 7.5. Then, 50 μL was mixed with 250 μL of Buffer GR containing: 100 mM sodium phosphate buffer pH 7.5, 0.1 mM EDTA, 0.3 mM NADPH, 0.6 U glutathione reductase, and 0.2 mM 5,5′-dithiobis (2-nitrobenzoic acid) (DTNB or Ellman's reagent, Sigma-Aldrich, St. Louis, MO, USA). Immediately after the start of the reaction, absorbance at 412 nm was recorded in the microplate reader every 20 s for a total of 5 min. For GSSG quantification, 50 μL of the neutralised supernatant was pre-treated with 1 μL of 1M 4-vinylpiridine for 45 min at room temperature as a GSH masking agent, and then mixed with 250 μL of GR buffer to complete the reaction. GSH content was obtained by subtracting GSSG content from total glutathione. Standard curves using commercial GSH (Sigma-Aldrich) were constructed for assay calibration. Data were normalised to protein concentrations, determined by the Bradford method [41].

4.6. Oxidised and Reduced Ascorbate

Reduced ascorbate (ASC) and dehydroascorbate (DHA) were measured according to the ferric reducing/antioxidant power assay [24,43], with modifications. Briefly, 100 mg of grounded sample was lysed with 600 μL of 0.1 M HCl, vortexed for 10 min at room temperature, and centrifuged at 17,800 g for 15 min at 4 °C. Then, for ASC quantification, 10 μL of supernatant was mixed with 290 μL of FRAP solution containing: Sodium acetate buffer 300 mM pH 3.6, 20 mM $FeCl_3$, and 10 mM 2,4,6-tripyridyl-s-triazine (TPTZ). Absorbance at 593 nm was immediately measured in the microplate reader. For total ascorbate (ASC + DHA) determination, 250 μL of the obtained supernatant was incubated with 2.5 μL of 100 mM dithiothreitol (DTT) for 1 h at room temperature. Then, 2.5 μL of 5% was *N*-ethylmaleimide applied to stop the reaction and 10 μL were used for the reaction with 290 μL of FRAP solution following the protocol described above. A standard curve using commercial L-ascorbate (Sigma-Aldrich) was constructed for assay calibration. Data were normalised to protein concentrations, determined by the Bradford method [41].

4.7. RNA Extraction

All samples were homogenised using mortars with liquid nitrogen. Then, RNA extraction was performed using 100 mg of ground tissue with the FavorPrep Plant Total RNA Purification Mini Kit (Favorgen, Biotech Corp, Ping-Tung, Taiwan), according the manufacturer's instructions. RNA purity was determined through electrophoresis gel and spectrophotometrically by 260/280 ratio, using a nanoplate within the microplate reader. RNA quantification was carried out by fluorescence using the

Quant-iT RiboGreen RNA Assay Kit (Invitrogen, Waltham, MA, USA) in a QFX Fluorometer (DeNovix, Wilmington, DE, USA).

4.8. cDNA Synthesis and qPCR

Synthesis of cDNA was carried out with 500 ng of total RNA using the ProtoScript First Stand cDNA Synthesis Kit (New England BioLabs, Ipswich, MA, USA), according to the manufacturer's guidelines. Then, for the qPCR reaction, 50 ng of cDNA (2 μL) were used together with 0.25 μM of each primer and the Brilliant II SYBR Green QPCR Master Mix (Agilent Technologies, Santa Clara, CA, USA) 1X (final volume of 20 μL). The qPCR program was: Initial denaturation at 95 °C for 5 min, 40 cycles of: 95 °C for 30 s, 55 °C for 30 s, 72 °C for 40 s, and final extension at 72°C for 10 min. All reactions were performed in a MIC qPCR Magnetic Induction Cycler (Bio Molecular Systems, Queensland, Australia). Primers were designed specifically for each gene using the GenScript online tool primer designer (https://www.genscript.com/tools/pcr-primers-designer) (Table 1), based on public transcriptome sequences from Rodríguez et al. [44]. The analysed transcripts corresponded to genes coding *CAT* (catalase), *SOD* (superoxide dismutase), *TRX* (thioredoxin), *APX* (ascorbate peroxidase), *DHAR* (dehydroascorbate reductase), and *GS* (glutathione synthase). Relative expression analyses were based in the $2^{-\Delta\Delta Ct}$ method using 18S rRNA as housekeeping gene [45].

Table 1. Genes and primers for qPCR analyses.

Gene	Encoded Product	Primer ID	Sequence 5′-3′	Start Position	Amplicon Size (bp)
CAT	Catalase	F1_CAT_Ulva	AGCGGAACAAGTCGGCGGCAA	717	208
		R1_CAT_Ulva	CAGCACCATGCGGCCAACGG	905	
SOD	Superoxide dismutase	F1_SOD2_Ulva	CCCTGCACCGCCGTCTGTCG	22	255
		R1_SOD2_Ulva	CGCGCTGCTGATCTTCGGAGCA	255	
TRX	Thioredoxin	F1_Trx2_Ulva	GCTGAGTGGCTCCTCTTCGCTGCAT	3	224
		R1_Trx2_Ulva	CGGCTGTGCTCACTGGTGCGT	206	
APX	Ascorbate peroxidase	F1c_Apx1_Ulva	AGGACGCGTGGCCCAAGTGC	131	297
		R1c_Apx1_Ulva	GTCTCCGCCGGATGGCCAAGG	408	
DHAR	Dehydroascorbate reductase	F1_Dhar_Ulva	CGCGGACTCCGGCGACATCT	309	327
		R1_Dhar_Ulva	AGCGCTGCGAGCTCCTCTGG	616	
GS	Glutathione synthetase	F1_GS_Ulva	TGGCGGCGAAGCTGCAGGAA	1322	245
		R1_GS_Ulva	GCCTGCCGCAACACCTCCCT	1547	
18S	18S rRNA subunit	18S-F-1183	AAT TTG ACT CAA CAC GGG	1183	448
		18S-R-1631	TAC AAA GGG CAG GGA CG	1631	

4.9. Statistical Analyses

Interactive effects in all parameters content were analysed using ANOVA. This test was performed for *U. compressa* including treatment (one-way) as fixed factor for the whole dataset (mean ± SE, $n = 3$). Each time period (6 h, 24 h, 48 h, and 6 d) included 9 different factors: T1, T2, T3, T4, T5, T6, T7, T8, and T9, assessed at a 95% confidence interval. Homogeneity of variance was checked with Cochran. Student Newman Keuls tests (SNK) were conducted after observing significant interactions in the ANOVA. All data complied with the requirements of normality and homogeneity of variance. Analyses were conducted with SPSS v.21 (IBM, USA). A PCO analysis was performed to detect patterns between the parameters on the basis of an Euclidean distance using PERMANOVA+ for PRIMER6 package [46]. PCO analyses were conducted for results in vivo of H_2O_2, TBARS, total glutathione ascorbate, as well as for the expression of the assessed genes. Finally, a PCO was also carried out considering the data contained in this investigation plus the results in the parallel article Celis-Plá et al. [1].

Supplementary Materials: Supplementary materials can be found at http://www.mdpi.com/1422-0067/20/18/4546/s1.

Author Contributions: Conceptualisation, F.R.-R., P.S.M.C.-P., A.M. and C.A.S. Methodology and investigation, F.R.-R., P.S.M.C.-P., L.M., F.M., P.T.M., M.G.L. and P.D. Formal analysis and writing original draft preparation,

F.R.-R., P.S.M.C.-P. and C.A.S. Writing review and editing, J.L.S.-L., M.T.B., A.M., F.R.-R., P.S.M.C.-P. and C.A.S. Supervision, C.A.S. Funding acquisition, C.A.S.

Funding: Project FONDECYT NO. 11160369 granted to C.A. Sáez

Conflicts of Interest: The authors declare no conflict of interest.

Abbreviations

MAPK	mitogen activated protein kinases
ERK	Extracellular Signal Regulated Kinases
JNK	c-Jun N-terminal kinases
p38	cytokinin specific binding protein
ROS	reactive oxygen species
GSH	reduced glutathione
GSSG	oxidised glutathione
ASC	reduced ascorbate
DHA	dehydroascorbate (form of oxidised ascorbate)
CAT	catalase
SOD	superoxide dismutase
TRX	thioredoxin
APX	ascorbate peroxidase
DAHR	dehydroascorbate reductase
GS	glutathione synthase

References

1. Celis-Plá, P.S.M.; Rodríguez-Rojas, F.; Méndez, L.; Moenne, F.; Muñoz, P.; Lobos, M.; Díaz, P.; Sánchez-Lizaso, J.L.; Brown, M.; Moenne, A.; et al. MAPK pathway under chronic copper excess in green macroalgae (Chlorophyta): Influence on metal exclusion/extrusion mechanisms and photosynthesis. *Int. J. Mol. Sci.* **2019**, in press.
2. Moenne, A.; González, A.; Sáez, C.A. Mechanisms of metal tolerance in marine macroalgae, with emphasis on copper tolerance in Chlorophyta and Rhodophyta. *Aquat. Toxicol.* **2016**, *176*, 30–37. [CrossRef] [PubMed]
3. Navarrete, A.; González, A.; Gómez, M.; Contreras, R.A.; Díaz, P.; Lobos, G.; Brown, M.T.; Sáez, C.A.; Moenne, A. Copper excess detoxification is mediated by a coordinated and complementary induction of glutathione, phytochelatins and metallothioneins in the green seaweed Ulva compressa. *Plant Physiol. Biochem.* **2019**, *135*, 423–431. [CrossRef] [PubMed]
4. Foyer, C.H.; Noctor, G. Ascorbate and glutathione: The heart of the redox hub. *Plant Physiol.* **2011**, *155*, 2–18. [CrossRef] [PubMed]
5. Sáez, C.A.; Roncarati, F.; Moenne, A.; Moody, J.A.; Brown, M.T. Copper-induced intra-specific oxidative damage and antioxidant responses in strains of the brown alga *Ectocarpus siliculosus* with different pollution histories. *Aquat. Toxicol.* **2015**, *159*, 81–89. [CrossRef] [PubMed]
6. Burkhead, J.L.; Reynolds, K.A.G.; Abdel-Ghany, S.E.; Cohu, C.M.; Pilon, M. Copper homeostasis. *New Phytol.* **2009**, *182*, 799–816. [CrossRef] [PubMed]
7. Cuypers, A.; Hendrix, S.; dos Reis, R.A.; De Smet, S.; Deckers, J.; Gielen, H.; Jozefczak, M.; Loix, C.; Vercampt, H.; Vangronsveld, J.; et al. Hydrogen peroxide, signaling in disguise during metal phytotoxicity. *Front. Plant Sci.* **2016**, *7*, 470. [CrossRef]
8. Buchanan, B.B. The path to thioredoxin and redox regulation beyond chloroplasts. *Plant Cell Physiol.* **2017**, *58*, 1826–1832. [CrossRef]
9. Ratkevicius, N.; Correa, J.A.; Moenne, A. Copper accumulation, synthesis of ascorbate and activation of ascorbate peroxidase in *Enteromorpha compressa* (L.) Grev. (Chlorophyta) from heavy metal-enriched environments in northern Chile. *Plant Cell Environ.* **2003**, *26*, 1599–1608. [CrossRef]
10. González, A.; Vera, J.; Castro, J.; Dennett, G.; Mellado, M.; Morales, B.; Correa, J.A.; Moenne, A. Co-occurring increases of calcium and organellar reactive oxygen species determine differential activation of antioxidant and defense enzymes in *Ulva compressa* (Chlorophyta) exposed to copper excess. *Plant Cell Environ.* **2010**, *33*, 1627–1640. [CrossRef]

11. González, A.; Cabrera Mde, L.; Henriquez, M.J.; Contreras, R.A.; Morales, B.; Moenne, A. Cross talk among calcium, hydrogen peroxide, and nitric oxide and activation of gene expression involving calmodulins and calcium-dependent protein kinases in *Ulva compressa* exposed to copper excess. *Plant Physiol.* **2012**, *158*, 1451–1462. [CrossRef] [PubMed]
12. Laporte, D.; Valdés, N.; González, A.; Sáez, C.A.; Zúñiga, A.; Navarrete, A.; Meneses, C.; Moenne, A. Copper-induced overexpression of genes encoding antioxidant system enzymes and metallothioneins involve the activation of CaMs, CDPKs and MEK1/2 in the marine alga *Ulva compressa*. *Aquat. Toxicol.* **2016**, *177*, 433–440. [CrossRef] [PubMed]
13. Mellado, M.; Contreras, R.A.; González, A.; Dennett, G.; Moenne, A. Copper-induced synthesis of ascorbate, glutathione and phytochelatins in the marine alga *Ulva compressa* (Chlorophyta). *Plant Physiol. Biochem.* **2012**, *51*, 102–108. [CrossRef] [PubMed]
14. Jervis, L.; Rees-Naesborg, R.; Brown, M. Biochemical responses of the marine macroalgae *Ulva lactuca* and *Fucus vesiculosus* to cadmium and copper-from sequestration to oxidative stress. *Biochem. Soc. Trans.* **1997**, *25*, 63S. [CrossRef] [PubMed]
15. Wu, T.M.; Lee, T.M. Regulation of activity and gene expression of antioxidant enzymes in *Ulva fasciata* Delile (Ulvales, Chlorophyta) in response to excess copper. *Phycologia* **2008**, *47*, 346–360. [CrossRef]
16. Song, Y.; Cheong, Y.K.; Kim, N.H.; Chung, H.T.; Kang, D.G.; Pae, H.O. Mitogen-activated protein kinases and reactive oxygen species: How can ROS activate MAPK pathways? *J. Signal Transduct.* **2011**, *2011*, 792639.
17. Sinha, A.K.; Jaggi, M.; Raghuram, B.; Tuteja, N. Mitogen-activated protein kinase signaling in plants under abiotic stress. *Plant Signal. Behav.* **2011**, *6*, 196–203. [CrossRef]
18. Janitza, P.; Ullrich, K.K.; Quint, M. Toward a comprehensive phylogenetic reconstruction of the evolutionary history of mitogen-activated protein kinases in the plant kingdom. *Front. Plant Sci.* **2012**, *3*, 271. [CrossRef]
19. Parages, M.L.; Capasso, J.M.; Niell, F.X.; Jimenez, C. Responses of cyclic phosphorylation of MAPK-like proteins in intertidal macroalgae after environmental stress. *J. Plant Physiol.* **2014**, *171*, 276–284. [CrossRef]
20. Charneco, G.O.; Parages, M.L.; Camarena-Gomez, T.; Jimenez, C. Phosphorylation of MAP Kinases crucially controls the response to environmental stress in *Dunaliella viridis*. *Environ. Exp. Bot.* **2018**, *156*, 203–213. [CrossRef]
21. Parages, M.L.; Figueroa, F.L.; Conde-Alvarez, R.M.; Jimenez, C. Phosphorylation of MAPK-like proteins in three intertidal macroalgae under stress conditions. *Aquat. Biol.* **2014**, *22*, 213–226. [CrossRef]
22. Ismail, G.A.; Ismail, M.M. Variation in oxidative stress indices of two green seaweeds growing under different heavy metal stresses. *Environ. Monit. Assess.* **2017**, *189*, 12. [CrossRef] [PubMed]
23. Foyer, C.H.; Noctor, G. Stress-triggered redox signalling: what's in pROSpect? *Plant Cell Environ.* **2016**, *39*, 951–964. [CrossRef] [PubMed]
24. Sáez, C.A.; González, A.; Contreras, R.A.; Moody, A.J.; Moenne, A.; Brown, M.T. A novel field transplantation technique reveals intra-specific metal-induced oxidative responses in strains of *Ectocarpus siliculosus* with different pollution histories. *Environ. Pollut.* **2015**, *199*, 130–138. [CrossRef] [PubMed]
25. Karuppanapandian, T.; Kim, W. Cobalt-induced oxidative stress causes growth inhibition associated with enhanced lipid peroxidation and activates antioxidant responses in Indian mustard (*Brassica juncea* L.) leaves. *Acta Physiol. Plant.* **2013**, *35*, 2429–2443. [CrossRef]
26. Shan, C.; Sun, H. Jasmonic acid-induced NO activates MEK1/2 in regulating the metabolism of ascorbate and glutathione in maize leaves. *Protoplasma* **2018**, *255*, 977–983. [CrossRef]
27. Qi, Q.; Guo, Z.; Liang, Y.; Li, K.; Xu, H. Hydrogen sulfide alleviates oxidative damage under excess nitrate stress through MAPK/NO signaling in cucumber. *Plant Physiol. Biochem.* **2019**, *135*, 1–8. [CrossRef] [PubMed]
28. Ahanger, M.A.; Ashraf, M.; Bajguz, A.; Ahmad, P. Brassinosteroids regulate growth in plants under stressful environments and crosstalk with other potential phytohormones. *J. Plant Growth Regul.* **2018**, *37*, 1007–1024. [CrossRef]
29. Jagodzik, P.; Tajdel-Zielinska, M.; Ciesla, A.; Marczak, M.; Ludwikow, A. Mitogen-activated protein kinase cascades in plant hormone signaling. *Front. Plant Sci.* **2018**, *9*, 1387. [CrossRef]
30. Yoo, S.J.; Kim, S.H.; Kim, M.J.; Ryu, C.M.; Kim, Y.C.; Cho, B.H.; Yang, K.Y. Involvement of the OsMKK4-OsMPK1 CASCADE AND ITS DOWNSTREAM TRANSCRIPTION FACTOR OsWRKY53 in the wounding response in rice. *Plant Pathol. J.* **2014**, *30*, 168–177. [CrossRef]
31. Zhao, F.Y.; Wang, K.; Zhang, S.Y.; Ren, J.; Liu, T.; Wang, X. Crosstalk between ABA, auxin, MAPK signaling, and the cell cycle in cadmium-stressed rice seedlings. *Acta Physiol. Plant.* **2014**, *36*, 1879–1892. [CrossRef]

32. Nazarian, R.; Shi, H.; Wang, Q.; Kong, X.; Koya, R.C.; Lee, H.; Chen, Z.; Lee, M.K.; Attar, N.; Sazegar, H.; et al. Melanomas acquire resistance to B-RAF(V600E) inhibition by RTK or N-RAS upregulation. *Nature* **2010**, *468*, 973–977. [CrossRef] [PubMed]
33. Villanueva, J.; Vultur, A.; Lee, J.T.; Somasundaram, R.; Fukunaga-Kalabis, M.; Cipolla, A.K.; Wubbenhorst, B.; Xu, X.; Gimotty, P.A.; Kee, D.; et al. Acquired resistance to BRAF inhibitors mediated by a RAF kinase switch in melanoma can be overcome by cotargeting MEK and IGF-1R/PI3K. *Cancer Cell* **2010**, *18*, 683–695. [CrossRef] [PubMed]
34. Paraiso, K.H.; Xiang, Y.; Rebecca, V.W.; Abel, E.V.; Chen, Y.A.; Munko, A.C.; Wood, E.; Fedorenko, I.V.; Sondak, V.K.; Anderson, A.R.; et al. PTEN loss confers BRAF inhibitor resistance to melanoma cells through the suppression of BIM expression. *Cancer Res.* **2011**, *71*, 2750–2760. [CrossRef] [PubMed]
35. Gopal, Y.N.; Deng, W.; Woodman, S.E.; Komurov, K.; Ram, P.; Smith, P.D.; Davies, M.A. Basal and treatment-induced activation of AKT mediates resistance to cell death by AZD6244 (ARRY-142886) in Braf-mutant human cutaneous melanoma cells. *Cancer Res.* **2010**, *70*, 8736–8747. [CrossRef] [PubMed]
36. Davies, M.A.; Kopetz, S. Overcoming resistance to MAPK pathway inhibitors. *J. Natl. Cancer Inst.* **2013**, *105*, 9–10. [CrossRef] [PubMed]
37. Abel, E.V.; Basile, K.J.; Kugel, C.H., 3rd; Witkiewicz, A.K.; Le, K.; Amaravadi, R.K.; Karakousis, G.C.; Xu, X.; Xu, W.; Schuchter, L.M.; et al. Melanoma adapts to RAF/MEK inhibitors through FOXD3-mediated upregulation of ERBB3. *J. Clin. Investig.* **2013**, *123*, 2155–2168. [CrossRef] [PubMed]
38. Newlaczyl, A.U.; Hood, F.E.; Coulson, J.M.; Prior, I.A. Decoding RAS isoform and codon-specific signalling. *Biochem. Soc. Trans.* **2014**, *42*, 742–746. [CrossRef] [PubMed]
39. Sáez, C.A.; Ramesh, K.; Greco, M.; Bitonti, M.B.; Brown, M.T. Enzymatic antioxidant defences are transcriptionally regulated in Es524, a copper-tolerant strain of *Ectocarpus siliculosus* (Ectocarpales; Phaeophyceae). *Phycologia* **2015**, *54*, 425–429. [CrossRef]
40. Junglee, S.; Urban, L.; Sallanon, H.; Lopez-Lauri, F. Optimized assay for hydrogen peroxide determination in plant tissue using potassium iodide. *Am. J. Anal. Chem.* **2014**, *5*, 730. [CrossRef]
41. Bradford, M.M. A rapid and sensitive method for the quantitation of microgram quantities of protein utilizing the principle of protein-dye binding. *Anal. Biochem.* **1976**, *72*, 248–254. [CrossRef]
42. Rahman, I.; Kode, A.; Biswas, S.K. Assay for quantitative determination of glutathione and glutathione disulfide levels using enzymatic recycling method. *Nat. Protoc.* **2006**, *1*, 3159. [CrossRef]
43. Benzie, I.F.; Strain, J. Ferric reducing/antioxidant power assay: Direct measure of total antioxidant activity of biological fluids and modified version for simultaneous measurement of total antioxidant power and ascorbic acid concentration. In *Methods in Enzymology*; Elsevier: Amsterdam, The Netherlands, 1999; Volume 299, pp. 15–27.
44. Rodríguez, F.E.; Laporte, D.; González, A.; Méndez, K.N.; Castro-Nallar, E.; Meneses, C.; Huidobro-Toro, J.P.; Moenne, A. Copper-induced increased expression of genes involved in photosynthesis, carotenoid synthesis and C assimilation in the marine alga *Ulva compressa*. *BMC Genom.* **2018**, *19*, 829. [CrossRef] [PubMed]
45. Livak, K.J.; Schmittgen, T.D. Analysis of relative gene expression data using real-time quantitative PCR and the 2− ΔΔCT method. *Methods* **2001**, *25*, 402–408. [CrossRef] [PubMed]
46. Anderson, M.; Gorley, R.N.; Clarke, K. *PERMANOVA+ for Primer: Guide to Software and Statistical Methods*; PRIMER-e: Plymouth, UK, 2008; p. 218.

© 2019 by the authors. Licensee MDPI, Basel, Switzerland. This article is an open access article distributed under the terms and conditions of the Creative Commons Attribution (CC BY) license (http://creativecommons.org/licenses/by/4.0/).

International Journal of
Molecular Sciences

Article

Insight into the Phytoremediation Capability of *Brassica juncea* (v. Malopolska): Metal Accumulation and Antioxidant Enzyme Activity

Arleta Małecka [1,*], Agnieszka Konkolewska [2], Anetta Hanć [3], Danuta Barałkiewicz [3], Liliana Ciszewska [2], Ewelina Ratajczak [4], Aleksandra Maria Staszak [5], Hanna Kmita [6] and Wiesława Jarmuszkiewicz [6]

1 Department of Biotechnology, Institute of Molecular and Biotechnology, Adam Mickiewicz University, Collegium Biologicum, Uniwersytetu Poznańskiego 6, 61-614 Poznań, Poland
2 Department of Biochemistry, Institute of Molecular Biology and Biotechnology, Adam Mickiewicz University, Collegium Biologicum, Uniwersytetu Poznańskiego 6, 61-614 Poznań, Poland
3 Department of Trace Element Analysis by Spectroscopy Method, Faculty of Chemistry, Adam Mickiewicz University, Uniwersytetu Poznańskiego 8, 61-614 Poznań, Poland
4 Institute of Dendrology, Polish Academy of Sciences, Parkowa 35, 62-035 Kórnik, Poland
5 Plant Physiology Department, Institute of Biology, University of Bialystok, Ciołkowskiego 1J, 15-245 Bialystok, Poland
6 Department of Bioenergetics, Institute of Molecular and Biotechnology, Adam Mickiewicz University, Collegium Biologicum, Uniwersytetu Poznańskiego 6, 61-614 Poznań, Poland
* Correspondence: arletam@amu.edu.pl

Received: 13 August 2019; Accepted: 31 August 2019; Published: 5 September 2019

Abstract: Metal hyperaccumulating plants should have extremely efficient defense mechanisms, enabling growth and development in a polluted environment. *Brassica* species are known to display hyperaccumulation capability. *Brassica juncea* (Indiana mustard) v. Malopolska plants were exposed to trace elements, i.e., cadmium (Cd), copper (Cu), lead (Pb), and zinc (Zn), at a concentration of 50 µM and were then harvested after 96 h for analysis. We observed a high index of tolerance (IT), higher than 90%, for all *B. juncea* plants treated with the four metals, and we showed that Cd, Cu, Pb, and Zn accumulation was higher in the above-ground parts than in the roots. We estimated the metal effects on the generation of reactive oxygen species (ROS) and the levels of protein oxidation, as well as on the activity and gene expression of antioxidant enzymes, including superoxide dismutase (SOD), catalase (CAT), and ascorbate peroxidase (APX). The obtained results indicate that organo-specific ROS generation was higher in plants exposed to essential metal elements (i.e., Cu and Zn), compared with non-essential ones (i.e., Cd and Pb), in conjunction with SOD, CAT, and APX activity and expression at the level of encoding mRNAs and existing proteins. In addition to the potential usefulness of *B. juncea* in the phytoremediation process, the data provide important information concerning plant response to the presence of trace metals.

Keywords: oxidative stress; antioxidative system; Brassicaceae family; heavy metals

1. Introduction

Trace metal element contamination in soils is one of the world's major environmental problems, posing significant risks to human health, as well as to ecosystems ([1]). Metals such as zinc (Zn), iron (Fe), and copper (Cu) are essential micronutrients required for a wide range of physiological processes in all plant organs, and the processes are based on the activities of various metal-dependent enzymes and proteins. However, they can also be toxic at elevated levels. Metals such as arsenic (As), mercury (Hg), cadmium (Cd), and lead (Pd) are nonessential and potentially highly toxic [2]. Trace metal element toxicity includes changes in the chlorophyll concentration in leaves and damage of the photosynthetic apparatus, inhibition of transpiration, and destruction of carbohydrate metabolism, as well as nutrition and oxidative stress, which collectively affect plant development and growth [3–7].

Biological organisms are incapable of degrading metals, so they persist in their body parts and environment, leading to health hazards [8]. Metal accumulation and other abiotic stresses cause excess reactive oxygen species (ROS) generation, leading to oxidative stress [7]. Plant cells are equipped with enzymatic mechanisms to eliminate or reduce oxidative damage that occurs under metal accumulation. The antioxidative defense system includes superoxide dismutase (SOD), catalase (CAT), and ascorbate peroxidase (APX), which are regarded as responsible for maintaining the balance between ROS production and scavenging [9].

The Brassicaceae family includes many genera abundant in metallophytes, such as *Thlaspi*, *Brassica*, and *Arabidopsis*. They accumulate a wide range of heavy metals, especially Zn, Cd, nickel (Ni), thallium (Tl), chromium (Cr), and selenium (Se) [10]. The term hyperaccumulator is used for plants that accumulate 1000 mg per kg of dry matter of any above-ground tissue when grown in their natural habitat [11,12]. As of 2013, approximately 500 metal hyperaccumulator plant species were described [13,14], and the number is increasing. *B. juncea* exhibits some traits of a metal hyperaccumulator—this species can take up significant quantities of Pb, Cd [15,16], Cr, Cu, Ni, Pb, and Zn [10,17], although its translocation ability is not as efficient as shown for other known hyperaccumulators. Metal hyperaccumulating plants should have extremely efficient defense mechanisms, enabling growth and development in a polluted environment. Therefore, the objective of the present study was to estimate the contribution of the *B. juncea* (v. Malopolska) enzymatic antioxidant system to combating the oxidative stress induced by essential (Cu, Zn) and non-essential (Pb, Cd) metal elements to allow survival under adverse environmental conditions. The analysis included trace metal accumulation, level of stress parameters, and antioxidant enzyme activity, as well as estimation of encoding mRNA and enzyme protein levels.

2. Results

2.1. Levels of Metal Accumulation

Research using laser ablation combined with plasma mass spectrometry (LA-ICP-MS) made it possible to determine the levels of metal accumulation in *B. juncea* organs (Figure 1). The analyses were performed for roots, stems, and leaves. In the case of roots, Pb constituted approximately 60% of all accumulated metals. In addition, approximately 4 times higher levels of accumulated Cu and Zn, as well as more than 140 times higher levels of Cd, were found in roots compared to control plant seedlings. In the stems and leaves, high levels of Cu and Zn were observed to be approximately 20 times higher than in control plants. The data allowed for calculation of the amount of accumulated Cu, Cd, Zn, and Pb in the above-ground parts, which were 58%, 55%, 52%, and 38% higher, respectively, than the amount in the roots. The results indicate that *B. juncea* is a good accumulator of trace metals, especially Cd.

2.2. Biomass and Morphological Changes

The metals used in the research did not dramatically increase *B. juncea* (v. Malopolska) seedling biomass (Figure 2). The highest inhibition of biomass growth was observed for seedlings exposed to Cu. After 96 h of treatment, the seedling biomass was approximately 34% lower than that of control plants. The weakest effect was observed for seedlings treated with Pb, as after 96 h of treatment, the seedlings were approximately 10% lighter compared to control plants. The metals used in the study also did not appreciably inhibit the increase in root length. The value of the index of tolerance (IT), based on average root length also did not change dramatically (Figure 2). After 96 h of treatment, we observed the lowest IT value for Pb (70%) and the highest IT value for Cd, i.e., 90,4%. We observed the occurrence of necrotic spots on leaves and the inhibition of leaf blade surface growth with respect to control seedlings in the above-ground parts of seedlings. Moreover, in Cd-treated seedlings, leaves were slightly twisted, whereas Cu caused strong chlorosis and shortening of the end of leaves. The smallest morphological changes were observed for seedlings treated with Zn.

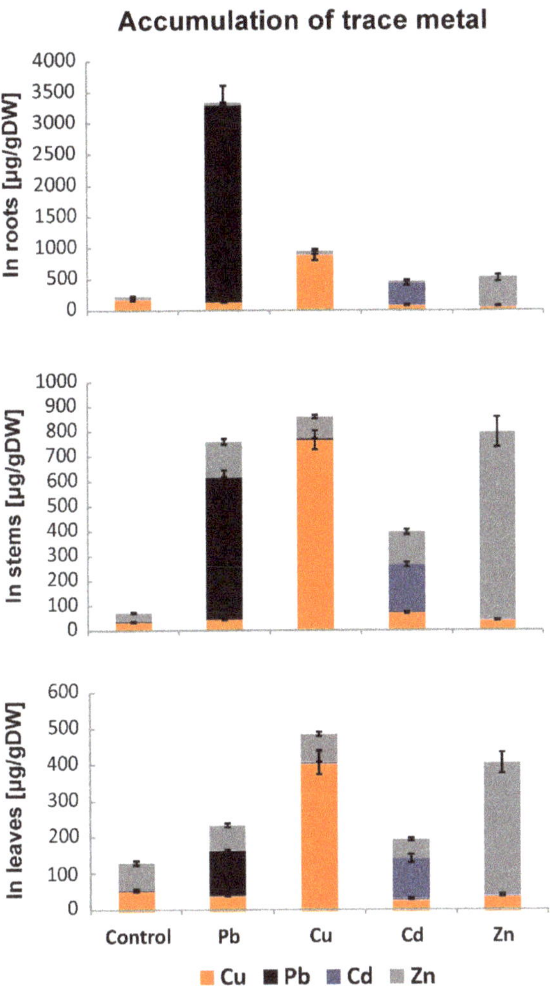

Figure 1. Accumulation of Pb, Cu, Cd, and Zn in the roots, stems, and leaves of *B. juncea* var. Malopolska seedlings grown in Hoagland's medium and treated with lead, cooper, cadmium, and zinc ions. Metal solutions $Pb(NO_3)_2$, $CuSO_4$, $CdCl_2$, and $ZnSO_4$ were applied at a 50 µM concentration. Mean values of three replicates (±SD).

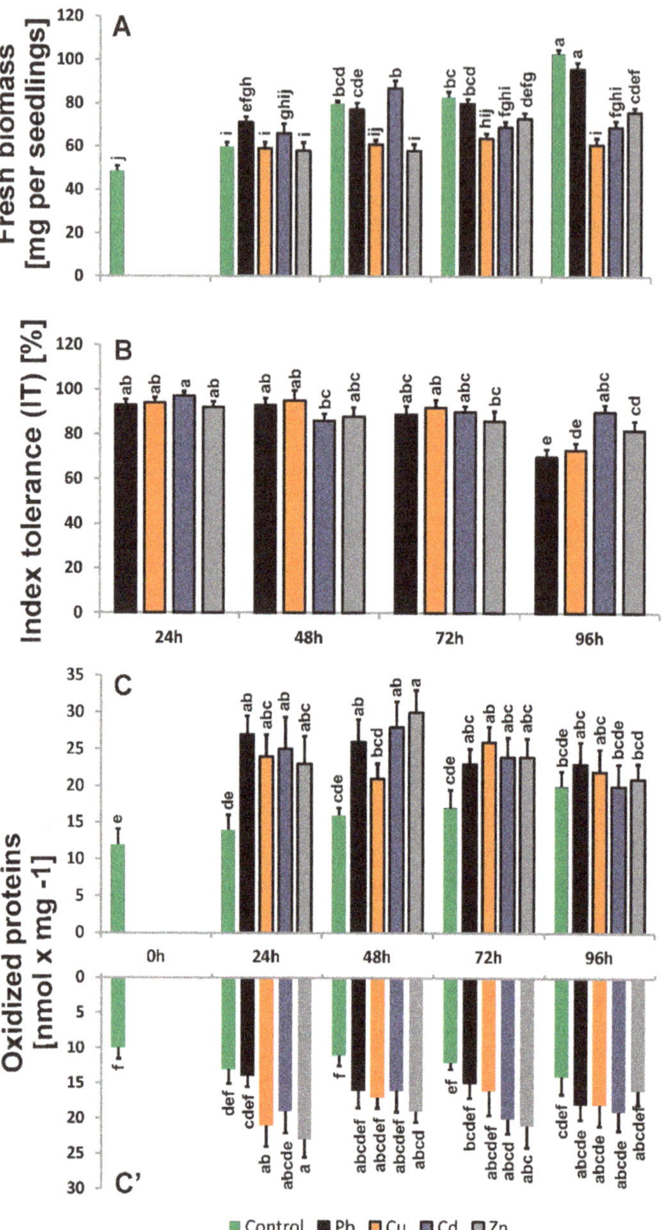

Figure 2. Stress parameters in *B. juncea* seedlings treated with trace metals: Pb, Cu, Cd, and Zn. The results are expressed as the mean ± standard deviation ($n = 3$). Metal solutions $Pb(NO_3)_2$, $CuSO_4$, $CdCl_2$, and $ZnSO_4$ were applied at a 50 μM concentration. Mean values of three replicates (±SD).

2.3. Production and Localization of ROS

The metal-treated seedlings increased $O_2^{\cdot-}$ production at levels comparable for shoots and roots compared to control seedlings, but the fluctuation in the production observed for control plants was maintained (Figure 3). In the roots, the highest values were mainly observed in the first 72 h (over 30%), whereas in the above-ground parts, the highest values were observed for 48 h (over 30–40%). After 96 h, the levels of $O_2^{\cdot -}$ decreased, which may indicate high activity of the SOD enzyme. The highest level of $O_2^{\cdot -}$ in roots was observed for plants treated with Zn compared with shoots treated with Zn and Cd.

The profile of the changes in the H_2O_2 level was similar for control roots and shoots, but the levels were distinctly higher in roots. The highest H_2O_2 amount was observed in roots treated with Cu, Cd, and Zn. For metal-treated samples, a significant increase in H_2O_2 occurred between 48 and 72 h of treatment, and the observed profile of H_2O_2 changes was more homogenous for shoots. We noticed a large difference in the level of H_2O_2 in roots after 96 h of treatment, reaching approximately 20–50% higher compared to the control. As in the case of $O_2^{\cdot -}$, H_2O_2 levels were also confirmed by confocal microscopy (Figure 4). The most intensive fluorescence DHE, indicating the presence of $O_2^{\cdot -}$, was observed for the *B. juncea* roots treated for 24 h with 50 µM Cd and Zn. The highest amount of H_2O_2 generated was observed in roots treated with 50 µM Cu, Cd, and Zn (Figure 5).

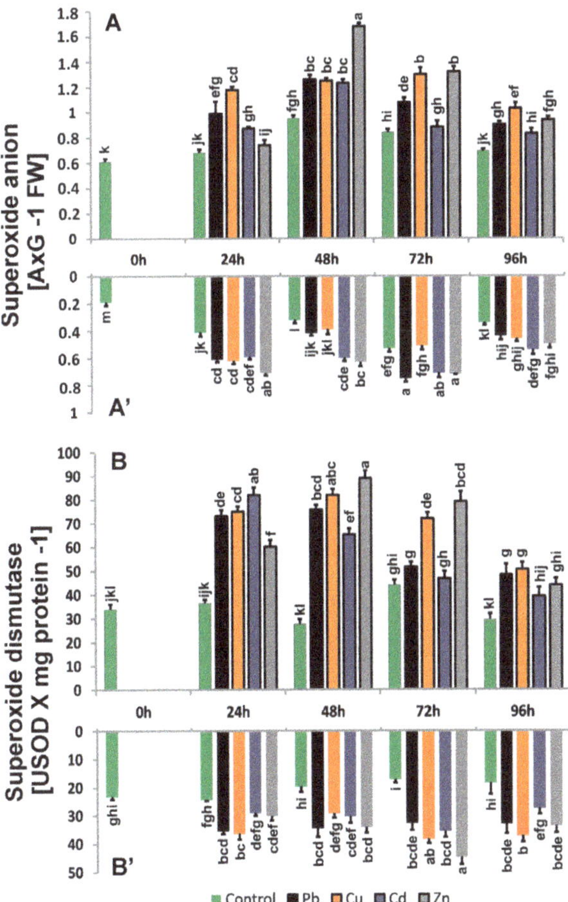

Figure 3. Superoxide anion (A580 g^{-1} FW) level and SOD (USOD mg^{-1} protein) activities in roots and above-ground parts of *B. juncea* var. Malopolska seedlings grown in Hoagland's medium and treated with lead, cooper, cadmium, and zinc ions. Metal solutions $Pb(NO_3)_2$, $CuSO_4$, $CdCl_2$, and $ZnSO_4$ were applied at a 50 µM concentration. Mean values of three replicates (±SD).

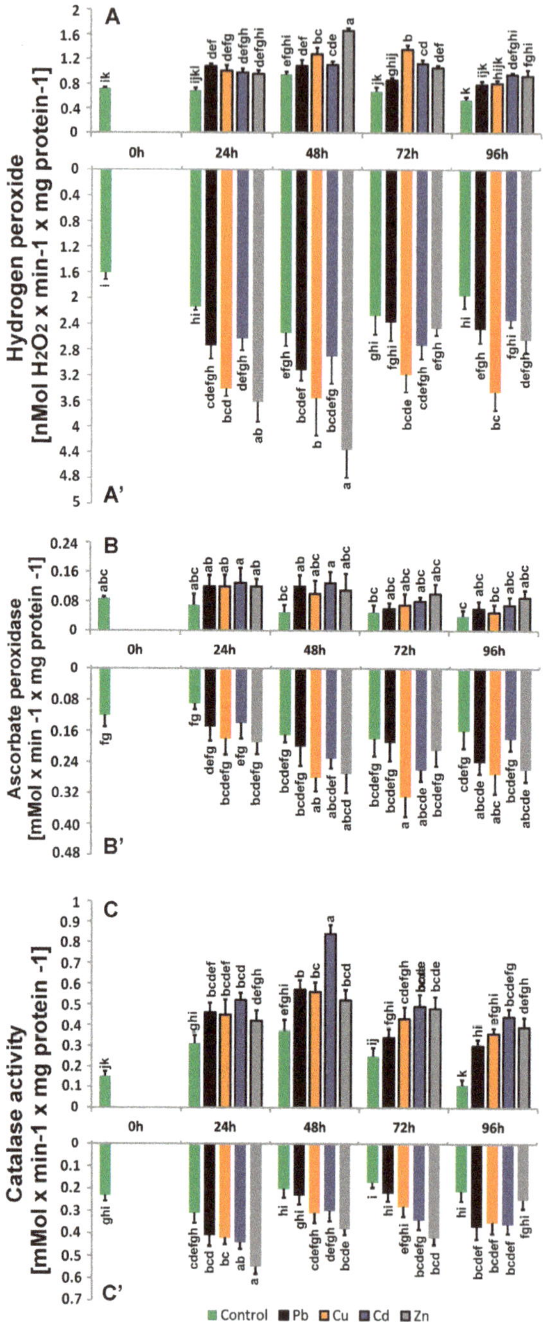

Figure 4. Hydrogen peroxide level (nMol $H_2O_2 \times min^{-1} \times$ mg protein^{-1}), CAT (μMol min^{-1} mg^{-1} protein), and APX (μMol $\times$ min^{-1} $\times$ mg protein^{-1}) activities in roots and above-ground parts of *B. juncea* var. Malopolska seedlings grown in Hoagland's medium and treated with lead, cooper, cadmium, and zinc ions. Metal solutions $Pb(NO_3)_2$, $CuSO_4$, $CdCl_2$, and $ZnSO_4$ were applied at a 50 μM concentration. Mean values of three replicates (±SD).

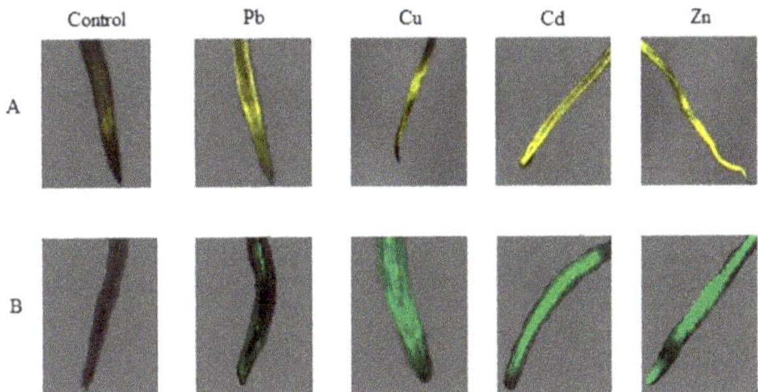

Figure 5. Trace metals-induced $O_2^{-\cdot}$ and H_2O_2 production in *B. juncea* var. Malopolska roots. Fluorescent images of *B. juncea* roots grown in Hoagland's medium in the presence of 50 μmol of $Pb(NO_3)_2$, $CuSO_4$, $CdCl_2$, and $ZnSO_4$ for 24 h and control roots of plants stained with DHE for 12 h (**A**) and DCFH-DA for 4 h (**B**). The bar indicates 1 μm.

2.4. Levels of Oxidized Proteins

The levels of protein oxidative modification imposed by the metal treatment were 12–44% higher for roots and above-ground parts compared to control plants (Figure 1). The level of oxidized proteins reached a maximum after 48 h and was 3-fold higher than in the shoots of control plants.

2.5. Enzyme Antioxidant Activity

SOD activities were 25–50% higher in the roots of plants treated with trace metals. In the above-ground parts, greater differences in SOD activity between research variants, ranging from 8 to 70%, were observed. However, the general activity of SOD was higher in roots and shoots compared to control seedlings (Figure 3) and changed differently for the seedling parts. In the case of roots, the activity level and profile were comparable to those of control seedlings, whereas for shoots, after the initial increase, the activity decreased significantly after 96 h. The generation of H_2O_2 caused a rapid increase in CAT activity within 24 h of cultivation, i.e., from 30% to 70% in the roots of plants treated with trace metals, especially in plants treated with Zn (Figure 4). In the next days, we observed a slight decrease (approximately 12–55%), but this decrease remained higher than that in control plants. The highest CAT activity was observed above ground in the first 48 h of cultivation (56%) in plants exposed to Cd. Activities of APX, a second enzyme involved in the dismutation of hydrogen peroxide, systematically increased in roots exposed to metals during the cultivation period, especially in plants grown in the presence of Cu and Zn, which had approximately 10–43% higher levels than those observed in the control (Figure 4). In the above-ground parts of *B. juncea* cultured in the presence of trace metals, we observed an increase in the intensity of APX during the first 48 h, reaching a maximum in plants treated with Cd for 48 h, approximately 62% higher than in the control, and then a slight decrease, but the activities were approximately 2-fold higher than those in control plants. The activity profiles of CAT and APX differed between the control roots and shoots (Figure 4). The metal treatment increased the activity of both enzymes, and the CAT activity profile appeared to be maintained in roots and shoots. However, the APX profile did not differ from that of the control plants with respect to treated shoots, whereas in treated roots, the APX activity profile was variable and metal-dependent, although comparable for Cu and Zn.

2.6. Levels of Gene Transcripts

To estimate possible changes at the level of CuZnSOD and MnSOD encoding gene transcripts, we used an electrophoretic separation technique and the CpAtlas program (Figure 6). In the case of CuZnSOD, a decrease in the expression of the gene encoding CuZnSOD was observed in the roots of plants treated with trace metals after 4 h and 24 h of cultivation, with the exception of the roots of

B. juncea-treated Cu. Induction of the gene in the above-ground parts was visible, with an approximate 2-fold increase in the level of the transcript in plants after 8 h of copper treatment and an approximate 2-fold decrease in plants after 4 h of zinc treatment. The results indicate that the presence of cadmium ions had no significant effect on the induction of CuZnSOD gene expression because no significant changes in the level of the transcripts was observed in either the roots or above-ground parts of *B. juncea* plants.

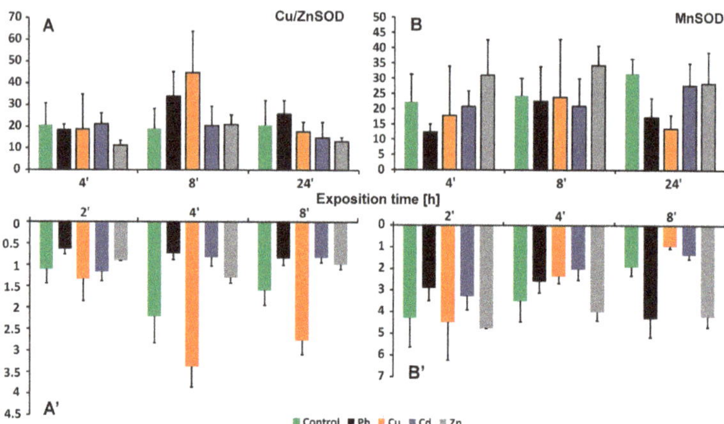

Figure 6. Transcriptional levels of genes encoding antioxidative enzymes in roots and above-ground parts of *B. juncea* var. Malopolska seedlings grown in Hoagland's medium and treated with lead, cooper, cadmium, and zinc ions. Metal solutions $Pb(NO_3)_2$, $CuSO_4$, $CdCl_2$, and $ZnSO_4$ were applied at a 50 µMol concentration. Enzymes chosen for the experiment were amplified using semi-quantitative RT-PCR with primers designed for *Arabidopsis thaliana* genes: *CSD1* for CuZnSOD and *MSD1* for MnSOD.

When analyzing changes in the expression of the gene encoding MnSOD, a decrease in the expression was observed in the roots and above-ground parts of plants after 4 h of treatment with lead ions; in the remaining research variants, there were no significant differences in transcript levels compared to control plants. An approximate 2-fold increase in the level of the transcript was found in plant roots after 24 h of Pb and Zn treatment in comparison to the control. The greatest decrease in expression was observed after 24 h in the above-ground parts of plants treated with Cu, which was almost 5-fold higher than that in the control (Figure 6).

2.7. Identification of Enzyme Forms

To distinguish between the enzyme forms, Western blot analysis was performed for protein extracts from roots and above-ground seedling parts in the absence and presence of the metal treatment (Figure 7). This allowed for the detection of MnSOD (25 kDa) and CuZnSOD (15 kDa and 20 kDa) subunits. The obtained signal was similar for both the treated and control seedlings. Thus, the metal presence likely did not change the levels of the CuZnSOD and MnSOD proteins.

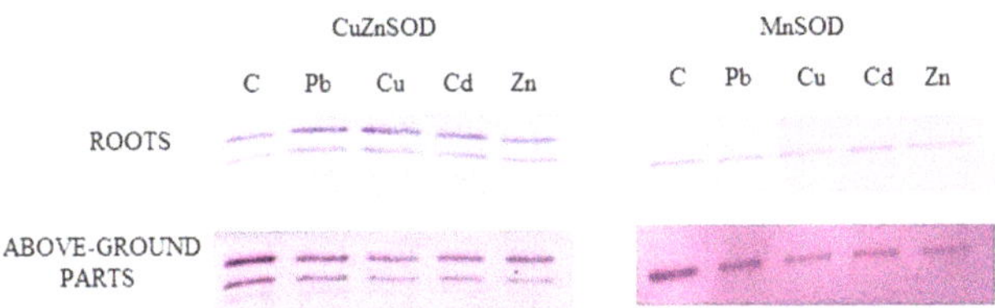

Figure 7. Effects of 50 µM Pb, Cu, Cd, and Zn for 24 h on the CuZnSOD and MnSOD of roots and above-ground parts of *B. juncea* var. Malopolska seedlings. The protein content was evaluated by Western blot using specific antibodies.

3. Discussion

Trace metals are one of the most important abiotic stress factors affecting the natural environment. As a result of anthropogenic activities, we can observe their increasing levels from year-to-year. Metal toxicity results in effects at physiological and cellular levels, leading to distorted metabolism, including plant metabolism [18]. Abiotic stresses, including the presence of trace metals in soil, are estimated to be the main cause of global crop yield reduction of ca. 70% and thus are considered a great constraint to crop production. This situation has worsened due to disturbed equilibrium between crop production and human population growth. Therefore, it is especially important to understand plant responses to such stress factors. This also applies to trace metals [12]. In the present study, this was clearly visible in the growth of plant biomass, which significantly decreased during the culture in the presence of heavy metals. Copper and zinc ions are essential for the normal growth and development of all organisms but can be toxic to plants at excessive levels. Lead and cadmium are nonessential elements and are toxic to plants even at low levels [8]. Essential and nonessential trace elements, when exceeding the threshold limits, can cause different physiological, morphological, and genetic plant anomalies, including reduced growth, mutations, and increased mortality [8]. Therefore, plants suitable for phytoremediation are, at present, of great importance.

In our study, we noticed that in the case of *B. juncea* v. Malopolska, all the mentioned metals used at 50 µM concentration displayed moderate phytotoxic properties. The biomass increments ranged between 96 mg for Pb-treated plants and 61 mg for Cu-treated plants, and the values were approximately 7% and 41% lower, respectively, than those in control plants. Several studies have shown that high concentrations of trace metals in the soil cause plant growth impairment [9,19]). In *Sesbania drummondii*, a reduction in seedling biomass was caused by Pb—21%, Cu—46.3%, Ni—31.5% and Zn—25.2% [20]. The inhibition of shoot growth by trace metals may be due to a decrease in photosynthesis, as trace metals disturb mineral nutrition and water balance, change hormonal status, and affect membrane structure and permeability [21]. Trace metals might cause an inhibition of root growth that alters water balance and nutrient absorption [12] and decrease calcium uptake in root tips, leading to a decrease in cell division or cell elongation [9,22,23]. According to Marshner [23], Cd-induced mineral stress can reduce plant dry weight accumulation. Other authors have shown a negative influence of Pb [24], Cu [25], Cd [26], and Zn [20]. Despite the inhibitory effect caused by trace metals on the growth of the biomass of *B. juncea*, we observed a high IT amounting to approximately 90% resistance of the plants to trace metals.

The bioaccumulation of trace metals is different for various plant species, reflected by their growth, reproduction, occurrence, and survival in metal-contaminated soil, because the mechanisms of elemental uptake by plants are not the same for all species. The capacity of plants to take up trace metals is different for different metals, and the same trace metal can be accumulated at different ratios in different plant species [27]. Metal bioavailability is also affected by the presence of organic compounds of that metal in plants [8]. The ICP-MS results we obtained indicate that the accumulation of trace metals was higher in above-ground parts than in roots, especially for cadmium, lead, and zinc. The metal concentrations followed an order of Pb > Cu > Zn > Cd in roots, Zn > Cu > Pb > Cd in the stem, and Zn > Cu > Cd > Pb in leaves [28]. Based on the obtained results, it can be concluded that *B. juncea* is a hyperaccumulator of Cd, Zn, and Pb. Cherif and co-authors [29] reported that Zn induced a decrease in Cd uptake and a simultaneous increase in Zn accumulation, indicating a strong competition between these two metals for the same membrane transporters. In our earlier study [28] in *B. juncea* plants treated with a binary combination of metals, namely, PbCu, PbCd, PbZn, CuZn, CuCd, and ZnCd, at a concentration of 25 μM of each, a synergistic response between Zn and Pb was observed, resulting in an increased accumulation of the two metals. The accumulation results obtained for plants treated with Cu are different from those of other researchers. Purakayastha and others [30] showed that Cu is accumulated mainly in above-ground parts of *B. juncea*. This difference may result from different exposure durations of the plant to the metal, other metal concentrations, and different plant ages at the time of analysis of the collected metal. Quaritacci et al. [31] reported that *B. juncea* was identified as a species able to take up and accumulate metals in its above-ground parts, such as Cd, Cu, Ni, Zn, Pb, and Se. It has been observed that this species concentrated Cu, Pb, and Zn in its above-ground parts in amounts much higher than those detected in the metal soluble fractions present in a soil contaminated by acidic water and pyritic slurry [31].

The accumulation of trace metals in organs is dangerous for plants. In an earlier study [32], we confirmed that plants are not adequately protected by the detoxification system because trace metals penetrate in areas with high metabolic activity, such as the cytoplasm, mitochondria, or cell membrane.

The occurrence of oxidation stress conditions in *B. juncea* treated with the trace metals Pb, Cu, Cd, and Zn was confirmed by the increase in the level of oxidized proteins in the roots (approximately 7–12%) and above-ground parts (approximately 13%). Several metals, including Cd, Pb, and Hg, have been shown to cause protein oxidation by depletion of protein thiol groups [33]. ROS cause protein modifications through the formation of carbonyl groups at certain amino acid residues. Such modifications were caused by the presence of heavy metals, e.g., cadmium [34], mercury lead, aluminum, zinc, copper, cobalt, nickel, and chromium [35].

ROS also act as signaling molecules involved in the regulation of many key physiological processes, such as root hair growth, stomatal movement, cell growth, and cell differentiation, when finely tuned and regulated by an antioxidative defense system [12]. We showed an increase in the level of ROS compared to control plants in all plants treated with heavy metals. The $O_2^{\cdot-}$ rate after 2 h of culture was 2 times higher than that observed in plants grown under control conditions. The high level of $O_2^{\cdot-}$ was the highest between 24 to 72 h of the treatment depending on the research variant. The highest value of O_2^{-} was measured in plants treated with Zn, while the highest H_2O_2 values were observed in plants treated with Cu and Cd. Similar results were obtained by other researchers. Markovska et al. [36] showed a 10-fold higher level of H_2O_2 in the leaves of *B. juncea* after 5 days of treatment with Cd ions at a concentration of 50 μM. Wang et al. [37] observed the highest levels of H_2O_2 in *B. juncea* roots treated with Cu ions for 4 days. In our research, the highest level of H_2O_2 was obtained after 4 days in plants treated with single metals. The reduction of $O_2^{\cdot-}$ and the H_2O_2 content in roots and above-ground parts of plants treated with trace metals during the cultivation period suggests that some antioxidative enzymes would work effectively in the removal of ROS. To detect ROS in plant cells, we used incubation with fluorescent labels such as 2′7′-difluoroscein and dihydroethidium and imaging under confocal microscopy. We observed increased generation of $O_2^{\cdot-}$ and H_2O_2 in the roots of *B. juncea* treated with trace metals—especially Cd and Zn for $O_2^{\cdot-}$, and Cu, Cd, and Zn for H_2O_2.

The increase in ROS production in metal-treated plants was precisely associated with changes in the activity of antioxidant enzymes. We always observed the induction of antioxidant enzyme activity in *B. juncea* roots and leaves, although there were no significant differences between the used metals. We observed increasing activity of antioxidant enzymes, i.e., 20–158% for SOD, 15–147% for CAT, and 6–68% for APX. The highest activity of SOD in both roots and shoots was observed in plants treated with Zn and Cu. The first line of defense against ROS-mediated toxicity is through SOD, which catalyzes the dismutation of superoxide anions to H_2O_2 and O_2. The stimulation of SOD activity has also been reported in several plants exposed to Pb, Cu, Cd, Zn, Ni, and as ions [20,25,38,39]. We noticed that in the roots of *B. juncea*, the most induced activity of CAT was for Zn, compared with Cd in the above-ground parts. APX was definitely lower than catalase, especially in the above-ground parts, which means that this enzyme complements CAT catalytic activity. APX activity was significantly elevated in the metal-treated plants, which suggests its role in the detoxification of H_2O_2. Enhanced CAT and APX activity has been observed in various plant species after application of trace metals: Pb, Cu, Cd, Zn, Ni, and As [20,25,38–40]. APX may be responsible for controlling the levels of H_2O_2 as signal molecules, and the CAT function is to remove large amounts of oxygen during oxidative stress. APX may be responsible for controlling the levels of H_2O_2 as signal molecules, and the CAT function is to remove large amounts of oxygen during oxidative stress [41]. Mohamed et al. [42] showed in *B. juncea* that the higher activity of antioxidant enzymes offers a greater detoxification efficiency, which provides better plant resistance against trace metal-induced oxidative stress. Yadav and co-authors [25] reported increases in the activities of antioxidant enzymes: SOD by 16.2%, DHAR by 27–58%, GR by 35.74%, GST and GPX by 19.19%, and APX by 42.75% in *B. juncea* plants treated with 0.0005 M Cu. The authors indicated that brassinostereoids can regulate the activity of the antioxidant system and help in scavenging overproduced ROS, and can provide tolerance by inducing the expression of regulatory genes such as respiratory burst oxidase homologue, mitogen-activated protein kinase-1, and mitogen-activated protein kinase 3, as well as activating genes involved in antioxidative defense and responses [25]. Other authors [12] have noted that brassinosteroids are a group of hormones that regulate ion uptake in plant cells and reduce trace metal accumulation in plants. An exogenous application of brassinosteroids is widely used to improve crop yield, as well as stress tolerance, in various plant species.

We previously demonstrated an increase in the activity of the antioxidant system at the physiological and biochemical levels. The next step was to determine whether trace metals influence the transcription level of genes encoding suitable defense proteins. ROS concentration at an appropriate level can promote plant development and reinforce resistance to stressors by modulating the expression of a set of genes and redox signaling pathways [12]. In our research, we observed differences in the expression induction depending on the exposure time and the metal used. We observed an increase in the level of the gene coding for CuZnSOD in plants treated with copper, zinc, and lead. The highest level of expression was obtained after 4 h in roots and 8 h in above-ground parts. Romero-Puertas and co-authors [34] noted a drastic reduction in the expression of genes coding for CuZnSOD and no changes in MnSOD in *Pisum sativum* under conditions of stress caused by the presence of Cd. Their results showed a reduction in CuZnSOD levels in the presence of Cd, while in our study, we did not observe significant differences in the level of transcript for plants treated with this metal in relation to control plants. We observed the induction of gene expression encoding MnSOD in *B. juncea* roots after 8 h of exposure to Zn and Pb ions, compared with lead ions in above-ground parts. Other authors did not observe any changes or a low expression of genes coding for SOD, e.g., Fidlago et al. [43] showed no differences in MnSOD-related mRNA accumulation in leaves and roots, but CuZnSOD-related transcripts decreased in leaves but did not change in roots in Cd-treated *Solanum nigrum* L. Others authors [44] indicated that Cd stress induced an upregulated expression of FeSOD, MnSOD, Chl CuZnSOD, Cyt CuZnSOD, APX, GPX, GR, and POD at 4–24 h after treatment began for *Lolium perenne* L., and their results suggested that the gene transcript profile was related to the enzyme activity under Cd stress. Romero-Puertas et al. [34] indicated two groups of genes in pea plants treated with Cd. First, some elements of the signal transduction cascade accentuated or attenuated the Cd effect on CAT, MDHAR, and CuZnSOD mRNA expression. The second was formed by the genes MnSOD, APX, and GR that were not affected by these modulators during the Cd treatment because their expression was not modified compared to control plants.

The effect of Cd on the expression of CuZnSOD was reversed by a nitric oxide (NO)· scavenger, indicating that NO· must be a key element in the regulation of this SOD, showing the existence of a relationship between an increase in ROS production and NO. NO-dependent downregulation was also observed for MnSOD, while the opposite effect was found for APX and GR. This suggests that protein phosphorylation is involved in the response to Cd stress [34]. Bernard and co-authors [45] indicate that molecular analysis (gene expression) is the first level of integration of environmental stressors, and it is supposed to respond to stressors earlier than biochemical markers.

Our results from Western blotting indicate that the presence of trace metals does not increase the synthesis of the proteins CuZnSOD and MnSOD in the organs of *B. juncea* plants, but induces an increase in their activity.

4. Materials and Methods

4.1. Plant Material

Brassica juncea v. Malopolska seeds were grown in Petri dishes for 7 days under optimal conditions. Next, seedlings were cultivated hydroponically on Hoagland's medium for 7 days in a growth room with a 16/8 h photoperiod, day/night at room temperature and light intensity of 82 μmol m^2 s^{-1}. Then, the applied medium was changed into 100×-diluted Hoagland's medium and a heavy metal solution in combination; Cu, Pb, Cd, and Zn ions at a concentration of 50 μM were applied. In the cultivation, a solution of $Pb(NO_3)_2$, $CuSO_4$, $CdCl_2$, Zn SO_4 was used. The roots and shoots were cut off after 0, 24, 48, 72, and 96 h of cultivation. The roots were dipped sequentially in cold solutions of 10 mM $CaCl_2$ and 10 mM EDTA for 5 min each to eliminate trace elements adsorbed at the root surface. Then, roots and shoots were rinsed three times with distilled water, frozen in liquid nitrogen, and stored at −80 °C until molecular analysis.

4.2. Phytotoxic Test

The index of tolerance (IT) was calculated according to Wilkins [46]:

$$IT = \frac{\text{average length of roots in tested solution}}{\text{average length of roots in control}} \times 100\% \quad (1)$$

The changes in fresh biomass of control plants and plants treated with metals were measured on a Radwag scale after 0, 24 28, 72, and 96 h of cultivation.

4.3. Accumulation of Trace Metals

The determination of trace metal accumulation was performed using inductively coupled plasma mass spectrometry (ICP-MS) model Elan DRC II, (Perkin Elmer Sciex, Concord, Ontario, Canada) connected with laser ablation (LA) model LSX-500 (CETAC Technologies, Omaha, NE, USA). Plant material (roots, stems, and leaves) was rinsed with distilled water, gently dried on blotting paper, weighed, and dried at 70 ± 2 °C. The dried samples were mineralized in an MDS-2000 microwave digestor oven (CEM Corporation Matthews, NC, USA). A three-stage dilution was conducted in a closed system using 5 mL of 65% HNO_3. After mineralization, samples were transferred to 10 mL flasks filled with deionized water. An ICP-MS was used to determine the concentration of elements in the mineralized plant tissues.

Plant roots, stems, and leaves were collected after 72 h of treatment for the analysis of metal distribution. Samples were cut into 3 mm long pieces and ablated along the pre-defined line across the cross-sections. Laser performance was optimized according to a detailed scheme [47] using a single variable method.

4.4. Superoxide Anion Determination

The superoxide anion content was determined according to Doke [48]. *B. juncea* roots (0.5 g) were placed in the test tubes that were filled with 7 mL of a mixture containing 50 mM phosphate buffer (pH 7.8), 0.05% NBT (nitro blue tetrazolium), and 10 mM of NaN_3. Next, the test tubes were incubated in the dark for 5 min, and then 2 mL of the solution was taken from the tubes, heated at 85 °C for 10–15 min, cooled on ice for 5 min, and the absorbance was measured at 580 nm against the control.

4.5. Hydrogen Peroxide Content

The hydrogen peroxide content was determined using the method described by Patterson et al. [49]. The decrease in absorbance was measured at 508 nm. The reaction mixture contained 50 mM phosphate buffer (pH 8.4) and reagents, 0.6 mM 4-(-2 pyridylazo)resorcinol, and 0.6 mM potassium-titanium oxalate (1:1). The corresponding concentration of H_2O_2 was determined against the standard curve of H_2O_2.

4.6. In Situ Detection of Superoxide Anion and Hydrogen Peroxide

The roots and shoots from plants exposed to metals for 24 h were submerged for 12 h in 100 µM of $CaCl_2$ containing 20 µM of dihydroethidium (DHE, pH 4.75; samples for superoxide anion radicals) or 4 µM dichlorodihydrofluorescein diacetate (DCFH-DA) (samples for hydrogen peroxide) in 5 mM dimethyl sulfoxide (DMSO). After rinsing with 100 µM of $CaCl_2$ or 50 mM phosphate buffer (pH 7.4), the roots and shoots were observed with a confocal microscope (Zeiss LSM 510, Axiovert 200 M, Jena, Germany) equipped with no. 10 filter set (excitation 450–490 nm, emission 520 nm or more).

4.7. Estimation of Protein Oxidation

For carbonyl quantification, the reaction with DNPH was used basically as described by Levine et al. [50]. For each determination, two replicates and their respective blanks were used. Roots and shoots (0.5 g) were incubated with isolation buffer containing 0.1 M Na-phosphate buffer, 0.2% (*v/v*) Triton X—100, 1 mM EDTA, and 1 mM PMSF. After centrifugation at 13,000× *g* for 15 min, supernatants (200 µL) were mixed with 300 µL of 10 mM DNPH in 2 M HCl. The blank was incubated in 2 M HCl. After 1 h incubation at room temperature, proteins were precipitated with 10% (*w/v*) trichloroacetic acid (TCA), and the pellets were washed three times with 500 µL of ethanol/ethylacetate (1:1). The pellets were finally dissolved in 6 M guanidine hydrochloride in 20 mM potassium phosphate buffer (pH 2.3), and the absorption was measured at 370 nm. Protein recovery was estimated for each sample by measuring the absorption at 280 nm. The carbonyl content was calculated using the molar absorption coefficient for aliphatic hydrazones, 22,000 M^{-1} cm^{-1}.

4.8. Determination of Antioxidant Enzyme Activities

The activity of SOD was assayed by measuring its ability to inhibit the photochemical reduction of NBT, adopting the method of Beauchamp and Fridovich [51]. The reaction mixture contained 13 µM riboflavin, 13 mM methionine, 63 mM NBT, and 50 mM potassium phosphate buffer (pH 7.8). Absorbance at 560 nm was then measured. One unit of SOD activity has been defined as the amount of enzyme that causes a 50% decrease in the inhibition of NBT reduction. The activity of CAT was determined by directly measuring the decomposition of H_2O_2 at 240 nm for 3 min as described by Aebi [52] in 50 mM phosphate buffer (pH 7.0) containing 5 mM H_2O_2 and enzyme extract). CAT activity was determined using the extinction coefficient of 36 mM^{-1} cm^{-1} for H_2O_2. The activity of APX was assayed using the method described by Nakano and Asada [53] by monitoring the rate of ascorbate oxidation at 290 nm (extinction coefficient of 2.9 mM^{-1} cm^{-1}) for 3 min. The reaction mixture consisted of 25–50 µL supernatant, 50 mM phosphate buffer (pH 7.0), 20 µM H_2O_2, 0.2 mM ascorbate, and 0.2 mM EDTA.

4.9. Isolation of Total RNA and RT-PCR

Roots and above-ground parts (100 mg) of *B. juncea* plants in the presence of trace metals and under control conditions were collected for total RNA isolation. The RNA was isolated with TRIzol reagent and tested spectrophotometrically for purity at 260 and 280 nm. Then, RNA was reverse-transcribed with oligo (dT) primers using the RevertAid Reverse Transcriptase Kit (Thermo Science, Lithuania, European Union) after DNA was treated with DNase I (Thermo Science).

Primer pair sequences were as follows (forward/reverse, gene accession number): gtgattgcttgca gggtttt/cagaatacggaagcaaatgtca, X54844.1 (TUB1), ggagcaagtttggttccatt/aaggttattcggccagattg, U30841.1 (MnSOD), gaacaatggtgaaggctgtg/gtgaccacctttcccaagat M63003.1 (CuZnSOD).

As a reference gene, the gene encoding tubulin was used. PCRs were performed with 30 (BJMnSOD) and 34 (BjCuZnSOD) cycles of denaturation, 95 °C for 30 s; annealing primers, 53 °C for 30 s; and elongation, 72 °C for 30 s using a 1:100 diluted cDNA template and REDAllegro*Taq* DNA Polymerase (Novazym, Poznań, Poland).

PCR products were separated by electrophoresis on a 1.3% agarose gel with ethidium bromide in TBE (445 mM Tris-HCL; 445 mM boric acid; 10 mM EDTA; pH 8.0), visualized under UV light and photographed using the Photo Print 215SD V.99 Vilber Lourmat Set. CP Atlas 2.0 were used for densitometric analysis of relative gene expression.

4.10. Western Blot

RIPA buffer (150 mM NaCl, 1% Triton X-100, 0.5% Na deoxycholate, 0.1% SDS, 50 mM Tris, pH 8.0) was used to lyse cells for protein extraction. The protein concentrations were determined using the Bradford method, and 20 µg of each extract was loaded onto a 12% SDS–PAGE (sodium dodecyl sulfate–polyacrylamide gel electrophoresis) gel. Separated proteins were transferred to polyvinylidene fluoride membrane (ImmobilonTM-P, Millipore, Burlington, MA, USA) at 350 mA for 1 h using the Mini Trans-BlotCell (Bio-Rad, Hercules, CA, USA). Membranes were blocked with 1% BSA and incubated with an antibody against CuZnSOD at a final dilution of 1:2500. The secondary antibody, goat anti-rabbit IgG conjugated with alkaline phosphatase (Sigma-Aldrich, St Louis, MO, USA), was used at a 1:3000 dilution to visualize protein bands by reaction with 5-bromo-4-chloro-3-indolyl phosphate/nitroblue tetrazolium (BCIP/NBT) (Sigma-Aldrich, St Louis, MO, USA/CALBIOCHEMV.S. and Canada) as a substrate.

4.11. Protein Quantification

Total soluble protein contents were determined according to the method of Bradford [54] using the Bio-Rad assay kit with bovine serum albumin as a calibration standard.

4.12. Statistical Analyses

Each experiment was performed in three biological and technical replicates. The mean values ± SE are given in the tables and figures. The data were analyzed statistically using IBM SPSS Statistics (Version 22 for Windows). Significant differences among treatments were analyzed by one-way ANOVA, taking $p < 0.05$ as the significance threshold, and the b-Tukey post-hoc test was conducted for pairwise comparisons between treatments.

5. Conclusions

This study was conducted to determine the interactive role of Pb, Cu, Cd, and Zn in metal uptake, plant growth, and the antioxidative system of *B. juncea*. Plants accumulated high amounts of trace metals, i.e., more than 40% in the roots, and in the above-ground parts, the values for Cu, Cd, Zn, and Pb were 58%, 55%, 52%, and 38%, respectively. The results suggest that *B. juncea* var. Malopolska is a good hyperaccumulator of trace metals, especially Cu, Cd, and Zn, and can be useful in phytoremediation. The presence of metals resulted in a considerable reduction in *B. juncea* biomass; the highest reduction was observed in plants treated with Cu and Cd. Despite the visible influence of trace metals on plant morphology, the IT coefficient was high and exceeded 90%, indicating the high resistance of *B. juncea* plants. Trace metals lead to the production of ROS, which causes an imbalance in the redox state in the plant cells and increases the level of oxidized proteins. We noticed that under the conditions of oxidative stress, the antioxidant system was activated: SOD, CAT, and APX. We observed that the presence of metals influenced the increase in the activity of antioxidant enzymes, while no significant differences were observed in the levels of CuZnSOD and MnSOD transcripts and proteins.

The results obtained indicate that *B. juncea* var. Malopolska has efficient defense mechanisms to cope with different metals.

Author Contributions: A.M. conceived and designed the experiments, performed part of analysis, wrote-original draft, review and editing, A.K. performed the molecular experiments, A.H. performed analysis LA-ICP-MS and ICP-MS, analyzed the data from ICP-MS D.B. analyzed the data from LA-ICP-MS, L.C. performed biochemical tests, E.R. perfomed immunological research, A.M.S. graphical analysis of results, H.K. manuscripts edition, W.J. substantive care.

Funding: This work was partially supported by the National Science Centre no. N N305 381138. AH was supported by the National Science Center in Poland under the grant number 2017/01/X/ST4/00373

Conflicts of Interest: The authors declare no conflicts of interest.

References

1. Chen, Z.; Zhao, Y.; Gu, L.; Wang, S.; Li, Y.; Dong, F. Accumulation and Localization of Cadmium in Potato (*Solanum tuberosum*) Under Different Soil Cd Levels. *Bull. Environ. Contam. Toxicol.* **2014**, *92*, 745–751. [CrossRef]
2. Dalvi, A.A.; Bhalerao, S.A. Response of Plants towards Heavy Metal Toxicity: An overview of Avoidance. Tolerance and Uptake Mechanism. *Ann. Plant Sci.* **2013**, *2*, 362–368.
3. Molas, J. Changes of chloroplast ultrastructure and total chlorophyll concentration in cabbage leaves caused by excess of organic Ni (II) complexes. *Environ. Exp. Bot.* **2002**, *47*, 115–126. [CrossRef]
4. Kramer, U.; Clemens, S. Functions and homeostasis of zinc, copper and nickel in plants. In *Molecular Biology of Metal Homeostasis and Detoxification*; Topics in Current Genetics; Springer: Berlin/Hiedelberg, Germany, 2005; Volume 14, p. 215.
5. Bhardwaj, P.; Chaturvedi, A.K.; Prasad, P. Effect of Enhanced Lead and Cadmium in soil on Physiological and Biochemical attributes of *Phaseolus vulgaris* L. *Nat. Sci.* **2009**, *7*, 63–75.
6. Bankaji, I.; Sleimi, N.; Lopez-Climent, M.F.; Perez-Clmente, R.M.; Gomez-Cadenas, A. Effects of combined abiotic stresses on grownth trace element accumulation and phytohormone regulation in two halophytic species. *J. Plant Growth Regul.* **2014**, *33*, 632–643. [CrossRef]
7. Malecka, A.; Kutrowska, A.; Piechalak, A. High Peroxide Level May Be a Characteristic Trait of a Hyperaccumulator. *Water Air Soil Pollut.* **2015**, *226*, 84. [CrossRef]
8. Khan, A.; Khan, S.; Khan, M.A.; Qamar, Z.; Waqas, M. The uptake and bioaccumulation of heavy metals by food plants, their effects on plants nutrients, and associated health risk: A review. *Environ. Sci. Pollut. Res.* **2015**, *22*, 13772–13799. [CrossRef]
9. Bankaji, I.; Cacador, I.; Sleimi, N. Physiological and biochemical responses of Sudaeda fruticosa to cadmium and copper stresses: Grownth, nutrient uptake, antioxidant enzymes, phytochelatin, and glutatione levels. *Environ. Sci. Pollut. Res.* **2015**, *22*, 13058–13069. [CrossRef]
10. Babula, P.; Adam, V.; Havel, L.; Kizek, R. Cadmium Accumulation by Plants of Brassicaceae Family and Its Connection with Their Primary and Secondary Metabolism. In *The Plant Family Brassicaceae*; Springer: Dordrecht, The Netherlands, 2012; pp. 71–97.
11. Eapen, S.; Souza, S.F.D. Prospects of genetic engineering of plants for phytoremediation of toxic metals. *Biotechnol. Adv.* **2005**, *23*, 97–114. [CrossRef]
12. Singh, S.; Parihar, P.; Singh, R.; Singh, V.P.; Prasad, S.M. Heavy Metals Tolerance in Plants: Role of Transcriptomics, Proteomics, Metabolomics, and Ionomics. *Front. Plant Sci.* **2016**, *6*, 1143. [CrossRef]
13. Rascio, N.; Navari-Izzo, F. Heavy metal hyperaccumulating plants: How and why do they do it? And what makes them so interesting? *Plant Sci.* **2011**, *180*, 169–181. [CrossRef]
14. Ent van der, A.; Baker, A.J.M.; Reeves, R.D.; Pollard, A.J.; Schat, H. Hyperaccumulators of metal and matalloid trace elements: Facts and fiction. *Plant Soil* **2013**, *362*, 319–334. [CrossRef]
15. Jiang, W.; Liu, D.; Hou, W. Hyperaccumulation of lead by roots, hypocotyls, and shoots of *Brassica juncea*. *Biol. Plant.* **2008**, *43*, 603–606. [CrossRef]
16. Meyers, D.E.R.; Auchterlonie, G.J.; Webb, R.I.; Wood, B. Uptake and localisation of lead in the root system of *Brassica juncea*. *Pollut. Environ.* **2008**, *153*, 323–332. [CrossRef]
17. Prasad, M.N.V.; Freitas, H. Metal hyperaccumulation in plants—Biodiversity prospecting for phytoremediation technology. *Electron. J. Biotechnol.* **2003**, 6. [CrossRef]

18. Hossain, M.A.; Piyatida, P.; da Silva, J.A.T.; Fujita, M. Molecular mechanism of heavy metal toxicity and tolerance in plants: Central role of glutathione in detoxification of reactive oxygen species and methylglyoxal and in heavy metal Chelation. *J. Bot.* **2012**. [CrossRef]
19. Malecka, A.; Piechalak, A.; Zielińska, B.; Kutrowska, A.; Tomaszewska, B. Response of the pea roots defense systems to the two-element combinations of metals (Cu, Zn, Cd, Pb). *Acta Biochim. Pol.* **2014**, *61*, 23–28. [CrossRef]
20. Israr, M.; Jewell, A.; Kumar, D.; Sahi, S.V. Interactive effects of lead, copper, nickel and zinc on growth, metal uptake and antioxidative metabolism of Sesbania drummondii. *J. Hazard. Mater.* **2011**, *186*, 1520–1526. [CrossRef]
21. Sharma, P.; Dubey, R.S. Lead toxicity in plants. *Braz. J. Plant Physiol.* **2005**, *17*, 35–52. [CrossRef]
22. Liu, D.; Zou, J.; Men, Q.; Zou, J.; Jiang, W. Uptake and accumulation and oxidative stress in garlic (*Allium sativum* L.) under lead phytotoxicity. *Ecotoxicology* **2009**, *18*, 134. [CrossRef]
23. Marshner, P. *Marschner's Mineral Nutration of Higher Plants*, 3rd ed.; Academic Press: London, UK, 2012.
24. Zaier, H.; Mudarra, A.; Kutcher, D.; Fernandez de la Campa, M.R.; Abdelly, C.; Sanz-Medel, A. Induced lead binding phytochelatins in Brassica juncea and Sesuvium portulacastrum investigated by orthogonal chromatography inductively coupled plasma-mass spectrometry and matrix assisted laser desorption ionization-time of flight-mass spectrometry. *Anal. Chim. Acta* **2010**, *671*, 48–54. [CrossRef]
25. Yadav, P.; Kaur, R.; Kanwar, M.K.; Bhardwaj, R.; Sirhindi, G.; Wijaya, L.; Alyemeni, M.N.; Ahmad, P. Ameliorative Role of Castasterone on Copper Metal Toxicity by Improving Redox Homeostsis in *Brassica juncea* L. *J. Plant Growth Regul.* **2018**, *37*, 575–590. [CrossRef]
26. Irfan, M.; Ahmad, A.; Shamsul, H. Effect of cadmium on the growth and antioxidant enzymes in two varieties of Brassica juncea. *Saudi, J. Biol. Sci.* **2014**, *21*, 125–131. [CrossRef]
27. Sinha, S.; Sinam, G.; Mishra, R.K.; Mallick, S. Metal accumulation, growth, antioxidants and oil yield of *Brassica juncea* L. exposed to different metals. *Ecotoxicol. Environm. Saf.* **2010**, *73*, 1352–1361. [CrossRef]
28. Kutrowska, A.; Małecka, A.; Piechalak, A.; Masiakowski, W.; Hanc, A.; Barałkiewicz, D.; Andrzejewska, B.; Zbierska, J.; Tomaszewska, B. Effects of binary metal combinations on zinc, copper, cadmium and lead uptake and distribution in *Brassica juncea*. *J. Trace Elem. Med. Biol.* **2017**, *44*, 32–39. [CrossRef]
29. Cherif, J.; Mediouni, C.; Ammar, W.B.; Jemal, F. Interactions of zinc and cadmium toxicity in their effects on growth and in antioxidative systems in tomato plants (*Solanum lycopersicum*). *J. Environ. Sci.* **2011**, *23*, 837–844. [CrossRef]
30. Purakayastha, T.J.; Viswanath, T.; Bhadraray, S.; Chhonkar, P.K.; Adhikari, P.P.; Suribabu, K. Phytoextraction of Zinc, Copper, Nickel and Lead from a Contaminated Soil by Different Species of Brassica. *Int. J. Phytoremed.* **2008**, *10*, 61–72. [CrossRef]
31. Quartacci, M.F.; Argilla, A.; Baker, A.J.M.; Navari-Izzo, F. Phytoextraction of metals from a multiply contaminated soil by Indian mustard. *Chemosphere* **2006**, *63*, 918–925. [CrossRef]
32. Hanć, A.; Małecka, A.; Kutrowska, A.; Bagniewska –Zadworna, A.; Tomaszewska, B.; Barałkiewicz, D. Direct analysis of elemental biodistribution in pea seedlings by LA-ICP-MS, EDAX and confocal microscopy: Imaging and quantification. *Microchem. J.* **2016**, *128*, 305–311. [CrossRef]
33. Sharma, R.; Jha, A.B.; Dubey, R.S.; Pessarakli, M. Reactive Oxygen Species, Oxidative Damage, and Antioxidative Defense Mechanism in Plants under Stressful Conditions. *J. Bot.* **2012**. [CrossRef]
34. Romero-Puertas, M.C.; Corpas, F.J.; Rodriguez-Serrano, M.; Gomez, M.; del Rıo, L.A.; Sandalio, L.M. Differential expression and regulation of antioxidative enzymes by cadmium in pea plant. *J. Plant Physiol.* **2007**, *164*, 1346–1357. [CrossRef]
35. Pena, L.B.; Tomaro, M.L.; Gallego, S.M. Effect of different metals on protease activity in sunflower cotyledons. *Electron. J. Biotechnol.* **2006**, *9*, 259–262. [CrossRef]
36. Markovska, Y.K.; Gorinova, N.J.; Nedkovska, M.P.; Miteva, K.M. Cadmium-induced oxidative damage and antioxidant responses in *Brassica juncea* plants. *Biol. Plant.* **2009**, *53*, 151–154. [CrossRef]
37. Wang, S.H.; Yang, Z.M.; Yang, H.; Lu, B.; Li, S.Q.; Lu, Y.P. Copper-induced stress and antioxidative responses in roots of *Brassica juncea* L. *Bot. Bull. Acad. Sin.* **2004**, *45*, 203–212.
38. Malecka, A.; Piechalak, A.; Mensinger, A.; Hanc, A.; Barałkiewicz, D.; Tomaszewska, B. Antioxidative defense system in Pisum sativum roots exposed to heavy metals (Pb, Cu, Cd, Zn). *Pol. J. Environ. Stud.* **2012**, *21*, 1721–1730.

39. Kanwar, M.K.; Poonam; Bhardwaj, R. Arsenuc induced modulation of antioxidative defense system and brassinosteroids in *Brassica juncea* L. *Ecotoxicol. Environ. Saf.* **2015**, *115*, 119–125. [CrossRef]
40. Wang, S.L.; Liao, W.B.; Lu, F.Q.; Liao, B.; Shu, W.S. Hyperaccumulation of lead, zinc and cadmium in plants growing on a lead/zinc outcrop in Yunnan Province, China. *Environ. Geol.* **2009**, *58*, 471–476. [CrossRef]
41. Pinto, A.P.; Alves, A.S.; Candeias, A.J.; Cardoso, A.L.; de Varennes, A.; Martins, L.L.; Mourato, M.P.; Goncales, M.L.S.; Mota, A.M. Cadmium accumulation and antioxidative defences in *Brassica juncea* L. Czer, *Nicotiana tabacum* L. and *Solanum nigrum* L. *Int. J. Environ. Anal. Chem.* **2009**, *89*, 661–676. [CrossRef]
42. Mohamed, A.A.; Castagna, A.; Ranieri, A.; Sanita di Toppi, L. Cadmium tolerance in Brassica juncea roots and shoots is affected by antioxidant status and phytochelatin biosynthesis. *Plant Physiol. Biochem.* **2012**, *22*, 13058–13069. [CrossRef]
43. Fidalgo, F.; Freita, R.; Ferreira, R.; Pessona, A.M.; Teixeira, J. *Solanum nigrum* L. antioxidant defence system isozymes are regulated transcriptionally and posttranslationally in Cd-induced stress. *Environ. Exp. Bot* **2011**, *72*, 312–319. [CrossRef]
44. Luo, H.; Li, H.; Zhang, X.; Fu, J. Antioxidant responses and gene expression in perennial ryegrass (*Lollium perenne* L.) under cadmium stress. *Ecotoxicology* **2011**, *20*, 770–778. [CrossRef]
45. Bernard, F.; Brulle, F.; Dumez, S.; Lemiere, S.; Platel, A.; Nesslany, F.; Cuny, D.; Deramz, A.; Vandenbulcke, F. Antioxidant responses of Annelids, Brassicaceae and Fabaceae to pollutants: A review. *Ecotoxicol. Environ. Saf.* **2015**, *114*, 273–303. [CrossRef]
46. Wilkins, D.A. A technique for the measurement of lead tolerance in plants. *Nature* **1957**, *180*, 37–38. [CrossRef]
47. Hanć, A.; Olszewska, H.; Barałkiewicz, D. Quantitative analysis of elements migration in human teeth with and without filling using LA-ICP-MS. *Microchem. J.* **2013**, *110*, 61–69. [CrossRef]
48. Doke, N. Invovement of superoxide anion generation in the hypersensitive response of potato tuber tissues to infection with an incompatible race of Phytophthora infestans and to the hyphal wall components. *Physiol. Mol. Plant Pathol.* **1983**, *23*, 345–355. [CrossRef]
49. Patterson, B.D.; Macrae, E.A.; Ferguson, I.B. Estimation of hydrogen peroxide in plant extracts using titanium(IV). *Anal. Biochem.* **1984**, *139*, 487–492. [CrossRef]
50. Levine, R.L.; Williams, J.A.; Stadtman, E.P.; Shacter, E. Carbonyl assays for determination of oxidatively modified proteins. *Methods Enzymol.* **1994**, *233*, 346–357. [CrossRef]
51. Beauchamp, C.; Fridovich, I. Superoxide dismutase: Improved assays and an assay applicable to acrylamide gels. *Anal. Biochem.* **1971**, *44*, 276–287. [CrossRef]
52. Aebi, H.E. Catalase in vitro. In *Methods of Enzymatic Analyses*; Bergmeyer, H.U., Ed.; Verlag Chemie: Weinheim, Germany, 1983; Volume 3, pp. 273–282.
53. Nakano, Y.; Asada, K. Hydrogen peroxide is scavenged by ascorbate-specific peroxidase in spinach chloroplasts. *Plant Cell Physiol.* **1981**, *22*, 867–880.
54. Bradford, M.M. A rapid and sensitive method for the quantitation of microgram quantitiesof protein utilizing the principle of protein- dye binding. *Anal. Biochem.* **1976**, *72*, 248–254. [CrossRef]

© 2019 by the authors. Licensee MDPI, Basel, Switzerland. This article is an open access article distributed under the terms and conditions of the Creative Commons Attribution (CC BY) license (http://creativecommons.org/licenses/by/4.0/).

International Journal of
Molecular Sciences

Article

Selenium Modulates the Level of Auxin to Alleviate the Toxicity of Cadmium in Tobacco

Yong Luo [1], Yuewei Wei [1], Shuguang Sun [2], Jian Wang [2], Weifeng Wang [3], Dan Han [1], Huifang Shao [1], Hongfang Jia [1,*] and Yunpeng Fu [1,*]

1 National Tobacco Cultivation & Physiology & Biochemistry Research Centre, College of Tobacco Science, Henan Agricultural University, Zhengzhou 450002, China
2 China Tobacco Hubei Industrial Co., Ltd., Wuhan 430040, China
3 Guangxi Zhuang Autonomous Region Provincial Branch of China National Tobacco Corporation, Nanning 530000, China
* Correspondence: jiahongfang@henau.edu.cn (H.J.); yunpengfu@henau.edu.cn (Y.F.); Tel.: +86-371-63555713

Received: 23 June 2019; Accepted: 30 July 2019; Published: 1 August 2019

Abstract: Cadmium (Cd) is an environmental pollutant that potentially threatens human health worldwide. Developing approaches for efficiently treating environmental Cd is a priority. Selenium (Se) plays important role in the protection of plants against various abiotic stresses, including heavy metals. Previous research has shown that Se can alleviate Cd toxicity, but the molecular mechanism is still not clear. In this study, we explore the function of auxin and phosphate (P) in tobacco (*Nicotiana tabacum*), with particular focus on their interaction with Se and Cd. Under Cd stress conditions, low Se (10 μM) significantly increased the biomass and antioxidant capacity of tobacco plants and reduced uptake of Cd. We also measured the auxin concentration and expression of auxin-relative genes in tobacco and found that plants treated with low Se (10 μM) had higher auxin concentrations at different Cd supply levels (0 μM, 20 μM, 50 μM) compared with no Se treatment, probably due to increased expression of auxin synthesis genes and auxin efflux carriers. Overexpression of a high affinity phosphate transporter NtPT2 enhanced the tolerance of tobacco to Cd stress, possibly by increasing the total P and Se content and decreasing Cd accumulation compared to that in the wild type (WT). Our results show that there is an interactive mechanism among P, Se, Cd, and auxin that affects plant growth and may provide a new approach for relieving Cd toxicity in plants.

Keywords: selenium; cadmium stress; auxin; root architecture; phosphate transporter; *Nicotiana tabacum*

---

1. Introduction

Cadmium (Cd) is a highly toxic heavy metal, which is widely distributed in the environment [1]. In recent years, industrial and agricultural production have discharged Cd to varying degrees, and it has become one of the most widely distributed agricultural pollutants [2]. Cd is not needed for plant growth and development but is more likely to accumulate in plants than other heavy metals [3]. Evidence suggests that a high concentration of Cd in soil affects the growth and development of plants through physiological and biochemical processes, including inhibition of plant enzyme and membrane activity [4], decreased cell division [5], reduced growth rate [6], damaged photosynthesis [7], inhibition of stomatal opening [8], and promotion of lipid peroxidation [9]. Cd bioaccumulates and in humans can cause diseases such as osteoporosis, anemia, hypertension, and kidney damage. Soil pollution by Cd has become a serious threat to the safety of agricultural produce. To address this issue, it is important to develop a comprehensive understanding of the mechanism of Cd uptake in plants. Selenium (Se) is not essential for plant nutrition, but it can play a beneficial role in plant health. A suitable dose of Se can enhance antioxidant capacity, delay aging, increase photosynthesis, boost auxin content, and promote plant growth [10,11]. By contrast, a high dose of Se can damage plants through reactive oxygen species (ROS) accumulation and inhibit plant growth [12]. Plants are able to utilize soil Se in its inorganic Se (IV) and Se (VI) forms (selenite and selenate, respectively). Studies have shown that plants absorb selenate through sulfate transporters and selenite through phosphate channels [13,14]. A previous study has suggested that Se (IV) uptake is mediated by Pi (inorganic phosphate) transporters [15]. Recent research suggests that auxin is involved in the interaction between Pi and Se in tobacco, which provides convincing evidence for understanding the molecular mechanism of how Se regulates plant growth [16].

Se inhibits Cd uptake, which can relieve the toxic effects of Cd in plants. Cary et al. (1981) first reported that Se fertilizer reduced the absorption of Cd in wheat and lettuce [17]. Increasing numbers of studies have shown that there is antagonism between Cd and Se in plants [18–20]. Cd stress has also been shown to inhibit root growth and leaf photosynthesis in winter wheat, although adding some doses of Se can alleviate the Cd toxicity by enhancing root growth [21]. Despite studies investigating the antagonistic relationship between Se and Cd, this relationship—especially the molecular mechanism of Se remission of toxic Cd effects—is still not clear.

Tobacco is an important economic and model crop. In this study, we aimed to clarify the molecular mechanism of Se remission of toxic Cd effects in tobacco. We chose *DR5::GUS* and *NtPT2* overexpressed transgenic tobacco as the test material. We studied the influence of Cd on growth; the antioxidant system; the auxin distribution; and the Se, Cd uptake of tobacco under various Se and Cd treatments. We aimed to (1) clarify the function of auxin in tobacco growth, under different doses of Se and Cd; (2) determine the mechanism by which Se enhances tolerance to Cd stress; and (3) reveal the function of phosphate transporter (*NtPT2*) in the interaction of Se and Cd in plants.

2. Results

2.1. Effects of Se on Tobacco Phenotype and Biomass under Cd Stress

To determine whether Se affects the growth of tobacco under different Cd treatments, we checked the characterization of tobacco plants. We found that Cd treatment had a substantial toxic effect on tobacco seedlings, with greater yellow leaf area and more poorly developed root architecture in Cd-treated seedlings than those in the Cd0 treatment (Figure 1A and Figure S1). Interestingly, we also found that low Se (Se10) could promote the growth of tobacco under Cd stress (Figure 1A); under Cd20 and Cd50 stress conditions, the biomass of shoots with Se10 treatments increased by 16.3% and 20.8%, respectively, compared to those with no Se added (Figure 1B), and the biomass of roots with Se10 treatments increased by 24.2% and 30.2%, respectively (Figure 1C). By contrast, high Se (Se50) levels suppressed tobacco growth under different Cd (Cd0, Cd20 and Cd50) treatments (Figure 1). These results show that Se has a dual effect on Cd stress, and the appropriate concentration of Se can alleviate Cd toxicity in tobacco.

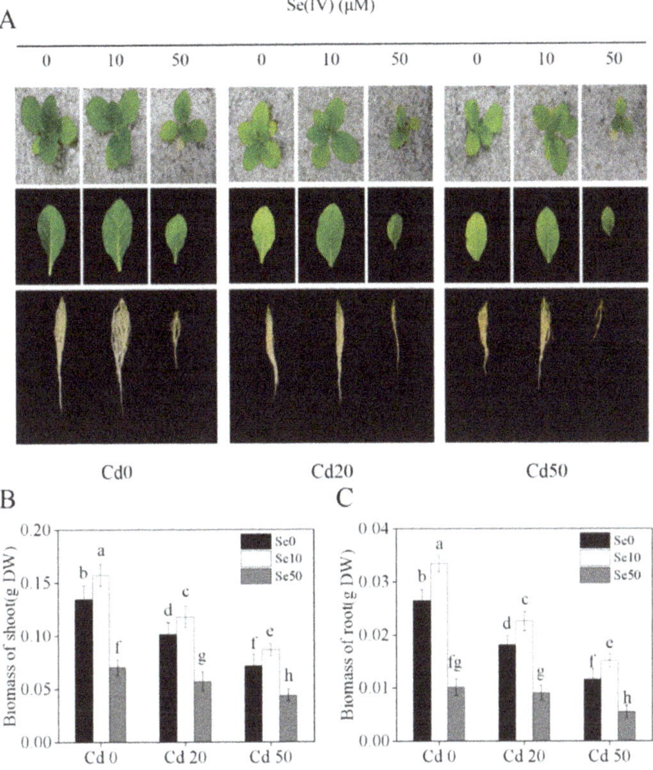

Figure 1. Characterization of tobacco under different Se and Cd concentration supply conditions for 21 days. (**A**) The phenotype of shoot, third young leaf and root under Se (IV) and Cd (II) treatments in tobacco; (**B**) The biomass of shoot under different Se (IV) and Cd (II) concentration supply conditions; (**C**) The biomass of root under different Se (IV) and Cd (II) concentration supply conditions. 14-days-old seedlings (wild-type, Yunyan 87) were grown in pots with sand under different Se (0, 10, 50 μM) and Cd (0, 20, 50 μM) concentrations for 21 days. **Se0**: no Se; **Se10**: Se, 10 μM; **Se50**: Se, 50 μM; **Cd0:** no Cd; **Cd20**: Cd, 20 μM; **Cd50**: 50 μM. Shown are mean ± standard deviation (SD) from five biological replicates. DW, dry weight. Different letters indicate significant differences ($p < 0.05$).

2.2. Effects of Se and Cd Interactions on Tobacco Antioxidant Capacity

Previous studies showed that Se can enhance the enzymatic and non-enzymatic anti-oxidation systems and improve plant resistance to abiotic stresses in plants [4–7]. To determine the function of Se under Cd stress conditions, we measured the antioxidant capacity and chlorophyll content of tobacco. Firstly, we determined the accumulation of H_2O_2 by nitroblue tetrazolium (NBT) staining. Low Se (Se10) obviously reduced the accumulation of H_2O_2 (Figure 2A), which was consistent with the results of Malondialdehyde (MDA) content under Cd stress conditions (Figure 2B). We also checked the chlorophyll content, and found that Se had notable effects on chlorophyll content in the leaves (Figure 2C). At no Se (Se0) and high Se (Se50) levels, increased Cd levels significantly reduced tobacco chlorophyll content. Notably, the chlorophyll content of the low Se (Se10) treatment showed a remarkable increase under different Cd levels, which implies that low Se can promote tobacco growth under Cd stress by improving the anti-oxidation activity of tobacco plants.

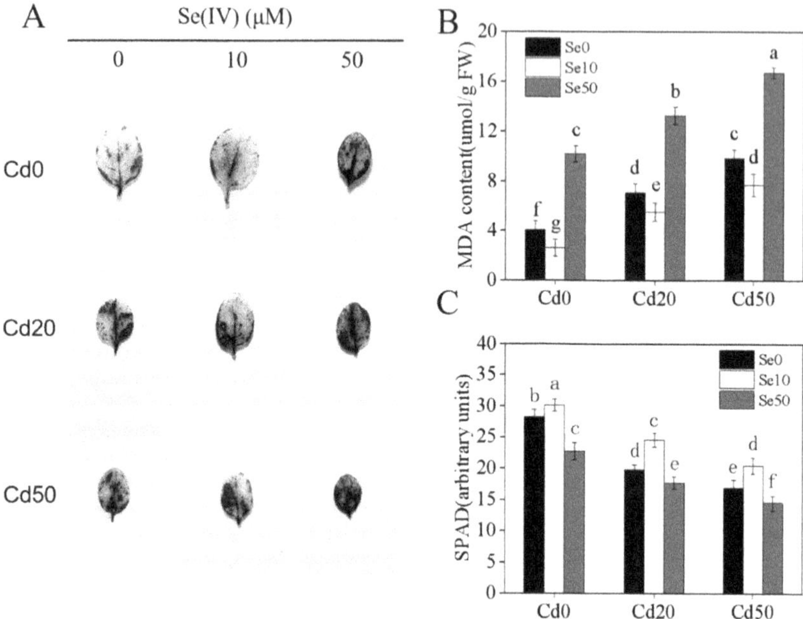

Figure 2. Effects of Se and Cd treatments on antioxidant capacity of tobacco. (**A**) nitroblue tetrazolium (NBT) staining of tobacco seedlings under different Se (IV) and Cd (II) concentration supply conditions; (**B**) the content of MDA of tobacco seedlings under different Se (IV) and Cd (II) concentration supply conditions; (**C**) the SPAD of the third young leaf of tobacco seedlings under different Se (IV) and Cd (II) concentration supply conditions; 14-days-old seedlings (wild-type, Yunyan 87) were grown in pot with sand under different Se (0, 10, 50 μM) and Cd (0, 20, 50 μM) concentrations for 21 days. **Se0**: no Se; **Se10**: Se, 10 μM; **Se50**: Se, 50 μM; **Cd0**: no Cd; **Cd20**: Cd, 20 μM; **Cd50**: Cd, 50 μM. Shown are mean ± SD from five biological replicates. Different letters indicate significant differences ($p < 0.05$).

2.3. Accumulation of Se and Cd in Tobacco

To investigate the accumulation of Cd in tobacco under low Se (Se10) and high Se (Se50) conditions, we monitored the Se and Cd content in the roots and shoots of tobacco seedlings (Figure 3). Under high Se and low Se conditions, the Se content of the shoots and roots increased in Cd20 and Cd50 treatments compared with the Cd0 treatment (Figure 3A,B). Under Cd stress (Cd20, Cd50) conditions, the low Se treatment significantly reduced the Cd content of tobacco shoots and roots, especially under the Cd20 treatment, where the root Cd content was 37.3% lower than that observed with the Se0 treatment (Figure 3C,D). We also found that high Se did not reduce the Cd content of tobacco plants, suggesting that variation in Se content can affect the uptake of Cd in tobacco.

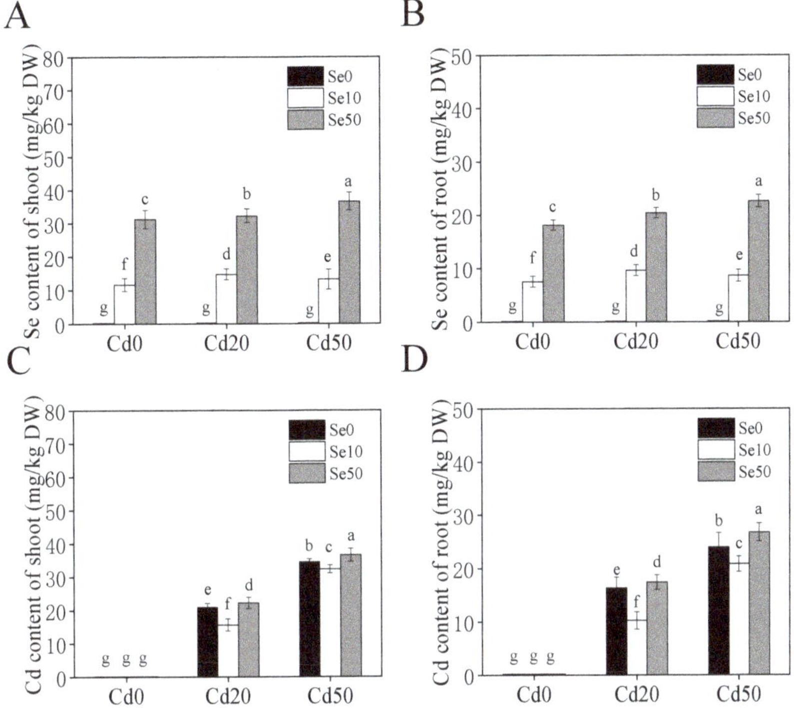

Figure 3. Content of Se and Cd in tobacco under different Se and Cd concentration supply conditions for 21 days. (**A**,**B**) Se content of the shoots and roots under different Se (IV) and Cd (II) concentration supply conditions; (**C**,**D**) Cd content of the shoots and roots under different Se (IV) and Cd (II) concentration supply conditions. 14-days-old seedlings (wild-type, Yunyan 87) were grown in pot with sand under different Se (0, 10, 50 μM) and Cd (0, 20, 50 μM) concentration for 21 days. **Se0**: no Se; **Se10**: Se, 10 μM; **Se50**: Se, 50 μM; **Cd0**: no Cd; **Cd20**: Cd, 20 μM; **Cd50**: Cd, 50 μM. Shown are mean ± SD from five biological replicates. Different letters indicate significant differences ($p < 0.05$).

2.4. Effects of Se on Auxin and Expression of Auxin-Related Genes in Tobacco under Cd Stress

To investigate whether auxin is involved in the growth of tobacco roots under Se and Cd treatment, we used *DR5::GUS* transgenic tobacco, which could reflect the distribution of auxin in the plant [22–24]. We detected *GUS* expression in the root tip under Se and Cd treatments (Figure 4A). Under Cd stress (Cd20, Cd50) conditions, the *GUS* expression in the root tip of plants was much lower than observed in the Cd0 treatment. Under low Se (Se10), the expression of *GUS* in the root tip was much greater than that in Se0 tobacco. Low Se could increase the expression of *GUS* under Cd stress. We also checked the auxin content of shoots and roots, which was consistent with the *GUS* expression results (Figure 4B,C).

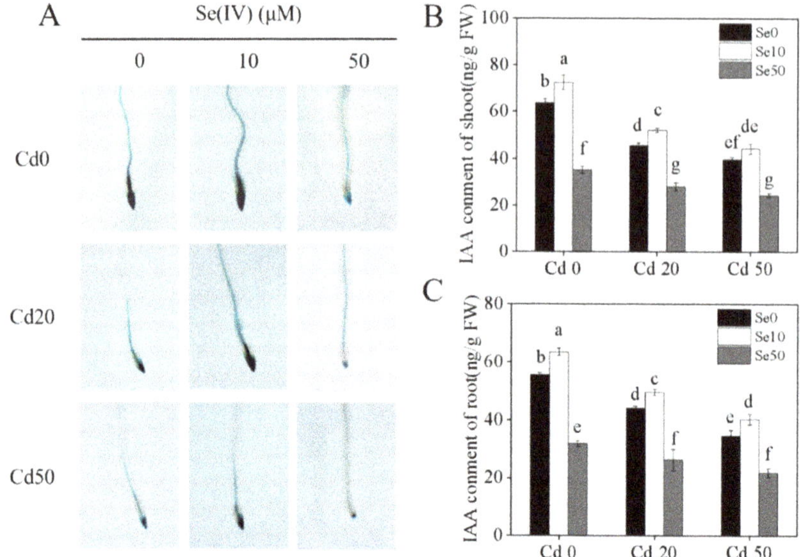

Figure 4. Histochemical localization of *DR5::GUS* and indole-3-acetic acid (IAA) contents of tobacco under different Se and Cd concentration supply conditions. (**A**) Histochemical localization of *DR5::GUS* in root tips of tobacco under different Se (IV) and Cd (II) concentration supply conditions; (**B**,**C**) IAA content of the shoots and roots under different Se (IV) and Cd (II) concentration supply conditions. 14-days-old seedlings (*DR5::GUS* transgenic tobacco) were grown in pot with sand under different Se (0, 10, 50 μM) and Cd (0, 20, 50 μM) concentrations for 21 days. **Se0**: no Se; **Se10**: Se, 10 μM; **Se50**: Se, 50 μM; **Cd0:** no Cd; **Cd20:** Cd, 20 μM; **Cd50**: Cd, 50 μM. Shown are mean ± SD from five biological replicates. Different letters indicate significant differences ($p < 0.05$).

Local auxin levels are determined by biosynthesis and intercellular transport in plant roots [25]. To detect whether Se and Cd treatments affect the auxin-signal pathway, we analyzed the expression of *YUCCAs* and *PINs* family genes, which are involved in auxin biosynthesis and transport in tobacco under different Se and Cd treatments. Under Cd stress conditions, the expression of *NtYUCCA 6*, *8*, and *9*, and *NtPIN 1a*, *1c*, and *4* was substantially higher under low Se treatments, which was consistent with the auxin content results (Figure 5A–F). All these results suggest that auxin may play a key role in growth under Se and Cd treatment conditions.

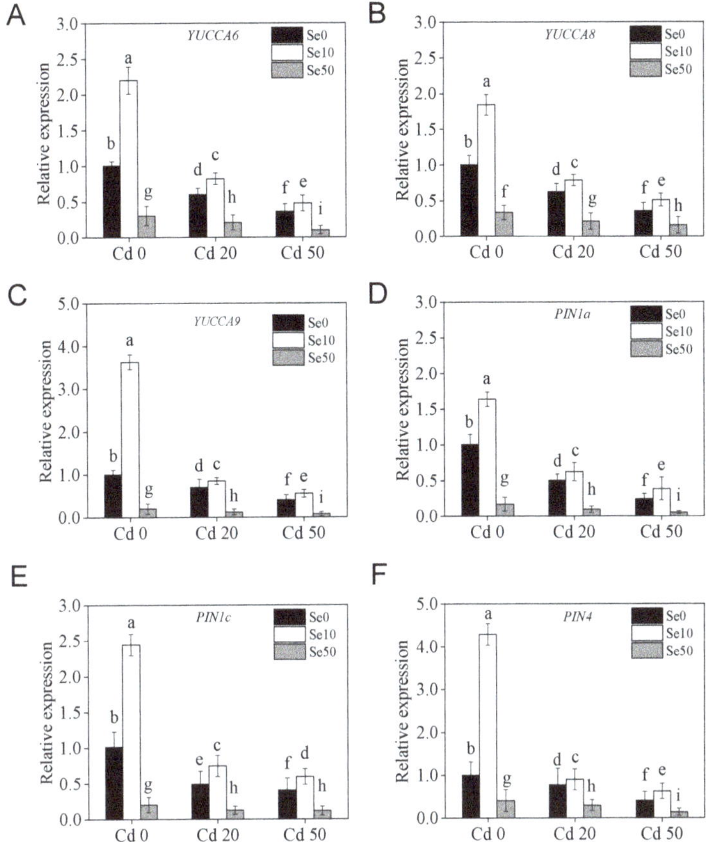

Figure 5. Expression of auxin-relative gene in tobacco under different Se and Cd concentration supply conditions. (**A**–C) Expression of three members (*YUCCA6, 8, 9*) of the tobacco *YUCCAs* family genes in shoots under different Se and Cd concentration supply conditions.; (**D**–**F**) Expression of three members (*PIN1a, 1c, 4*) of the tobacco *PINs* family genes in roots under different Se and Cd concentration supply conditions. 14-days-old seedlings (*DR5::GUS* transgenic tobacco) were grown in pot with sand under different Se (0, 10, 50 μM) and Cd (0, 20, 50 μM) concentrations for 21 days. The tobacco housekeeping gene *L25* was used as an internal control. The relative expression levels are shown compared with the expression under Cd0 and Se0 (Cd0 + Se0) conditions as 1 expression. **Se0**: no Se; **Se10**: Se, 10 μM; **Se50**: Se, 50 μM; **Cd0**: no Cd; **Cd20**: Cd, 20 μM; **Cd50**: Cd, 50 μM. Shown are mean ± SD from five biological replicates. Different letters indicate significant differences ($p < 0.05$).

2.5. Overexpression of NtPT2 Could Enhance the Tolerance of Cd Stress under Low Se Conditions

Phosphate transporters are not only involved in Pi uptake, but also in selenite uptake [14,26]. Recently, we reported that the expression of a high-affinity phosphate transporter (*NtPT2*) involved in Se uptake in tobacco is induced in the roots and shoots under low Pi conditions [16]. In this study, we used *NtPT2* overexpression (*NtPT2-Oe*, two independent transgenic lines: Oe1 and Oe2) in transgenic tobacco to clarify the molecular mechanism by which Se can alleviate Cd toxicity. We found that *NtPT2* overexpression in plants enhanced their tolerance to Cd stress, with plants exhibiting better roots and shoots than the WT under Cd stress conditions. This was consistent with the auxin content differences between the *NtPT2-Oe* plant and the WT (Figure 6A,B and Figure S2). We also checked the P, Se, and Cd content in both *NtPT2-Oe* and WT plants; the total P and Se content of *NtPT2-Oe* plants was higher than that of the WT plants under Cd stress conditions (Figure 6C,D), suggesting that *NtPT2* is involved in P and Se uptake in tobacco. We also found that Cd content was significantly reduced in *NtPT2-Oe* plants under Cd stress conditions (Figure 6E), confirming that overexpression of *NtPT2* could enhance the tolerance of tobacco plants to Cd stress.

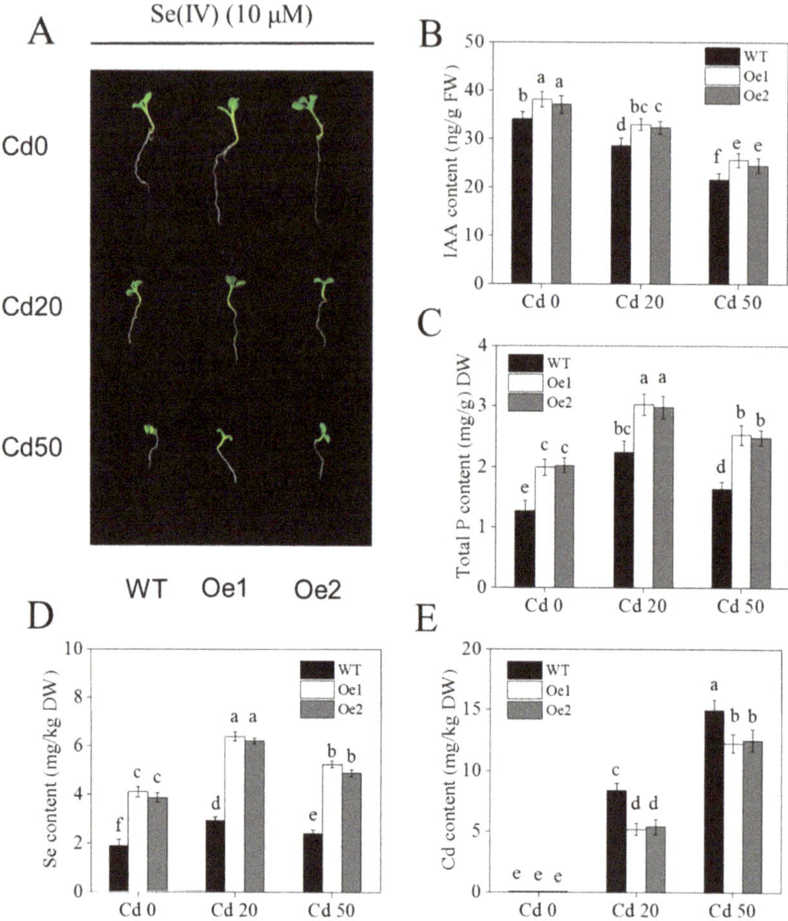

Figure 6. Effects of Cd on *NtPT2-Oe* transgenic tobacco under different Se and Cd concentration supply conditions. (**A**,**B**) The phenotype and IAA content of *NtPT2-Oe* transgenic tobacco seedlings in Se (10 μM) and different Cd (0, 20, 50 μM) concentrations supply conditions; (**C**–**E**) Total P, Se and Cd content of the whole transgenic plant. Tobacco seeds were grown in 1/2MS culture under Se (10 μM) and different Cd (0, 20, 50 μM) concentrations for 14 days. **Se0**: no Se; **Se10**: Se, 10 μM; **Se50**: Se, 50 μM; **Cd0**: no Cd; **Cd20**: Cd, 20 μM; **Cd50**: Cd, 50 μM. Shown are mean ± SD from five biological replicates. Different letters indicate significant differences ($p < 0.05$).

3. Discussion

3.1. Se Affects the Growth of Tobacco Roots by Changing Auxin Concentration under Cd Stress

Cd has high levels of biological toxicity and can inhibit the growth and development of plants [27]. Biomass is an important indicator of plant growth and development. In this study, we showed that under Cd stress conditions, the biomass of tobacco, especially in the roots, was significantly lower than plants not exposed to Cd stresses. This indicates that the roots and shoots were damaged by Cd stress, which is consistent with the results of Li et al. [28]. We also found that low Se levels stimulated growth in tobacco and could effectively alleviate the toxic effects of Cd stress (Figure 1), which is consistent with the results of previous studies [21,29]. Previous studies have shown that low Se enhances the antioxidation of enzymatic and non-enzymatic systems, changes root growth, and promotes absorption of nutrients [30,31], thus, improving the ability of plants to resist abiotic stresses. Low Se increased anti-oxidation activity (Figure 2), reduced the content of Cd (Figure 3) and changed root development by increasing auxin concentration in the roots (Figure 4), which may have increased growth. Further analyses on auxin-related genes showed that the expression levels of *YUCCAs* and *NtPINs* family genes were markedly higher under low Se conditions (Figure 5), indicating that low Se affects the growth of roots by changing auxin concentration under Cd stress.

Se often exerts a dual effect on plant growth. High levels of Se cause toxicity in plants, such as accumulation of ROS and inhibition of plant development [12,30,31]. A recent study showed that high Se levels inhibit root growth [16]. In this study, we also found high Se significantly decreased the biomass and auxin content of roots. Notably, under high Se conditions, the Cd treatments significantly decreased the biomass and auxin content in the shoots and roots. To investigate whether auxin is a major regulator of plant growth and development under Cd stress conditions, we also checked the characterization of root in *DR5::GUS* transgenic tobacco under different Se and Cd concentration supply conditions by adding IAA (100 nM). The results showed that exogenous IAA could increase the root biomass and the length of primary root, which implies that IAA plays a key roles in regulating the root architecture under Cd stress conditions (Figure S3).

Altogether, our data suggest that, under Cd stress conditions, Se might affect the growth and development of plants by changing auxin concentration. Not only the auxin pathway, but also other endogenous hormones (e.g., cytokinin, ethylene and gibberellin) may have an effect on plant root development. Recent studies showed that Se increases primary root length through alteration of the auxin and ethylene balance in rice, growth inhibition in Se-treated *Arabidopsis* is associated with an incomplete mobilization of starch, and high concentrations of selenite-induced enhancement of ethylene biosynthesis may result in plant cell death [32–34]. Therefore, the function of cytokinin, ethylene, gibberellin and other hormones in the development of tobacco roots under the treatment of Se and Cd need further study.

3.2. NtPT2 *Might be a Potential Candidate Gene for Breeding Cd-Tolerant Plants*

A large number of genes encoding Pi transporters have been identified in different plant species. Pi transporters are generally classified into the *Pht1*, *Pht2*, *Pht3*, and PT gene families [35,36]. Most of the high-affinity Pi transporter (*Pht1*) family genes had induced expression under low Pi stress conditions, suggesting that *Pht1* plays a crucial role in both Pi uptake and translocation under Pi deficiency [37,38]. Recent studies have suggested that Se and Pi share similar uptake mechanisms and that Pi transporters are involved in Se uptake in plants, which might occur via a Se-H^+ symport process in the plant cell membrane [14,16,26].

Although a large number of *Pht1* family genes have been studied in rice and *Arabidopsis*, the function of key Pi transporters in plants need further research. For instance, recently, two key phosphate transporter genes *OsPT2* and *OsPT8* were found to play a key role in As and Se uptake and translation in rice, implying that some phosphate transporters are involved in the uptake of heavy metals or trace elements [39]. At present, studies on the *Pht1* family gene in tobacco are few. In our previous work, we showed that *NtPT2* is the most closely related to *OsPT2*, and that *NtPT2* has similar expression patterns to *OsPT2* (low Pi-induced expression in roots) [16]. In this study, we used *NtPT2* overexpressing transgenic tobacco to further clarify whether Se could alleviate the toxicity of Cd. Our results showed that when 10 μM Se was supplied, the *NtPT2* overexpression significantly increased biomass, total P, axuin, and Se content. In contrast, the Cd content in transgenic tobacco obviously reduced under Cd treatments compared with WT (Figure 6), implying that a suitable level of Se modulates the level of auxin, enhancing the tolerance of tobacco to Cd stress.

With the improvement of health awareness, how to reduce the Cd content in crop production and food chain has become a research hotspot in recent years. It has a potential impact on reducing the accumulation of Cd in tissues such as liver, kidney, lungs and bones, and avoiding the induction of various diseases such as lung cancer, hypertension and cardiomyopathy [40,41]. Our results suggest that *NtPT2* might be a potential candidate gene for breeding Cd-tolerant plants, which also need further research.

Based on this study, we propose a possible model for revealing the mechanism controlling the Se, Cd stress, P, and auxin response in tobacco (Figure 7). Under Cd stress conditions, (1) plant growth is inhibited. (2) Under low Se and Cd stress conditions, the Pi transporter is involved in the uptake of Se (IV) and P in the root, and low Se (accumulation of a small amount Se in plant) changes the auxin-related gene expression, which increases auxin content to promote plant growth. (3) Low Se not only increases the biomass of the root but also enhances the antioxidant capacity. We conclude that low Se alleviates Cd toxicity in tobacco. However, the function of other hormones in the development of tobacco roots under the treatment of Se and Cd needs further study.

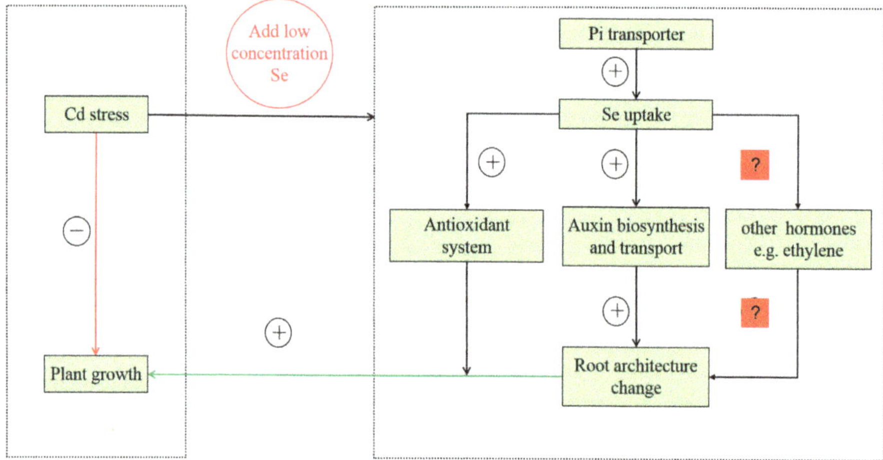

Figure 7. The model for the interaction mechanism between Se, Cd stress, *p* and auxin response in tobacco. The model is based on the results presented here. + indicates positive regulations and – indicates negative regulations. ? indicates that requires further research.

4. Materials and Methods

4.1. Plant Material and Experimental Conditions

The tested materials were the wild-type (*Nicotiana tabacum* cv,Yunyan 87), *DR5::GUS* and *NtPT2-Oe* transgenic seeds (T2 generation) (Figure S4). The generation of transgenic tobacco material and the construction of *pDR5::GUS* had been detailed in the previous studies [16,26,42].

Tobacco seeds were sterilized in solution of 75% (*v/v*) ethanol for 30 s and 10% (*v/v*) sodium hypochlorite for 7 min, then followed by washing 6 times with sterile distilled water. The seeds were then transferred to seedling tray (3 days) kept in the culture room at 28 °C in dark for proper germination. Germinated seedlings were placed in greenhouse for 10 days. The culture during the first five days used one-fourth-strength Hoagland's nutrient solution, and half-strength nutrient solution was used in the second five days. The tobacco seedlings were transferred in pot with sand and half-strength nutrient solution was used for 4 days in seedling recovering stage. Then they were exposed to Se (Na_2SeO_3) and Cd ($CdCl_2{\cdot}2.5H_2O$) for 21 days. The experiment of three Se levels, i.e., 0, 10 and 50 μM, and three Cd levels, i.e., 0, 20 and 50 μM were designed. There was a total of nine treatments: Cd0+Se0, Cd0+Se10, Cd0+Se50, Cd20+Se0, Cd20+Se10, Cd20+Se50, Cd50+Se0, Cd50+Se10, and Cd0+Se50. Se and Cd were added into the nutrient solution in the forms of Na_2SeO_3 and $CdCl_2{\cdot}2.5H_2O$, respectively.

The tobacco seedings were harvested after 21 days of treatment. Shoots and roots were washed with deionized water for further analysis. (1) We observed and recorded the phenotype of tobacco plants. (2) Some seedings were used to measure the IAA, chlorophyll and MDA content. (3) Some of the leaves and roots were used for NBT and *DR5::GUS* staining. (4) The other tobacco seedings were oven-dried and used to measure the contents of Se and Cd. (5) Some seedings were used for detecting the gene expression and enzyme activity.

For IAA (Indole-3-acetic acid, dissolved in 1 M NaOH) treatments, 100 nM IAA was added to the nutrient solution under different Se and Cd treatments, respectively. The nine treatment plants, namely Cd0+Se0+IAA, Cd0+Se10+IAA, Cd0+Se50+IAA, Cd20+Se0+IAA, Cd20+Se10+IAA, Cd20+Se50+IAA, Cd50+Se0+IAA, Cd50+Se10+IAA and Cd50+Se50+IAA. The tobacco seedings were grown in a growth chamber for 7 days. The plants were then harvested for next stage of analysis. (1) To observe the phenphtype of tobacco plants. (2) To record the histochemical localization of GUS.

4.2. GUS Staining and Nitroblue Tetrazolium (NBT) Staining of Plant Tissues

Plants were stained for GUS activity for 24 h at 37 °C, and then seedlings were immersed in 95% ethanol to eliminate chlorophyll pigmentation. Plants were stained in NBT solution for 5 h at 30 °C, and 95% ethanol was used until decolorization was complete. The stereo microscope (Olympus SZX16, Olympus, Tokyo, Japan) equipped with a colorcharge coupled device (CCD) camera were used to photograph stained plant tissues.

4.3. Chlorophyll and MDA Measurement

The relative amount of chlorophyll in plants was determined by a SPAD-502 chlorophyll meter (Konica, Tokyo, Japan) [43]. Five sites of one leaf were measured, and the results were averaged. The content of MDA was determined by the thio-barbituric acid method (TBA) which was calculated by using the difference in absorbance of the extract at 532 nm and 600 nm [44,45].

4.4. IAA Measurement

The IAA contents of shoots and roots in tobacco seedlings were measured as described by Sun et al. (2014) and Jia et al. (2018) [16,46]. The samples were grinded with appropriate amount of the antioxidant butyleret hydroxytoluen (BHT) and 80% pre-cooled methanol for 12–16 h. We collected and concentrated the extracted fluid by a rotary evaporator to 10 mL at 40 mL, and then the fluid was extracted with petroleum ether of the same volume. Under a layer liquid it was adjusted to pH 8.5 and added 0.2 g polyvinylpyrrolidone (PVP) then vibrated for 30 min, and then filtered through a 0.45 μm filter over an OASIS HLB (St. Louis, Mo, USA), and chromatographic conditions were described by: Waters 600–2487; Hibar column RT 250 × 4.6 mm; Purospher STARRP-18 (5 μm); column temperature 45 °C; fluid phase: methanol:1% acetic acid (*v/v*, 40/60), isocratic elution; fluid rate: 0.6 mL min^{-1}; ultraviolet (UV) detector, l = 269 nm; injection volume 20 μL. A 0.22 μm filter was used for filtration of both the buffer and the samples before high-performance liquid chromatography (HPLC) analysis.

4.5. Reverse Transcription Polymerase Chain Reaction (qRT-PCR) Analysis

Trizol reagent was used to prepared Total RNAs from the roots and shoots of tobacco seeding. DNase I-treated total RNAs were used for RT by Superscript II. Triplicate quantitative assays were performed with SYBR Premix Ex Taq™ II (Perfect Real Time) kit (TaKaRa Biotechnology, Dalian, China) on the Step One Plus RealTime PCR Systems (Applied Biosystems, Bio-Rad, Berkeley, CA USA). The gene-specific primers for *YUCCAs* and *PINs* family genes of tobacco were used to perform reverse transcription polymerase chain reaction (qRT-PCR) analysis. The primers were shown in Supplemental Table S1. The analysis of relative expression levels used *NtL25* (L18908.1) as internal reference gene and presented as $2^{-\Delta\Delta Ct}$.

4.6. Determination of Total P in Plant

Dry samples of about 0.05 g were digested with 5 mL of 98% H_2SO_4 and 3 mL of 30% hydrogen peroxide. Then, total P content was analyzed by the molybdate blue method [47].

4.7. Measurement of Se and Cd Contents

The comminuted tobacco samples were digested with concentrated HNO_3 and $HClO_4$ (*v/v*, 4:1) [48]. Se and Cd contents were determined by inductively coupled plasma mass spectrometry (ICP-MS 7500A, Agilent, Palo Alto, CA, USA). The accuracy of elemental analysis was verified using standard reference materials from the China Standard Reference Center.

4.8. Statistical Analysis

Two-way analysis of variance (ANOVA) and Tukey's multi-comparisons test ($p \leq 0.05$) were applied to all data. The results were expressed as the means and the corresponding standard errors. All statistical analyses were completed using the Origin2018 (Origin Lab, Northampton, MA, USA) software.

5. Conclusions

This study showed that proper Se supply effectively alleviates the toxicity of Cd in tobacco. Selenium affected the growth of tobacco in the Se–Cd interaction by regulating the expression of the auxin-related genes and enhancing the tolerance to Cd stress by increasing the content of auxin in tobacco. Overexpression of a high-affinity phosphate transporter *NtPT2* increased the content of P and Se and decreased the accumulation of Cd. This study reveals the interaction mechanism of P and auxin in plant growth under the action of Se and Cd and provides new ideas for the safe cultivation of crops in Cd-contaminated soil.

Supplementary Materials: Supplementary materials can be found at http://www.mdpi.com/1422-0067/20/15/3772/s1.

Author Contributions: H.J. and Y.F. conceived the research project. Y.L. performed most experiments and wrote the manuscript. Y.W. and W.W. conducted the transgenic tobacco. S.S. and J.W. checked the content of Se and Cd; D.H. and H.S. revised the manuscript. All authors saw and commented on the manuscript.

Funding: This work was supported by the grants from National Natural Science Foundation of China (grant no.31301837) and the Foundation of Henan Educational Committee (grant no.15A210029).

Conflicts of Interest: The authors declare no conflict of interest.

References

1. Raiesi, F.; Razmkhah, M.; Kiani, S. Salinity stress accelerates the effect of cadmium toxicity on soil N dynamics and cycling: Does joint effect of these stresses matter? *Ecotox. Environ. Safe.* **2018**, *153*, 160–167. [CrossRef]
2. Chai, M.W.; Shi, F.C.; Li, R.L.; Liu, L.M.; Liu, Y.; Liu, F.C. Interactive effects of cadmium and carbon nanotubes on the growth and metal accumulation in a halophyte Spartina alterniflora (Poaceae). *Plant Growth Regul.* **2013**, *71*, 171–179. [CrossRef]
3. Xue, Y.; Wang, Y.Y.; Yao, Q.H.; Song, K.; Zheng, X.Q.; Yang, J.J. Research progress of plants resistance to heavy metal Cd in soil. *Ecol. Environ. Sci.* **2014**, *23*, 528–534.
4. He, X.; Richmond, M.E.; Williams, D.V.; Zheng, W.; Wu, F. Exogenous Glycinebetaine Reduces Cadmium Uptake and Mitigates Cadmium Toxicity in Two Tobacco Genotypes Differing in Cadmium Tolerance. *Int. J. Mol. Sci.* **2019**, *20*, 1612. [CrossRef]
5. Zhang, S.S.; Zhang, H.M.; Qin, R.; Jiang, W.S.; Liu, D.H. Cadmium induction of lipid peroxidation and effects on root tip cells and antioxidant enzyme activities inVicia fabaL. *Ecotoxicology* **2009**, *18*, 814–823. [CrossRef]

6. Dias, M.C.; Monteiro, C.; Moutinho-Pereira, J.; Correia, C.; Gonçalves, B.; Conceição, S. Cadmium toxicity affects photosynthesis and plant growth at different levels. *Acta Physiol. Plant* **2013**, *35*, 1281–1289. [CrossRef]
7. Parmar, P.; Kumari, N.; Sharma, V. Structural and functional alterations in photosynthetic apparatus of plants under cadmium stress. *Bot. Stud.* **2013**, *54*, 45. [CrossRef]
8. Fasahat, P.; Fasahat, P. Advances in Understanding of Cadmium Toxicity and Tolerance in Rice. *Emir. J. Food Agr.* **2014**, *27*, 94–105. [CrossRef]
9. Pinto, F.R.; Mourato, M.P.; Sales, J.R.; Moreira, I.N.; Martins, L.L. Oxidative stress response in spinach plants induced by cadmium. *J. Plant Nutr.* **2017**, *40*, 268–276. [CrossRef]
10. Kaur, N.; Sharma, S.; Kaur, S.; Nayyar, H. Selenium in agriculture: A nutrient or contaminant for crops? *Arch. Agron Soil. Sci.* **2014**, *60*, 1593–1624. [CrossRef]
11. Çakir, Ö.; Turgut-Kara, N.; Ari, Ş. Selenium induced selenocysteine methyltransferase gene expression and antioxidant enzyme activities in Astragalus chrysochlorus. *Acta Bot. Croat.* **2016**, *75*, 11–16. [CrossRef]
12. Hartikainen, H.; Xue, T.; Piironen, V. Selenium as an anti-oxidant and pro-oxidant in ryegrass. *Plant Soil.* **2000**, *225*, 193–200. [CrossRef]
13. Sors, T.G.; Ellis, D.R.; Salt, D.E. Selenium uptake, translocation, assimilation and metabolic fate in plants. *Photosynth. Res.* **2005**, *86*, 373–389. [CrossRef]
14. Zhang, L.H.; Hu, B.; Li, W.; Che, R.H.; Deng, K.; Li, H.; Yu, F.Y.; Ling, H.Q.; Li, Y.J.; Chu, C.C. OsPT2, a phosphate transporter, is involved in the active uptake of selenite in rice. *New Phytol.* **2014**, *201*, 1183–1191. [CrossRef]
15. Li, H.F.; Steve, P.M.; Zhao, F.J. Selenium uptake, translocation and speciation in wheat supplied with selenate or selenite. *New Phytol.* **2008**, *178*, 92–102. [CrossRef]
16. Jia, H.F.; Song, Z.P.; Wu, F.Y.; Ma, M.; Li, Y.T.; Han, D.; Yang, Y.X.; Zhang, S.T.; Cui, H. Low selenium increases the auxin concentration and enhances tolerance to low phosphorous stress in tobacco. *Environ. Exp. Bot.* **2018**, *153*, 127–134. [CrossRef]
17. Cary, E.E. Effect of Selenium and Cadmium Additions to Soil on Their Concentrations in Lettuce and Wheat 1. *Agron. J.* **1981**, *73*, 703–706. [CrossRef]
18. He, P.P.; Lv, X.Z.; Wang, G.Y. Effects of Se and Zn Supplementation on the Antagonism against Pb and Cd in Vegetables. *Environ. Int.* **2004**, *30*, 167–172. [CrossRef]
19. Saidi, I.; Chtourou, Y.; Djebali, W. Selenium alleviates cadmium toxicity by preventing oxidative stress in sunflower (Helianthus annuus) seedlings. *J. Plant Physiol.* **2014**, *171*, 85–91. [CrossRef]
20. Wan, Y.N.; Yu, Y.Y.; Wang, Q.; Qiao, Y.H.; Li, H.F. Cadmium uptake dynamics and translocation in rice seedling: Influence of different forms of selenium. *Ecotox. Environ. Safe.* **2016**, *133*, 127–134. [CrossRef]
21. Qin, X.M.; Nie, Z.J.; Liu, H.G.; Zhao, P.; Qin, S.Y.; Shi, Z.W. Influence of selenium on root morphology and photosynthetic characteristics of winter wheat under cadmium stress. *Environ. Exp. Bot.* **2018**, *150*, 232–239. [CrossRef]
22. Ulmasov, T.; Murfett, J.; Hagen, G.; Guilfoyle, T.J. Aux/IAA proteins repress expression of reporter genes containing natural and highly active synthetic auxin response elements. *Plant Cell.* **1997**, *9*, 1963–1971.
23. Miura, K.; Lee, J.; Gong, Q.Q.; Ma, S.S.; Jin, J.B.; Yoo, C.Y.; Miura, T.; Sato, A.; Bohnert, H.J.; Hasegawa, P.M. SIZ1 Regulation of Phosphate Starvation-Induced Root Architecture Remodeling Involves the Control of Auxin Accumulation. *Plant Physiol.* **2011**, *155*, 1000–1012. [CrossRef]
24. Sabatini, S.; Beis, D.; Wolkenfelt, H.; Murfett, J.; Guilfoyle, T.; Malamy, J.; Benfey, P.; Leyser, O.; Bechtold, N.; Weisbeek, P.; et al. An auxin-dependent distal organizer of pattern and polarity in the Arabidopsis root. *Cell* **1999**, *99*, 463–472. [CrossRef]
25. Ding, Z.; Jirí, F. Auxin regulates distal stem cell differentiation in Arabidopsis roots. *Proc. Natl. Acad. Sci. USA* **2010**, *107*, 12046–12051. [CrossRef]
26. Song, Z.P.; Shao, H.F.; Huang, H.G.; Shen, Y.; Wang, L.Z.; Wu, F.Y.; Han, D.; Song, J.Y.; Jia, H.F. Overexpression of the phosphate transporter gene, OsPT8, improves the Pi and selenium contents in, Nicotiana tabacum. *Environ. Exp. Bot.* **2017**, *137*, 158–165. [CrossRef]
27. Nedjimi, B.; Daoud, Y. Cadmium accumulation in Atriplex halimus subsp. schweinfurthii and its influence on growth, proline, root hydraulic conductivity and nutrient uptake. *Flora (Jena)* **2009**, *204*, 316–324. [CrossRef]
28. Li, H.; Li, X.; Xiang, L.; Zhao, H.M.; Li, Y.W.; Cai, Q.Y.; Zhu, L.; Mo, C.H.; Wong, M.H. Phytoremediation of soil co-contaminated with Cd and BDE-209 using hyperaccumulator enhanced by AM fungi and surfactant. *Sci. Total Environ.* **2017**, *613–614*, 447–455. [CrossRef]

29. Li, L.; Zhou, W.H.; Dai, H.X.; Cao, F.B.; Zhang, G.P.; Wu, F.B. Selenium reduces cadmium uptake and mitigates cadmium toxicity in rice. *J. Hazard. Mater.* **2012**, *235–236*, 343–351.
30. Han, D.; Xiong, S.L.; Tu, S.X.; Liu, J.C.; Chen, C. Interactive effects of selenium and arsenic on growth, antioxidant system, arsenic and selenium species of Nicotiana tabacum L. *Environ. Exp. Bot.* **2015**, *117*, 12–19. [CrossRef]
31. Ding, Y.Z.; Feng, R.W.; Wang, R.G.; Guo, J.K.; Zheng, X.Q. A dual effect of Se on Cd toxicity: Evidence from plant growth, root morphology and responses of the antioxidative systems of paddy rice. *Plant Soil.* **2014**, *375*, 289–301. [CrossRef]
32. Rafael, S.P.M.; Lucas, C.C.; Rodrigo, T.Á.; Thaline, M.P.; Lubia, S.T.; Fred, A.L.B.; Agustín, Z.; Wagner, L.A.; Ribeiro, D.M. Selenium downregulates auxin and ethylene biosynthesis in rice seedlings to modify primary metabolism and root architecture. *Planta* **2019**, *250*, 333–345.
33. Ribeiro, D.M.; Silva, J.; Dalton, D.; Cardoso, F.B.; Martins, A.O.; Silva, W.A.; Nascimento, V.L.; Araújo, W.L. Growth inhibition by selenium is associated with changes in primary metabolism and nutrient levels inArabidopsis thaliana. *Plant Cell Environ.* **2016**, *39*, 2235–2246. [CrossRef]
34. Lehotai, N.; Kolbert, Z.; Peto, A.; Feigl, G.; Ordog, A.; Kumar, D.; Tari, I.; Erdei, L. Selenite-induced hormonal and signalling mechanisms during root growth of Arabidopsis thaliana L. *J. Exp. Bot.* **2012**, *63*, 5677–5687. [CrossRef]
35. Raghothama, K.G. Phosphate transport and signaling. *Curr. Opin. Plant Biol.* **2000**, *3*, 182–187. [CrossRef]
36. Rausch, C.; Bucher, M. Molecular mechanisms of phosphate transport in plants. *Planta (Berl.)* **2002**, *216*, 23–37. [CrossRef]
37. Yu, J.; Hu, S.; Wang, J.; Wong, G.K.; Li, S.; Liu, B.; Deng, Y.; Dai, L.; Zhou, Y.; Zhang, X.; et al. A Draft Sequence of the Rice Genome (Oryza sativa L. ssp. japonica). *Science* **2002**, *296*, 92–100. [CrossRef]
38. Ai, P.H.; Sun, S.B.; Zhao, J.N.; Fan, X.R.; Xin, W.J.; Guo, Q.; Yu, L.; Shen, Q.R.; Wu, P.; Miller, A.J.; et al. Two rice phosphate transporters, OsPht1;2 and OsPht1;6, have different functions and kinetic properties in uptake and translocation. *Plant J.* **2009**, *57*, 798–809. [CrossRef]
39. Wang, P.; Zhang, W.; Mao, C.; Xu, G.; Zhao, F.J. The role of OsPT8 in arsenate uptake and varietal difference in arsenate tolerance in rice. *J. Exp. Bot.* **2016**, *67*, 6051–6059. [CrossRef]
40. Xie, L.H.; Xu, Z.R. The toxicity of heavy metal Cadmium to animals and humans. *Acta Agric. Zhejianggensis* **2003**, *6*, 52–57. (In Chinese)
41. Prozialeck, W.C.; Edwards, J.R.; Nebert, D.W.; Woods, J.M.; Barchowsky, A.; Atchison, W.D. The Vascular System as a Target of Metal Toxicity. *Toxicol. Sci.* **2007**, *102*, 207–218. [CrossRef]
42. Jia, H.F.; Zhang, S.T.; Wang, L.Z.; Yang, Y.X.; Zhang, H.Y.; Cui, H.; Shao, H.F.; Xu, G.H. OsPht1;8, a phosphate transporter, is involved in auxin and phosphate starvation response in rice. *J. Exp. Bot.* **2017**, *68*, 5057–5068. [CrossRef]
43. Li, Z.; Tan, X.F.; Lu, K.; Zhang, L.; Long, H.X.; Lv, J.B.; Lin, Q. Influence of drought stress on the growth, leaf gas exchange, and chlorophyll fluorescence in two varieties of tung tree seedlings. *Acta Ecol. Sin.* **2017**, *37*, 1515–1524.
44. Feng, R.W.; Wei, C.Y.; Tu, S.X.; Sun, X. Interactive effects of selenium and arsenic on their uptake by Pteris vittata L. under hydroponic conditions. *Environ. Exp. Bot.* **2009**, *65*, 363–368. [CrossRef]
45. Han, S.; Fang, L.; Ren, X.J.; Wang, W.L.; Jiang, J. MPK6 controls H_2O_2-induced root elongation by mediating Ca^{2+} influx across the plasma membrane of root cells in Arabidopsis seedlings. *New Phytol.* **2015**, *205*, 695–706. [CrossRef]
46. Sun, H.W.; Tao, J.Y.; Liu, S.J.; Huang, S.J.; Chen, S.; Yoneyama, Y.; Zhang, Y.L.; Xu, G.H. Strigolactones are involved in phosphate- and nitrate-deficiency-induced root development and auxin transport in rice. *J. Exp. Bot.* **2014**, *65*, 6735–6746. [CrossRef]
47. Chen, A.Q.; Hu, J.; Sun, S.B.; Xu, G.H. Conservation and divergence of both phosphate- and mycorrhiza-regutelad physiological responses and expression patterns of phosphate transporters in solanaceous species. *New Phytol.* **2007**, *173*, 817–831. [CrossRef]
48. Wei, C.Y.; Sun, X.; Wang, C.; Wang, W.Y. Factors influencing arsenic accumulation by Pteris vittata: A comparative field study at two sites. *Environ. Pollut.* **2006**, *141*, 488–493. [CrossRef]

© 2019 by the authors. Licensee MDPI, Basel, Switzerland. This article is an open access article distributed under the terms and conditions of the Creative Commons Attribution (CC BY) license (http://creativecommons.org/licenses/by/4.0/).

International Journal of
Molecular Sciences

Article

Ectopic Expression of Poplar ABC Transporter PtoABCG36 Confers Cd Tolerance in *Arabidopsis thaliana*

Huihong Wang †, Yuanyuan Liu †, Zaihui Peng, Jianchun Li, Weipeng Huang, Yan Liu, Xuening Wang, Shengli Xie, Liping Sun, Erqin Han, Nengbiao Wu, Keming Luo and Bangjun Wang *

Chongqing Key Laboratory of Plant Resource Conservation and Germplasm Innovation, Key Laboratory of Eco-Environments in Three Gorges Reservoir Region (Ministry of Education), College of Life Sciences, Southwest University, Chongqing 400715, China

* Correspondence: bangjunwang@swu.edu.cn; Tel.: +86-23-6825-3235

† These authors contributed equally to this work.

Received: 17 May 2019; Accepted: 1 July 2019; Published: 4 July 2019

Abstract: Cadmium (Cd) is one of the most toxic heavy metals for plant growth in soil. ATP-binding cassette (ABC) transporters play important roles in biotic and abiotic stresses. However, few ABC transporters have been characterized in poplar. In this study, we isolated an ABC transporter gene *PtoABCG36* from *Populus tomentosa*. The *PtoABCG36* transcript can be detected in leaves, stems and roots, and the expression in the root was 3.8 and 2 times that in stems and leaves, respectively. The *PtoABCG36* expression was induced and peaked at 12 h after exposure to Cd stress. Transient expression of *PtoABCG36* in tobacco showed that PtoABCG36 is localized at the plasma membrane. When overexpressed in yeast and Arabidopsis, PtoABCG36 could decrease Cd accumulation and confer higher Cd tolerance in transgenic lines than in wild-type (WT) lines. Net Cd^{2+} efflux measurements showed a decreasing Cd uptake in transgenic Arabidopsis roots than WT. These results demonstrated that PtoABCG36 functions as a cadmium extrusion pump participating in enhancing tolerance to Cd through decreasing Cd content in plants, which provides a promising way for making heavy metal tolerant poplar by manipulating ABC transporters in cadmium polluted areas.

Keywords: Cd; *PtoABCG36*; tolerance; poplar; accumulation; efflux

---

1. Introduction

Cadmium (Cd) is a highly toxic pollutant in the environment. Cadmium is nephrotoxic, and it can lead to serious human diseases, including kidney disorders, bone damage and neurotoxicity [1]. For example, high environmental exposure in Japan resulting from a stable diet of cadmium contaminated rice caused itai-itai disease [2]. Cadmium can inactivate or denature proteins by binding to the sulfhydryl groups, leading to cellular damage by displacing co-factors from a variety of proteins including transcription factors and enzymes, and by indirectly generating reactive oxygen species [3,4]. Heavy metal pollution in agricultural soils has become a serious problem. Therefore, it is essential to prevent cadmium from getting into the food chain and make the best use of cadmium contaminated soil.

Plants are able to tolerate heavy metal stress to a certain extent, with the participation of some transporters. These transporters can enhance heavy metal tolerance by pumping heavy metals into vacuoles or out of cells. Previous studies showed that two type 1(B) heavy metal-transporting subfamily of the P-type ATPases AtHMA2 and AtHMA4 are localized at the plasma membrane and can transport excessive zinc and cadmium to the outside of the cell in *Arabidopsis thaliana*, which are important players in the plant detoxification process [5]. The members of the cation diffusion facilitator (CDF) family, natural resistance-associated macrophage protein (Nramp) and Zrt/IRT-like protein (ZIP) families of transporters are also involved in the transport of heavy metals in a variety of organisms [6,7]. In addition, ATP-binding cassette (ABC) transporters are essential for plant growth and development. ABC transporters are driven by ATP hydrolysis and can act as exporters as well as importers. The Arabidopsis nuclear genome encodes for more than 100 ABC transporters, which are divided into eight subfamilies (ABCA, ABCB, ABCC, ABCD, ABCE, ABCF, ABCG and ABCI), largely exceeding that of animal. Most plant ABC transporters are present in cell membranes and are involved in detoxification processes, organ growth, plant nutrition, plant development and response to abiotic and biotic stresses [8]. Some ABC transporters are closely related to the detoxification of heavy metals. In *Saccharomyces cerevisiae*, an ABCC-like heavy metal transporter ScYCF1 (yeast cadmium factor 1) has been found to contribute to detoxifying cadmium by pumping it into vacuoles [9], and overexpression of *ScYCF1* in Arabidopsis can improve cadmium tolerance [10]. Similarly, a half-size ABC transporter HMT1 (heavy metal tolerance 1) from *Schizosaccharomyces pombe* can transport phytochelatin–Cd complexes into the vacuole, which is considered to be the first transporter to transport heavy metal-phytochelatin complexes [11]. AtABCB25/AtATM3, a close homolog of SpHMT1, contributes to Cd resistance and can transport glutamine synthetase conjugated Cd (II) across the mitochondrial membrane [12]. Full-size ABC transporters AtABCC1 and AtABCC2 have been demonstrated to be major vacuolar phytochelatins (PCs) transporters to participate in arsenic (As), mercury (Hg) and Cd resistance in Arabidopsis [13], and their homologous rice ABCC transporter OsABCC1 is involved in the As detoxification and reduces As accumulation in the rice grains [14]. Another homologous ABCC transporter PtABCC1 can enhance tolerance to Cd in poplar [15]. It has reported that AtABCC3 can complement the Cd sensitive phenotype of the *ycf1* mutant in *Saccharomyces cerevisiae* [16]. The level of expression of *AtABCC6/AtMRP6* can be up-regulated in response to cadmium (Cd) treatment [17]. Furthermore, some ABCG subfamily transporters are also involved in heavy metal resistance. AtABCG36/PDR8, localized at the plasma membrane, plays an important role in Cd extrusion from root cells [18]. The transcription of cucumber genes *CsPDR8/CsABCG36* can be up-regulated under Cd stress [19]. The rice OsABCG43/PDR5 is a Cd inducible-transporter and confers high Cd resistance in yeast cells [20].

Poplar is a woody plant with established genetic transformation system and abundant biomass. In China, there is a very large number of *Populus tomentosa*, taking up very large land resources. The previous research on poplar mainly focuses on insect resistance, herbicide resistance, biomass traits, stress tolerance, disease resistance, hormone modification, flowering modification and phytoremediation [21]. Recently, a study found that exogenous abscisic acid (ABA) stimulated the expression level of poplar ABCG40 transporter involved in lead (Pb) uptake, transport and detoxification [22]. The two multidrug and toxic compound extrusion (MATE) family genes *PtrMATE1* and *PtrMATE2* from poplar induced by aluminum (Al) can enhance aluminum resistance in acidic soils [23]. The *YCF1*-expressing transgenic poplar plants exhibited enhanced growth, reduced toxicity symptoms, and increased Cd content in the aerial tissue compared to the non-transgenic plants [24]. However, there are few studies on poplar ABCG transporters involved in Cd resistance. Therefore, the engineering of *Populus tomentosa* by manipulating ABC transporters is a significant step towards the effective utilization of Cd contaminated soil.

In the present study, we cloned a novel ABC transporter gene *PtoABCG36* (GenBank accession: MH660448) by BLAST search in the poplar database using *AtABCG36* as a query sequence. Yeast and Arabidopsis overexpressing *PtoABCG36* were measured in terms of their Cd tolerance and Cd content after Cd treatment. The results showed that overexpressing *PtoABCG36* is effective in enhancing Cd tolerance through decreasing Cd content in plants, indicating that PtoABCG36 transporter functions as a cadmium extrusion pump to participate in Cd stress in plants, which provides a reasonable way to make heavy metal tolerant poplar by manipulating ABC transporters in the areas with cadmium pollution.

2. Results

2.1. Structural and Phylogenetic Analysis of PtoABCG36

PtoABCG36 was isolated from full-length cDNA of leaves of six-month-old *Populus tomentosa* and submitted to GenBank (accession number: MH660448). The sequence encoded 1478 amino acid residues and contained two putative transmembrane domains (TMD) and two putative nucleotide-binding domains (NBD) (Figure 1A). Each NBD domain has about 200 amino acid residues, and it contains a Walker A motif (GXXGXGKS/T), a Walker B motif (hhhhD) and an ABC signature motif (LSGGQQ/R/KQR) [25]. Some ABCG subfamily transporters have been identified in many plant species, including *Arabidopsis thaliana, Glycine Max, Ricinus conmunis, Vitis vinifera, Gossypium arboretum* and *Oryza sativa*. The two NBD domains are highly conserved (Figure 1A).

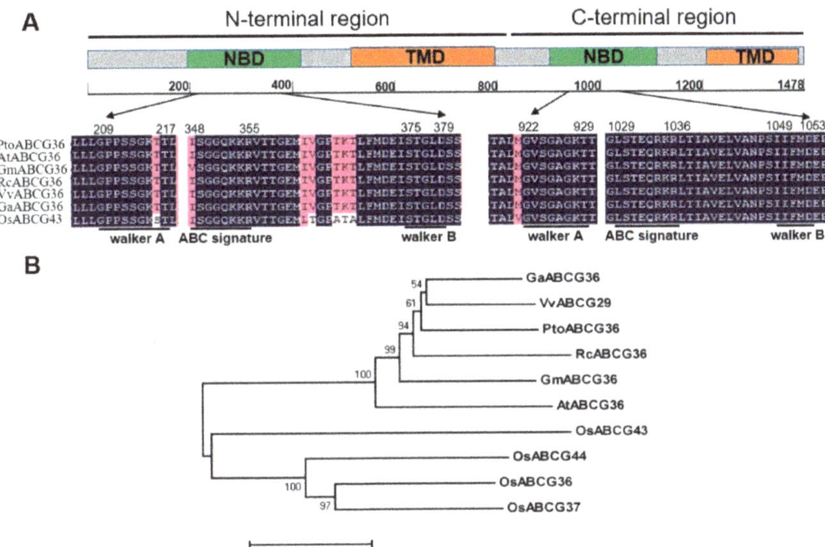

Figure 1. Amino acid sequence alignment and phylogenetic analysis. (**A**) Structure analysis and amino acid multi-alignment of the nucleotide-binding domains (NBD) of ABCG proteins from different plant species. ABCG domains are marked as two green and orange blocks. TMD, transmembrane domain; Walker A, ATP-binding cassette (ABC) signature; Walker B, NBD associated motifs. Blue indicates identical amino acids; pink indicates similar amino acids. (**B**) Phylogenetic analysis of ABCG proteins from *Populus trichocarpa* (PtoABCG36, MH660448); *Arabidopsis thaliana* (AtABCG36, NP_176196); *Oryza sativa* (OsABCG36, XP_015648358; OsABCG37, XP_015648329; OsABCG43, XP_015646575; OsABCG44, XP_015650488); *Glycine Max* (GmABCG36, XP_006585572); *Ricinus conmunis* (RcABCG36, XP_002515970); and *Vitis vinifera* (VvABCG29, XP_010654625); *Gossypium arboretum* (GaABCG36, XP_017606959). The numbers beside the branches represent bootstrap values based on 1000 replications.

To investigate the homology between PtoABCG36 and other plant species, ten plant ABCG transporters were analyzed. PtoABCG36 had 81.9%, 81.4%, 78.4%, 78.2% and 74.1% amino acid sequence similarity to GaABCG36 (XP_017606959), VvABCG29 (XP_010654625), GmABCG36 (XP_006585572), RcABCG36 (XP_002515970) and AtABCG36 (NP_176196), respectively. Phylogenetic analysis also revealed that PtoABCG36 was homologous with the ABCG proteins from dicotyledons such as *Vitis vinifera, Gossypium arboretum, Ricinus conmunis, Glycine Max* and *Arabidopsis thaliana*, as well as monocotyledons such as *Oryza sativa* (Figure 1B).

2.2. The PtoABCG36 Gene Is Highly Expressed in Response to Cd Stress in Poplar

To confirm the function of the PtoABCG36 transporter, we measured its gene expression level. *PtoABCG36* transcript can be detected in leaves, stems and roots, and the expression in the root was 3.8 and 2 times that of the stems and leaves, respectively. The higher expression level in the roots indicated that PtoABCG36 mainly functioned in the roots (Figure 2A). In addition, to confirm the function of PtoABCG36 in response to Cd stress, we performed induced expression using quantitative real-time PCR after the six-month-old poplars were immersed in woody plant medium (WPM) supplemented with different concentrations of $CdCl_2$ for 12 h. Poplar gene-specific primers were used for qRT-PCR analysis of *PtoABCG36*. The results showed that the expression of *PtoABCG36* was significantly increased in roots with increasing cadmium concentration and reached the highest level when treated with 100 μM $CdCl_2$ for 12 h. *PtoABCG36* expression was also significantly increased in stems and leaves but not as highly as that in roots. However, when treated with 150 or 200 μM $CdCl_2$, the expression of the *PtoABCG36* gradually declined, but it could still be induced in roots, stems and leaves (Figure 2B). Furthermore, temporal spatial expression analysis upon treatment with 100 μM $CdCl_2$ for 24 h showed that *PtoABCG36* transcript increased overtime and peaked at 12 h, with a level seven times that of the control, then gradually decreased (Figure 2C). These results further determined that *PtoABCG36* could be induced and participate in resisting Cd stress.

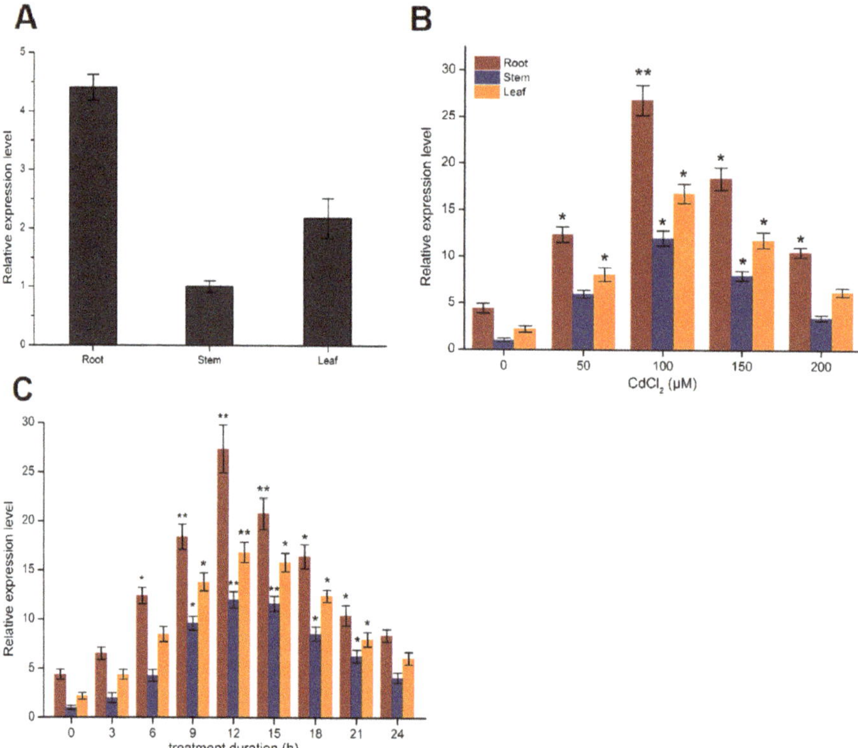

Figure 2. Expression analysis of *PtoABCG36* gene. (**A**) Relative expression level of *PtoABCG36* gene in roots, stems, leaves of *Populus tomentosa*. (**B**) Expression of *PtoABCG36* in poplar roots, stems and leaves under different concentrations of Cd^{2+} for12 h. (**C**) Time course of *PtoABCG36* expression in poplar roots, stems and leaves in response to 100 μM Cd^{2+} treatment. The results are shown as the mean expression ± standard deviation (SD) of three independent experiments. Poplar ubiquitin (*UBQ*) expression was used as a control and gene-specific primers were used for qRT-PCR analysis of *PtoABCG36* gene. Student's t-test, * $p < 0.05$, ** $p < 0.01$.

2.3. The PtoABCG36 Transporter is Localized at the Plasma Membrane

In order to determine the subcellular localization of PtoABCG36, the 35S:*PtoABCG36*-GFP construct, in which the *PtoABCG36*-GFP fusion gene was driven by the CaMV 35S promoter, was transiently expressed in the leaves of three-week-old *Nicotiana benthamiana*. Compared with the control where GFP was observed at the plasma membrane (PM), endoplasmic reticulum (ER) and nucleus (NU) in the epidermal cells (Figure 3A–D), the PtoABCG36 signal was observed only at the plasma membrane (Figure 3E–H), indicating that PtoABCG36 is localized at the plasma membrane to function as transporter, consistent with the localization pattern of AtABCG36 in *Arabidopsis thaliana*.

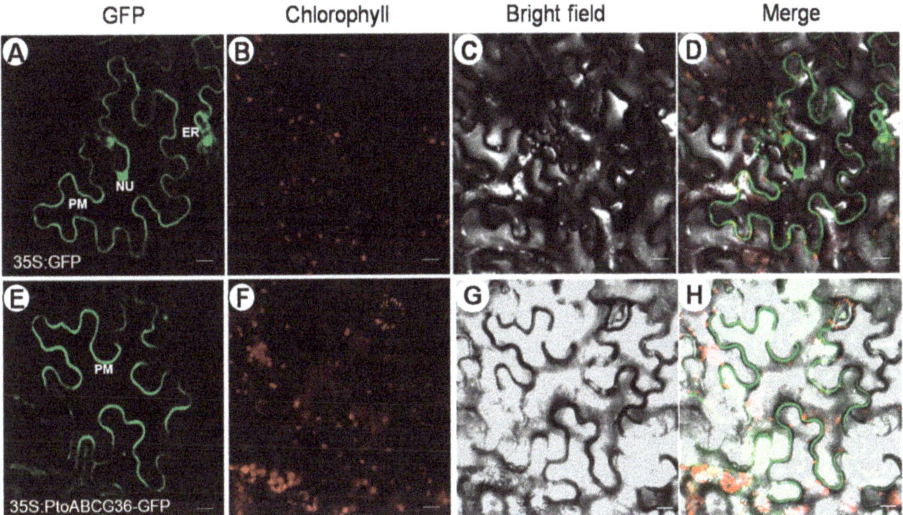

Figure 3. Subcellular localization of PtoABCG36 in epidermal cells of *Nicotiana benthamiana*. The fluorescence of green fluorescent protein (GFP) or PtoABCG36-GFP signal in tobacco leaf cell (**A**,**E**). Chlorophyll autofluorescence (**B**,**F**). Bright field (**C**,**G**). The overlap images of bright field and fluorescence images (**D**,**H**). Scale bars = 20 μm. NU, nucleus; PM, plasma membrane; ER, endoplasmic reticulum.

2.4. Heterologous Expression of PtoABCG36 Confers Cd Tolerance in Yeast

To investigate whether PtoABCG36 is involved in Cd tolerance, pDR-*PtoABCG36* was produced and transformed into the yeast Cd sensitive mutant strain Δ*yap1* and wild-type strain Y252. We found that on the SD-Ura medium, growth was similar between the yeast cells carrying the empty vector and those expressing *PtoABCG36*. However, on the SD-Ura medium containing 100 μM or 200 μM $CdCl_2$, the Δ*yap1* or Y252 with pDR-*PtoABCG36* exhibited stronger Cd tolerance than mutants or wild-type with the empty vector (Figure 4A). Yeast growth in liquid SD-Ura medium containing 40 μM $CdCl_2$ was analyzed overtime. In the absence of Cd, there was no growth difference between the *PtoABCG36*-carrying yeast and the control (Figure 4B). However, upon $CdCl_2$ exposure, the growth of *PtoABCG36*-carrying Δ*yap1* and Y252 were better than the yeast cells carrying the empty vector. Additionally, complementary strains partially restored their tolerance to Cd (Figure 4C), further confirming heterologous expression of *PtoABCG36* could confer Cd tolerance in yeast.

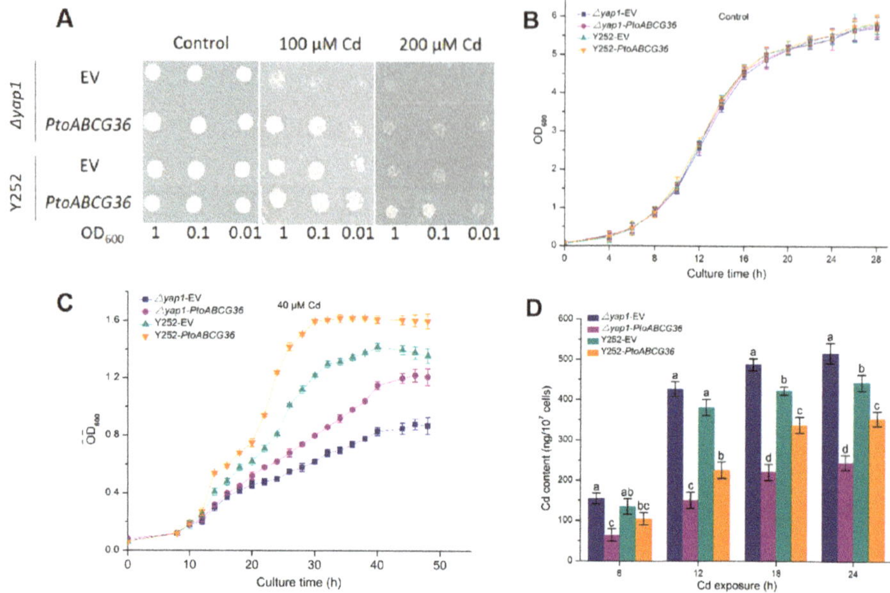

Figure 4. PtoABCG36 enhances cadmium tolerance in yeasts. (**A**) Δ*yap1* and the wild-type Y252 were transformed with EV (pDR196 empty vector) and pDR196-*PtoABCG36*, and grown on SD plates with indicated concentrations of $CdCl_2$ for 7 d. (**B**,**C**) Growth curves of yeast cells Δ*yap1*-EV (square), Δ*yap1*-*PtoABCG36* (circle), Y252- EV (up-triangle) and Y252- *PtoABCG36* (down-triangle) under control (**B**) and 40 μM $CdCl_2$ condition (**C**) for indicated time. (**D**) Accumulation of cadmium in Δ*yap1*-EV (navy blue), Δ*yap1*-*PtoABCG36* (purple), Y252-EV (dark cyan) and Y252-*PtoABCG36* (orange) yeasts. Yeast cells (1×10^7) were exposed to 40 μM Cd treatment for 6, 12, 18 or 24 h at 30 °C. Cd concentrations in the yeast cells were measured by ICP-OES. Error bars indicate standard deviation ($n = 3$). Different letters indicated significant differences ($p < 0.05$).

Previous studies have shown that yeast could resist cadmium by transporting it into the vacuoles or out of the cells. We tested Cd concentration in the yeast cells culturing in liquid SD-Ura medium containing 40 μM $CdCl_2$. As shown in Figure 4D, after 24 h of treatment, the accumulation of Cd in *PtoABCG36*-carrying Δ*yap1* and Y252 was significantly less (52.5% and 20.3% less, respectively) than that in mutant and wild-type. These results indicated that PtoABCG36 can contribute to Cd resistance by transporting it out of the yeast cells.

2.5. Overexpression of PtoABCG36 Increases Tolerance to Cd and Decreases Cd Accumulation in Plants

In order to investigate the function of PtoABCG36 in plants, the construct 35S:*PtoABCG36* was introduced into Arabidopsis. The *PtoABCG36* transcript levels were detected by qRT-PCR for further analysis (Supplementary Figure S2). Arabidopsis transgenic plants T4, wild-type and mutant seeds were analyzed after treatment without Cd and with 20 μM, 40 μM or 60 μM $CdCl_2$ for 2 weeks. There was no growth difference among these lines in the absence of Cd, while the growth of Arabidopsis was significantly inhibited when grown on half MS agar plates containing 20 μM, 40 μM and 60 μM $CdCl_2$. The *abcg36* mutants displayed shortest roots. However, the transgenic plants had longer roots and grew better than wild-type plants (Figure 5A,B), indicating that PtoABCG36 was also involved in mediating tolerance to Cd in plants. Quantitative analysis showed that the roots of overexpression lines (OX-2 and OX-3) were significantly longer than those of wild-type plants in the presence of 40 μM $CdCl_2$ (44% and 48% longer, respectively) and 60 μM $CdCl_2$ (116.7% and 112.5% longer, respectively). These results further indicated that PtoABCG36 enhanced tolerance to Cd in plants.

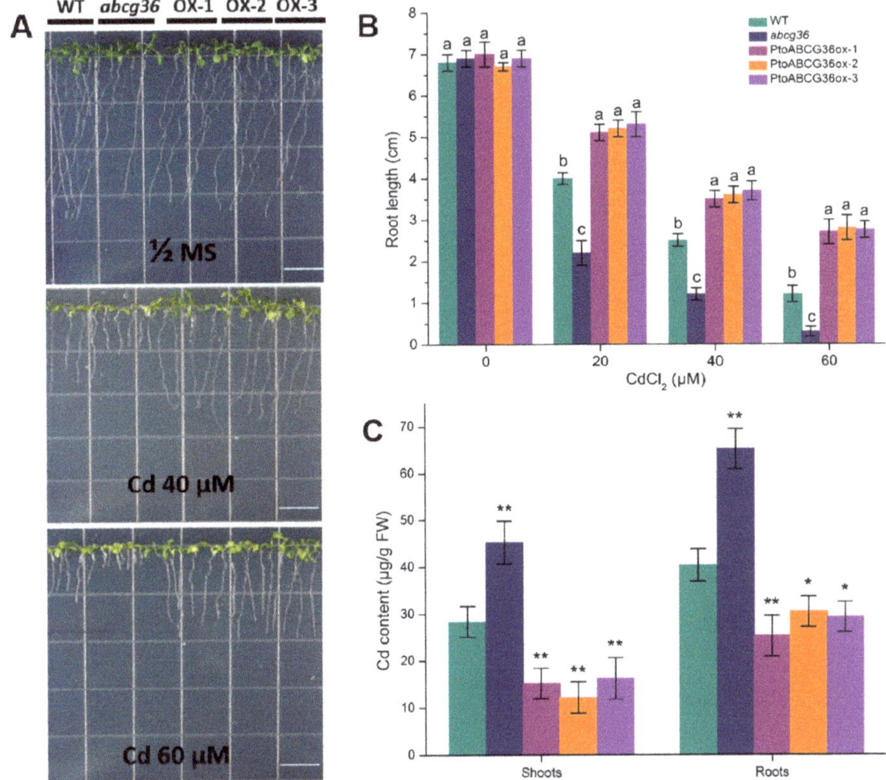

Figure 5. PtoABCG36 enhances cadmium tolerance in Arabidopsis. Arabidopsis seeds were grown on half-strength MS medium containing 0, 40 or 60 μM $CdCl_2$ for two weeks (**A**) and primary root length (**B**) were analyzed. (**C**) Accumulation of cadmium in plants after treatment with the half MS liquid medium containing 100 μM $CdCl_2$ for 24 h. For root lengths, n = 120–124 from three independent experiments. WT, wild-type; *abcg36*, *abcg36* mutant SALK_1422526; OX-1 and OX-2, OX-3 *PtoABCG36*-overexpressing Arabidopsis lines. White bars = 15 mm. Cd concentrations in plants were measured by ICP-OES. Error bars indicate standard deviation. Different letters indicated significant differences ($p < 0.05$). Student's t-test, * $p < 0.05$, ** $p < 0.01$.

To explain the detoxification mechanism of PtoABCG36 in plants, we tested the cadmium content in the mutants, wild-type and transgenic plants after treatment with the half MS liquid medium containing 100 μM $CdCl_2$ for 24 h. We found that transgenic plants OX-1, OX-2 and XO-3 had lower Cd content than wild-type in the shoots (46.68%, 57% and 42.91% lower, respectively) and roots (37.42%, 22.3% and 27.37% lower, respectively). In contrast, the mutant *abcg36* had higher Cd content than the wild-type in the roots (61.65% higher) and shoots (59.24% higher). More importantly, the levels of cadmium reduction in the roots of transgenic plants were much greater than those in the shoots (Figure 5C), suggesting that PtoABCG36 contributed to Cd tolerance by pumping it out of the plants and reducing Cd toxicity in plant roots.

To further determine the function of PtoABCG36 in plant roots, we investigated the Cd^{2+} uptake in root tips of *abcg36* mutants, WT and plants overexpressing *PtoABCG36* through a non-invasive micro-test (NMT) technique. In the presence of 50 μM $CdCl_2$, the net Cd^{2+} influxes of OX-1, OX-2 and OX-3 lines were lower than WT plants (62.39%, 54.50% and 53.30% lower, respectively) (Figure 6). In contrast, the mutant *abcg36* had higher Cd net Cd^{2+} influx than the WT plants. These results indicated that a decreasing Cd uptake capacity existed in lines overexpressing *PtoABCG36* than the WT plants.

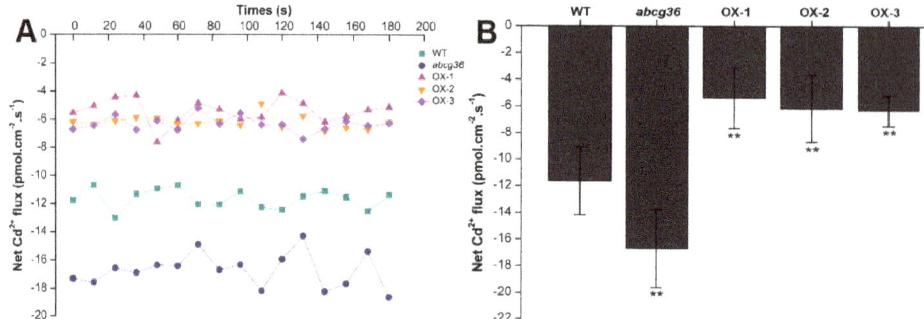

Figure 6. Net Cd^{2+} fluxes. Net Cd^{2+} fluxes in the roots of WT, *abcg36* mutant and transgenic plants (OX-1, OX-2, and OX-3) treated with $CdCl_2$ stress (**A**). The average 180 s net Cd^{2+} fluxes are illustrated to highlight the trend differences (**B**). Bars indicate means ± SD. Student's t-test, * $p < 0.05$, ** $p < 0.01$.

3. Discussion

To date, how to effectively use soil containing cadmium has become a worldwide problem. Previous studies have showed that several transporters, including the P-type ATPases AtHMA2 and AtHMA4, the CDF, Nramp and ZIP families of transporters and ABC transporters could be involved in the heavy metal tolerance [5–7,9–11].

In this study, we identified the ABC transporter ABCG36 of *Populus tomentosa*. Protein sequence analysis showed that it contained conserved Walker A, Walker B, and ABC signal (Figure 1A). In previous studies, Walker A, Walker B, ABC signal of NBD were demonstrated to function as ABC transporters motifs [26]. Phylogenetic tree analysis showed that PtoABCG36 in poplar is an ortholog of Arabidopsis AtABCG36, which acts as transporter involved in biotic or abiotic stress [18,27] (Figure 1B). Expression pattern showed that the accumulation of *PtoABCG36* transcript was mainly detected in the roots (Figure 2A). In line with our results (Figure 2B,C), it has been also reported that transcript levels of ABCG transporters were induced rapidly by biotic or abiotic stress [28–32]. Interestingly, *PtoABCG36* expression was induced by Cd, peaking at 12 h after Cd treatment. Additionally, the expression of *PtoABCG36* was significantly higher in poplar roots than that in shoots under Cd treatment, which is different from its ortholog in other species (Figure 2C).

It is important for plants to cope with heavy metal stress. In this study, first, we found that ectopic expression of *PtoABCG36* in yeast and Arabidopsis all significantly increased Cd tolerance (Figures 3 and 5). Interestingly, our data showed that the growth of *PtoABCG36*-carrying *Δyap1* yeast stain, which has a lower level of Cd, was not better than that of the wild-type Y252 (Figure 4C,D). It is known that Yap1 increased cellular tolerance to cadmium by activating the expression of *ScYCF1* as a transcription factor. Yeast wild-type Y252 can resist Cd stress through ABC transporter ScYCF1 localized at vacuolar membrane and plasma membrane pumping Cd into vacuoles or out from the cells [10]. The expression of *YCF1* in Y252-*PtoABCG36* could pump Cd into vacuoles, while inhibition of *YCF1* in *Δyap1-PtoABCG36* could decrease the transport of heavy metals to vacuoles. Therefore, Y252-*PtoABCG36* has higher accumulation of Cd compared to *Δyap1-PtoABCG36* (Figure 4D). In addition, our data indicated that Arabidopsis *PtoABCG36*-overexpressing lines could enhance Cd tolerance (Figure 5). The *abcg36* plants are sensitive to Cd, whereas the *PtoABCG36*-overexpressing plants are tolerant (Figure 5). *PtoABCG36*-overexpressing plants have reduced cadmium content in their shoots and roots, but *abcg36* plants were the opposite. The wild-type plants accumulate 1.2 to 1.5 times as much Cd in roots and shoots as the transgenic plants (Figure 5C), suggesting that the overexpression of *PtoABCG36* could expel heavy metals from plants. Non-invasive micro-test (NMT) technique showed that overexpressing *PtoABCG36* can decrease Cd uptake capacity in plants (Figure 6). The detoxification mechanism of PtoABCG36 might be similar to that of its homologous AtABCG36 located at the plasma membrane, which can transport Cd out from the cells.

Taken together, our study provided the evidence for the biological functions of PtoABCG36 as a transporter in regulating Cd resistance in plants. Additionally, it plays a crucial role in reducing Cd accumulation in plants, providing a theoretical basis to make heavy metal tolerant poplar by manipulating ABC transporters in cadmium polluted areas. The present study has also provided insight on the roles of ABCG transporters in economic forest cultivation.

4. Materials and Methods

4.1. Materials and Growth Conditions

Arabidopsis seeds of wild-type (ecotype Columbia-0), *abcg36* (a loss-of-function mutant of *AtABCG36*, SALK_1422526) [18], and transgenic plants OX-1, OX-2, OX-3 were vernalized in the dark at 4 °C for 2 days, and then grew on half-strength MS agar medium plates containing 1.5% sucrose in a controlled environment with a 16 h light with 120 µmol m^{-2} s^{-1} light intensity and 8 h dark at 22 °C/18 °C for the indicated duration.

P. tomentosa Carr. (clone 741) (Chinese white poplar), kindly provided by Institute of Resources Botany, Southwest University, and transgenic poplars were cultivated in a greenhouse at 24 °C under a 14 h/10 h light/dark cycle with 45 µmol m^{-2} s^{-1} of light and maintained in sterile woody plant medium (WPM) containing 0.8% (w/v) agar. Gene expression patterns were analyzed in leaves, roots and stems from 6-month-old plants.

4.2. Gene Cloning, Expression Vector Construction, Structural and Phylogenetic Analysis of PtoABCG36

Total RNA was extracted from the leaves of 6-month-old *P. tomentosa Carr.* by using the Trizol Reagent (Tiangen, China), then revers transcribed to cDNA by using the RT-AMV transcriptase Kit (TaKaRa, Dalian, China). The *PtoABCG36* specific fragment was amplified by PCR using specific primers (Supplementary Table S1). Cycling conditions were: 98 °C for 3 min followed by 34 cycles of 98 °C for 30 s, 56.6 °C for 30 s and 72 °C for 2 min 58 s, adding a final prolongation step at 72 °C for 10 min. The amplification products were cloned into the *Bam*HI site of the plant binary vector pCAMBIA-1300-GFP [33] as well as the *Spe*I and *Xma*I sites of the yeast vector pDR196 [34], to construct pCAMBIA-1300-*PtoABCG36* and pDR196-*PtoABCG36*.

Prediction and analysis of the structure of PtoABCG36 protein was performed with the Simple Modular Architecture Research Tool (SMART, http://smart.embl-heidelberg.de). The homologous amino acid sequences of PtoABCG36 in other species were downloaded from NCBI (http://www.ncbi.nlm.nih.gov), and aligned with DNAMAN 8.0 (Lynnon Biosoft, San Ramon, CA, USA). The phylogenetic analysis of amino acid sequences was carried out with MEGA 5.0 software by using neighbor-joining (NJ).

4.3. Transformation and Selection for Yeast and Arabidopsis

The yeast expression vectors pDR196 and pDR196-*PtoABCG36* were transformed into the Cd sensitive-yeast mutant Δ*yap1* (*MATa ura3 lys2 ade2 trp1 leu2 yap1::leu2*) and the wild-type Y252 (*MATa ura3 lys2 ade2 trp1 leu2*) [35], kindly provided by Ji-Ming Gong (Shanghai Institutes for Biological Sciences, Chinese Academy of Sciences, shanghai, China) for metal sensitivity assay, as described [36]. Yap1 is a transcription factor that increases the tolerance of cells to cadmium by activating *YCF1* expression [37].

pCAMBIA-1300-*PtoABCG36* was transformed into the *Agrobacterium tumefaciens* strain GV3101, then transformed into wild-type Arabidopsis by the floral dip method [38]. The selection of putative transgenic plants was performed on half MS medium with 40 mg/L hygromycin and 200 mg/L cefotaxime, and further confirmed by PCR analysis (Supplementary Figure S1) and qRT-PCR analysis (Supplementary Figure S2).

4.4. The Metal Assay of Yeast Cells and Plants

For phenotypic analysis, yeast cells were cultured in SD-Ura liquid medium to log phase and diluted to the corresponding concentration after collection, then spotted onto SD-Ura plates containing 100 μM and 200 μM $CdCl_2$. Plates were kept at 30 °C for 7 days before being photographed. Yeast cells were also cultured in liquid medium containing 40 μM $CdCl_2$ for 12 h and OD_{600} was measured at indicated time [35].

For phenotypic analysis, Arabidopsis transgenic plants T4, wild-type and mutant seeds were grown on half MS agar plates in the absence or presence of 20, 40 and 60 μM $CdCl_2$ for 2 weeks before being photographed and the averages of root lengths were measured in different experiments. Four untreated seedlings, each with a distinctive genotype, were grown in the half MS liquid medium with 100 μM $CdCl_2$ for 24 h, and were used for determination of cadmium content. Three technical replicates were performed.

For induced expression experiment, 6-month-old poplars were immersed in WPM medium supplemented with different concentrations of $CdCl_2$ for 12 h. Meanwhile, poplars treated with WPM medium without Cd were used as control. For the temporal spatial expression analysis, 6-month-old poplars were immersed in WPM medium supplemented with 100 μM $CdCl_2$. Roots, stems and leaves were collected every 3 h for real-time quantitative PCR. Three technical replicates were performed.

4.5. Subcellular Localization of PtoABCG36

PtoABCG36 was ligated into pCAMBIA1300-GFP vector to produce 35S:*PtoABCG36*-GFP, which was transiently expressed in the leaves of 3-week-old *Nicotiana benthamiana* to examine the subcellular localization of PtoABCG36 after 72 h of infiltration. The 35S:*PtoABCG36*-GFP construct was transformed into GV3101 cells. The cells were grown at 28 °C to OD_{600} of 0.8, resuspended in infiltration buffer (10 mM MES, pH=5.7, 10 mM $MgCl_2$, and 100 μM acetosyringone) to adjust the OD_{600} to 0.6 and infiltrated into 3-week-old *Nicotiana benthamiana* leaves. Analysis was carried out with a confocal microscope (Olympus FV1200, Tokyo, Japan). Conditions for imaging were set as 488-nm excitation, collecting bandwidth at 500 to 552 nm for GFP, 633-nm excitation, collecting bandwidth at 650 to 750 nm for chlorophyll autofluorescence.

4.6. Quantitative Real-Time PCR Analysis

Total RNA was extracted from different plant tissues by using the RNA RNeasy Plant Mini Kit (Qiagen, Duesseldorf, Germany). First-strand cDNA synthesis was performed using the PrimeScript™ RT reagent kit (Perfect Real Time; Takara, Dalian, China). qRT-PCR was performed to detect the transcript of *PtoABCG36* in Arabidopsis and poplar by using the SYBR Green-based qPCR Master Mix (Promega, Madison, WI, USA). The gene-specific primers for qRT-PCR are listed in the supplementary Table S1. The poplar reference gene *UBQ* (FJ438462) was used as an internal control to normalize the expression data. The PCR conditions and relative gene expression calculations were conducted as previously described [14]. Three biological replicates and three technical replicates were performed.

4.7. Determination of Cadmium Content in Yeasts and Plants

Cells of each line (1×10^7) were added to 30 mL of liquid SD-Ura medium containing 40 μM $CdCl_2$ and then cultured for different durations (6, 12, 18 or 24 h) at 30 °C. The cells were then collected and washed twice with distilled water and digested with HNO_3 and H_2O_2 (3:1) at 140 °C for 10 min, 200 °C for 20 min and 140 °C for 10 min. The 2-week-old plants were immersed in half MS medium supplemented with 100 μM $CdCl_2$ for 24 h. Then, shoots and roots were digested with HNO_3 and H_2O_2 (3:1) at 140 °C for 10 min, 200 °C for 20 min and 140 °C for 10 min [39]. All of samples were analyzed for total Cd detection by using Inductively Coupled Plasma Optical Emission Spectrometer (ICP-OES; ThermoFisher ICAP 6300, Waltham, MA, USA). All analyses were repeated three times.

4.8. Net Cd^{2+} Efflux Measurements

Fifteen-day-old seedlings were treated with 50 μM $CdCl_2$ for 24 h and soaked in testing buffer (0.1 mM KCl, 0.1 mM $CaCl_2$, 0.05 mM $CdCl_2$, 0.3 mM 2-(*N*-morpholino) ethane sulfonic acid, pH 5.8) for 15 min. Roots were immobilized on the bottom of a measuring dish in fresh testing buffer. The measuring site was 800 μm from the root apex, and the net flux of Cd^{2+} was detected using a non-invasive micro-test technique (NMT; BIO-001A, Younger United States Science and Technology Corp, Beijing, China). The ion flux of Cd^{2+} was calculated according to Fick's law of diffusion, $J_0 = -D \times (d_C/d_X)$, where J_0 is the net ion flux (in $\mu mol \cdot cm^{-2}$ per second), D is the self-diffusion coefficient for the ion (in $cm^2 \cdot s^{-1}$), d_C is the difference in the ion concentrations between the two positions, and d_X is the 10 μm excursion over which the electrode moved in these experiments.

4.9. Statistical Analysis

The experimental data related to roots length, Cd content, OD_{600} of yeast, and quantitative RT-PCR were analyzed by the statistical software SPSS 9.0. One-way analysis of variance (ANOVA) with Duncan's multiple range tests was considered as significance test. Different letters represented significant differences ($p < 0.05$). Values represented means ± standard deviation. Quantitative difference between two groups of data for comparison in each experiment was found to be statistically significant (* $p < 0.05$; ** $p < 0.01$).

Supplementary Materials: Supplementary materials can be found at http://www.mdpi.com/1422-0067/20/13/3293/s1.

Author Contributions: Data curation, H.W., Y.L., X.W., S.X., L.S. and E.H.; Investigation, Y.L.; Methodology, B.W.; Project administration, J.L., Y.L., Z.P. and W.H.; Supervision, N.W. and B.W.; Validation, K.L.; Writing–original draft, H.W.; Writing–review & editing, B.W. All authors reviewed the manuscript.

Funding: This work was supported by the National Natural Science Foundation of China (Grant No. 31571584 and 31370317), the Ministry of Science and Technology of China (Grant No. 2016YFD0100504), the Natural Science Foundation of Chongqing (Grant No. cstc2013jcyjA80016 and cstc2016shmszx20008), and Fundamental Research Funds for the Central Universities (XDJK2013B032), the National Undergraduate Training Programs for Innovation and entrepreneurship of China (Grant No. 201810635034).

Acknowledgments: We thank Ye Ning and Dexin Zhou for fruitful discussions and critical reading of the manuscript. We thank Ji-Ming Gong for kindly providing the yeast strains.

Conflicts of Interest: The authors declare no conflicts of interest.

References

1. Järup, L.; Åkesson, A. Current status of cadmium as an environmental health problem. *Toxicol. Appl. Pharmacol.* **2009**, *238*, 201–208. [CrossRef]
2. Kazantzis, G. Cadmium, osteoporosis and calcium metabolism. *Biometals* **2004**, *17*, 493–498. [CrossRef] [PubMed]
3. Goyer, R.A. Toxic and essential metal interactions. *Annu. Rev. Nutr.* **1997**, *17*, 37–50. [CrossRef] [PubMed]
4. Schützendübel, A.; Schwanz, P.; Teichmann, T.; Gross, K.; Langenfeld Heyser, R.; Godbold, D.L.; Polle, A. Cadmium-induced changes in antioxidative systems, hydrogen peroxide content, and differentiation in Scots pine roots. *Plant Physiol.* **2001**, *127*, 887–898. [CrossRef] [PubMed]
5. Hussain, D.; Haydon, M.J.; Wang, Y.; Wong, E.; Sherson, S.M.; Young, J.; Camakaris, J.; Harper, J.F.; Cobbett, C.S. P-type ATPase heavy metal transporters with roles in essential zinc homeostasis in Arabidopsis. *Plant Cell* **2004**, *16*, 1327–1339. [CrossRef]
6. Clemens, S. Molecular mechanisms of plant metal tolerance and homeostasis. *Planta* **2001**, *212*, 475–486. [CrossRef] [PubMed]
7. Williams, L.E.; Pittman, J.K.; Hall, J.L. Emerging mechanisms for heavy metal transport in plants. *Biochim. Biophys. Acta. Biomembr.* **2000**, *1465*, 104–126. [CrossRef]

8. Do, T.H.T.; Martinoia, E.; Lee, Y. Functions of ABC transporters in plant growth and development. *Curr. Opin. Plant Biol.* **2018**, *41*, 32–38. [CrossRef] [PubMed]
9. Szczypka, M.S.; Wemmie, J.A.; Moye Rowley, W.S.; Thiele, D.J. A yeast metal resistance protein similar to human cystic fibrosis transmembrane conductance regulator (CFTR) and multidrug resistance-associated protein. *J. Biol. Chem.* **1994**, *269*, 22853–22857.
10. Song, W.Y.; Sohn, E.J.; Martinoia, E.; Lee, Y.J.; Yang, Y.Y.; Jasinski, M.; Forestier, C.; Hwang, I.; Lee, Y. Engineering tolerance and accumulation of lead and cadmium in transgenic plants. *Nat. Biotechnol.* **2003**, *21*, 914. [CrossRef]
11. Ortiz, D.F.; Ruscitti, T.; Mccue, K.F.; Ow, D.W. Transport of metal-binding peptides by HMT1, a fission yeast ABC-type vacuolar membrane protein. *J. Biol. Chem.* **1995**, *270*, 4721–4728. [CrossRef] [PubMed]
12. Kim, D.Y.; Bovet, L.; Kushnir, S.; Noh, E.W.; Martinoia, E.; Lee, Y. AtATM3 is involved in heavy metal resistance in Arabidopsis. *Plant physiol.* **2006**, *140*, 922–932. [CrossRef] [PubMed]
13. Park, J.; Song, W.Y.; Ko, D.; Eom, Y.; Hansen, T.H.; Schiller, M.; Lee, T.G.; Martinoia, E.; Lee, Y. The phytochelatin transporters AtABCC1 and AtABCC2 mediate tolerance to cadmium and mercury. *Plant J.* **2012**, *69*, 278–288. [CrossRef] [PubMed]
14. Song, W.Y.; Yamaki, T.; Yamaji, N.; Ko, D.; Jung, K.H.; Fujii Kashino, M.; An, G.; Martinoia, E.; Lee, Y.; Ma, J.F. A rice ABC transporter, OsABCC1, reduces arsenic accumulation in the grain. *Proc. Natl. Acad. Sci. USA* **2014**, *111*, 15699–15704. [CrossRef] [PubMed]
15. Sun, L.; Ma, Y.; Wang, H.; Huang, W.; Wang, X.; Han, L.; Sun, W.; Han, E.; Wang, B. Overexpression of PtABCC1 contributes to mercury tolerance and accumulation in Arabidopsis and poplar. *Biochem. Biophys. Res. Commun.* **2018**, *497*, 997–1002. [CrossRef] [PubMed]
16. Tommasini, R.; Vogt, E.; Fromenteau, M.; Hörtensteiner, S.; Matile, P.; Amrhein, N.; Martinoia, E. An ABC-transporter of *Arabidopsis thaliana* has both glutathione-conjugate and chlorophyll catabolite transport activity. *Plant J.* **1998**, *13*, 773–780. [CrossRef] [PubMed]
17. Gaillard, S.; Jacquet, H.; Vavasseur, A.; Leonhardt, N.; Forestier, C. AtMRP6/AtABCC6, an ATP-Binding Cassette transporter gene expressed during early steps of seedling development and up-regulated by cadmium in *Arabidopsis thaliana*. *BMC Plant Biol.* **2008**, *8*, 22. [CrossRef]
18. Kim, D.Y.; Bovet, L.; Maeshima, M.; Martinoia, E.; Lee, Y. The ABC transporter AtPDR8 is a cadmium extrusion pump conferring heavy metal resistance. *Plant J.* **2007**, *50*, 207–218. [CrossRef]
19. Migocka, M.; Papierniak, A.; Rajsz, A. Cucumber PDR8/ABCG36 and PDR12/ABCG40 plasma membrane proteins and their up-regulation under abiotic stresses. *Biol. Plant.* **2017**, *61*, 115–126. [CrossRef]
20. Oda, K.; Otani, M.; Uraguchi, S.; Akihiro, T.; Fujiwara, T. Rice ABCG43 Is Cd Inducible and Confers Cd Tolerance on Yeast. *Bios. Biotechnol. Biochem.* **2011**, *75*, 1211–1213. [CrossRef]
21. Melnikova, N.V.; Borkhert, E.V.; Snezhkina, A.V.; Kudryavtseva, A.V.; Dmitriev, A.A. Sex-Specific Response to Stress in *Populus*. *Front. Plant Sci.* **2017**, *8*, 1827. [CrossRef] [PubMed]
22. Shi, W.G.; Liu, W.; Yu, W.; Zhang, Y.; Ding, S.; Li, H.; Mrak, T.; Kraigher, H.; Luo, Z.B. Abscisic acid enhances lead translocation from the roots to the leaves and alleviates its toxicity in *Populus × canescens*. *J. Hazard. Mater.* **2019**, *362*, 275–285. [CrossRef] [PubMed]
23. Li, N.; Meng, H.; Xing, H.; Liang, L.; Zhao, X.; Luo, K.; Raines, C. Genome-wide analysis of MATE transporters and molecular characterization of aluminum resistance in *Populus*. *J. Exp. Bot.* **2017**, *68*, 5669–5683. [CrossRef] [PubMed]
24. Shim, D.; Kim, S.; Choi, Y.I.; Song, W.Y.; Park, J.; Youk, E.S.; Jeong, S.C.; Martinoia, E.; Noh, E.W.; Lee, Y. Transgenic poplar trees expressing yeast cadmium factor 1 exhibit the characteristics necessary for the phytoremediation of mine tailing soil. *Chemosphere* **2013**, *90*, 1478–1486. [CrossRef] [PubMed]
25. Schneider, E.; Hunke, S. ATP-binding-cassette (ABC) transport systems: Functional and structural aspects of the ATP-hydrolyzing subunits/domains. *FEMS Microbiol. Rev.* **1998**, *22*, 1–20. [CrossRef] [PubMed]
26. Hyde, S.C.; Emsley, P.; Hartshorn, M.J.; Mimmack, M.M.; Gileadi, U.; Pearce, S.R.; Gallagher, M.P.; Gill, D.R.; Hubbard, R.E.; Higgins, C.F. Structural model of ATP-binding proteing associated with cystic fibrosis, multidrug resistance and bacterial transport. *Nature* **1990**, *346*, 362–365. [CrossRef] [PubMed]
27. Kobae, Y.; Sekino, T.; Yoshioka, H.; Nakagawa, T.; Martinoia, E.; Maeshima, M. Loss of AtPDR8, a Plasma Membrane ABC Transporter of *Arabidopsis thaliana*, Causes Hypersensitive Cell Death Upon Pathogen Infection. *Plant Cell Physiol.* **2006**, *47*, 309–318. [CrossRef] [PubMed]

28. Ji, H.; Peng, Y.; Meckes, N.; Allen, S.; Stewart, C.N.; Traw, M.B. ATP-Dependent Binding Cassette Transporter G Family Member 16 Increases Plant Tolerance to Abscisic Acid and Assists in Basal Resistance against *Pseudomonas syringae* DC3000. *Plant Physiol.* **2014**, *166*, 879–888. [CrossRef]
29. Khare, D.; Choi, H.; Huh, S.U.; Bassin, B.; Kim, J.; Martinoia, E.; Sohn, K.H.; Paek, K.H.; Lee, Y. Arabidopsis ABCG34 contributes to defense against necrotrophic pathogens by mediating the secretion of camalexin. *Proc. Natl. Acad. Sci. USA* **2017**, *114*, 5712–5720. [CrossRef]
30. Kuromori, T.; Sugimoto, E.; Shinozaki, K. Arabidopsis mutants of AtABCG22, an ABC transporter gene, increase water transpiration and drought susceptibility. *Plant J.* **2011**, *67*, 885–894. [CrossRef]
31. Moons, A. Osgstu3 and osgtu4, encoding tau class glutathione S-transferases, are heavy metal- and hypoxic stress-induced and differentially salt stress-responsive in rice roots. *FEBS Lett.* **2003**, *553*, 427–432. [CrossRef]
32. Ogawa, I.; Nakanishi, H.; Mori, S.; Nishizawa, N.K. Time course analysis of gene regulation under cadmium stress in rice. *Plant Soil.* **2009**, *325*, 97–108. [CrossRef]
33. Songbiao, C.; Pattavipha, S.; Jianli, L.; Guo Liang, W. A versatile zero background T-vector system for gene cloning and functional genomics. *Plant Physiol.* **2009**, *150*, 1111–1121.
34. Rentsch, D.; Laloi, M.; Rouhara, I.; Schmelzer, E.; Delrot, S.; Frommer, W.B. NTR1 encodes a high affinity oligopeptide transporter in Arabidopsis. *FEBS Lett.* **1995**, *370*, 264–268. [CrossRef]
35. Peng, J.S.; Ding, G.; Meng, S.; Yi, H.Y.; Gong, J.M. Enhanced metal tolerance correlates with heterotypic variation in SpMTL, a metallothionein-like protein from the hyperaccumulator *Sedum plumbizincicola*. *Plant Cell Environ.* **2017**, *40*, 1368–1378. [CrossRef] [PubMed]
36. Gietz, R.D.; Schiestl, R.H. Frozen competent yeast cells that can be transformed with high efficiency using the LiAc/SS carrier DNA/PEG method. *Nat. Protoc.* **2007**, *2*, 1–4. [CrossRef] [PubMed]
37. Wemmie, J.A.; Szczypka, M.S.; Thiele, D.J.; Moye Rowley, W.S. Cadmium tolerance mediated by the yeast AP-1 protein requires the presence of an ATP-binding cassette transporter-encoding gene, *YCF1*. *J. Biol. Chem.* **1994**, *269*, 32592–32597. [PubMed]
38. Clough, S.J.; Bent, A.F. Floral dip: A simplified method for Agrobacterium-mediated transformation of *Arabidopsis thaliana*. *Plant J.* **1998**, *16*, 735–743. [CrossRef]
39. Brunetti, P.; Zanella, L.; De Paolis, A.; Di Litta, D.; Cecchetti, V.; Falasca, G.; Barbieri, M.; Altamura, M.M.; Costantino, P.; Cardarelli, M. Cadmium-inducible expression of the ABC-type transporter AtABCC3 increases phytochelatin-mediated cadmium tolerance in Arabidopsis. *J. Exp. Bot.* **2015**, *66*, 3815–3829. [CrossRef]

© 2019 by the authors. Licensee MDPI, Basel, Switzerland. This article is an open access article distributed under the terms and conditions of the Creative Commons Attribution (CC BY) license (http://creativecommons.org/licenses/by/4.0/).

International Journal of
Molecular Sciences

Article

Comparative Transcriptomic Studies on a Cadmium Hyperaccumulator *Viola baoshanensis* and Its Non-Tolerant Counterpart *V. inconspicua*

Haoyue Shu [1], Jun Zhang [2], Fuye Liu [1], Chao Bian [3], Jieliang Liang [4], Jiaqi Liang [2], Weihe Liang [2], Zhiliang Lin [1], Wensheng Shu [4], Jintian Li [4], Qiong Shi [3,*] and Bin Liao [1,*]

1 State Key Laboratory of Biocontrol, School of Life Sciences, Sun Yat-Sen University, Guangzhou 510275, China; shuhy@mail2.sysu.edu.cn (H.S.); liufy6@126.com (F.L.); linzhliang@126.com (Z.L.)
2 School of Biosciences and Biopharmaceutics, Guangdong Province Key Laboratory for Biotechnology Drug Candidates, Guangdong Pharmaceutical University, Guangzhou 510006, China; zhangj80@aliyun.com (J.Z.); liangjq_phil@163.com (J.L.); TTlweihe@163.com (W.L.)
3 Shenzhen Key Lab of Marine Genomics, Guangdong Provincial Key Lab of Molecular Breeding in Marine Economic Animals, BGI Academy of Marine Sciences, BGI Marine, BGI, Shenzhen 518083, China; bianchao@genomics.cn
4 School of Life Sciences, South China Normal University, Guangzhou 510631, China; liangjl@m.scnu.edu.cn (J.L.); shuws@mail.sysu.edu.cn (W.S.); lijtian@mail.sysu.edu.cn (J.L.)
* Correspondence: shiqiong@genomics.cn (Q.S.); lsslb@mail.sysu.edu.cn (B.L.); Tel.: +86-189-0229-9133 (B.L.)

Received: 3 March 2019; Accepted: 16 April 2019; Published: 17 April 2019

Abstract: Many *Viola* plants growing in mining areas exhibit high levels of cadmium (Cd) tolerance and accumulation, and thus are ideal organisms for comparative studies on molecular mechanisms of Cd hyperaccumulation. However, transcriptomic studies of hyperaccumulative plants in Violaceae are rare. *Viola baoshanensis* is an amazing Cd hyperaccumulator in metalliferous areas of China, whereas its relative *V. inconspicua* is a non-tolerant accumulator that resides at non-metalliferous sites. Here, comparative studies by transcriptome sequencing were performed to investigate the key pathways that are potentially responsible for the differential levels of Cd tolerance between these two *Viola* species. A cascade of genes involved in the ubiquitin proteosome system (UPS) pathway were observed to have constitutively higher transcription levels and more activation in response to Cd exposure in *V. baoshanensis*, implying that the enhanced degradation of misfolded proteins may lead to high resistance against Cd in this hyperaccumulator. Many genes related to sucrose metabolism, especially those involved in callose and trehalose biosynthesis, are among the most differentially expressed genes between the two *Viola* species, suggesting a crucial role of sucrose metabolism not only in cell wall modification through carbon supply but also in the antioxidant system as signaling molecules or antioxidants. A comparison among transcriptional patterns of some known transporters revealed that several tonoplast transporters are up-regulated in *V. baoshanensis* under Cd stress, suggesting more efficient compartmentalization of Cd in the vacuoles. Taken together, our findings provide valuable insight into Cd hypertolerance in *V. baoshanensis*, and the corresponding molecular mechanisms will be useful for future genetic engineering in phytoremediation.

Keywords: Cadmium; hyperaccumulator; *Viola baoshanensis*; transcriptome; detoxification

1. Introduction

Cadmium (Cd) is a non-essential trace element that can cause toxic reactions in plants and can be easily transferred into vegetative cover and ultimately enter the food chain, becoming a threat to human health [1]. Therefore, Cd is considered a primary cause of soil pollution, and the control of risks related to Cd exposure has become a worldwide concern. The development of efficient phytoremediation and strategies to reduce Cd concentrations in crops are the two most promising strategies for preventing health risks from Cd contamination [2,3]. Indeed, both strategies require deep understanding of the molecular mechanisms involved in Cd absorbance, internal translocation and detoxification. A class of rare plants called hyperaccumulators, which possess extremely high tolerance and high accumulation of heavy metals, have recently evoked considerable interest as model plants for studying plant responses to heavy metal stress, and they are potential genetic resources for the development of future genetic engineering technologies [4].

To date, there have been reports of over 500 hyperaccumulators, while only a small fraction of them have been recognized as Cd hyperaccumulators with the distribution being among the Brassicaceae, Crassulaceae and Violaceae families [5–8]. Several basic physiological processes, possibly shared by most hyperaccumulators, have already been proposed for the Cd hyperaccumulation and hypertolerance, such as transporter-mediated absorbance and root-to-shoot transportation [9], sequestration of Cd chelates and Cd^{2+} to the vacuole or cell wall [10,11], and scavenging of reactive oxygen species (ROS) [12]. The critical role of the *HMA4* gene (encoding a heavy metal ATPase) in enhanced root-to-shoot translocation within the xylem has a particularly interesting cause, which is likely a highly conserved mechanism in different hyperaccumulators across several phyla [13]. However, most current molecular biological studies focus on the most well-known Cd hyperaccumulators *Arabidopsis halleri* and *Noccaea caerulescens* from the family Brassicaceae by taking advantage of the genetic resources developed for *A. thaliana* [14]. It is generally accepted that most of these hyperaccumulators have independent origins [15], indicating the existence of different genetic architectures for hyperaccumulation traits among these various hyperaccumulators. Therefore, expanding the scope of taxa may provide new insights into the molecular mechanisms of the hyperaccumulation traits [13].

Previous studies have suggested that tolerance and accumulation are separated traits that are mediated by genetically and physiologically distinct mechanisms, and thus, the examined plants can be assigned into four categories: tolerant accumulators, non-tolerant accumulators, non-tolerant non-accumulators and tolerant non-accumulators [16]. Although the progression of more detailed transcriptomic descriptions of hyperaccumulators has benefited from next-generation sequencing (NGS) technologies, some notable questions are still present in the majority of comparative analyses between tolerant accumulators and non-tolerant non-accumulators. For example, comparative studies on *A. halleri* versus *A. thaliana* [17], *N. caerulescens* versus *A. thaliana* [14], and *Sedum plumbizincicola* versus *S. alfredii* [18] may have generated confusing results, since they could not clearly distinguish the genes responsible for metal tolerance versus those responsible for metal accumulation. When we compare the tolerant accumulators and non-tolerant accumulators, this misleading information can be avoided by controlling the accumulative traits by setting them as constant variables while focusing on the mechanisms responsible for differential metal tolerance. However, transcriptome-based comparative analyses between tolerant accumulators and non-tolerant accumulators are still lacking.

A large number of hyperaccumulators in Violaceae have been reported in Asian and European countries, for example, *V. baoshanensis* for the hyperaccumulation of Cd [19,20], *V. calaminaria* [20] and *V. lutea* [21] for zinc (Zn), and *Rinorea niccolifera* [22] for nickel (Ni). Previous studies on the responses of *Viola* species to Cd stress were mostly limited to ecological and physiological levels, and only few studies were done at the transcriptional level, while comparative analyses with closely related non-metalliferous species are lacking. *V. baoshanensis*, growing on Baoshan lead/zinc mines in Hunan Province of China was identified by us as both a Cd-hyperaccumulator and a strong accumulator of Pb (lead) and Zn [19,23]; whereas, interestingly, *V. inconspicua* from non-contaminated sites showed less tolerance to Cd than *V. baoshanensis*, but with similar amounts of Cd accumulation in roots and shoots. Accordingly, the striking phenotypic difference between these two *Viola* species provides an exceptional opportunity to elucidate the molecular mechanisms for their differential levels of Cd tolerance on the basis of controlling the Cd accumulation traits. In this study, we sequenced and compared the transcriptomes of *V. baoshanensis* and *V. inconspicua* on an Illumina sequencing platform to investigate the gene transcription patterns of these two distinct *Viola* species under Cd exposure.

We attempted to identify candidate genes involved in the Cd detoxification regulatory networks and also tested whether those known mechanisms from Brassicaceae are conserved among Violaceae.

2. Results

2.1. Metal Accumulation and Tolerance in the Two Viola Species from Hydroponic Experiments

In order to evaluate the different capabilities for Cd accumulation and tolerance between these two *Viola* species, we performed hydroponic Cd stress experiments. We compared the Cd concentrations in the roots (Figure 1a) and shoots (Figure 1b) and calculated the tolerance indices (ratios of total dry biomass in the Cd treatments and controls) in the roots and shoots of the two Cd-treated *Viola* species (Figure 1c). Finally, we observed that the tolerance indices of *V. baoshanensis* from the contaminated sites were >1 in the 25 and 50 μM Cd treatments, while they were almost equal to 1 in the 100 μM Cd treatments (see Figure 1c); however, all these values were significantly higher than those in *V. inconspicua samples*, which were collected from the non-contaminated sites (see more details in Figure 1c). Interestingly, among the 25, 50 and 100 μM Cd treatments, *V. baoshanensis* took up similar levels of Cd compared with *V. inconspicua* in both the roots (Figure 1a) and the shoots (Figure 1b). In summary, our results indicate that the two *Viola* species exhibit a great difference in Cd tolerance, while they have a similar capacity to accumulate exogenous Cd.

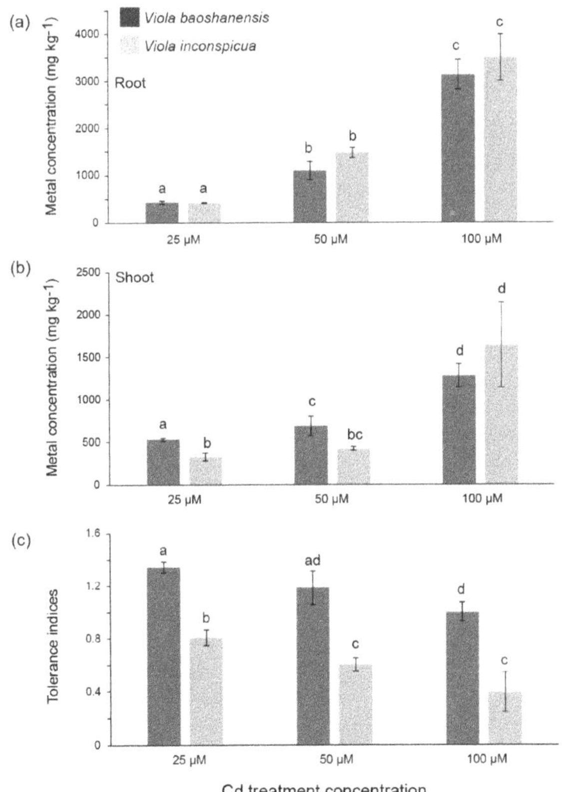

Figure 1. Concentrations of Cd in *Viola baoshanensis* and *V. inconspicua* after different Cd treatments. (**a**) roots, (**b**) shoots, and (**c**) Tolerance indices. The different letters above the columns ($n = 4$), using ANOVA analysis, indicate significant differences ($p < 0.05$) among the Cd treatments.

2.2. Summary of the Data for Transcriptome Sequencing, De Novo Assembling, and Annotation

To investigate the potential molecular mechanisms of Cd hypertolerance in *V. baoshanensis*, we sequenced and compared the transcriptome data between *V. baoshanensis* and *V. inconspicua* under Cd stressed conditions. The roots and shoots of both *Viola* species were harvested from controls or 100 μM Cd treated groups (with three biological replicates), resulting in a total of 24 cDNA libraries that produced a total of 2500 million paired reads (Table S1). In order to improve the quality of the sequences, we trimmed each read by using quality scores (only those higher than 20 at the 3′-end were kept) and removed those reads with excess non-sequenced (N) bases. Subsequently, these clean reads were used separately for further *de novo* assembling by Trinity and TGICI (see more details in Section 4.5). As a result, a total of 105,280 transcripts (>300 bp) were assembled for *V. baoshanensis* and 101,616 contigs for *V. inconspicua* (before filtering; Table 1).

Table 1. Statistics of the assembling and annotation data for the transcriptomes of two *Viola* species.

Parameter	*Viola baoshanensis*	*Viola inconspicua*
No. of contigs (before filtering)	105,280	101,616
No. of contigs	82,854	80,059
Maximum length of contigs (bp)	67,368	42,149
Average length of contigs (bp)	1984	1348
Contig N50 (bp)	2778	1709
GC content (%)	43.8	42.5
Annotated in KEGG	31,772	31,141
Annotated in GO	49,354	45,646

After filtering of low coverage and short ORFs (open reading frames), we obtained the final assemblies with 82,854 protein-coding transcripts for *V. baoshanensis* and 80,059 for *V. inconspicua* (Table 1). To assess the completeness of our transcriptome assemblies, we submitted the two final transcript sets to a BUSCO (Benchmarking Universal Single-Copy Orthologs) evaluation [24], which revealed that the majority of the Eudicotyledon core genes had been successfully recovered in the two assemblies (Table S2). These data indicate that our two *Viola* transcriptome assemblies were of high quality.

To compare inter-species differences in transcription level, we identified 19,794 putative orthologous transcripts between *V. baoshanensis* and *V. inconspicua* by the RBH method with an E-value cutoff of 1.0×10^{-10} and a coverage threshold of 50% (see more details in Section 4.5). Despite the different quality of the two *Viola* assemblies reflected by their N50 sizes and the BUSCO completeness assessment, the length distribution of orthologous contigs demonstrated that the orthologous transcript sets of *V. baoshanensis* and *V. inconspicua* have similar length distribution (Figure S1b) and similar contig N50 sizes (2223 versus 2067 bp). That is to say, the quality of both orthologous transcript sets is comparable across the two *Viola* species, suggesting the reliability of our interspecies comparison [25].

The GO (Gene Ontology), KEGG (Kyoto Encyclopedia of Genes and Genomes), and InterproScan databases were employed to annotate the two transcript sets (Table 1). A total of 49,354 (46.8%) genes were mapped to 129,359 GO terms, and 45,776 (45.1%) genes were mapped to 128,847 GO terms for *V. baoshanensis* and *V. inconspicua*, respectively. Detailed proportions of the GO annotation for individual assembly are shown in Figure 2, indicating that molecular functions, biological processes, and cellular components were well represented. By the way, we observed a high GO distribution similarity between these two *Viola* species (see Figure 2).

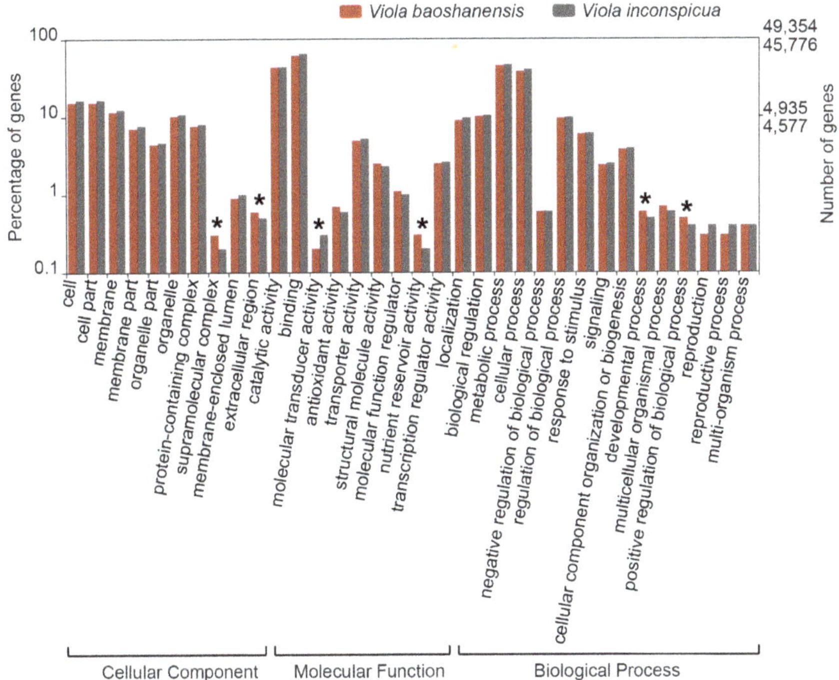

Figure 2. Gene Ontology (GO) distributions for *Viola baoshanensis* (red) and *V. inconspicua* (grey). Annotation results were mapped to categories in the second level of GO terms, respectively. Those GO terms that contain less than 0.1 % of the total genes were excluded from this graph. The asterisks represent significant differences (p <0.05) between the two *Viola* species. See more details in Table S14.

In the cellular component category, cell part (15.2% versus 16.3%) and membrane-related functions (11.5% versus 12.3%) were the most abundant. The highest proportions of the mapped GO terms for "Molecular Function" for the two *Viola* species were related to binning (61.2% versus 64.9%) and catalytic activity (43.3% versus 44.1%). Under the "Biological Process", metabolic process and cellular process were the most enriched terms, comprising 22,446 (45.5%) and 18,950 (38.4 %) genes in *V. baoshanensis* and 21,267 (46.5%) and 18,493 (40.4%) genes in *V. inconspicua*, respectively. We also found that several GO terms significantly differed between these two *Viola* species. For example, under the "Cellular Component" category, *V. baoshanensis* had a higher percentage of genes in the extracellular region and the supramolecular complex (p = 0.019) when compared with *V. inconspicua*. In the "Molecular Function" category, we observed that genes in *V. baoshanensis* were significantly over-represented in terms of nutrient reservoir activity (p = 0.003), of which seed storage proteins were the main component, the seed storage proteins accumulate significantly in the developing seed, whose main function is to act as a storage reserve for nitrogen, carbon, and sulfur. while molecular transducer activity was less represented (p = 0.008) in *V. baoshanensis*. In the last major category, *V. baoshanensis* had more genes related to developmental process (p = 0.004) but less genes related to positive regulation of biological process (p = 0.004).

2.3. Analyses of Differential Expression (DE) and Functional Enrichment

After quantification of transcripts with RNA-Seq by RSEM (see more details in Section 4.6), we employed edgeR to compare the transcriptional changes between *V. baoshanensis* and *V. inconspicua* in response to Cd and also identified the differentially expressed genes (DEGs) in the two species to gain a comprehensive insight into the molecular mechanisms underlying the differences in Cd tolerance.

To determine the genes that responded to the Cd treatment in both *Viola* species, mRNA profiles of Cd treated and untreated (control) roots and shoots were compared (Figure 3 and Figure S2). In *V. baoshanensis*, a total of 5564 transcripts were determined to be DEGs (with q values < 0.01 and log2 (fold change) > 2) between the Cd treatment and the control in roots, in which 3,867 were upregulated and 1697 were downregulated in response to the Cd treatment (Figure 3e). A total of 3192 DEGs were determined in the shoots; among them, 1514 were upregulated and 1678 were downregulated respectively in response to the Cd treatment (Figure 3f). However, in *V. inconspicua*, a total of 8380 and 3299 differentially expressed transcripts were identified in the roots (4410 upregulated and 3970 downregulated; Figure 3g) and the shoots (1275 upregulated and 2024 downregulated; Figure 3h) in response to the Cd treatment. A Venn diagram was used to show the specific upregulated transcripts in *V. baoshanensis* (Figure S3). In total, 570 and 403 transcripts in the orthologous gene sets were induced differently in response to Cd between the two *Viola* species in roots and shoots, respectively (see detailed information in Tables S9 and S10). It seems that the two *Viola* species have more divergent responses in the shoot compared with the root under Cd stress, due to a noticeably smaller fraction of commonly induced transcripts in shoots than roots (9.97% versus 36.7%).

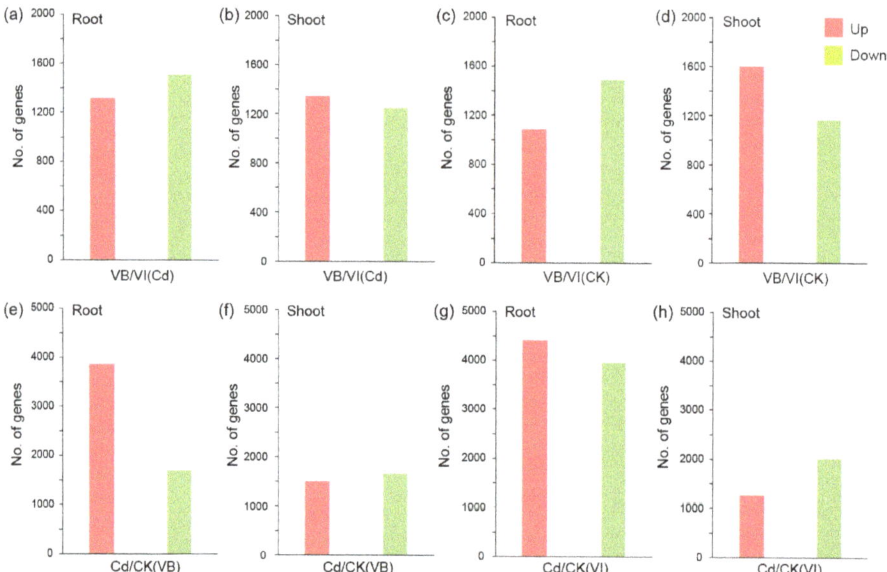

Figure 3. Differentially expressed genes (DEGs) between Cd-treated and control samples in both *Viola* species. Numbers of the genes with significantly different transcription levels (up- or downregulated) between two species in roots (**a**) and shoots (**b**) under the Cd stressed condition, or in the roots (**c**) and shoots (**d**) of the control samples, were summarized for comparison. DEGs between the Cd treated and control samples in VB roots (**e**), VB shoots (**f**), VI roots (**g**), and VI shoots (**h**). CK, control; VB, *Viola baoshanensis*; VI, *Viola inconspicua*.

For the interspecies comparisons, read counting and DE analysis were restricted to the orthologous genes annotated in *V. baoshanensis* and *V. inconspicua*, which were identified by using the reciprocal best blast hits method. After the DE analysis, we identified 2823 and 2602 DEGs between the two species in the roots (Figure 3a,c) and the shoots (Figure 3b,d), respectively. In the Cd treated roots, 1316 and 1507 genes were upregulated in *V. baoshanensis* (Figure 3a) and *V. inconspicua* (Figure 3a), respectively.

Within the Cd treated shoots, 1347 and 1255 genes were upregulated in *V. baoshanensis* (Figure 3b) and *V. inconspicua* (Figure 3b), respectively.

To determine the functional significance of these interspecies and intraspecies variances in response to Cd treatments, we implemented GO classifications and enrichment analysis for these DEGs (for more information see Figures S4 and S5). A similar pattern was observed between *V. baoshanensis* and *V. inconspicua* in response to Cd exposure (Figure S5); the overrepresented GO terms of the two phenotypes presented a significant overlap for shoots (Figure S5c,d), and the genes related to antioxidant metabolism (such as reactive oxygen species metabolic process and antibiotic metabolic process) were enriched in both species (Figure S5c). However, genes related to metal internal translocation (such as divalent metal ion transport, inorganic ion transmembrane transport, metal ion transmembrane transporter activity, and divalent inorganic cation transmembrane transporter activity) were enriched specifically in *V. baoshanensis* (Figure S5c,d).

For the roots in *V. baoshanensis*, the most representative categories were regulation of cellular biosynthetic processes, sulfur metabolism-related GO terms (including sulfate transport and sulfate reduction) and ubiquitin-related GO terms (such as ubiquitin-protein transferase activity and ubiquitin-like protein transferase activity; Figure S5b). However, we noted that the ubiquitin related GO terms is not enriched in *V. inconspicua* roots (Figure S5b), suggesting that these genes may be related to the differential Cd tolerance between the two *Viola* species.

Interestingly, for the interspecies comparisons (Figure S4), we observed that DEGs with higher transcription levels in *V. baoshanensis* roots than in *V. inconspicua* roots were overrepresented by oxidation–reduction processes, tetrapyrrole binding, and transition metal ion transmembrane transporter activity (Figure S4c). In *V. baoshanensis* shoots, however, upregulated DEGs were enriched by microtubule motor activity, heme binding, antioxidant activity, and structural constituents of the cell wall (Figure S4d).

2.4. Ubiquitin Proteosome System (UPS) Pathway-Related Genes: Response to Cd Stress

The GO enrichment analysis showed that ubiquitin-protein transferase activity was specifically and significantly enriched in *V. baoshanensis* (Figure S5b), suggesting that the UPS pathway (Figure 4) may play important roles in the differential Cd tolerance between the two *Viola* species. These DEGs in the UPS pathway were further compared between the two phenotypes (Figure 4; Tables S3, S4 and S11). Interestingly, we identified differential expression of transcripts at almost all steps of this pathway (Figure 4), and *V. baoshanensis* apparently has more upregulated genes than *V. inconspicua* for responding to the Cd stress (Tables S3 and S4). Totals of 125 and 36 UPS-related genes were upregulated by Cd in *V. baoshanensis* roots and shoots, respectively; however, fewer upregulated DEGs were observed in *V. inconspicua* roots and shoots, especially regarding the Fbox and Ubox E3 ligase gene families (see more details in Table S11).

For the interspecies comparisons, we observed that 88 UPS-related genes had significantly higher transcription levels in *V. baoshanensis* compared with *V. inconspicua*, including 5 in the E2 gene family, 27 in the Fbox gene family, 2 in the HECT gene family, 42 in the RING gene family, 8 in the Ubox gene family, and 4 in the proteasome gene family (Tables S3 and S11). Interestingly, significantly fewer UPS-related genes were transcribed at significantly higher rates in *V. inconspicua* than in *V. baoshanensis* (Table S4).

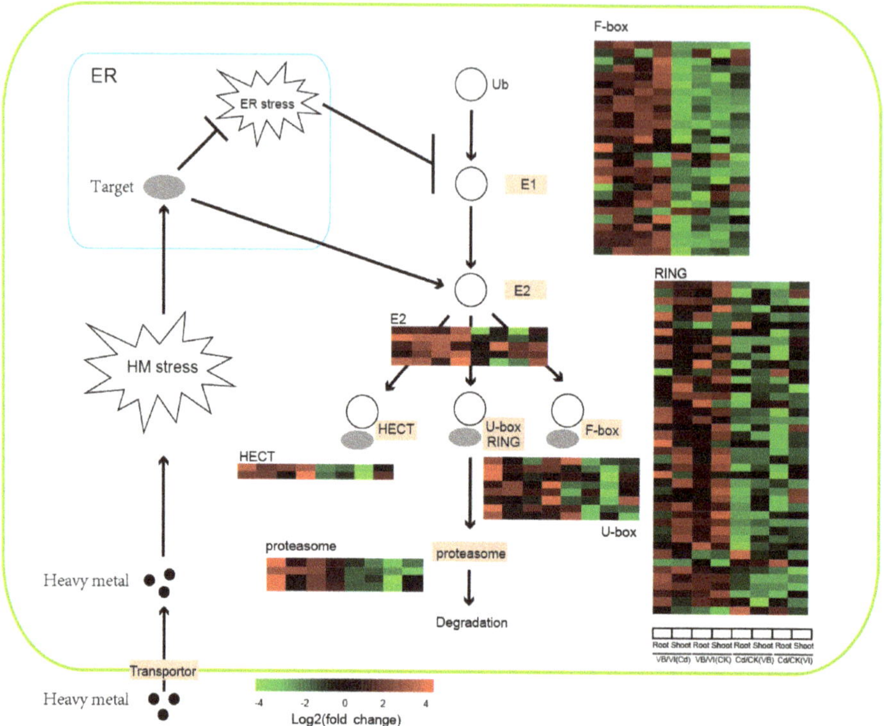

Figure 4. DEGs involved in the ubiquitin proteasome system (UPS) pathway with constitutively higher transcription levels in *Viola baoshanensis*. The heat maps for the relative transcription of DEGs were calculated by log2-fold changes. VB/VI(Cd), VB/VI(CK), Cd/CK(VB), and Cd/CK(VI) represent ratios of transcription levels between *V. baoshanensis* (VB) and *V. inconspicua* (VI) in the Cd treatment (Cd) and control (CK), and ratios of transcription levels of Cd/CK in VB and VI, respectively. E1, Ubiquitin-activating enzymes; E2, ubiquitin conjugases; HECT, U-box, F-box and RING, subfamilies of ubiquitin ligases. See more details about activation of the UPS pathway by ER stress in Table S11, Sections 2.4 and 3.1 (under the Discussion).

2.5. DEGs Involved in Sucrose Metabolism

Our KEGG enrichment analysis demonstrated that the constitutive DEGs in *V. baoshanensis* were enriched in the sucrose metabolism pathway (Figure 5a) in both roots and shoots (Figures S6 and S7), implying that this important pathway may be responsible in part for the phenotypic difference between the two *Viola* species. For the interspecies comparisons, we found that a total of 40 genes, encoding 19 enzymes in the starch and sucrose metabolism pathway, presented higher transcription levels in *V. baoshanensis* than *V. inconspicua* (see more details in Tables S5 and S6).

Nine genes encode 1,3-beta-glucan synthases (key enzymes for callose biosynthesis), and none of the members of this gene family had a higher transcription level in *V. inconspicua* (Figure 5; Tables S5 and S6). For a comparison in the Cd treated samples and controls, many genes were upregulated in response to Cd stress in the roots (41 versus 50) and shoots (22 versus 33) of *V. baoshanensis* and *V. inconspicua*, respectively (see more details in Table S12), suggesting a common activation of the sucrose metabolism pathway by Cd in both *Viola* species.

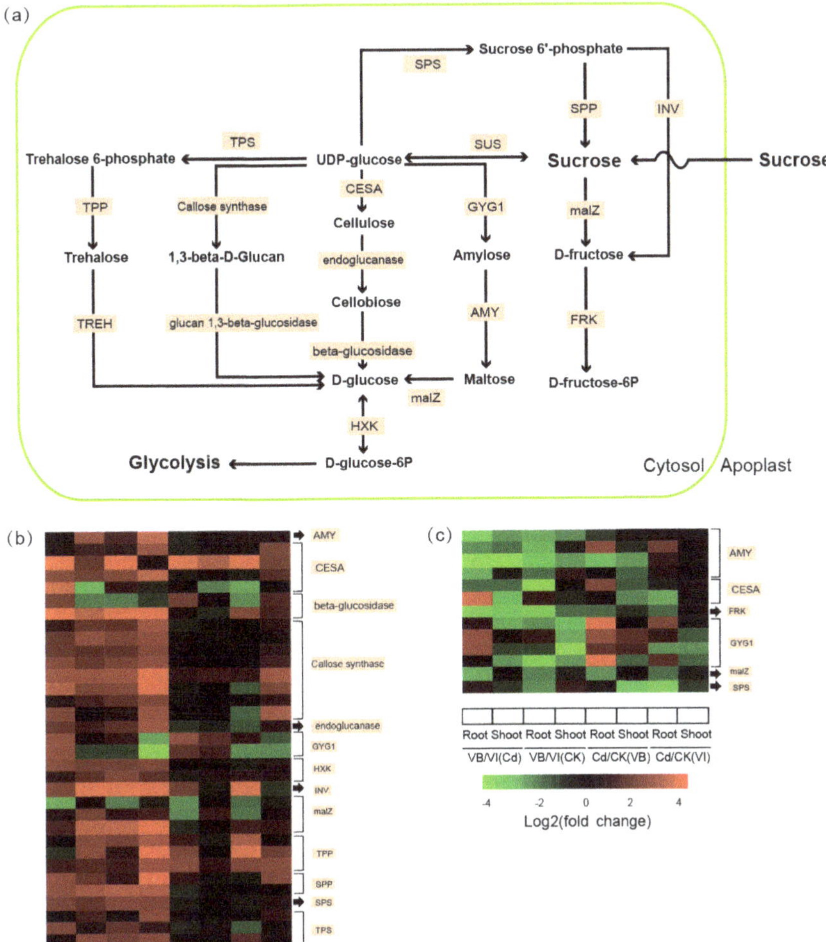

Figure 5. DEGs involved in sucrose metabolism with differential transcription levels in the two *Viola* species. (**a**) Schematic diagram of the sucrose metabolism pathway in *Viola*. (**b**) Heat maps of genes involved in sucrose metabolism with constitutively higher transcription levels in *Viola baoshanensis*. (**c**) Heat maps of genes with constitutively higher transcription levels in *V. inconspicua*. AMY, alpha-amylase; CESA, cellulose synthase; FRK, Fructokinase; HXK, Hexokinase; GYG1, glycogenin; INV, beta-fructofuranosidase; malZ, alpha-glucosidase; SPP, sucrose-6-phosphatase; SPS, sucrose-phosphate synthase; TPP, trehalose 6-phosphate phosphatase; TPS, trehalose 6-phosphate synthase; TREH, alpha trehalase. See more details in Table S12, Sections 2.5 and 3.2.

2.6. DEGs Encoding Metal Transporter Proteins

Metal transporters have been widely regarded as the critical components in hyperaccumulators due to their Cd accumulation, root-to-shoot transportation and internal sequestration [26]. Among the known 10 transporter families, 43 transcripts (24 ABCs, 2 MTPs, 1 CTR, 6 HIPPs, 2 HMAs, and 1 MATE, 1 NRAMP, and 5 ZIPs; see Figure 6) were identified as orthologous DEGs with constitutively higher transcription levels in *V. baoshanensis*; meanwhile, less transcripts from these transporter families (especially in ZIPs, ABCs, and HMAs) showed constitutively higher transcription levels in *V. inconspicua* (Tables S7 and S8).

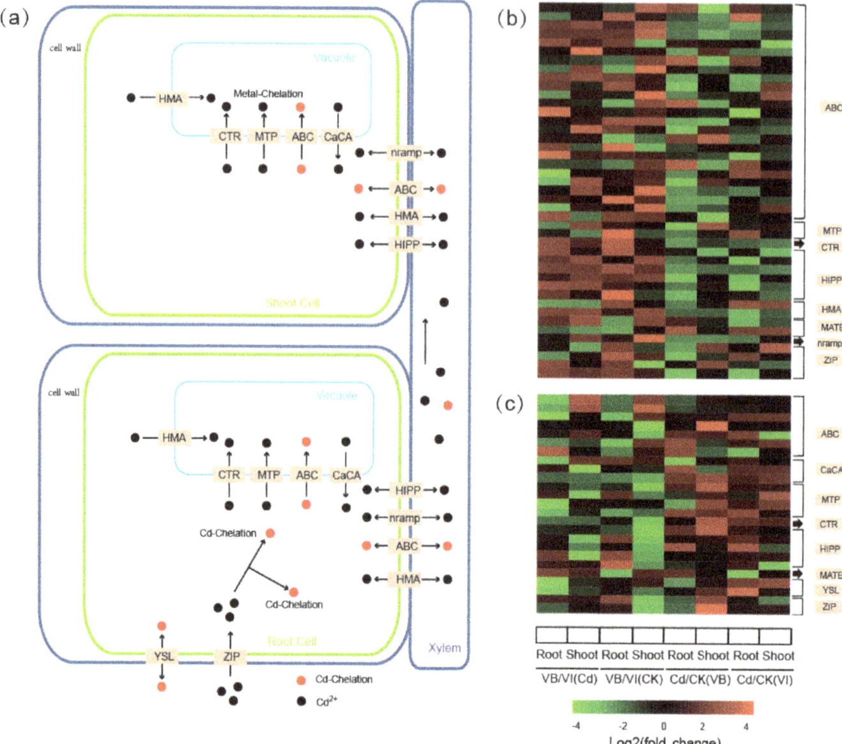

Figure 6. DEGs encoding putative metal transporters. (**a**) A schematic representation of the main processes for differential Cd uptake and internal translocation in *Viola* species. (**b**) Heat maps of genes encoding putative metal transporters with constitutively higher transcription levels in *V. baoshanensis*. (**c**) Heat maps of genes encoding putative metal transporters with constitutively higher transcription levels in *V. inconspicua*. ABC, ATP-binding cassette transporter; CaCA, the Ca^{2+}: cation antiporter protein; CTR, the copper transporter protein; HIPP, heavy metal-associated isoprenylated plant protein; HMA, heavy metal ATPase; MATE, multi-antimicrobial extrusion protein; MTP, metal tolerance protein; Nramp, natural resistance-associated macrophage protein; YSL, yellow stripe-like protein; ZIP, the Zinc/Iron permease protein. See more details in Table S13, Sections 2.6 and 3.3.

Within the 24 ABC DEGs, 12, 6, 5, and 1 belong to the ABCG, ABCC, ABCB, and ABCA subfamilies respectively. Two HMA genes, *HMA4* and *HMA5*, were shown to be transcribed at higher levels in *V. baoshanensis*; in particular, the *HMA4* gene presented a ~100-fold higher level in *V. baoshanensis* than in *V. inconspicua* in both the Cd treated and control samples (Figure 6b; Table S13). For the comparisons between the Cd treated and control samples, 14 genes encoding ZIP proteins were upregulated in *V. baoshanensis* shoots in response to the Cd stress, while only 2 were upregulated in *V. inconspicua* shoots (Figure 6b; Table S13). On the other hand, we also found that several genes related to the YSL family and CaCA family were more activated in *V. inconspicua* (Figure 6c; Table S13). Our findings may reveal the distinct strategies in *V. baoshanensis* and *V. inconspicua* for Cd transportation and sequestration as a consequence of different surroundings.

3. Discussion

Despite the consecutive reporting of hyperaccumulators in Violaceae, progress in understanding the hypertolerance abilities of metallicolous plants in this class has been limited to metal mobilizing in cells at a physiological level [21], while transcriptome-level studies are rare. Here, we compared the transcriptomes of two *Viola* species, a tolerant hyperaccumulator *V. baoshanensis* and a non-tolerant accumulator *V. inconspicua*, from metalliferous and nonmetalliferous sites respectively. Several key pathways are proposed as being associated with their differential Cd tolerance.

3.1. How Does the UPS Pathway Enhance Heavy-metal Resistance in Plants?

As shown in Figure 3, a cascade of genes involved in the UPS pathway displayed constitutively higher transcription levels and greater activation in response to Cd stress in *V. baoshanensis* than in *V. inconspicua*. In plants, the UPS pathway acts through the sequential actions of a cascade of enzymes (see more details in Figure 4) to add multiple copies of the protein ubiquitin (ub) to a substrate protein that is then targeted for degradation by the 26S proteasome [27]. Both transcriptome and proteome studies in various plant species have shown that UPS pathway related genes can be activated by heavy metals, such as Cd, Cu (copper), Hg (mercury), and Pb. For instance, the ubiquitin-dependent proteolysis pathway in yeast was activated in response to Cd exposure [28], and the polyubiquitin genes in common bean and rice were strongly stimulated by Hg [29] and low concentrations of Cd [30] respectively.

However, there are two distinct theories to explain how the overexpression of genes in the UPS cascade enhances the tolerance of heavy metals. The first one focuses on the UPS pathway, since it is a rapid and effective method for precise degradation of misfolded proteins that are induced by heavy metal ions [31–33]. Previous studies demonstrated that heavy metal ions inhibit the refolding of chemically denatured proteins in vitro, obstruct protein folding in vivo, and stimulate the aggregation of nascent proteins in living cells [34]. Together with evidences that yeast mutants in the proteasome are hypersensitive to Cd, it was suggested that heavy metal tolerance can be mediated by degradation of abnormal proteins [28]. Another theory depends on the ubiquitination process, which may indirectly mediate the tolerance of heavy metals by the regulation of heavy metal transporters [35–37]. For example, rice *OsHIR1* E3 ligase protein is able to control metal uptake through regulation of the *OsTIP4-1* protein via ubiquitination [36]. In the present study, for the first time, we found that the UPS pathway related genes have constitutively higher transcription levels in tolerant hyperaccumulators than closely related non-tolerant accumulator species in both roots and shoots; however, it is hard to infer which theory or whether both of them may mediate the Cd hypertolerance in *V. baoshanensis*. The detailed mechanisms are worthy of in-depth investigation. Moreover, it seems that the effects of the UPS pathway on heavy metal stress may not be generalized, since the UPS system can be applied to develop biotechnological tools not only to reduce metal concentrations for food safety but also to strengthen metal accumulations for phytoremediation [37].

3.2. Relations of Sucrose Metabolism with Heavy-Metal Stress in Plants

Sucrose and starch metabolism play pivotal roles in development, the stress response and synthesis of essential components (including proteins, cellulose, and starch) in higher plants [38]. For example, the addition of dialdehyde starch derivatives in heavy metal-contaminated soils limited the negative impact of these metals both in terms of yield and heavy metal content in maize [39]. In this study, we observed a significant difference in transcription levels at almost all steps of sucrose and starch metabolism (Figure 5a) between *V. baoshanensis* and *V. inconspicua* (Figure 5b). These data are consistent with previous descriptions of the transcriptome in response to heavy metal stress in other phyla [40–42]; however, detailed descriptions and discussion of this pathway are still limited.

Hexokinases (HXKs) and fructokinases (FRKs) are two families of enzymes with the capacity to catalyze the essential irreversible phosphorylation of glucose and fructose, and therefore, they may play central roles in the regulation of plant sugar metabolism. It has been suggested that in vivo, HXKs probably mainly phosphorylate glucose, whereas fructose is phosphorylated primarily by FRKs [43]. Interestingly, 2 HXKs homologous to *Arabidopsis HXK1* were found to have constitutively higher transcription levels in *V. baoshanensis* than in *V. inconspicua*, while 2 FRKs homologous to *Arabidopsis*

FRK1 and *FRK2* were constitutively higher in *V. inconspicua*. A previous study has shown that the reduction of *FRK2* activity in aspen (*Populus tremula*) led to thinner fiber cell walls with a reduction in the proportion of cellulose by decreased carbon flux to cell wall polysaccharide precursors [44], indicating that the cell wall biosynthesis may be repressed in *V. baoshanensis* by low expression of FRK and thus may reduce the capacity of cell walls to act as barriers against Cd translocation. HXKs have been demonstrated to play potential roles in the uptake of Zn in roots, since the HXK-dependent transporter *ZIP11* is unrelated to sugar sensing but may be related to sugar metabolism downstream of HXK [45]. These findings are in accordance with our present observation that *ZIP11* transcribed with significantly higher levels in *V. baoshanensis* roots than in *V. inconspicua* roots under Cd stress.

Uridine diphosphate glucose (UDP-Glc), one product of the sucrose cleavage reactions, is the substrate for biosynthesis of callose, which is associated with the plasma membrane. In this study, we found that *V. baoshanensis* had constitutively higher transcription levels of two related enzymes, callose synthases (β-1,3-glucan synthases) and β-1,3-glucanases, which can produce and break down callose respectively [46]. Of the 8 upregulated genes encoding callose synthases and the 2 genes encoding glucan 1,3-beta-glucosidase in *V. baoshanensis*, 7 CALSs (2 *CALS10*, 2 *CALS7*, *CALS5*, *CAL3*, and *CLA9*) and 2 *pdBG1s* have been reported to localize in the plasmodesmata and to regulate callose and plasmodesmatal permeability [47–49], suggesting an alteration of the plasmodesmatal permeability between the two *Viola* species. With the evidence that accumulation of callose accompanies a reduction in plasmodesmatal permeability leading to reduced growth and depletion of the stem cell population [50], the decreased callose levels in response to heavy-metal stress may buffer the negative effects on primary root growth, and thus to increase heavy metal trafficking through the plasmodesamata [51] and subsequently, increase resistance to heavy metals [52]. Hence, we speculate that *V. baoshanensis* may have decreased the plasmodesmatal permeability to a greater extent than *V. inconspicua*, thereby enhancing Cd tolerance.

In addition to cell wall modification, which is regulated by sucrose metabolism through the carbon supply, sugars also play protective functions against various abiotic stresses in several physiological processes by acting as signaling molecules in plants [53,54]. Six genes, encoding trehalose-6-phosphate phosphatase (TPP) and trehalose-6-phosphate synthase (TPS), which are involved in trehalose biosynthesis in the sucrose metabolism pathway, were found to have significantly elevated transcription levels in *V. baoshanensis*, implying that trehalose-6-phosphate (T6P) and trehalose may play significant roles in the phenotypic difference in Cd tolerance between the two *Viola* species. Several previous studies support this hypothesis. For instance, enhanced endogenous trehalose levels in rice seedlings significantly mitigated the toxic effects of excessive Cu^{2+} by inhibiting Cu uptake and regulating the antioxidant and glyoxalase systems [55]. Another case is that overexpression of *Arabidopsis* trehalose-6-phosphate synthase in tobacco plants (*AtTPS1*) was shown to lead to better acclimation to Cd and excess Cu than in the wild-type [56]. Regarding the mechanisms involved in the use of trehalose biosynthesis to mediate responses to heavy metals, the production of antioxidants or antioxidant enzymes to alleviate excessive heavy metals has been proven to be induced by trehalose-6-phosphate [53,55–57]. In line with our expectations, several genes with significantly higher transcription levels in *V. baoshanensis* were enriched in the GO term of peroxidase activity. Moreover, trehalose can act directly as an antioxidant on excess ROS [58]. Interestingly, the effects of trehalose on heavy metal uptake seem to be contradictory. Reduced Cu uptake was reported in rice seedlings with trehalose treatment [55], while higher transcription levels of *AtTPS1* in tobacco resulted in more Cd and Cu accumulation than in a transgenic line [56].

3.3. Contributions of Transporter Proteins to Heavy Metal Tolerance in Plants

It is well known that the enhancement of transporters for essential elements (such as Fe^{2+}, Zn^{2+}, and Ca^{2+}) may be involved in non-essential metal uptake and transport in hyperaccumulators [59]. To test whether these mechanisms of heavy metal hyperaccumulation are conserved in the *Viola* phyla, we identified and compared several related gene families between the two *Viola* transcriptome sets. ZIP family genes are the most important Zn/Fe plasma membrane transporters, and in this study 5 ZIP transporters (*ZIP1, 3, 4, 5,* and *11*) displayed higher levels of transcription in *V. baoshanensis* than in *V. inconspicua*. Previous studies have shown that *ZIP4* is necessary for the enhanced accumulation of metal ions, and the metal accumulating capacity correlates with higher expression levels of ZIP4 in a known hyperaccumulator *N. caerulescens* [60].

P1B-type ATPases (heavy metal transporting ATPases, HMAs) have been shown to be involved in root-to-shoot long-distance transportation of heavy metals. In particular, *HMA4* is responsible for efficient xylem loading of Cd, and it has been regarded as a key gene for Zn/Cd accumulation in shoots by overexpression in a hyperaccumulator [61]. We found that transcription levels of the *HMA4* gene in *V. baoshanensis* can reach ~100 fold higher than in *V. inconspicua*, although both *Viola* species can accumulate high levels of Cd in the aboveground parts. Thus, we propose that the *HMA4* protein may have other functions besides its role in root-to-shoot metal transportation.

We also observed a member of the Nramp family (*Nramp1*) with constitutively higher transcription levels in *V. baoshanensis*, which was reported to be involved in the influx of Cd across endodermal plasma membrane for root-to-shoot transportation [62]. Although *V. baoshanensis* and *V. inconspicua* have similar abilities to accumulate Cd, we still observed contrasting transcriptional patterns of transporters in relation to metal uptake and root-to-shoot transport, suggesting that the movement of metals from roots to shoots in the two *Viola* species may be regulated by different pathways.

Different from the transporters responsible for heavy metal uptake and root-to-shoot transport, tonoplast transporters usually sequester heavy metals in leaf cellular vacuoles with a central role in the heavy metal homeostasis of plants [63]. Several vacuolar transporters, such as *MTP3* and *COPT5* (using divalent heavy-metal irons as substrates) and 25 ABC transporters (using the heavy metal–phytochelatin (HM-PC) complex as the substrate) were found to have higher transcription levels in *V. baoshanensis* than in *V. inconspicua*, implying that the enhancement of vacuolar compartmenta-lization in *V. baoshanensis* may have contributed to its higher Cd tolerance than *V. inconspicua*.

However, our comparative transcriptome analysis showed constitutively higher transcription levels and more activation of the CaCA gene family and YSL gene family (8 genes homologs to *AtYSL3* and 2 genes homologs to *AtYSL1*), suggesting that they were induced by Cd stress in *V. inconspicua*, especially in the roots. The *YSL* gene family encodes plasma-localized transporters to deliver various heavy metal–nicotianamine (HM–NA) complexes containing Fe(II), Cu, Zn, and Cd, and *YSL3* has been received extensive attention as it may be responsible for internal metal transport in hyperaccumulators in response to metal stress. Nevertheless, the function of *YSL3* may not be conserved across various phyla. Overexpression of *TcYSL3* and *SnYSL3* from the hyperaccumulators *N. caerulescens* and *Solanum nigrum* has been reported to function in metal hyperaccumulation in shoots [41,64], while high expression of *AhYSL3.1* from peanut and rice plants with excess Cu, resulted in a low concentration of Cu in young leaves [65], implying a potential capacity of *YSL3* to reduce metal toxicity by metal efflux. The contradictory expression patterns of HM-NA transporter YSL genes and HM-PC transporter ABC genes in the two *Viola* species may represent two distinct evolutionary lines. We suggest that high expression of MTP, CTR, and ABC genes may enhance vacuolar mobilization in both root and shoot cells to support *V. baoshanensis* with a greater Cd tolerance.

4. Materials and Methods

4.1. Collection of Plant Samples

Two *Viola* species were investigated in this study. *V. baoshanensis* was collected from anthropogenically contaminated soils from local Baoshan Pb/Zn mines in Guiyang City, Hunan Provinces China, whereas *V. inconspicua* was collected from non-metalliferous sites in Guangzhou City, Guangdong Provinces China.

4.2. Hydroponic Experiments

For hydroponic experiments, tissue-cultured seedlings of *V. baoshanensis* were prepared as described in our previous report [66]. *V. inconspicua* seedlings were collected from natural soils at Sun

Yat-Sen University (Guangzhou, China). These seedlings were subsequently cultured in 0.1-strength Hoagland solutions [67]. After 4 weeks of growth, various amounts of Cd were added into the culturing solutions, which were refreshed at 2-day intervals. In the Cd treatments, Cd was supplied as $CdCl_2$ at concentrations of 25, 50, or 100 μM; meanwhile, the treatment without Cd addition was regarded as the control (CK). The hydroponic experiments were conducted at 25 °C with an illumination of LD 16/8 in a glasshouse. The root and shoot samples for transcriptome sequencing (RNA-seq, 100 μM Cd) and the Cd elemental analysis were harvested after one month of the Cd exposure tests. A schematic diagram of the experimental design is summarized in Figure 7. The Cd tolerance indices of *V. baoshanensis* and *V. inconspicua* were calculated as the ratios of the total dray biomass in the Cd treatment and the control.

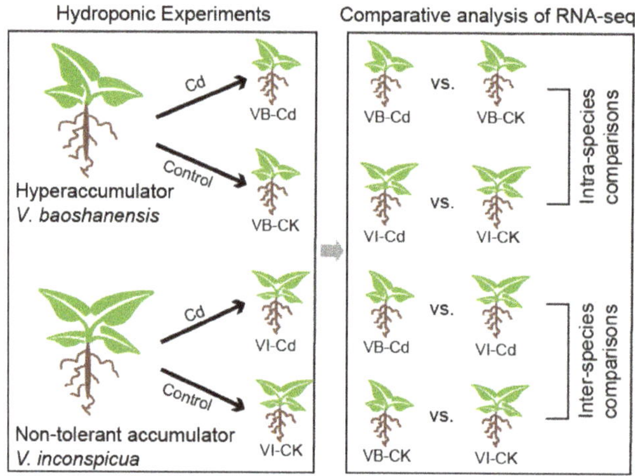

Figure 7. A schematic view of the experimental design for subsequent RNA-seq. VB, *V. baoshanensis*; VI, *V. inconspicua*; CK, control samples.

4.3. Elemental Analysis of the Hydroponic Samples

The plants collected from the hydroponic experiments were digested with a mixture of concentrated HNO_3 and $HClO_4$ at 5:1 (*v*/*v*) [68]. The concentrations of Cd in plant samples were measured by an Inductively Coupled Plasma Optical Emission Spectroscopy (ICP-OES) (PerkinElmer, Shelton, CT, USA). Quality control was performed using standard reference materials (GBW 07604) purchased from the China Standard Materials Research Center and including blanks in the digestion batches. The recovery rates for all samples were around 90 ± 10%.

4.4. RNA Extraction and cDNA Library Preparation

Total RNA was isolated using a HiPure Plant RNA mini kit (Magen, Guangzhou, China). The concentrations and integrity of the isolated RNA samples were evaluated by 1.2% agarose gel electrophoresis and ultraviolet spectrophotometry before cDNA synthesis. The poly(A) RNA was enriched from 2 μg of total RNA using magnetic oligo (dT) beads. The harvested poly(A) RNA samples were fragmented into small pieces using divalent cations under 94 °C for 8 min. cDNA synthesis was performed using the Illumina TruSeq RNA Sample Preparation v2 Kit (Illumina Inc, San Diego, CA, USA). The cleaved RNA fragments were converted into first-strand cDNA using reverse transcriptase and random primers. Second-strand cDNA synthesis was conducted using DNA Polymerase I and RNase H. The synthesized short cDNA fragments were further processed by end repair, adapter ligation, and agarose gel separation. Finally, the correct-sized fragments were selected as templates for PCR amplification with the sequencing primer pairs.

4.5. Transcriptome Sequencing, Assembling, and Annotation

Paired-end sequencing (2 × 150 bp) was performed on the Illumina HiSeq2000 sequencing platform at a sequencing depth of 30–50 million reads per library. Preprocessing of those raw short reads by custom script for quality control included the following three steps: (1) elimination of reads with adapter contamination; (2) removal of reads with excess of non-sequenced bases (N; >5% of each read); (3) trimming of low-quality reads (quality value < 20 at the 3′-ends). Quality control of both raw and processed reads was performed with customized Perl scripts. The Transcriptome Shotgun Sequencing project has been deposited at NCBI Sequence Read Archive under the accession number PRJNA524759.

Given the lack of an available reference genome, we subjected the filtered reads to de novo assembling using Trinity [69] with default parameters. Two reference transcriptomes were produced from *V. baoshanensis* and *V. inconspicua*, respectively, by assembling the reads across all tissues (roots and shoots) and the biological replicates obtained from the constructed libraries. To obtain sets of non-redundant transcripts, TGICL-2.1 [70] was employed to reassemble highly similar transcripts with an identity threshold of 0.94. Subsequently, we applied TransDecoder (https://transdecoder.github.io/) to the ORFs for each transcript and removed those transcripts with short ORFs (<100 bp in length) and low transcription (average Transcripts Per Million bases (TPM) < 1). To provide comprehensive descriptions of the final transcript sets, we employed several public databases including Swissprot/Uniprot [71], KEGG (Kyoto Encyclopedia of Genes and Genomes) and InterproScan [72] to annotate these unigenes. We used WEGO [73] to visualize and compare the GO (Gene Ontology) annotation results of *V. baoshanensis* and *V. inconspicua*. Based on the WEGO manual, the Chi-square test of independence was applied to determine whether there were significant differences in the frequencies of genes within GO terms between the two species.

4.6. Differential Expression and Statistical Analysis

Paired-end reads from each library were individually mapped to their respective transcriptome assemblies using Bowtie2 [74]. RSEM [75] was used to estimate the transcription and raw counts of each transcript. For intra-species comparisons, all expressed transcripts were used for differential expression analysis between Cd-treated samples and control samples. For interspecies comparisons, in order to directly compare the transcription levels between *V. baoshanensis* and *V. inconspicua*, the differential expression analysis was narrowed down to only those constitutive transcripts that were orthologously presented in the two species. We identified orthologous genes between the two transcriptome assemblies by reciprocal best hits (RBH) BLAST [76]. Differential expression analysis was realized using the Bioconductor package edgeR [77] with parameters (minimum fold change = 4, p-value cutoff = 0.01 after FDR correction). The differentially expressed transcripts were then subjected to enrichment analyses, using GOeast [78] for GO enrichment and clusterProfile [79] for KEGG enrichment. We used the entire transcript annotation as the background set when conducting the DEG functional enrichment analysis, which resulted from intraspecies comparisons. For interspecies comparisons, however, we used annotations of orthologous genes as a background set for DEG functional enrichments analysis. The significance was assessed using a hypergeometric test with FDR p-value correction ($p < 0.05$). Comparisons of the Cd accumulation in tissues and the tolerance indices between *V. baoshanensis* and *V. inconspicua* were performed using the statistical package SPSS 13.0 for Windows (IBM, Armonk, NY, USA). The data were examined using one-way ANOVA, followed by multiple comparisons using the least significant difference (LSD) test. The level of significance was set at $p < 0.05$ (two-tailed).

5. Conclusions

In summary, we sequenced and assembled high-quality transcriptomes for the Cd hyperaccumulator *V. baoshanensis* and its non-tolerant counterpart *V. inconspicua*, with an important contribution to the accumulation of the genetic resources of hyperaccumulators with more diverse taxa. Furthermore, intraspecies and interspecies differential expression genes and related functional enrichments were revealed by comparative analyses. Our results suggest an integrated strategy of Cd detoxification, mediated by UPS-dependent proteolysis, sucrose metabolism and vacuolar mobilization, as being responsible for the Cd hypertolerance of *V. baoshanensis*. The transcriptomic data presented in this study also provide genetic support for deep investigations on the conservation of these candidate genes and pathways in other plants.

Supplementary Materials: Supplementary materials can be found online at http://www.mdpi.com/1422-0067/20/8/1906/s1. Table S1. Statistics of the transcriptome sequencing data. Table S2. Summary of the BUSCO evaluation. Table S3. Statistics of upregulated DEGs in the UPS pathway. Table S4. Statistics of downregulated DEGs in the UPS pathway. Table S5. Statistics of upregulated DEGs in the sucrose metabolism pathway. Table S6. Statistics of downregulated DEGs in the sucrose metabolism pathway. Table S7. Statistics of upregulated DEGs encoding transporters. Table S8. Statistics of downregulated DEGs encoding transporters. Table S9. *V. baoshanensis* specific upregulated transcripts in roots in response to Cd stress. Table S10. *V. baoshanensis* specific upregulated transcripts in shoots in response to Cd stress. Table S11. Fold changes of DEGs involved in the UPS pathway between *V. baoshanesis* and *V. inconspicua*. Table S12. Fold changes of DEGs involved in sucrose metabolism between *V. baoshanesis* and *V. inconspicua*. Table S13. Fold changes of DEGs encoding transporters between *V. baoshanesis* and *V. inconspicua*. Table S14. Detailed proportions of GO terms in *V. baoshanensis* and *V. inconspicua*. Figure S1. Length distribution of entire transcripts and orthologous transcripts in *V. baoshanensis* and *V. inconspicua*. Figure S2. Volcano plots of gene transcription-level changes in the two *Viola* species with different Cd treatments (from 0 μM to 100 μM). Figure S3. A Venn diagram of specific and common up-regulated transcripts in *V. baoshanensis* and *V. inconspicua* in response to Cd stress. Figure S4. Summary of significantly upregulated genes in *V. baoshanensis* compared with *V. inconspicua*. Figure S5. GO enrichment of upregulated DEGs in *V. baoshanensis* and *V. inconspicua*. Figure S6. KEGG enrichment of upregulated DEGs in roots and shoots of *V. baoshanensis* compared with *V. inconspicua*. Figure S7. KEGG enrichment of up-regulated DEGs in *V. baoshanensis* and *V. inconspicua*.

Author Contributions: B.L., Q.S. and W.S. designed the project; F.L., W.L., Z.L. and J.L. (Jiaqi Liang) performed the experiments; H.S., C.B. and J.L. (Jieliang Liang) analyzed the data; H.S. prepared the manuscript; B.L., Q.S., J.Z. and J.L. (Jintian Li) revised the manuscript.

Funding: The research was supported by National Natural Science Foundation of China (Nos. 31570506, 41622106, 31570500 and U1501232) and Natural Science Foundation of Guangdong Province (No. 2016A030312003).

Acknowledgments: We thank Guangdong Magigene Biotechnology Co. Ltd. (Guangzhou, China) for assistance in transcriptome sequencing and data analysis.

Conflicts of Interest: The authors declare no conflicts of interest.

Abbreviations

ABC	ATP-binding cassette transporters
AMY	alpha-amylase
bglX	beta-glucosidase
BUSCO	Benchmarking Universal Single-Copy Orthologs
Cd	cadmium
CaCA	the Ca^{2+}: cation antiporter Family
CESA	cellulose synthase
CTR	the Copper transporter family
DEG	differential expression gene
FRK	fructokinase
GO	Gene Ontology
GYG1	glycogenin
HMA	heavy metal ATPase
HIPP	heavy metal-associated isoprenylated plant protein
HXK	hexokinase
INV	beta-fructofuranosidase
KEGG	Kyoto Encyclopedia of Genes and Genomes
malZ	alpha-glucosidase
MATE	multi-antimicrobial extrusion protein
MTP	metal tolerance protein
Ni	nickel

Nramp	natural resistance-associated macrophage protein
OM	organic matter
Pb	lead
SPP	sucrose-6-phosphatase
SPS	sucrose-phosphate synthase
TPP	trehalose 6-phosphate phosphatase
TPS	trehalose 6-phosphate synthase
TREH	alpha, alpha-trehalase
UPS	ubiquitin proteosome system
YSL	yellow stripe-like family
ZIP	the Zinc/Iron permease family
Zn	zinc

References

1. Buha, A.; Matovic, V.; Antonijevic, B.; Bulat, Z.; Curcic, M.; Renieri, E.; Tsatsakis, A.; Schweitzer, A.; Wallace, D. Overview of cadmium thyroid disrupting effects and mechanisms. *Int. J. Mol. Sci.* **2018**, *19*, 1501. [CrossRef] [PubMed]
2. Clemens, S.; Aarts, M.G.; Thomine, S.; Verbruggen, N. Plant science: The key to preventing slow cadmium poisoning. *Trends Plant Sci.* **2013**, *18*, 92–99. [CrossRef] [PubMed]
3. Ali, H.; Khan, E.; Sajad, M.A. Phytoremediation of heavy metals-Concepts and applications. *Chemosphere* **2013**, *91*, 869–881. [CrossRef] [PubMed]
4. Rascio, N.; Navari-Izzo, F. Heavy metal hyperaccumulating plants: How and why do they do it? And what makes them so interesting? *Plant Sci.* **2011**, *180*, 169–181. [CrossRef] [PubMed]
5. Nathalie, V.; Christian, H.; Henk, S. Molecular mechanisms of metal hyperaccumulation in plants. *New Phytol.* **2009**, *181*, 759–776. [CrossRef]
6. Krämer, U. Metal Hyperaccumulation in Plants. *Annu. Rev. Plant Biol.* **2010**, *61*, 517–534. [CrossRef] [PubMed]
7. Li, J.T.; Gurajala, H.K.; Wu, L.; Antony, V.D.E.; Qiu, R.L.; Baker, A.J.M.; Tang, Y.T.; Yang, X.; Shu, W. Hyperaccumulator plants from China: A synthesis of the current state of knowledge. *Environ. Sci. Technol.* **2018**, *52*, 11980–11994. [CrossRef]
8. Kusznierewicz, B.; Bączek-Kwinta, R.; Bartoszek, A.; Piekarska, A.; Huk, A.; Manikowska, A.; Antonkiewicz, J.; Namieśnik, J.; Konieczka, P. The dose-dependent influence of zinc and cadmium contamination of soil on their uptake and glucosinolate content in white cabbage (*Brassica oleracea* var. *capitata* f. *alba*). *Environ. Toxicol. Chem.* **2012**, *31*, 2482–2489. [CrossRef]
9. Shahid, M.; Dumat, C.; Khalid, S.; Schreck, E.; Xiong, T.; Niazi, N.K. Foliar heavy metal uptake, toxicity and detoxification in plants: A comparison of foliar and root metal uptake. *J. Hazard Mater.* **2016**, *325*, 36–58. [CrossRef]
10. Cobbett, C.S. Phytochelatins and their roles in heavy metal detoxification. *Plant Physiol.* **2000**, *123*, 825–832. [CrossRef] [PubMed]
11. Luigi, P.; Gea, G.; Kjell, S.; Giampiero, C.; Jean-Francois, H. Target or barrier? The cell wall of early- and later-diverging plants vs cadmium toxicity: Differences in the response mechanisms. *Front. Plant Sci.* **2015**, *6*, 133. [CrossRef]
12. Hossain, M.A.; Piyatida, P.; da Silva, J.A.T.; Fujita, M. Molecular mechanism of heavy metal toxicity and tolerance in plants: Central role of glutathione in detoxification of reactive oxygen species and methylglyoxal and in heavy metal chelation. *J. Bot.* **2012**, *2012*, 872875. [CrossRef]
13. Verbruggen, N.; Hanikenne, M.; Clemens, S. A more complete picture of metal hyperaccumulation through next-generation sequencing technologies. *Front. Plant Sci.* **2013**, *4*, 388. [CrossRef] [PubMed]
14. Halimaa, P.; Blande, D.; Aarts, M.G.; Tuomainen, M.; Tervahauta, A.; Karenlampi, S. Comparative transcriptome analysis of the metal hyperaccumulator Noccaea caerulescens. *Front. Plant Sci.* **2014**, *5*, 213. [CrossRef]
15. Cappa, J.J.; Pilon-Smits, E.A.H. Evolutionary aspects of elemental hyperaccumulation. *Planta* **2014**, *239*, 267–275. [CrossRef]

16. Goolsby, E.W.; Mason, C.M. Toward a more physiologically and evolutionarily relevant definition of metal hyperaccumulation in plants. *Front. Plant Sci.* **2015**, *6*, 33. [CrossRef]
17. Weber, M.; Trampczynska, A.; Clemens, S. Comparative transcriptome analysis of toxic metal responses in Arabidopsis thaliana and the Cd(2+)-hypertolerant facultative metallophyte Arabidopsis halleri. *Plant Cell Environ.* **2010**, *29*, 950–963. [CrossRef]
18. Peng, J.-S.; Wang, Y.-J.; Ding, G.; Ma, H.-L.; Zhang, Y.-J.; Gong, J.-M. A pivotal role of cell wall in cadmium accumulation in the Crassulaceae hyperaccumulator Sedum plumbizincicola. *Mol. Plant* **2017**, *10*, 771–774. [CrossRef] [PubMed]
19. Liu, W.; Shu, W.; Lan, C. Viola baoshanensis, a plant that hyperaccumulates cadmium. *Chin. Sci. Bull.* **2004**, *49*, 29–32. [CrossRef]
20. Tonin, C.; Vandenkoornhuyse, P.; Joner, E.J.; Straczek, J.; Leyval, C. Assessment of arbuscular mycorrhizal fungi diversity in the rhizosphere of Viola calaminaria and effect of these fungi on heavy metal uptake by clover. *Mycorrhiza* **2001**, *10*, 161–168. [CrossRef]
21. Sychta, K.; Słomka, A.; Suski, S.; Fiedor, E.; Gregoraszczuk, E.; Kuta, E. Suspended cells of metallicolous and nonmetallicolous Viola species tolerate, accumulate and detoxify zinc and lead. *Plant Physiol. Biochem.* **2018**, *132*, 666–674. [CrossRef] [PubMed]
22. Fernando, E.S.; Quimado, M.O.; Doronila, A.I. Rinorea niccolifera (Violaceae), a new, nickel-hyperaccumulating species from Luzon Island, Philippines. *PhytoKeys* **2014**, *37*, 1–13. [CrossRef] [PubMed]
23. Wu, C.A.; Liao, B.; Wang, S.L.; Zhang, J.; Li, J.T. Pb and Zn accumulation in a Cd-hyperaccumulator (Viola baoshanensis). *Int. J. Phytoremediat.* **2010**, *12*, 574–585. [CrossRef]
24. Simão, F.A.; Waterhouse, R.M.; Ioannidis, P.; Kriventseva, E.V.; Zdobnov, E.M. BUSCO: Assessing genome assembly and annotation completeness with single-copy orthologs. *Bioinformatics* **2015**, *31*, 3210–3212. [CrossRef]
25. Parekh, S.; Vieth, B.; Ziegenhain, C.; Enard, W.; Hellmann, I. Strategies for quantitative RNA-seq analyses among closely related species. *bioRxiv* **2018**, 297408. [CrossRef]
26. Song, Y.; Jin, L.; Wang, X. Cadmium absorption and transportation pathways in plants. *Int. J. Phytoremediat.* **2017**, *19*, 133–141. [CrossRef] [PubMed]
27. Ari, S.; Mark, B.; Richard, E.; Jack, L.; Stuart, N. The ubiquitin-proteasome system: Central modifier of plant signalling. *New Phytol.* **2012**, *196*, 13–28. [CrossRef]
28. Jungmann, J.; Reins, H.A.; Schobert, C.; Jentsch, S. Resistance to cadmium mediated by ubiquitin-dependent proteolysis. *Nature* **1993**, *361*, 369. [CrossRef]
29. Chai, T.Y.; Zhang, Y.X. Expression analysis of polyubiquitin genes from bean in response to heavy metals. *Acta Bot. Sin.* **1999**, *41*, 1052–1057.
30. Oono, Y.; Yazawa, T.; Kanamori, H.; Sasaki, H.; Mori, S.; Handa, H.; Matsumoto, T. Genome-Wide Transcriptome Analysis of Cadmium Stress in Rice. *BioMed Res. Int.* **2016**, *2016*, 1–9. [CrossRef]
31. Flick, K.; Kaiser, P. Protein degradation and the stress response. *Cell Dev. Biol.* **2012**, *23*, 515–522. [CrossRef] [PubMed]
32. Amm, I.; Sommer, T.; Wolf, D.H. Protein quality control and elimination of protein waste: The role of the ubiquitin-proteasome system. *BBA-Mol. Cell. Res.* **2014**, *1843*, 182–196. [CrossRef]
33. Hasan, M.K.; Cheng, Y.; Kanwar, M.K.; Chu, X.Y.; Ahammed, G.J.; Qi, Z.Y. Responses of Plant Proteins to Heavy Metal Stress—A Review. *Front. Plant Sci.* **2017**, *8*, 1492. [CrossRef]
34. Sharma, S.K.; Goloubinoff, P.; Christen, P. *Cellular Effects of Heavy Metals*; Springer: New York, NY, USA, 2011; pp. 263–274.
35. Chen, C.C.; Chen, Y.Y.; Tang, I.C.; Liang, H.M.; Lai, C.C.; Chiou, J.M.; Yeh, K.C. Arabidopsis SUMO E3 Ligase SIZ1 Is Involved in Excess Copper Tolerance. *Plant Physiol.* **2011**, *156*, 2225–2234. [CrossRef]
36. Lim, S.D.; Jin, G.H.; Han, A.R.; Yong, C.P.; Lee, C.; Yong, S.O.; Jang, C.S. Positive regulation of rice RING E3 ligase OsHIR1 in arsenic and cadmium uptakes. *Plant Mol. Biol.* **2014**, *85*, 365. [CrossRef] [PubMed]
37. Dametto, A.; Buffon, G.; dos Reis Blasi, Ã.A.; Sperotto, R.A. Ubiquitination pathway as a target to develop abiotic stress tolerance in rice. *Plant Signal. Behav.* **2015**, *10*, e1057369. [CrossRef] [PubMed]
38. Yongling, R. Sucrose metabolism: Gateway to diverse carbon use and sugar signaling. *Annu. Rev. Plant Biol.* **2014**, *65*, 33–67. [CrossRef]
39. Antonkiewicz, J.; Para, A. The use of dialdehyde starch derivatives in the phytoremediation of soils contaminated with heavy metals. *Int. J. Phytoremediat.* **2016**, *18*, 245–250. [CrossRef]

40. Baccio, D.D.; Galla, G.; Bracci, T.; Andreucci, A.; Barcaccia, G.; Tognetti, R.; Sebastiani, L. Transcriptome analyses of Populus × euramericana clone I-214 leaves exposed to excess zinc. *Tree Physiol.* **2011**, *31*, 1293–1308. [CrossRef]
41. Feng, J.; Jia, W.; Lv, S.; Bao, H.; Miao, F.; Zhang, X.; Wang, J.; Li, J.; Li, D.; Zhu, C. Comparative transcriptome combined with morpho-physiological analyses revealed key factors for differential cadmium accumulation in two contrasting sweet sorghum genotypes. *Plant Biotechnol. J.* **2017**, *16*, 558–571. [CrossRef]
42. Huang, Y.Y.; Shen, C.; Chen, J.X.; He, C.T.; Zhou, Q.; Tan, X.; Yuan, J.; Yang, Z. Comparative transcriptome analysis of two Ipomoea aquatica Forsk. cultivars targeted to explore possible mechanism of genotype dependent accumulation of cadmium. *J. Agric. Food Chem.* **2016**, *64*, 5241–5250. [CrossRef] [PubMed]
43. Granot, D. Role of tomato hexose kinases. *Funct. Plant Biol.* **2007**, *34*, 564–570. [CrossRef]
44. Melissa, R.; Lorenz, G.; David, S.; András, G.; Mattias, H.M.; Manoj, K.; Marie Caroline, S.; Regina, F.; Geoffrey, D.; Mark, S. Fructokinase is required for carbon partitioning to cellulose in aspen wood. *Plant J.* **2012**, *70*, 967–977. [CrossRef]
45. Laurence, L.; Judith, W.; Marjorie, P.; Joanna Marie-France, C.; Pascal, T.; Alain, G. Oxidative pentose phosphate pathway-dependent sugar sensing as a mechanism for regulation of root ion transporters by photosynthesis. *Plant Physiol.* **2008**, *146*, 2036–2053. [CrossRef]
46. Chen, X.-Y.; Kim, J.-Y. Callose synthesis in higher plants. *Plant Signal. Behav.* **2009**, *4*, 489–492. [CrossRef] [PubMed]
47. Bo, X.; Xiaomin, W.; Maosheng, Z.; Zhongming, Z.; Zonglie, H. CalS7 encodes a callose synthase responsible for callose deposition in the phloem. *Plant J.* **2011**, *65*, 1–14. [CrossRef]
48. Vatén, A.; Dettmer, J.; Shuang, W.; Stierhof, Y.D.; Miyashima, S.; Yadav, S.R.; Roberts, C.; Campilho, A.; Bulone, V.; Lichtenberger, R. Callose Biosynthesis Regulates Symplastic Trafficking during Root Development. *Dev. Cell* **2011**, *21*, 1144–1155. [CrossRef]
49. Levy, A.; Erlanger, M.; Rosenthal, M.; Epel, B.L. A plasmodesmata-associated β-1,3-glucanase in Arabidopsis. *Plant J.* **2010**, *49*, 669–682. [CrossRef] [PubMed]
50. Jens, M.; Theresa, T.; Marcus, H.; Janine, T.; Moore, K.L.; Gerd, H.; Dhurvas Chandrasekaran, D.; Katharina, B.; Steffen, A. Iron-dependent callose deposition adjusts root meristem maintenance to phosphate availability. *Dev. Cell* **2015**, *33*, 216–230. [CrossRef]
51. Mira, H.; Martinez-Garcia, F.; Penarrubia, L. Evidence for the plant-specific intercellular transport of the Arabidopsiscopper chaperone CCH. *Plant J.* **2010**, *25*, 521–528. [CrossRef]
52. Zhang, H.; Shi, W.L.; You, J.F.; Bian, M.D.; Qin, X.M.; Hui, Y.U.; Liu, Q.; Ryan, P.R.; Yang, Z.M. Transgenic *Arabidopsis thaliana* plants expressing a β-1,3-glucanase from sweet sorghum (*Sorghum bicolor* L.) show reduced callose deposition and increased tolerance to aluminium toxicity. *Plant Cell Environ.* **2015**, *38*, 1178–1188. [CrossRef] [PubMed]
53. Keunen, E.; Peshev, D.; Vangronsveld, J.; Van, D.E.W.; Cuypers, A. Plant sugars are crucial players in the oxidative challenge during abiotic stress: Extending the traditional concept. *Plant Cell Environ.* **2013**, *36*, 1242–1255. [CrossRef] [PubMed]
54. Sami, F.; Yusuf, M.; Faizan, M.; Faraz, A.; Hayat, S. Role of sugars under abiotic stress. *Plant Physiol. Biochem.* **2016**, *109*, 54–61. [CrossRef]
55. Mostofa, M.G.; Hossain, M.A.; Fujita, M.; Tran, L.S. Physiological and biochemical mechanisms associated with trehalose-induced copper-stress tolerance in rice. *Sci. Rep.* **2015**, *5*, 11433. [CrossRef]
56. Martins, L.L.; Mourato, M.P.; Baptista, S.; Reis, R.; Carvalheiro, F.; Almeida, A.M.; Fevereiro, P.; Cuypers, A. Response to oxidative stress induced by cadmium and copper in tobacco plants (Nicotiana tabacum) engineered with the trehalose-6-phosphate synthase gene (AtTPS1). *Acta Physiol. Plant.* **2014**, *36*, 755–765. [CrossRef]
57. Ali, Q.; Ashraf, M. Induction of Drought Tolerance in Maize (*Zea mays* L.) due to Exogenous Application of Trehalose: Growth, Photosynthesis, Water Relations and Oxidative Defence Mechanism. *J. Agron. Crop Sci.* **2011**, *197*, 258–271. [CrossRef]
58. Benaroudj, N.; Lee, D.; Goldberg, A. Trehalose accumulation during cellular stress protects cells and cellular proteins from damage by oxygen radicals. *J. Biol. Chem.* **2001**, *276*, 24261–24267. [CrossRef]
59. Lux, A.; Martinka, M.; Vaculík, M.; White, P.J. Root responses to cadmium in the rhizosphere: A review. *J. Exp. Bot.* **2011**, *62*, 21–37. [CrossRef] [PubMed]

60. Pence, N.S.; Larsen, P.B.; Ebbs, S.D.; Letham, D.L.; Lasat, M.M.; Garvin, D.F.; Eide, D.; Kochian, L.V. The molecular physiology of heavy metal transport in the Zn/Cd hyperaccumulator Thlaspi caerulescens. *Proc. Natl. Acad. Sci. USA* **2000**, *97*, 4956–4960. [CrossRef]
61. Hanikenne, M.; Talke, I.N.; Haydon, M.J.; Lanz, C.; Nolte, A.; Motte, P.; Kroymann, J.; Weigel, D.; Kramer, U. Evolution of metal hyperaccumulation required cis-regulatory changes and triplication of HMA4. *Nature* **2008**, *453*, 391–395. [CrossRef]
62. Li, J.; Wang, L.; Zheng, L.; Wang, Y.; Chen, X.; Zhang, W. A Functional Study Identifying Critical Residues Involving Metal Transport Activity and Selectivity in Natural Resistance-Associated Macrophage Protein 3 in Arabidopsis thaliana. *Int. J. Mol. Sci.* **2018**, *19*, 1430. [CrossRef]
63. Sharma, S.S.; Dietz, K.J.; Mimura, T. Vacuolar compartmentalization as indispensable component of heavy metal detoxification in plants. *Plant Cell Environ.* **2016**, *39*, 1112–1126. [CrossRef] [PubMed]
64. Gendre, D.; Czernic, P.; Conéjéro, G.; Pianelli, K.; Briat, J.F.; Lebrun, M.; Mari, S. TcYSL3, a member of the YSL gene family from the hyper-accumulator Thlaspi caerulescens, encodes a nicotianamine-Ni/Fe transporter. *Plant J.* **2010**, *49*, 1–15. [CrossRef]
65. Dai, J.; Wang, N.; Xiong, H.; Qiu, W.; Nakanishi, H.; Kobayashi, T.; Nishizawa, N.K.; Zuo, Y. The Yellow Stripe-Like (YSL) Gene Functions in Internal Copper Transport in Peanut. *Genes* **2018**, *9*, 635. [CrossRef]
66. Li, J.T.; Deng, D.M.; Peng, G.T.; Deng, J.C.; Zhang, J.; Liao, B. Successful micropropagation of the cadmium hyperaccumulator Viola baoshanensis (Violaceae). *Int. J. Phytoremediat.* **2010**, *12*, 761–771. [CrossRef]
67. Hoagland, D.R.; Arnon, D.I. The water-culture method for growing plants without soil. *Circ. Calif. Agric. Exp. Stn.* **1950**, *347*, 357–359.
68. Allen, S.E.; Grimshaw, H.M.; Parkinson, J.A.; Quarmby, C. Chemical analysis of ecological materials. *J. Appl. Ecol.* **1974**, *13*, 650.
69. Grabherr, M.G.; Haas, B.J.; Yassour, M.; Levin, J.Z.; Thompson, D.A.; Amit, I.; Xian, A.; Lin, F.; Raychowdhury, R.; Zeng, Q. Trinity: Reconstructing a full-length transcriptome without a genome from RNA-Seq data. *Nat. Biotechnol.* **2011**, *29*, 644–652. [CrossRef]
70. Pertea, G.; Huang, X.; Liang, F.; Antonescu, V.; Sultana, R.; Karamycheva, S.; Lee, Y.; White, J.; Cheung, F.; Parvizi, B. TIGR Gene Indices clustering tools (TGICL): A software system for fast clustering of large EST datasets. *Bioinformatics* **2003**, *19*, 651–652. [CrossRef] [PubMed]
71. Consortium, U.P. The Universal Protein Resource (UniProt) 2009. *Nucleic Acids Res.* **2009**, *37*, 169–174. [CrossRef]
72. Zdobnov, E.M.; Apweiler, R. InterProScan—An integration platform for the signature-recognition methods in InterPro. *Bioinformatics* **2001**, *17*, 847–848. [CrossRef] [PubMed]
73. Ye, J.; Zhang, Y.; Cui, H.; Liu, J.; Wu, Y.; Cheng, Y.; Xu, H.; Huang, X.; Li, S.; Zhou, A. WEGO 2.0: A web tool for analyzing and plotting GO annotations, 2018 update. *Nucleic Acids Res.* **2018**, *46*, W71–W75. [CrossRef]
74. Langmead, B.; Salzberg, S.L. Fast gapped-read alignment with Bowtie 2. *Nat. Methods* **2012**, *9*, 357–359. [CrossRef] [PubMed]
75. Bo, L.; Dewey, C.N. RSEM: Accurate transcript quantification from RNA-Seq data with or without a reference genome. *BMC Bioinform.* **2011**, *12*, 323. [CrossRef]
76. Moreno-Hagelsieb, G.; Latimer, K. Choosing BLAST options for better detection of orthologs as reciprocal best hits. *Bioinformatics* **2008**, *24*, 319–324. [CrossRef] [PubMed]
77. Robinson, M.D.; Mccarthy, D.J.; Smyth, G.K. edgeR: A Bioconductor package for differential expression analysis of digital gene expression data. *Bioinformatics* **2010**, *26*, 139–140. [CrossRef] [PubMed]
78. Zheng, Q.; Wang, X.-J. GOEAST: A web-based software toolkit for Gene Ontology enrichment analysis. *Nucleic Acids Res.* **2008**, *36*, 358–363. [CrossRef] [PubMed]
79. Yu, G.; Wang, L.G.; Han, Y.; He, Q.Y. clusterProfiler: An R package for comparing biological themes among gene clusters. *Omics* **2012**, *16*, 284–287. [CrossRef]

© 2019 by the authors. Licensee MDPI, Basel, Switzerland. This article is an open access article distributed under the terms and conditions of the Creative Commons Attribution (CC BY) license (http://creativecommons.org/licenses/by/4.0/).

International Journal of
Molecular Sciences

MDPI

Article

Exogenous Glycinebetaine Reduces Cadmium Uptake and Mitigates Cadmium Toxicity in Two Tobacco Genotypes Differing in Cadmium Tolerance

Xiaoyan He [1,2], Marvin E.A. Richmond [1], Darron V. Williams [1], Weite Zheng [1] and Feibo Wu [1,*]

1 Department of Agronomy, College of Agriculture and Biotechnology, Zijingang Campus, Zhejiang University, Hangzhou 310058, China; hexiaoyan@zju.edu.cn (X.H.); m.richmond25@hotmail.com (M.E.R.); darronw3@gmail.com (D.V.W.); 21116045@zju.edu.cn (W.Z.)

2 College of Agronomy, Qingdao Agricultural University, Qingdao 266109, China

* Correspondence: wufeibo@zju.edu.cn

Received: 20 February 2019; Accepted: 29 March 2019; Published: 31 March 2019

Abstract: Greenhouse hydroponic experiments were conducted using Cd-sensitive (*cv.* Guiyan1) and Cd-tolerant (*cv.* Yunyan2) tobacco cultivars to study the ameliorative effects of exogenous glycinebetaine (GB) upon 5 μM Cd stress. The foliar spray of GB markedly reduced Cd concentrations in plants and alleviated Cd-induced soil plant analysis development (SPAD) value, plant height and root length inhibition, with the mitigation effect being more obvious in Yunyan2. External GB markedly reduced Cd-induced malondialdehyde (MDA) accumulation, induced stomatal closure, ameliorated Cd-induced damages on leaf/root ultrastructure, and increased the chlorophyll content and fluorescence parameters of Fo, Fm, and Fv/Fm in both cultivars and Pn in Yunyan2. Exogenous GB counteracted Cd-induced alterations of certain antioxidant enzymes and nutrients uptake, e.g., the depressed Cd-induced increase of superoxide dismutase (SOD) and peroxidase (POD) activities, but significantly elevated the depressed catalase (CAT) and ascorbate peroxidase (APX) activities. The results indicate that alleviated Cd toxicity by GB application is related to the reduced Cd uptake and MDA accumulation, balanced nutrients and antioxidant enzyme activities, improved PSII, and integrated ultrastructure in tobacco plants.

Keywords: cadmium; glycinebetaine; photosynthesis; ultrastructure; tobacco (*Nicotiana tabacum* L.)

1. Introduction

A high cadmium (Cd) content in soil results in the inhibition of plant growth and yield reduction [1,2]. Tobacco (*Nicotiana tabacum* L.) is one of the most economically important crops worldwide. However, it is more acclimated to Cd uptake than other crops and preferentially enriches Cd in leaves, readily causing a risk for human health through the inhalation of smoke from cigarettes [3]. It has been demonstrated that Cd-exposed populations through smoking has a 2–3 folds higher morbidity risk of peripheral arterial disease than those of nonsmokers [4]. Accordingly, smoking has become one of the most important absorptive pathways of Cd in humans [5]. Thus, there is an urgent need to develop reliable approaches to prevent Cd accumulation in tobacco. The application of chemical regulators to alleviate Cd toxicity and to reduce plant Cd uptake in medium or slightly contaminated farmlands might offer a cost-effective and practically acceptable strategy for the complete utilization of natural resources and safe tobacco production.

Cd interferes the biochemical and physiological processes, such as mineral uptake [6], photosynthesis [7], oxidative stress [8], stomatal conductance, and transpiration [9]. Cd injury is probably attributed to an alteration in the oxidant level in plants, as Cd may cause reactive oxygen species (ROS), leading to oxidative injury [10]. Correspondingly, the plant internal metabolites and scavenging enzymes of active oxygen are relatively changed, which is beneficial to the development of a defense system. The different influence patterns of Cd toxicity on ROS-scavenging enzymes activities, such as superoxide dismutase (SOD), peroxidase (POD), catalase (CAT), and ascorbate peroxidase (APX), were found [11–13]. Therefore, in order to verify the hypothesis that some antioxidants may also be sensitive targets of Cd toxicity besides their function in the detoxification in plants, the changes in Cd-induced oxidant stress and antioxidant systems is imperative to determine.

Glycinebetaine (GB) (Figure 1), being a nitrogenous compound (quaternary amine) that behaves as zwitterion, is known to perform two main functions: osmotic adjustment and cellular compatibility in plants. The natural accumulation of GB in plants is correlated with abiotic stress tolerance [14,15]. A genetic transformation with GB synthesizing enzyme gene(s) in naturally non-accumulating plants has resulted in an enhanced tolerance against a variety of abiotic stresses [16,17]. The application of GB, foliar spray or genetic modification, improves stress tolerance in different plant species and enhances antioxidant defense systems in plant responses to various oxidative stresses [18,19]. Banu et al. [20] found that GB induces the expression of ROS scavenging antioxidant defense genes and suppresses ROS accumulation and cell death in cultured tobacco cells exposed to NaCl stress. As to Cd stress, Duman et al. [21] observed that exogenous GB and trehalose reduced the deleterious effects of Cd stress in duckweed (*Lemna gibba*). Islam and colleagues [22,23] reported that exogenous proline and glycinebetaine increase ascorbate-glutathione cycle and antioxidant enzyme activities and confer tolerance to Cd stress in cultured tobacco cells. However, the mechanism by which GB confers tolerance to plants against heavy metal stress, including Cd stress, is still poorly understood. Therefore, the question arises whether GB participates in Cd tolerance and whether GB application could reduce Cd accumulation in tobacco plants. It is also of great significance to understand whether exogenous GB can be used as a regulator of Cd stress or as an antioxidant intervention strategy in preventing oxidative stress in responses to Cd stress so as to better understand how plants adapt to adverse environments.

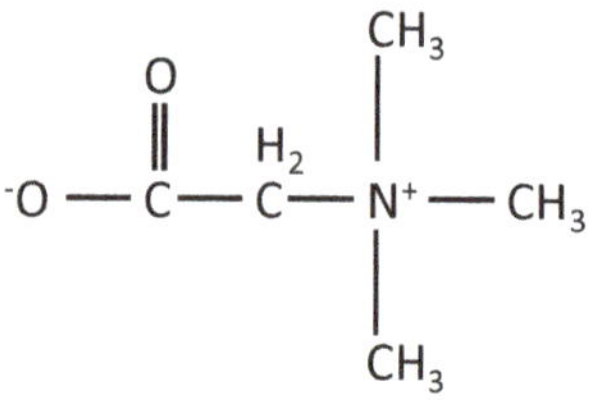

Figure 1. The chemical structure of glycinebetaine.

The present study was conducted via a hydroponic experiment to investigate the potential role of exogenous GB in alleviating Cd-induced changes in antioxidative metabolism, ultrastructure, photosynthesis, and chlorophyll fluorescence of two tobacco cultivars of Cd-sensitive (*cv.* Guiyan1) and Cd-tolerant (*cv.* Yunyan2). We aimed to provide a basis for developing strategies to reduce the risks associated with Cd toxicity and maintaining a sustainable tobacco production.

2. Results

2.1. Effect of Exogenous GB on Cd-Induced Suppression in Plant Growth

After 15 days of Cd exposure, Cd toxicity markedly hindered the soil plant analysis development (SPAD) value, plant height, root length, and biomass (Table 1). The tolerant cultivar Yunyan2 was less affected in terms of the abovementioned growth traits, whereas the sensitive Guiyan1 was more affected. Cd + GB treatment apparently alleviated the Cd-induced SPAD value, plant height, and root length inhibition. The alleviating effects were evaluated using the formula-based integrated relation. There is a positive correlation between the alleviating effects and the integrated scores. According to the integrated scores, Yunyan2 had a better mitigation effect under Cd + GB with a higher score than Guiyan1 (Table 1). A parallel experiment was performed to evaluate the vigor of root cells using flurescein diacetate-propidium iodide (FDA-PI) assay (Figure 2). After 15 days of Cd treatment, bright and red fluorescence were observed in roots of tobacco seedlings grown in Cd media, and Yunyan2 showed relatively few red fluorescence compared with Guiyan1. Cd + GB treatment markedly decreased the red fluorescence intensity but increased the green fluorescence compared with Cd alone treatment. The red fluorescence was very low, but detectable levels of red fluorescence were observed in the Cd + GB roots.

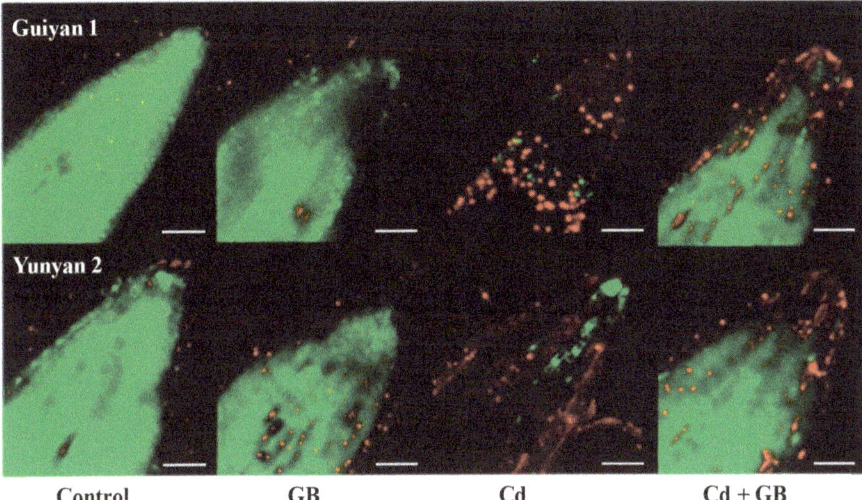

Figure 2. The effects of Cd and exogenous GB on the cell vigor in the root tips of Guiyan1 and Yunyan2 detected by an flurescein diacetate-propidium iodide (FDA-PI) dual fluorescent dye after 15 days of Cd exposure: The green and red fluorescence indicate the fluorescent dye in living and dead cells, respectively. Control, GB, Cd, and Cd + GB correspond to BNS + foliar spray of deionized water, BNS + foliar spray of 500 μM GB, BNS + foliar spray of deionized water + 5 μM $CdCl_2$, and BNS + foliar spray of 500 μM GB + 5 μM $CdCl_2$, respectively. Scale bars = 250 μm. The figure is representative of three different experiments.

Table 1. The effects of exogenous GB on the growth of Cd-treated tobacco seedlings after 15 days of Cd exposure.

Treatment	SPAD Value		Plant Height (cm)		Root Length (cm)		Fresh Weight ($g \cdot plant^{-1}$)				Dry Weight ($g \cdot plant^{-1}$)				Integrated Score *
							Shoot		Root		Shoot		Root		
							Guiyan1								
Control	41.46 ± 2.76	a	39.13 ± 0.58	a	37.40 ± 1.35	a	71.34 ± 1.99	a	19.98 ± 0.95	a	7.57 ± 0.61	a	1.15 ± 0.09	a	43.61
GB	42.16 ± 1.84	a	37.03 ± 1.18	b	35.11 ± 1.56	a	69.09 ± 1.62	a	15.84 ± 1.20	b	7.43 ± 0.40	a	1.11 ± 0.10	a	41.55
Cd	34.02 ± 3.72	b	20.66 ± 1.25	d	22.95 ± 1.81	c	39.47 ± 1.08	b	9.86 ± 0.69	c	5.08 ± 0.47	b	0.61 ± 0.11	b	26.53
Cd + GB	39.72 ± 3.80	a	25.03 ± 1.03	c	26.03 ± 0.32	b	39.76 ± 1.43	b	9.91 ± 0.76	c	5.41 ± 0.36	b	0.68 ± 0.01	b	29.31
							Yunyan2								
Control	45.50 ± 3.41	a	40.30 ± 1.42	a	34.75 ± 1.44	a	65.11 ± 1.68	a	15.57 ± 0.51	a	7.52 ± 0.41	a	1.19 ± 0.07	a	41.99
GB	45.88 ± 2.28	a	38.52 ± 0.86	a	34.07 ± 1.65	a	67.86 ± 2.30	a	15.13 ± 0.77	a	7.43 ± 0.36	a	1.15 ± 0.10	a	42.01
Cd	37.60 ± 2.98	b	28.60 ± 0.77	c	27.69 ± 1.54	b	46.27 ± 2.09	b	10.72 ± 1.56	b	4.78 ± 0.59	b	0.70 ± 0.00	c	31.27
Cd + GB	43.80 ± 2.77	a	31.10 ± 1.00	b	33.51 ± 1.14	a	48.12 ± 0.79	b	11.13 ± 0.65	b	4.95 ± 0.54	b	0.83 ± 0.02	b	34.69

The data were the means of three independent replicates. The different letters in each column indicate the significant differences ($p < 0.05$) among the 4 treatments within each cultivar. Control, GB (glycinebetaine), Cd, and Cd + GB correspond to the basic nutrition solution (BNS) + foliar spray of deionized water, BNS + foliar spray of 500 μM GB, BNS + foliar spray of deionized water + 5 μM $CdCl_2$, and BNS + foliar spray of 500 μM GB + 5 μM $CdCl_2$, respectively. * Integrated score = the absolute values of (the soil plant analysis development (SPAD) value × 0.2 + shoot height × 0.2 + root length × 0.2 + fresh weight × 0.2 + dry weight × 0.2).

2.2. Effect of Exogenous GB on Cd and Nutrient Elements Contents in Tobacco Seedlings

The 2 tobacco cultivars had clear differences in Cd concentration: Yunyan2 recorded 2%, 21%, and 8% more than Guiyan1 in S1 (three top leaves with stem), S2 (middle part of shoot), and S3 (three bottom leaves with stem), but Guiyan1 recorded 85%, 23%, and 1% more than Yunyan2 in R1 (three cm of root tip), R2 (middle part of root), and R3 (three cm of root bottom) (Table 2). Exogenous GB markedly suppressed Cd concentration compared with the Cd alone treatment, i.e., in S1, S2, and S3, the reductions were 14%, 20%, and 17% in Guiyan1 and 9%, 21%, and 7% in Yunyan2, and in R1, R2, and R3, these were 15%, 7%, and 8% and 15%, 13%, and 7%, respectively (Table 2). Cd treatment markedly decreased the shoot and root Zn and Mn in Guiyan1 and Yunyan2 and decreased the root Cu in Guiyan1 and shoot Cu in Yunyan2 but increased the shoot Fe and Cu and root Fe in Guiyan1 and root Fe and Cu in Yunyan2. However, to a degree, exogenous GB application recovered the shoot/root Fe and Cu in Guiyan1 and the shoot Zn and root Fe in Yunyan2 (Table 2).

Table 2. The effects of Cd and exogenous GB on the element concentrations in the shoots and roots of tobacco seedlings at day 15.

Elements (mg·kg^{-1} DW)		Guiyan1								Yunyan2							
		Control		GE		Cd		Cd + GB		Control		GB		Cd		Cd + GB	
Shoot																	
Cd	S1	0.05 ± 0.01	d	0.03 ± 0.01	d	11.13 ± 0.09	a	9.56 ± 0.24	c	0.07 ± 0.00	d	0.08 ± 0.00	d	11.36 ± 0.75	a	10.33 ± 0.46	b
	S2	0.04 ± 0.01	d	0.04 ± 0.01	d	10.65 ± 0.59	b	8.52 ± 0.30	c	0.07 ± 0.01	d	0.06 ± 0.03	d	12.89 ± 0.77	a	10.15 ± 0.26	b
	S3	0.05 ± 0.02	d	0.03 ± 0.02	d	12.93 ± 0.71	b	10.74 ± 0.42	c	0.08 ± 0.01	d	0.06 ± 0.01	d	14.05 ± 0.25	a	13.02 ± 0.36	b
Zn	S1	13.39 ± 0.54	d	12.85 ± 0.68	d	10.43 ± 0.42	e	10.26 ± 0.29	e	22.68 ± 0.52	b	25.79 ± 0.61	a	16.35 ± 0.76	c	15.83 ± 0.76	c
	S2	15.36 ± 0.62	bc	17.73 ± 0.90	a	9.41 ± 0.47	d	9.08 ± 0.76	d	15.96 ± 0.25	b	15.87 ± 0.48	b	14.71 ± 0.15	c	17.82 ± 0.22	a
	S3	26.69 ± 0.64	a	25.70 ± 0.11	b	10.51 ± 0.38	e	9.12 ± 0.13	f	22.49 ± 0.09	c	22.68 ± 0.72	c	16.72 ± 0.33	d	16.15 ± 0.67	d
Mn	S1	12.02 ± 0.31	c	11.10 ± 0.54	d	9.73 ± 0.27	e	9.68 ± 0.40	e	15.68 ± 0.17	a	14.90 ± 0.47	b	7.53 ± 0.60	f	7.42 ± 0.26	f
	S2	12.76 ± 0.56	bc	12.30 ± 0.47	c	10.58 ± 0.63	d	10.53 ± 0.56	d	14.47 ± 0.49	a	13.60 ± 0.77	ab	10.59 ± 0.41	d	10.00 ± 0.20	d
	S3	23.85 ± 0.38	a	22.66 ± 0.91	b	14.09 ± 0.98	e	13.80 ± 0.47	e	20.45 ± 0.57	c	19.21 ± 0.23	d	12.48 ± 0.19	f	10.51 ± 0.72	g
Cu	S1	2.74 ± 0.25	c	2.70 ± 0.26	c	5.60 ± 0.18	a	5.29 ± 0.25	a	5.32 ± 0.47	a	5.52 ± 0.15	a	4.45 ± 0.05	b	4.18 ± 0.18	b
	S2	2.67 ± 0.25	e	3.31 ± 0.22	cd	4.01 ± 0.45	b	2.70 ± 0.24	de	5.24 ± 0.19	a	4.93 ± 0.46	a	4.93 ± 0.30	a	3.70 ± 0.54	bc
	S3	4.32 ± 0.76	bcd	3.84 ± 0.52	cde	5.61 ± 0.32	a	3.08 ± 0.93	e	4.52 ± 0.34	bc	5.03 ± 0.44	ab	3.48 ± 0.45	de	4.42 ± 0.00	bc
Fe	S1	13.54 ± 0.42	b	8.18 ± 0.39	d	10.17 ± 0.94	c	8.97 ± 0.34	d	13.21 ± 0.79	b	13.32 ± 0.74	b	19.28 ± 0.55	a	10.18 ± 0.79	c
	S2	12.16 ± 0.41	cd	10.22 ± 0.81	ef	11.16 ± 1.32	de	8.56 ± 0.48	g	12.66 ± 0.67	c	19.38 ± 0.40	a	17.95 ± 0.71	b	9.86 ± 0.29	f
	S3	11.94 ± 0.73	d	10.65 ± 0.94	e	17.12 ± 0.75	c	8.37 ± 0.67	f	21.10 ± 0.34	b	24.47 ± 0.68	a	20.99 ± 0.30	b	12.10 ± 0.55	d
Root																	
Cd	R1	0.07 ± 0.00	e	0.07 ± 0.01	e	68.45 ± 1.76	a	58.38 ± 1.15	b	0.05 ± 0.01	e	0.05 ± 0.01	e	36.99 ± 1.04	c	31.38 ± 0.60	d
	R2	0.06 ± 0.01	e	0.08 ± 0.01	e	62.76 ± 0.22	a	58.11 ± 0.39	b	0.06 ± 0.02	e	0.04 ± 0.01	e	50.59 ± 1.70	c	44.00 ± 2.20	d
	R3	0.05 ± 0.01	c	0.07 ± 0.01	c	62.01 ± 1.18	a	57.28 ± 0.59	b	0.05 ± 0.02	c	0.03 ± 0.00	c	61.10 ± 2.57	a	56.79 ± 0.81	b
Zn	R1	73.37 ± 3.05	a	74.63 ± 1.69	a	65.56 ± 1.94	b	65.37 ± 1.84	b	47.67 ± 1.77	c	47.80 ± 1.72	c	41.75 ± 1.03	d	40.87 ± 1.18	d
	R2	71.08 ± 4.45	a	66.17 ± 0.78	b	48.77 ± 2.14	c	43.90 ± 1.41	d	37.90 ± 1.04	e	40.27 ± 2.27	de	33.55 ± 1.54	f	33.47 ± 2.07	f
	R3	52.01 ± 1.71	a	48.22 ± 2.29	b	34.05 ± 0.82	c	31.29 ± 1.68	d	24.20 ± 1.66	e	23.93 ± 0.87	e	17.60 ± 1.55	f	16.47 ± 0.28	f
Mn	R1	118.52 ± 1.06	b	79.37 ± 1.47	c	52.20 ± 0.26	f	45.33 ± 2.88	f	133.03 ± 8.22	a	122.98 ± 6.39	b	70.68 ± 3.49	d	62.96 ± 4.57	e
	R2	181.20 ± 9.78	a	95.07 ± 3.84	c	64.91 ± 4.49	e	60.10 ± 6.05	e	121.38 ± 7.01	b	115.42 ± 7.24	b	90.21 ± 7.95	cd	83.33 ± 4.54	d
	R3	122.53 ± 9.39	a	93.55 ± 3.43	b	52.38 ± 5.78	d	52.03 ± 6.06	d	84.68 ± 3.30	b	74.60 ± 4.00	c	50.24 ± 4.40	d	51.72 ± 2.47	d
Cu	R1	25.13 ± 1.72	b	24.49 ± 0.88	b	14.17 ± 2.62	d	36.83 ± 1.14	a	19.19 ± 0.30	c	12.24 ± 1.10	d	12.89 ± 3.83	d	12.95 ± 0.96	d
	R2	16.98 ± 0.55	a	15.48 ± 1.34	b	17.86 ± 0.85	a	13.31 ± 1.30	c	12.48 ± 0.35	cd	11.20 ± 0.41	de	10.85 ± 0.30	e	10.40 ± 0.55	e
	R3	18.00 ± 0.76	a	14.32 ± 0.14	b	12.19 ± 0.36	c	10.81 ± 0.14	de	11.06 ± 0.36	d	7.51 ± 0.88	f	9.95 ± 0.85	e	9.89 ± 0.47	e
Fe	R1	96.49 ± 3.96	d	102.10 ± 7.09	cd	169.59 ± 8.86	a	162.76 ± 8.29	a	77.71 ± 1.77	e	97.06 ± 2.83	d	112.7 ± 9.10	bc	121.08 ± 10.54	b
	R2	94.15 ± 2.89	b	81.92 ± 4.16	c	114.01 ± 6.46	a	110.04 ± 1.41	a	75.35 ± 7.13	c	72.21 ± 5.45	c	119.32 ± 7.25	a	117.29 ± 10.22	a
	R3	77.49 ± 3.92	bc	72.34 ± 3.08	cd	106.11 ± 4.93	a	84.46 ± 8.94	b	61.62 ± 3.97	e	64.90 ± 2.69	de	110.39 ± 9.31	a	62.37 ± 4.96	e

The data were the means of three independent replicates. The different letters in each line indicate the significant differences ($p < 0.05$) among the 4 treatments within two cultivars. S1, S2, and S3 correspond to the top part of the shoot (three top leaves with stem), the middle part of the shoot, and the bottom part of the shoot (three bottom leaves with stem), respectively; R1, R2, and R3 correspond to the top part of the root (three cm of root top), the middle part of the root, and the bottom part of the root (three cm of root bottom), respectively. Control, GB, Cd, and Cd + GB correspond to BNS + foliar spray of deionized water, BNS + foliar spray of 500 μM GB, BNS + foliar spray of deionized water + 5 μM $CdCl_2$, and BNS + foliar spray of 500 μM GB + 5 μM $CdCl_2$, respectively.

2.3. *Effect of Cd and Exogenous GB on MDA Accumulation and Certain Antioxidant Enzyme Activities*

Cd treatment displayed a significant accumulation of MDA in tobacco seedlings, which was decreased with the application of exogenous GB (Supplemental Figure S1). Averaged over the 3 sampling dates (5, 10, and 15 days of Cd exposure), the shoot/root MDA contents in Cd + GB were significantly decreased by 8%/6% and 15%/6% in Guiyan1 and Yunyan2, respectively, compared with Cd alone treatment.

Cd stress significantly promoted shoot and root SOD and POD activities in Guiyan1 and Yunyan2. Cd + GB treatment decreased the SOD and POD activities at different levels (Figure 3; Supplemental Figure S2). For example, the root POD activities were significantly decreased by 36% and 26% in Guiyan1 after 5 and 10 day of GB + Cd treatment, respectively, while were decreased by 11% in Yunyan2 after 15 day of GB + Cd treatment compared with Cd alone treatment (Figure 3). Averaged over the 3 sampling dates, the CAT/APX activities of Guiyan1 and Yunyan2 under Cd alone treatment were 16%/14% and 19%/12% in the shoots and 29%/16% and 25%/14% in the roots, respectively, lower than the controls. GB addition improved the CAT and APX activities compared with Cd alone treatment. Averaged over the 3 sampling dates, the CAT activity in the shoots/roots increased by 7%/5% and 13%/9% in Guiyan1 and Yunyan2 under Cd + GB, respectively, and some of the CAT activities in Cd + GB were even closed to the control and GB alone groups (Figure 3; Supplemental Figure S2).

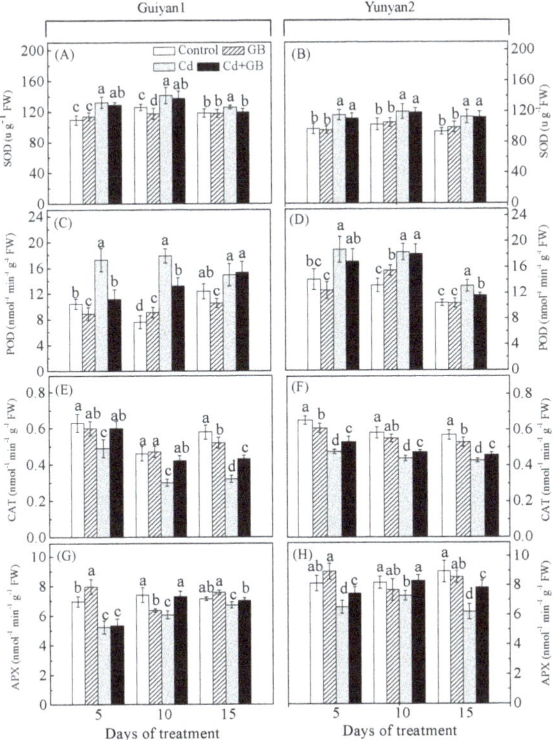

Figure 3. The effects of Cd and exogenous GB application on the superoxide dismutase (SOD) (**A**,**B**), peroxidase (POD) (**C**,**D**), catalase (CAT) (**E**,**F**), and ascorbate peroxidase (APX) (**G**,**H**) activity in the roots of two tobacco cultivars exposed to 5 μM Cd for 5, 10, and 15 days: The error bars represent the SD values ($n = 3$). The different letters indicate the significant differences ($p < 0.05$) among the 4 treatments within each sampling date. Control, GB, Cd, and Cd + GB correspond to BNS + foliar spray of deionized water, BNS + foliar spray of 500 μM GB, BNS + foliar spray of deionized water + 5 μM $CdCl_2$, and BNS + foliar spray of 500 μM GB + 5 μM $CdCl_2$, respectively.

2.4. Effect of Cd and Exogenous GB on Chlorophyll Fluorescence and Photosynthetic Parameters

Cd markedly reduced Pn, gs, Ci, and Tr by 23%, 34%, 9%, and 36% in Guiyan1 and 16%, 48%, 12%, and 30% in Yunyan2, respectively, compared with the controls. GB (Cd + GB) increased Pn and Ci of Yunyan2 by 9% each compared with Cd alone treatment (Supplemental Figure S3). Meanwhile, compared with the controls, Cd significantly decreased Fo, Fm, and Fv/Fm in Guiyan1 and Yunyan2, whereas, GB application improved Fo, Fm, and Fv/Fm by 11%, 10%, and 17% and by 8%, 10%, and 6% in Guiyan1 and Yunyan2, respectively, compared with the Cd alone treatment (Figure 4). False color images visually depicted the changes in Fv/Fm. The leaf color changed from blue to green with a decrease in the Fv/Fm ratio under Cd stress, especially for Guiyan1, but it was recovered obviously after GB application (Figure 4).

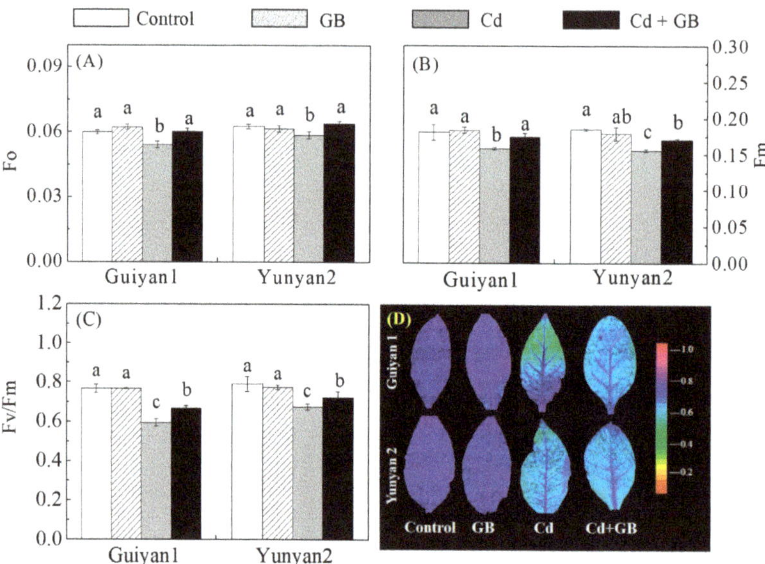

Figure 4. The effects of Cd and exogenous GB on the fluorescence parameters (**A**–**C**) and the Fv/Fm false color images (**D**) of Guiyan1 and Yunyan2 after 15 days of Cd exposure. Fo, minimal fluorescence yield (**A**); Fm, maximal fluorescence yield (**B**); and Fv/Fm, maximum quantum yield of PSII (**C**). The error bars represent the SD values ($n = 3$). The different letters indicate the significant differences ($p < 0.05$) among the 4 treatments within each cultivar. Control, GB, Cd, and Cd + GB correspond to BNS + foliar spray of deionized water, BNS + foliar spray of 500 µM GB, BNS + foliar spray of deionized water + 5 µM $CdCl_2$ and BNS + foliar spray of 500 µM GB + 5 µM $CdCl_2$, respectively. The figure is representative of three different experiments.

2.5. Effect of Cd and Exogenous GB on the Ultrastructure of Roots and Leaves

In control and GB-alone conditions, there were oval chloroplasts with a regular arrangement of thylakoid membranes of the stroma and of the grana in the spongy mesophyll cells and large starch grains and few osmiophilic plastoglobuli in the chloroplasts. After 15 days of Cd exposure, an irregular outline of chloroplasts could be found, and thylakoid membranes were dissolved and rarely visible; the cell wall was also distorted, and the osmiophilic plastoglobulis were much bigger than controls, especially in Guiyan1. The foliar application of GB obviously alleviated Cd-induced chloroplast damage in Yunyan2 but had little effect on Guiyan1 (Figure 5). After applying Cd to basic nutrition solution (BNS), some changes were also noted in the root cells (Supplemental Figure S4). An irregular outline of the cell structure and nuclear membrane was observed, nucleoli and karyoplasms were loosened, plasmolysis was evident, and a number of vacuoles accumulated along with the emergence of many great electron dense granules (EDG). Adding GB showed that cell structure and nuclei were better formed than those of the Cd alone treatment in Yunyan2. However, exogenous GB had a very little mitigation on the Cd-induced root cell changes in Guiyan1 (Supplemental Figure S4). Observations of the leaf surface of two tobacco cultivars by SEM revealed that stomatal movement could be regulated by Cd and exogenous GB application in Guiyan1 and Yunyan2 (Supplemental Figure S5). In the control condition, most of stomata were open; however, almost all the stomata were closed under Cd + GB treatment.

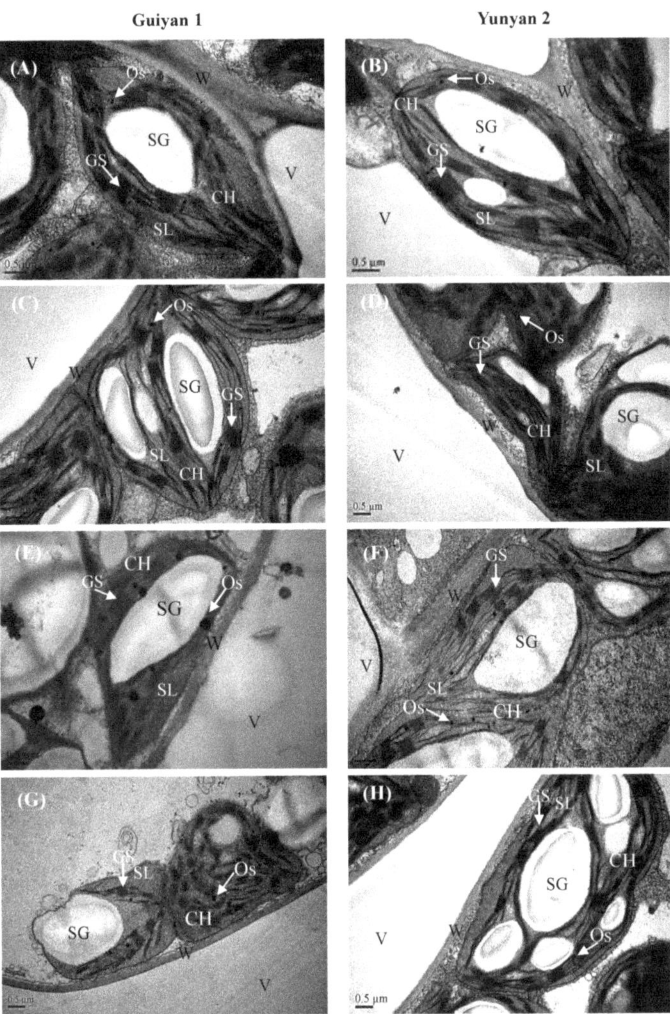

Figure 5. Transmission electron micrograph images of the leaf mesophyll cells of Guiyan1 (left panel) and Yunyan2 (right panel) after 15 days of Control (**A**,**B**), GB (**C**,**D**), Cd (**E**,**F**), and Cd + GB (**G**,**H**) treatments, respectively. Control, GB, Cd and Cd + GB correspond to BNS + foliar spray of deionized water, BNS + foliar spray of 500 μM GB, BNS + foliar spray of deionized water + 5 μM $CdCl_2$, and BNS + foliar spray of 500 μM GB + 5 μM $CdCl_2$, respectively. Labels: CH, chloroplast; GS, grana stack; SL, stroma lamella; Os, osmiophilic plastolobuli; SG, starch grain; V, vacuole; W, cell wall. Bar = 0.5 μm.

3. Discussion

3.1. Exogenous GB Reduces Cd Concentration in Tobacco Plants and Alleviates Cd-Induced Suppression in Plant Growth and Counteracts Nutrient Elements Changes

With tobacco being one of the susceptible plants to Cd stress, strategies for reducing Cd accumulation and the related health risks are urgently desired [3,4]. GB is water-soluble, nontoxic, and environmentally safe. It can easily be procured as a relatively inexpensive by-product from sugar beets. A number of reports have demonstrated that the exogenous application of GB improves stress tolerance [22,23]. In this study, Cd concentrations in the shoots and roots of both tobacco cultivars were reduced by GB application, and the mitigation effect was more obvious in Yunyan2. These results indicate the pronounced role of exogenous GB in protecting Cd toxicity and its potential use for reducing Cd concentrations in tobacco plants (Figure 2, Table 2). Many studies on the interaction between the uptake and distribution of Cd and essential mineral elements in crops have been reported, but the results are inconsistent. Previous studies have shown that Zn, Mn, and Cu interfere with the uptake of Cd and the translocation from roots to shoots [24,25]. Similar to our current study, Cd application significantly decreased leaf/root Zn, Mn, and Cu contents but increased Fe content, which led to an imbalance of the mineral content in plants through a disturbed metal uptake and allocation (Table 2). Exogenous GB application, to a degree, recovered shoot/root Fe, Cu, and Zn contents caused by Cd stress, indicating that GB can mitigate Cd toxicity partly through balancing the metabolism of elements.

3.2. Exogenous GB Counteracts Cd-Induced Alterations in the Antioxidant System

Oxidative stress is a central part of abiotic and biotic stresses. This mechanism is caused by a serious cell imbalance between the production of ROS and antioxidative enzymes, which leads to dramatic physiological disorders. In our study, the inhibition in plant growth by Cd was linked with Cd-induced MDA accumulation (Supplemental Figure S1), which indicates that Cd may cause oxidative damage to plants. However, the exogenous application of GB resulted in a significant reduction in MDA levels, and a higher reduction ratio was detected in Yunyan2 shoot. Results of cell viability in roots were consistent with the measurements on Cd stress-generated MDA. This result indicated that GB has a potential mitigation role on Cd-induced oxidative stress through decreasing the MDA content in tobacco seedlings.

The enzymatic ROS scavenging system plays an important role in maintaining the structure and function of membrane and cellular redox equilibrium [26]. In the present study, Cd treatment promoted SOD and POD activities but suppressed CAT and APX activities in the shoots and roots of both tobacco cultivars; however, the pattern of alterations in their activities induced by Cd stress was counteracted in the presence of GB (Figure 3; Supplemental Figure S2). Previous investigations observed that the application of GB could alleviate the oxidative stress induced by heavy metals [27,28]. The high SOD and POD decreasing ratio in Guiyan1 roots could be attributable to the excessive production of ROS, resulting in a greater growth inhibition under Cd + GB. However, Yunyan2 showed a relatively better ROS scavenging capacity under Cd stress than Guiyan1 because of its higher promoted activities of CAT under Cd + GB and lower decrease of APX activities under the Cd alone treatment. All these results indicated that the greater tolerance of Yunyan2 to Cd might be brought about by its strong antioxidant enzyme system compared with that of Guiyan1 and that GB can harmonize the activities of antioxidant enzymes and protect cells and tissues against oxidative damage caused by oxidative stress.

3.3. Exogenous GB Offsets Cd-Induced Inhibition in Photosynthesis and ChlorophyII Fluorescence Characteristics

Photosynthetic and chlorophyll fluorescence parameters are considered powerful tools for studying the physiological responses of plants in response to metal stress and for providing a direct method for evaluating photosynthetic activities [29,30]. In this study, Cd stress negatively influenced various photosynthetic parameters like Pn, Ci, and Tr (Supplemental Figure S3), and the application of GB alleviated the Cd-induced decrease of Pn in Yunyan2. However, Guiyan1 had higher gs and Tr than Yunyan2 under Cd + GB, suggesting that a Cd-induced photosynthetic system impairment was rehabilitated by exogenous GB fortification, possibly through maintaining the photosynthetic capacity by regulating stomatal conductance and decreasing the transpiration rate, which in turn countered the uptake of Cd.

Photosystem II (PSII) is believed to play a key role in the response of leaf photosynthesis to environmental perturbations [31]. Reducing PSII is considered the main target of Cd toxicity stress in plants [17]. Similar results were found in the present study, i.e., Cd significantly reduced Fo, Fm, and Fv/Fm in 2 tobacco cultivars (Figure 4). Reduced values of Fv/Fm also indicate that Cd-induced stress impaired the maximum quantum efficiency of PSII. A greater decrease of Fo and Fv/Fm in Guiyan1 demonstrated that the PSII of Guiyan1 was more sensitive to Cd stress, suggesting that the high tolerance of Yunyan2 was partly attributable to the higher protective capacity of PSII. Wang et al. [32] found that the enhanced drought tolerance was due to the accelerated recovery of the PSII from a photoinactivated state by GB. In this study, the foliar application of GB significantly alleviated the alteration of PSII caused by Cd stress. The results provided evidence that the application of GB can protect of PSII, contributing to Cd tolerance in tobacco plants.

3.4. Exogenous GB Mitigates Cd-Induced Damage in Cell Ultrastructure

The structure of the chloroplast has a relationship with the photochemistry activity. Thylakoid membrane leakage under Cd stress might be responsible for the reduced photosynthetic parameters, and it is the first limiting factor for photosynthesis [33]. In our current study, abnormal-shaped chloroplasts, dissolved thylakoid membranes, and bigger osmiophilic plastolobuli were observed under Cd stress, especially in Guiyan1, and the foliar spray of GB alleviated these damages particularly in Yunyan2 (Figure 5). This explains the higher photosynthetic activity in Yunyan2 under the Cd alone and Cd + GB treatments than Guiyan1. On the other hand, the stomatal movement is traditionally considered to be tightly regulated and to reflect the level of abiotic stress [34]. Although both the cultivars did not exhibit significant differences in stomatal movements (Supplemental Figure S5), the application of GB led to the osmotically driven changes in cell turgor which mediates stomatal movements with closed stomata, leading to a decreased stomatal conductance and resulting in an improved plant growth under Cd stress.

The nucleus is the genetic center for all eukaryotes. Experiencing the toxicity of Cd, the damage to the root nucleus was serious in 2 tobacco cultivars, reflected in the irregular cell structure, loose nucleoli and karyoplasms, large vacuoles, great electron dense granules, more mitochondrion, and plasmolysis (Supplemental Figure S4). The application of GB to Yunyan2 alleviated most of the disorder caused by Cd stress, but exogenous GB had very little effect on mitigating Guiyan1 disorders, as even bigger and larger vacuoles were formed under Cd + GB condition. The vacuoles of plants are considered to be the organelles in which the nourishment is accumulated and stored. However, when heavy metals such as Cd exist in the cell, the superfluous Cd in the cytoplasm could be stockpiled in the vacuoles [35]. The large and well-shaped vacuoles indicate vacuole compartmentation might be a probable mechanism of Cd detoxification in Guiyan1, explaining the high Cd content in its roots.

4. Materials and Methods

4.1. Plant Material and Experimental Designs

The greenhouse hydroponic experiment was carried out on Zijingang Campus, Zhejiang University, Hangzhou, China. Healthy tobacco seeds of Guiyan1 (Cd-sensitive) and Yunyan2 (Cd-tolerant) (Lab of Prof. Guoping Zhang, Department of Agronomy of Zhejiang University, Hangzhou China) were germinated in sterilized, moist vermiculite in a growth chamber at 25 °C/20 °C (day/night). Uniform healthy 4-leaf stage (50 day old) seedlings were transplanted to 5-L containers filled with 4.5-L basal nutrient solution (BNS), and the containers were placed in a greenhouse. The composition of BNS was the same as in the study by Liu et al. [25]. The solution pH was adjusted to 5.8 ± 0.1 with HCl or NaOH as required. After 15 days of transplanting, 4 treatments were performed: control (BNS+ foliar spray of deionized water), GB (BNS + foliar spray of 500 μM GB), Cd (BNS + foliar spray of deionized water + 5 μM $CdCl_2$), and Cd + GB (BNS + foliar spray of 500 μM GB + 5 μM $CdCl_2$). The foliar spray of deionized water and GB were conducted one day before Cd treatment and 1 and 3 days after Cd treatment. The experiment was laid in a split-plot design with treatment as the main plot and genotype as the subplot with three replicates for each treatment. The nutrient solution was continuously aerated with pumps and renewed every 5 d. After 5, 10, and 15 days of treatment, the fresh plant samples were immediately frozen in liquid nitrogen and stored frozen at −80 °C for the determination of antioxidative enzyme activities and MDA contents. Meanwhile, the growth and photosynthesis parameters were measured for the plants after 15 d treatment, and leaf/root ultrastructure were also performed.

4.2. Chlorophyll Content and Growth Measurement and Metal Analysis

After 15 days of treatment, the upper second fully opened leaves were selected to measure the SPAD values (chlorophyll meter readings) with three replicates using a chlorophyll meter Minolta SPAD-502 (Minolta, Tokyo, Japan). After measuring the plant heights and root lengths, the roots were soaked in 20 mM Na_2-EDTA for 3 h and rinsed thoroughly with deionized water. Then plants were separated into roots and shoots, and the fresh weights were measured simultaneously. The roots and shoots were dried at 80 °C and weighed. The dried roots and shoots were ground and ashed at 550 °C for 8 h and then digested with 30% HNO_3. The Cd and metal concentrations were determined using flame atomic absorption spectrometry (Shimadzu AA-6300, Shimadzu, Kyoto, Japan).

4.3. Assay of MDA Content and Enzyme Activities

To determine the enzyme activity, after 5, 10, and 15 days of treatments, the plant fresh roots and shoots were homogenized in 8 mL of a 50 mM sodium phosphate buffer (PBS, pH 7.8) using a prechilled mortar and pestle and subsequently centrifuged at 10,000× *g* for 20 min at 4 °C. The supernatant was used for the assays of all enzyme activities. The SOD, POD, CAT, and APX activities were determined as described by Zeng [36]. To analyze the MDA content, the thiobarbituric acid reaction was measured, which reflected the level of lipid peroxidation. The plant fresh roots and shoots were homogenized and extracted in 10 mL of 0.25% thiobarbituric acid. The extract was heated at 95 °C for 30 min and then quickly cooled on ice. After centrifugation at 10,000× *g* for 10 min, the absorbance of the supernatant was measured at 532 nm. A correction of the nonspecific turbidity was made by subtracting the absorbance value taken at 600 nm [13].

4.4. Chlorophyll Fluorescence and Photosynthetic Parameters

The chlorophyll fluorescence parameters were determined using an IMAGING-PAM chlorophyll fluorometer (Walz, Effeltrich, Germany) [37]. The leaves were kept in the dark for 20 min before measurement. The minimal fluorescence level (Fo) was measured by measuring light (<0.05 μmol m^{-2} s^{-1} PAR), and the maximal fluorescence level (Fm) was determined by a saturating pulse (2500 μmol m^{-2} s^{-1} PAR). The actinic light intensity was set as 280 μmol mol^{-2} s^{-1} PAR. Variable fluorescence was calculated as Fv = Fm − Fo; the maximal quantum yield of PSII photochemistry was Fv/Fm = (Fm − Fo)/Fm. The Fv/Fm false color images were created by ImagingWin software (IMAGING-PAM, Walz, Effeltrich, Germany). The net photosynthetic rate (Pn), transpiration rate (Tr), stomatal conductance (gs), and intracellular CO_2 concentration (Ci) were measured by a portable photosynthesis system LI-6400 (LI-COR, Lincoln, NE, USA). All the measurements were carried out with the upper second fully expanded leaves after 15 days of Cd treatments.

4.5. Determination of Cell Viability

To evaluate cell viability, after 15 days of Cd exposure, the root tips were rinsed 3 times with deionized water and blotted dry gently, then treated by staining with fluorescein diacetate and propidium iodide (FDA-PI) for 40 min, and washed 3 times with deionized water for 5 min. Red and green fluorescence and concurrent differential interference contrast images were obtained with a Zeiss LSM 780 fluorescent microscope (Zeiss, Oberkochen, Germany) with an excitation at 488 nm and an emission at 514 nm. The nonfluorescent esterase substrate FDA was cleaved by esterases in viable cells, releasing fluorescein which stains the cells green, while the characteristics of PI were totally opposite with FDA, which may interact with DNA/RNA in cells, leaving the red fluorescence of dead cells [38].

4.6. Ultrastructural Studies Using Electron Microscopy

Small sections (1 mm^2) from the middle of the upper second fully expanded leaves and root tips (1–3 mm) were used for TEM studies. The samples were fixed in 2.5% glutaraldehyde in a phosphate buffer (pH 7.2) for 6–8 h, postfixed in 1% OsO_4 (Osmium (VIII) oxide) for 1 h at 4 °C, and then thoroughly washed 3 times with the same buffer. Dehydration was carried out in a graded ethanol series, and then, the samples were infiltrated and embedded in Spurr's resin. Also ultrathin sections (about 70–90 nm) were cut on an Ultracut E ultramicrotome (Reichert-Jung, Vienna, Austria) and were placed on Formvar-coated copper grids, then were stained with uranyl acetate and lead citrate, and finally examined and photographed with a JEM-1230 (JEOL, Tokyo, Japan) electron microscope.

For the SEM studies, small portions (1 mm^2) from the middle of the upper second fully expanded leaves were selected and treated in the same way as described above for the TEM analysis. After a graded dehydration, the samples were transferred to the mixture of alcohol and isoamyl acetate (v/v = 1/1) for about 30 min, then transferred to pure isoamyl acetate for about 1–2 h, and followed by dehydration in a Hitachi ModelHCP-2 critical point dryer with liquid CO_2. In the end, the specimen was coated with gold-palladium in a Eiko Model IB5 ion coater for 4–5 min and then observed in a scanning electron microscope TM-1000 (Hitachi, Tokyo, Japan).

4.7. Statistic Analysis

The statistical analyses were performed by Data Processing System statistical software package with ANOVA followed by Duncan's multiple range tests to evaluate the significant treatment effects at a significance level of $p \leq 0.05$. To evaluate the alleviating effects, the following formula-based integrated relation was used: The absolute values of (SPAD value × 0.2 + shoot height × 0.2 + root length × 0.2 + fresh weight × 0.2 + dry weight × 0.2) was adopted [39].

5. Conclusions

In conclusion, exogenous GB application significantly alleviated Cd-induced growth inhibition in 2 tobacco cultivars especially for the tolerant cultivar Yunyan2. The alleviation mechanism of GB to Cd stress is associated with (1) a decreased Cd accumulation and a balanced nutrient status; (2) the amelioration of Cd-induced damages on leaf and root ultrastructure, an improved membrane-stabilizing/integrity, an increased SPAD value, and chlorophyll fluorescence including Fv/Fm, Fo, and Fm in both cultivars and Pn in Yunyan2; (3) the partially reversed Cd-induced changes in antioxidant enzyme activities and depressed MDA content compared with Cd alone treatment. The results suggest a potential role for GB as a potent alleviator in plants in responses to Cd stress. Ultimately, the decrease of Cd by GB is expected to have contributions to the reduction of Cd toxicity to humans.

Supplementary Materials: Supplementary materials can be found at http://www.mdpi.com/1422-0067/20/7/1612/s1. **Figure S1.** The effects of Cd and exogenous GB application on MDA accumulation in the shoots (**A**,**B**) and roots (**C**,**D**) of two tobacco cultivars (Left: Guiyan1; Right: Yunyan2) exposed to 5 μM Cd for 5, 10, and 15 days: The error bars represent the SD values (n = 3). The different letters indicate the significant differences ($p < 0.05$) among the 4 treatments within each sampling date. Control, GB, Cd, and Cd + GB correspond to BNS

+ foliar spray of deionized water, BNS + foliar spray of 500 μM GB, BNS + foliar spray of deionized water + 5 μM $CdCl_2$, and BNS + foliar spray of 500 μM GB + 5 μM $CdCl_2$, respectively. **Figure S2.** The effects of Cd and exogenous GB application on the SOD, POD, CAT, and APX activities in the shoots of two tobacco cultivars (Left: Guiyan1; Right: Yunyan2) exposed to 5 μM Cd for 5, 10, and 15 days: The error bars represent the SD values ($n = 3$). The different letters indicate the significant differences ($p < 0.05$) among the 4 treatments within each sampling date. Control, GB, Cd, and Cd + GB correspond to BNS + foliar spray of deionized water, BNS + foliar spray of 500 μM GB, BNS + foliar spray of deionized water + 5 μM $CdCl_2$, and BNS + foliar spray of 500 μM GB + 5 μM $CdCl_2$, respectively. **Figure S3.** The effects of Cd and exogenous GB application on the photosynthesis parameters of two tobacco cultivars after 15 days of Cd exposure: The error bars represent the SD values ($n = 3$). The different letters indicate the significant differences ($p < 0.05$) among the 4 treatments within each cultivar. Control, GB, Cd, and Cd + GB correspond to BNS + foliar spray of deionized water, BNS + foliar spray of 500 μM GB, BNS + foliar spray of deionized water + 5 μM $CdCl_2$, and BNS + foliar spray of 500 μM GB + 5 μM $CdCl_2$, respectively. Pn = net photosynthetic rate, gs = stomatal conductance, Tr = transpiration rate, and Ci = intercellular CO_2 concentration. **Figure S4.** Transmission electron micrograph images of the root tip cells of Guiyan1 (left panel) and Yunyan2 (right panel) after 15 days of Control (**A**,**B**), GB (**C**,**D**), Cd (**E**,**F**), and Cd + GB (**G**,**H**) treatments, respectively. Control, GB, Cd, and Cd + GB correspond to BNS + foliar spray of deionized water, BNS + foliar spray of 500 μM GB, BNS + foliar spray of deionized water + 5 μM $CdCl_2$, and BNS + foliar spray of 500 μM GB + 5 μM $CdCl_2$, respectively. Labels: N, nuclear; NL, nucleolus; V, vacuole; W, cell wall; M, mitochondrion, PL, plasmolysis. The arrows indicate the electron dense granules (EDG). Bar = 1 μm. **Figure S5.** The effects of Cd and GB on the stomatal aperture of Guiyan1 (left panel) and Yunyan2 (right panel) after 15 days of Control (**A**,**B**), GB (**C**,**D**), Cd (**E**,**F**), and Cd + GB (**G**,**H**) treatments, respectively. Control, GB, Cd, and Cd + GB correspond to BNS + foliar spray of deionized water, BNS + foliar spray of 500 μM GB, BNS + foliar spray of deionized water + 5 μM $CdCl_2$, and BNS + foliar spray of 500 μM GB + 5 μM $CdCl_2$, respectively. The photos were detected by a scanning electron microscope. Labels: OS, open stomata with aperture width ≥ 2 μm; CS, close stomata with aperture width < 2 μm; EH, epidermal hair. Bar = 20 μm.

Author Contributions: Conceptualization, F.W. and W.Z.; methodology, F.W.; formal analysis, X.H. and F.W.; investigation, X.H., W.Z., M.E.R., and D.V.W.; writing—original draft preparation, X.H.; writing—review and editing, X.H. and F.W.; supervision, F.W.

Funding: This research received no external funding.

Acknowledgments: We thank Guoping Zhang from the Department of Agronomy of Zhejiang University for providing the tobacco seeds used for this experiment.

Conflicts of Interest: The authors declare no conflict of interest.

References

1. Deckert, J. Cadmium toxicity in plants: Is there any analogy to its carcinogenic effect in mammalian cells? *Biometals* **2005**, *18*, 475–481. [CrossRef] [PubMed]
2. Dong, J.; Mao, W.H.; Zhang, G.P.; Wu, F.B. Root excretion and plant tolerance to cadmium toxicity—A review. *Plant Soil Environ.* **2007**, *53*, 193–200. [CrossRef]
3. Hecht, E.M.; Landy, D.C.; Ahn, S.; Hlaing, W.W.M.; Hennekens, C.H. Hypothesis: Cadmium explains, in part, why smoking increases the risk of cardiovascular disease. *J. Cardiovasc. Pharm. T.* **2013**, *18*, 550–554. [CrossRef]
4. Ju, Y.R.; Chen, W.Y.; Liao, C.M. Assessing human exposure risk to cadmium through inhalation and seafood consumption. *J. Hazard. Mater.* **2012**, *227*, 357–361. [CrossRef] [PubMed]
5. Galazyn-Sidorczuk, M.; Brzoska, M.M.; Moniuszko-Jakoniuk, J. Estimation of polish cigarette contamination with cadmium and lead, and exposure to these metals via smoking. *Environ. Monit. Assess.* **2008**, *137*, 481–493. [CrossRef] [PubMed]
6. Li, S.; Yu, J.; Zhu, M.; Zhao, F.; Luan, S. Cadmium impairs ion homeostasis by altering K^+ and Ca^{2+} channel activities in rice root hair cells. *Plant Cell Environ.* **2012**, *35*, 1998–2013. [CrossRef] [PubMed]
7. Gill, S.S.; Khan, N.A.; Tuteja, N. Cadmium at high dose perturbs growth, photosynthesis and nitrogen metabolism while at low dose it up regulates sulfur assimilation and antioxidant machinery in garden cress (*Lepidium sativum* L.). *Plant Sci.* **2012**, *182*, 112–120. [CrossRef] [PubMed]
8. Gallego, S.M.; Pena, L.B.; Barcia, R.A.; Azpilicueta, C.E.; Iannone, M.F.; Rosales, E.P.; Zawozenik, M.S.; Groppa, M.D.; Benavides, M.P. Unravelling cadmium toxicity and tolerance in plants: Insight into regulatory mechanisms. *Environ. Exp. Bot.* **2012**, *83*, 33–46. [CrossRef]
9. Souza, V.L.; de Almeida, A.A.F.; Lima, S.G.C.; de M. Cascardo, J.C.; Silva, C.D.; Mangabeira, P.A.O.; Gomes, F.P. Morphophysiological responses and programmed cell death induced by cadmium in *Genipa americana* L. (Rubiaceae). *Biometals* **2011**, *24*, 59–71. [CrossRef] [PubMed]

10. Cho, U.H.; Seo, N.H. Oxidative stress in *Arabidopsis thaliana* exposed to cadmium is due to hydrogen peroxide accumulation. *Plant Sci.* **2005**, *168*, 113–120. [CrossRef]
11. Hegedus, A.; Erdei, S.; Horvath, G. Comparative studies of H_2O_2 detoxifying enzymes in green and greening barley seedlings under cadmium stress. *Plant Sci.* **2001**, *16*, 1085–1093. [CrossRef]
12. Schutzendubel, A.; Schwanz, P.; Teichmann, T.; Gross, K.; Langenfeld-Heyser, R.; Godbold, D.L.; Polle, A. Cadmium-induced changes in antioxidative systems, hydrogen peroxide content, and differentiation in scots pine roots. *Plant Physiol.* **2001**, *127*, 887–898. [CrossRef] [PubMed]
13. Wu, F.B.; Zhang, G.P.; Dominy, P. Four barley genotypes respond differently to cadmium: Lipid peroxidation and activities of antioxidant capacity. *Environ. Exp. Bot.* **2003**, *50*, 67–78. [CrossRef]
14. Sakamoto, A.; Murata, N. The role of glycine betaine in the protection of plants from stress: Clues from transgenic plants. *Plant Cell Environ.* **2002**, *25*, 163–171. [CrossRef] [PubMed]
15. Chen, T.H.H.; Murata, N. Glycinebetaine protects plants against abiotic stress: Mechanisms and biotechnological applications. *Plant Cell Environ.* **2011**, *34*, 1–20. [CrossRef] [PubMed]
16. Park, E.J.; Jeknic, Z.; Pino, M.T.; Murata, N.; Chen, T.H. Glycinebetaine accumulation is more effective in chloroplasts than in the cytosol for protecting transgenic tomato plants against abiotic stress. *Plant Cell Environ.* **2007**, *30*, 994–1005. [CrossRef]
17. Yang, X.H.; Wen, X.G.; Gong, H.M.; Lu, Q.T.; Yang, Z.P.; Tang, Y.L.; Liang, Z.; Lu, C.M. Genetic engineering of the biosynthesis of glycinebetaine enhances thermo tolerance of photosystem II in tobacco plants. *Planta* **2007**, *225*, 719–733. [CrossRef] [PubMed]
18. Ma, Q.Q.; Wang, W.; Li, Y.H.; Li, D.Q.; Zou, Q. Alleviation of photoinhibition in drought-stressed wheat (*Triticum aestivum*) by foliar-applied glycinebetaine. *J. Plant Physiol.* **2006**, *163*, 165–175. [CrossRef]
19. Ashraf, M.; Foolad, M.R. Roles of glycine betaine and proline in improving plant abiotic stress resistance. *Environ. Exp. Bot.* **2007**, *59*, 206–216. [CrossRef]
20. Banu, M.N.A.; Hoque, M.A.; Watanabe-Sugimoto, W.M.; Natsuoke, K.; Nakamura, Y.; Shimoishi, Y.; Murata, Y. Proline and glycinebetaine induce antioxidant defense gene expression and suppress cell death in cultured tobacco cells under salt stress. *J. Plant Physiol.* **2009**, *30*, 146–158. [CrossRef]
21. Duman, F.; Aksoy, A.; Aydin, Z.; Temizgul, R. Effects of exogenous glycinebetaine and trehalose on cadmium accumulation and biological responses of an aquatic plant (*Lemna gibba* L.). *Water Air Soil Poll.* **2011**, *217*, 545–556. [CrossRef]
22. Islam, M.M.; Hoque, M.A.; Okuma, E.; Banu, M.N.A.; Shimoishi, Y.; Nakamura, Y.; Murata, Y. Exogenous proline and glycinebetaine increase antioxidant enzyme activities and confer tolerance to cadmium stress in cultured tobacco cells. *J. Plant Physiol.* **2009**, *166*, 1587–1597. [CrossRef] [PubMed]
23. Islam, M.M.; Hoque, M.A.; Okuma, E.; Jannat, R.; Banu, M.N.A.; Jahan, M.S.; Nakamura, Y.; Murata, Y. Proline and glycinebetaine confer cadmium tolerance on tobacco bright yellow-2 cells by increasing ascorbate-glutathione cycle enzyme activities. *Biosci. Biotech. Bioch.* **2009**, *73*, 2320–2323. [CrossRef] [PubMed]
24. Wu, F.B.; Zhang, G.P. Genotypic differences in effect of Cd on growth and mineral concentrations in barley seedlings. *Bull. Environ. Contam. Toxicol.* **2002**, *69*, 219–227. [CrossRef] [PubMed]
25. Liu, W.X.; Shang, S.H.; Feng, X.; Zhang, G.P.; Wu, F.B. Modulation of exogenous selenium in cadmium-induced changes in antioxidative metabolism, cadmium uptake, and photosynthetic performance in the 2 tobacco genotypes differing in cadmium tolerance. *Environ. Toxicol. Chem.* **2015**, *34*, 92–99. [CrossRef] [PubMed]
26. Muhammad, D.; Cao, F.B.; Jahangir, M.M.; Zhang, G.P.; Wu, F.B. Alleviation of aluminum toxicity by hydrogen sulfide is related to elevated ATPase, and suppressed aluminum uptake and oxidative stress in barley. *J. Hazard. Mater.* **2012**, *209*, 121–128.
27. Bharwana, S.A.; Ali, S.; Farooq, M.A.; Iqbal, N.; Hameed, A.; Abbas, F.; Ahmad, M.S.A. Glycine betaine-induced lead toxicity tolerance related to elevated photosynthesis, antioxidant enzymes suppressed lead uptake and oxidative stress in cotton. *Turk. J. Bot.* **2014**, *38*, 281–292. [CrossRef]
28. Rasheed, R.; Ashraf, M.A.; Hussain, I.; Haider, M.Z.; Kanwal, U.; Iqbal, M. Exogenous proline and glycinebetaine mitigate cadmium stress in two genetically different spring wheat (*Triticum aestivum* L.) cultivars. *Braz. J. Bot.* **2014**, *37*, 399–406. [CrossRef]
29. Kupper, H.; Kroneck, P.M.H. Heavy metal uptake by plants and cyanobacteria. In *Metal Ions in Biological Systems*; Sigel, A., Sigel, H., Sigel, R.K.O., Eds.; Marcel Dekker Inc.: New York, NY, USA, 2005; pp. 97–142.

30. Kupper, H.; Parameswaran, A.; Leitenmaier, B.; Trtilek, M.; Setlik, I. Cadmium-induced inhibition of photosynthesis and long-term acclimation to cadmium stress in the hyperaccumulator *Thlaspicaerulescens*. *New Phytol.* **2007**, *175*, 655–674. [CrossRef] [PubMed]
31. Murata, N.; Takahashi, S.; Nishiyama, Y.; Allakhverdiev, S.I. Photoinhibition of photosystem II under environmental stress. *Biochim. Biophys. Acta* **2007**, *1767*, 414–421. [CrossRef] [PubMed]
32. Wang, G.P.; Tian, F.X.; Zhang, M.; Wang, W. The overaccumulation of glycinebetaine alleviated damages to PSII of wheat flag leaves under drought and high temperature stress combination. *Acta Physiol. Plant.* **2014**, *36*, 2743–2753. [CrossRef]
33. Najeeb, U.; Jilani, G.; Ali, S.; Sarwar, M.; Xu, L.; Zhou, W.J. Insight into cadmium induced physiological and ultra-structural disorders in *Juncuseffusus* L. and its remediation through exogenous citric acid. *J. Hazard. Mater.* **2011**, *186*, 565–574. [CrossRef] [PubMed]
34. Roelfsema, M.R.G.; Hedrich, R. In the light of stomatal opening: New insights into 'the Watergate'. *New Phytol.* **2005**, *167*, 665–691. [CrossRef] [PubMed]
35. Daud, M.K.; Sun, Y.Q.; Dawood, M.; Hayat, Y.; Variatha, M.T.; Wu, Y.X.; Raziuddin; Mishkat, U.; Salahuddin; Najeeb, U.; Zhu, S.J. Cadmium-induced functional and ultrastructural alterationsin roots of two transgenic cotton cultivars. *J. Hazard. Mater.* **2009**, *161*, 463–473. [CrossRef] [PubMed]
36. Zeng, F.R.; Zhao, F.S.; Qiu, B.Y.; Ouyang, Y.N.; Wu, F.B.; Zhang, G.P. Alleviation of chromium toxicity by silicon addition in rice plants. *Agric. Sci. China* **2011**, *10*, 1188–1196. [CrossRef]
37. Cai, Y.; Cao, F.; Cheng, W.; Zhang, G.P.; Wu, F.B. Modulation of exogenous glutathione in phytochelatins and photosynthetic performance against Cd stress in the two rice genotypes differing in Cd tolerance. *Biol. Trace Elem. Res.* **2011**, *143*, 1159–1173. [CrossRef] [PubMed]
38. Lin, L.; Zhou, W.H.; Dai, H.X.; Cao, F.B.; Zhang, G.P.; Wu, F.B. Selenium reduces cadmium uptake and mitigates cadmium toxicity in rice. *J. Hazard. Mater.* **2012**, *235*, 343–351. [CrossRef] [PubMed]
39. Cao, F.B.; Wang, N.B.; Zhang, M.; Dai, H.X.; Muhammad, D.; Zhang, G.P.; Wu, F.B. Comparative study of alleviating effects of GSH, Se and Zn under combined contamination of cadmium and chromium in rice (Oryza sativa). *Biometals* **2013**, *26*, 297–308. [CrossRef] [PubMed]

© 2019 by the authors. Licensee MDPI, Basel, Switzerland. This article is an open access article distributed under the terms and conditions of the Creative Commons Attribution (CC BY) license (http://creativecommons.org/licenses/by/4.0/).

Article

Transcriptome Analysis Reveals Cotton (*Gossypium hirsutum*) Genes That Are Differentially Expressed in Cadmium Stress Tolerance

Mingge Han [1], Xuke Lu [1], John Yu [2], Xiugui Chen [1], Xiaoge Wang [1], Waqar Afzal Malik [1], Junjuan Wang [1], Delong Wang [1], Shuai Wang [1], Lixue Guo [1], Chao Chen [1], Ruifeng Cui [1], Xiaoming Yang [1] and Wuwei Ye [1,*]

1 Institute of Cotton Research of Chinese Academy of Agricultural Science, State Key Laboratory of Cotton Biology, Key Laboratory for Cotton Genetic Improvement, Anyang 455000, Henan, China; h13707663917@163.com (M.H.); 15824990556@163.com (X.L.); pycxg-007@163.com (X.C.); wangxiaoge1990@126.com (X.W.); Waqarviqi244@gmail.com (W.A.M.); Wjj2004liyuan@sina.com (J.W.); wdl_21@126.com (D.W.); wangshuai_19871201@163.com (S.W.); guolixue0114@163.com (L.G.); cc1218@163.com (C.C.); xiaocui0126@126.com (R.C.); xmyang152@163.com (X.Y.)

2 USDA-ARS Southern Plains Agricultural Research Center, College Station, TX 77845, USA; john.yu@ars.usda.gov

* Correspondence: yew158@163.com

Received: 29 January 2019; Accepted: 19 March 2019; Published: 24 March 2019

Abstract: High concentrations of heavy metals in the soil should be removed for environmental safety. Cadmium (Cd) is a heavy metal that pollutes the soil when its concentration exceeds 3.4 mg/kg. Although the potential use of cotton to remediate heavy Cd-polluted soils is known, little is understood about the molecular mechanisms of Cd tolerance. In this study, transcriptome analysis was used to identify Cd tolerance genes and their potential mechanisms in cotton. We exposed cotton plants to excess Cd and identified 4627 differentially expressed genes (DEGs) in the root, 3022 DEGs in the stem and 3854 DEGs in the leaves through RNA-Seq analysis. Among these genes were heavy metal transporter coding genes (ABC, CDF, HMA, etc.), annexin genes and heat shock genes (HSP), amongst others. Gene ontology (GO) analysis showed that the DEGs were mainly involved in the oxidation–reduction process and metal ion binding. The DEGs were mainly enriched in two pathways, the influenza A and pyruvate pathway. GhHMAD5, a protein containing a heavy-metal binding domain, was identified in the pathway to transport or to detoxify heavy metal ions. We constructed a *GhHMAD5* overexpression system in *Arabidopsis thaliana* that showed longer roots compared to control plants. GhHMAD5-silenced cotton plants showed more sensitivity to Cd stress. The results indicate that *GhHMAD5* is involved in Cd tolerance, which gives a preliminary understanding of the Cd tolerance mechanism in upland cotton. Overall, this study provides valuable information for the use of cotton to remediate soils polluted with Cd and potentially other heavy metals.

Keywords: cotton (*Gossypium hirsutum* L.); transcriptome; Cd stress; *GhHMAD5*; overexpression; VIGS (virus induced gene silence)

---

1. Introduction

Cadmium (Cd), one of the most common heavy metals with the strongest toxicity (exceeding 3.4 mg/kg in soil) [1], causes significant pollution to farmland, and actively transfers in the soil–plant system [2]. If the concentration of Cd in the plant reaches a certain level, toxic symptoms appear, such as crinkled and turned yellowing of plant leaves, degraded chloroplasts, closed stomata, and an imbalance of moisture metabolism, and by inhibiting the functional enzyme activity, it facilitates the decomposition of ascorbic acid and damages the chlorophyll. These factors can lead to a decline in crop yield and quality. The Cd absorbed by the plant will enter into the food chain, affecting the metabolism of calcium and phosphorus in the human body, and even increasing the possibility of teratogenicity and cancer [3]. Thus elimination of Cd from the soil and reduction of its harmful effects on human life, are of great importance.

The traditional methods to counteract heavy metal pollution are physical and chemical, but these are time-consuming and laborious, and can cause secondary pollution (i.e., primary pollutants under physical or chemical conditions result in new pollutants) [4]. At present, phytoremediation, a novel strategy for the removal of heavy metals from the soil using plants, has been introduced to resolve Cd pollution. A large number of plants were identified as able to remove Cd but had small biomass and slow growth, and therefore cannot be widely used to treat contaminated soil [5]. The most effective way for remediation of Cd polluted soil is to cultivate plants that can accumulate the maximum amount of Cd in their specific organs, which will not only eliminate Cd from the soil, but also maintain the rational use of the land to achieve sustainable production.

Cotton, one of the major economic crops grown all over the world has a large planted area of about 3.39×10^7 hectares worldwide [6]. It is reported that cotton, strongly tolerant of Cd, can be used as a restoration plant for Cd-contaminated soil [7]. It is widely believed that cotton is suitable for planting in industrially polluted areas [8]. Daud et al. found that the treatment of cotton seeds with a low concentration of Cd (10–100 μm) significantly increased their germination rate, which decreased with 1 mM concentration of Cd [9]. The activity of SOD, APX and other enzymes decreased in cotton under Cd stress [10]. Lawali et al. found significant differences in the toxicity of Cd in different cotton varieties [11]. The order of absorption of Cd in all parts of the reproductive organs of cotton was kernel > bell shell > fiber, with a significant difference among varieties. Cotton was reported as a remediation crop in heavy metal contaminated areas with a resilient absorptive capacity for heavy metals and tolerance against Cd stress [12], and with few effects on cotton fibre quality [13]. The previous studies of Cd stress in cotton mainly concentrated on physiological and biochemical aspects. We believe it is of great significance to uncover the Cd tolerance genes in cotton, and to analyze the regulation network of cotton Cd tolerance for the remediation of Cd-contaminated soil by molecular technology.

RNA-Seq can be adopted to analyze gene expression of plants under various biotic and abiotic stresses [14,15], and has been successfully used under Cd stress on many plants, such as ramie [16], maize [17], and rice [18]. Previous research has also reported a lot of proteins and gene families related to Cd stress, such as ABC transporters, Nramp family proteins, Zinc finger transporter families, YSL family proteins [19,20], P-type ATPase family proteins [21], and the CDF families [22].

Cd has been reported to enter organs through the cytomembrane via the Ca^{2+} pathway [23], inducing many free radicals and reactive oxygen species (ROS) to produce oxidative stress [4,24]. By adding Ca^{2+} or Mg^{2+} ion into the Cd stress solution, the degree of damage to soybean from Cd can be alleviated [25]. The Ca^{2+} signal transduction pathway plays a key role under Cd stress [26]. Overexpression of *SpHMA3* in *Sedum alfredii* could enhance Cd tolerance [27]. Previous studies indicated that there was a close relationship between the transport of glutathione reductase (GR) and the accumulation of jasmonic acid in *Lycium barbarum* under Cd stress [28]. The ascorbic acid (ASA) and glutathione (GSH) in the leaf significantly reduced in maize seedlings under Cd stress [29]. Three bHLH transcription factors (*FIT, AtbHLH38,* and *AtbHLH39*) of *Arabidopsis* were reported to be involved in the plant's response to Cd stress; the transgenic plants were more tolerant to Cd than the wild type plants [30]. In this study, DEGs were explored and the regulation network under Cd stress was constructed by cotton transcriptome sequencing, and the gene function of *GhHMAD5* was validated. This study would provide more information for understanding the mechanism of tolerance to Cd in cotton, and lay the foundation to repair heavy metal contaminated soil through molecular breeding methods.

2. Results

2.1. Phenotypic Analysis of Cotton under Cd Stress

Han 242, cotton cultivar (*Gossypium hirsutum* L.), was treated with 4 mM $CdCl_2$ compared with the control, which was subjected to the same amount of pure water. Phenotypic characters on cotton roots, stems, and leaves are shown in Figure 1A. The stems turned black, the leaves turned yellow and the veins lost their green pigment in Cd-treated plants compared to the controls. Over time, the cotton leaves became dry, and the stems turned black. The phenotypic traits of different cotton varieties was shown (Figure S1) under the same Cd concentration stress. Under Cd stress, remarkable accumulation of Cd was found in their root, stem and leaf tissues after 9 h. The results also indicated that the root and leaf tissues showed the highest and the lowest Cd content, respectively, compared to the stem (Figure 1B). The Cd content of the cotton also indicated that cotton has a strong absorptive capacity for Cd.

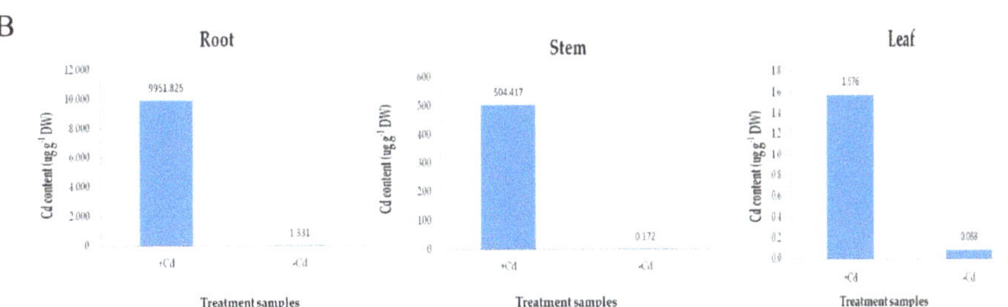

Figure 1. Phenotypic symptoms (**A**) and Cd content (**B**) in roots, stems and leaves of Han242 cotton seedlings with or without 4 mM Cd stress for 9 h.

2.2. Quality Analysis of the Transcriptome Sequence

The samples of roots, stems, and leaves of treated plants and controls were collected separately, and three biological replicates were conducted for both groups (treated versus control). Totally 18 qualified libraries were established. The raw reads were filtered and low quality reads were removed to get clean reads. Approximately 909 million clean valid reads were done, which contained 136.49 Gb of sequence data. Over 97.77% of the clean reads at a Q20 level and over 89.02% of the clean reads at a Q30 level were obtained. The GC content of the sequence data reached 44% (Table 1). Pearson correlation coefficient (PCC) analysis was performed on 18 established libraries to check the correlation between several tissues of all the samples (Figure 2). A dendrogram of raw RNA-Seq reads from all samples was acquired (Figure S7). A close correlation in root tissue was observed between the treated and control groups. The reason may be because of tissue specificity or root tissue being the first tissue to respond to Cd stress. Principal component analysis (PCA) showed that the repeatability was satisfactoryaccordance among the three samples (Figure S5). Above all, the results of transcriptome sequencing were reliable. Purified and valid reads were mapped into the *G.hirsutum* reference genome from CottonGen with a high proportion rate (Figure S6). The raw sequence data of RNA-Seq were submitted to NCBI with an accession number of GSE126671.

Table 1. Overview of the transcriptome sequencing of upland cotton. R_T1: Root Cd treatment 1, R_T2: Root Cd treatment 2, R_T3: Root Cd treatment 3, R_C1: Root water control 1, R_C2: Root water control 2, R_C3: Root water control 3, S_T1: Stem Cd treatment 1, S_T2: Stem Cd treatment 2, S_T3: Stem Cd treatment 3, S_C1: Stem water control 1, S_C2: Stem water control 2, S_C3: Stem water control 3, L_T1: Leaf Cd treatment 1, L_T2: Leaf Cd treatment 2, L_T3: Leaf Cd treatment 3, L_C1: Leaf water control 1, L_C2: Leaf water control 2, L_C3: Leaf water control 3.

Sample	Number of Raw Reads	Number of Valid Reads	Q20 Percentage (%)	Q30 Percentage (%)	GC Content (%)
R_T1	62,264,174	61,471,556	98.04	91.00	45
R_T2	55,526,318	54,811,464	97.93	91.32	45
R_T3	57,133,152	56,462,772	98.27	92.14	45
R_C1	55,124,734	54,473,502	97.93	90.83	45
R_C2	65,102,420	64,390,908	98.45	92.15	44
R_C3	52,774,982	52,144,004	97.78	90.97	45
S_T1	54,098,272	53,474,704	99.35	95.05	44
S_T2	43,663,718	43,152,570	99.19	95.13	45
S_T3	43,207,572	42,631,528	99.43	96.20	44
S_C1	57,188,916	56,544,630	98.99	94.40	45
S_C2	42,417,384	41,057,536	98.01	89.91	44
S_C3	50,332,770	48,940,728	98.28	90.06	45
L_T1	42,697,218	41,184,508	98.22	89.03	44
L_T2	45,204,982	44,593,254	98.25	89.42	45
L_T3	48,167,202	47,153,586	98.30	90.14	44
L_C1	63,540,928	62,794,552	99.67	96.60	44
L_C2	39,136,086	38,791,254	99.66	95.22	44
L_C3	46,343,822	45,925,388	99.74	95.38	44

Figure 2. Pearson correlation between samples. R_T1: Root Cd treatment 1, R_T2: Root Cd Table 2. R_T3: Root Cd treatment 3, R_C1: Root water control 1, R_C2: Root water control 2, R_C3: Root water control 3, S_T1: Stem Cd treatment 1, S_T2: Stem Cd treatment 2, S_T3: Stem Cd treatment 3, S_C1: Stem water control 1, S_C2: Stem water control 2, S_C3: Stem water control 3, L_T1: Leaf Cd treatment 1, L_T2: Leaf Cd treatment 2, L_T3: Leaf Cd treatment 3, L_C1: Leaf water control 1, L_C2: Leaf water control 2, L_C3: Leaf water control 3. The colors of the box represent the degree of correlation; red color represents the highest degree of correlation and white color represents the lowest degree of correlation.

2.3. Analysis of Differential Expression Genes under Cd Stress in Cotton

Comparative analysis between the controls and the treated samples to check the transcriptional changes in response to Cd stress was conducted using *cuffdiff* software. Gene expression profiles were comprised of 135,162 genes that consisted of 117,377 annotated genes and 17,785 novel genes (Table S5).

To effectively analyze and interpret the DEGs of the transcriptome, the *P-value* < 0.05 and $|\log_2$ *fold change*$| \geq 2$ was used, and an overall distribution of the DEGs was visualized in every tissue (Figure 3A). Overall 4627 DEGs were found in the root section including 2467 up-regulated genes and 2160 down-regulated genes. Similarly, 3022 DEGs were found involved in stem portions that consisted of 1324 up-regulated and 1698 down-regulated genes, whereas a total of 3854 DEGs were observed in leaves including 1879 up-regulated genes and 1975 down-regulated genes. Furthermore, a Venn diagram was developed to show the statistical analysis of DEGs in each tissue (Figure 3B). For the up-regulated mechanism, it showed 316 DEGs in the root–stem group, 183 DEGs in the stem–leaf group and 167 DEGs in the root–leaf group. Whereas 167 down-regulated DEGs were found in the root–stem, 122 in the stem–leaf and 80 in the root–stem groups. Results indicated that there were 3953 DEGs (up-regulated: 2032; down-regulated: 1921) of roots, 2471 DEGs (up-regulated: 873; down-regulated: 1598) of stems and 3177 DEGs (up-regulated: 1396; down-regulated: 1781) of leaves under tissue specific expression.

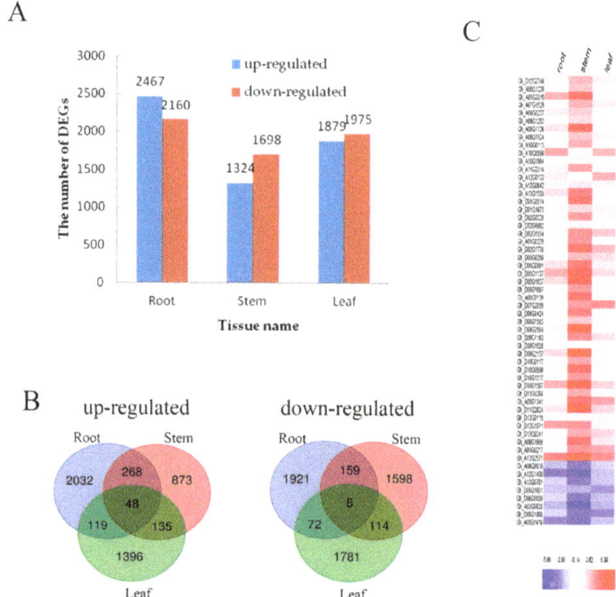

Figure 3. DEGs in roots, stems and leaves. (**A**) The number of genes up- or down-regulated by *fold change* ≥ 2 ($p < 0.05$) in roots, stems and leaves under Cd stress. (**B**) Venn diagrams showing the unique and shared regulated genes in cotton roots, stems and leaves under Cd stress; (**C**) The relative expression analysis of 56 genes that are expressed in roots, stems and leaves. The fold-change ratios of the genes are indicated by the different colors. The red color represents the highest expression, the blue color represents the lowest expression.

Under Cd stress, 56 DEGs (up-regulated: 48; down-regulated: 8) were identified in the root, stem and leaf (Figure 3B). In order to analyze the expression intensity of these 56 DEGs in roots, stems and leaves, the result was illustrated by heat map (Figure 3C). A large proportion of these 56 DEGs were evidently expressed in the stem, which maintained its consistency with continued Cd stress. The number of DEGs in roots, stems and leaves was ranked as root > stem > leaf. This data supported the fact that the root responded first and strongly to Cd stress.

2.4. Expression of Heavy Metal Transport Proteins in Cotton under Cd Stress

Heavy metal transporters play a vital role in the plants metabolism under Cd Stress. There are a large number of transporters in cotton (Figure 4), such as Nramp, P-type ATPase, CE, PC, MT, bHLH, Zip, ABC, YSL, MATE, CAX, OPT, HSP and ferredoxin. The P-type ATPase and PC did not show a significant expression in the stems, while Nramp, P-type ATPase and PC also did not express in the leaves, which was due to tissue specificity. Overexpressed *NRAMP* gene has been reported to enhance Cd tolerance in *Arabidopsis thaliana* [31], and overexpressed *PC* and *MT* genes showed increased Cd tolerance in *Escherichia coli* and *Arabidopsis thaliana* [32,33].

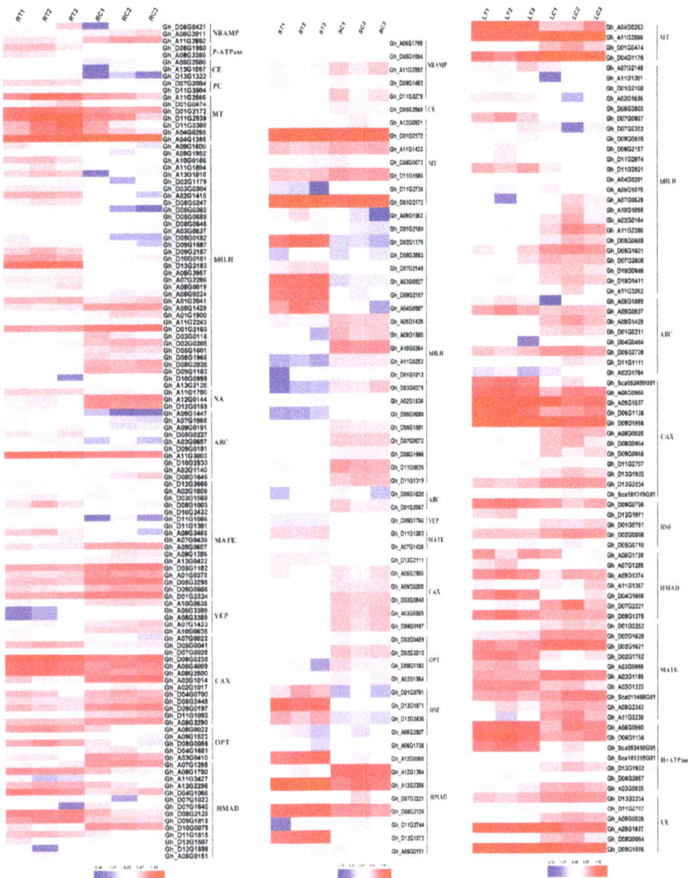

Figure 4. Metal transporters expressed in roots, stems and leaves under Cd stress. RT1: Root Cd treatment 1, RT2: Root Cd treatment 2, RT3: Root Cd treatment 3, RC1: Root water control 1, RC2: Root water control 2, RC3: Root water control 3, ST1: Stem Cd treatment 1, ST2: Stem Cd treatment 2, ST3: Stem Cd treatment 3, SC1: Stem water control 1, SC2: Stem water control 2, SC3: Stem water control 3, LT1: Leaf Cd treatment 1, LT2: Leaf Cd treatment 2, LT3: Leaf Cd treatment 3, LC1: Leaf water control 1, LC2: Leaf water control 2, LC3: Leaf water control 3. The fold-change ratios of the genes are indicated by the different colors. The red color represents the highest expression, the blue color represents the lowest expression. CE: cation efflux family, ABC: ATP-binding cassette transporter, CAX: calcium exchanger, NRAMP: natural resistance-associated macrophage protein, NA: nicotianamine, OPT: peptide transporter, PC: phytochelation, MT: metallothionein, MATE: multidrug resistance-associated protein, YEP: yellow stripe protein, bHLH: basic helix-loop-helix DNA-binding protein, HMAD: heavy metal transporter. HSF: heat shock factor.

2.5. Functional Classification of DEGs

The functional classification of DEGs were determined by using gene ontology (GO) terms based on their corresponding biological processes, cellular components, and molecular functions. Similar molecular biological processes and molecular functions were observed in cotton roots, stems and leaves under Cd stress.

GO enrichment analysis was performed by *p*-value, and the top 20 functional terms with the minimum *p*-value and the most specific genes were selected for statistical analysis (Figure 5). It was found that the biological functions of the DEGs in root, stem and leaf tissues was enriched in the oxidation–reduction process responding to salt stress. In roots, the DEGs were found enriched in metal ion binding, hydrolase activity, and copper ion binding. Among the most abundant genes in stems were enriched in oxidoreductase activity, calmodulin binding, and pyridoxal phosphate binding. The DEGs of leaves were found most abundant in metal ion binding, oxidoreductase activity, and catalytic activity. The DEGs enriched in cellular components were found in the cell nucleus, cytoplast, and plasma membrane in cotton.

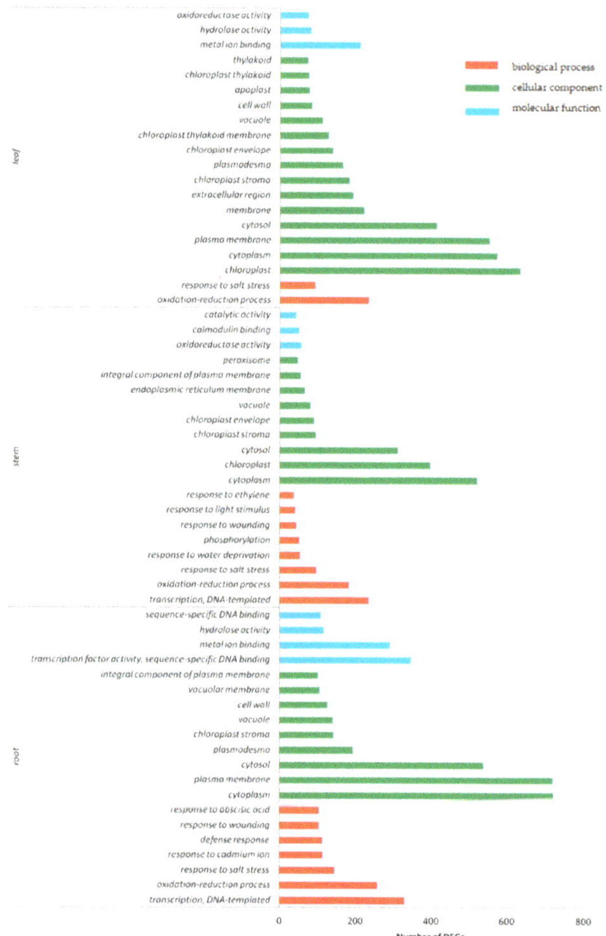

Figure 5. GO (gene ontology) function classification of DEGs in roots, stems, and leaves under Cd stress. Red color represents the biological process, green color represents cellular component, and blue color represents molecular function.

One GO term (GO: 0046686) was classified as responding to Cd stress based on GO annotation analysis. Figure 6A shows the total number of genes responding to Cd stress in roots, stems and leaves by Venn diagram. There were 111 DEGs in the roots, 44 DEGs in the stems and 69 DEGs in the leaves, with 150 DEGs (root: 81, stem: 29, leaf: 40) under tissue specific expression. Among these 150 DEGs, 2 DEGs were identified as up-regulated genes in roots, stems, and leaves, which were Gh_A12G0132

(Aldolase-type TIM barrel family protein) and Gh_D12G1971 (mitochondrion-localized small heat shock protein 23.6).

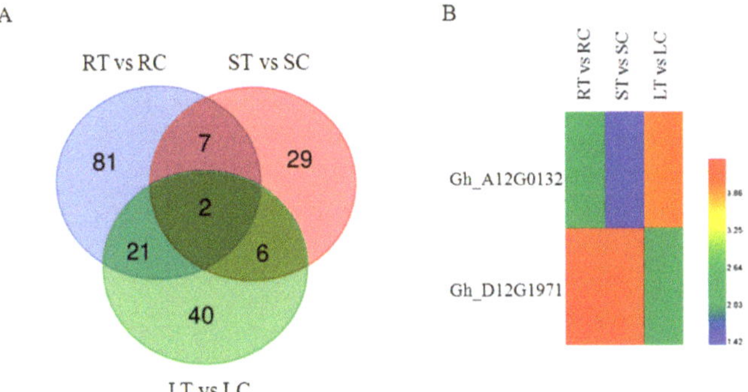

Figure 6. (**A**) Venn diagrams of DEGs in the GO term (GO: 0046686) under Cd stress; (**B**) Cluster map of DEGs in the GO term under Cd stress (GO: 0046686). The red color represents the highest expression, the blue color represents the lowest expression. The samples treated with $CdCl_2$ were called RT, ST and LT. The samples not treated with $CdCl_2$ were called RC, SC, and LC, respectively. Then 'R' indicates the root tissue, 'S' indicates the stem tissue, 'L' indicates the leaf tissue, 'T' indicates the treated, 'C' indicates the control.

2.6. KEGG Analysis of DEGs

To determine the pathway for DEGs under Cd stress, KOBAS was used for gene annotation. We obtained the enriched pathway map of DEGs in roots, stems, and leaves (Figure 7). The graph showed that DEGs in the roots played a role in influenza A, carbon metabolism, and biosynthesis of amino acids. In KEGG pathway analysis, DEGs enriched in the stems were important in carbon metabolism, the MAPK signaling pathway and drug metabolism–cytochrome P450. The DEGs enriched in the leaves played a role in carbon metabolism, biosynthesis of amino acids, and glycolysis/gluconeogenesis.

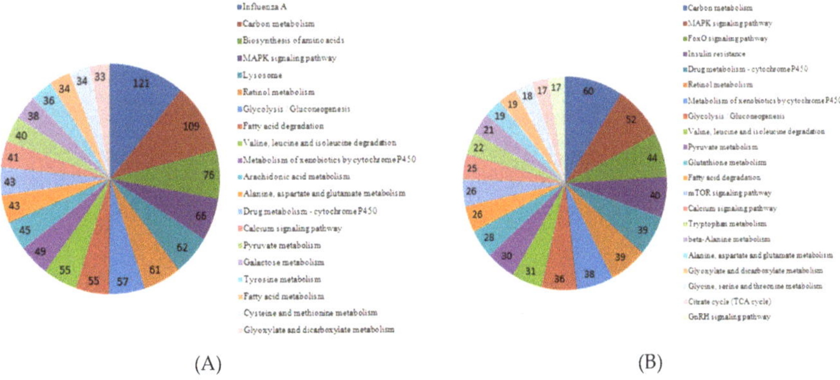

Figure 7. *Cont.*

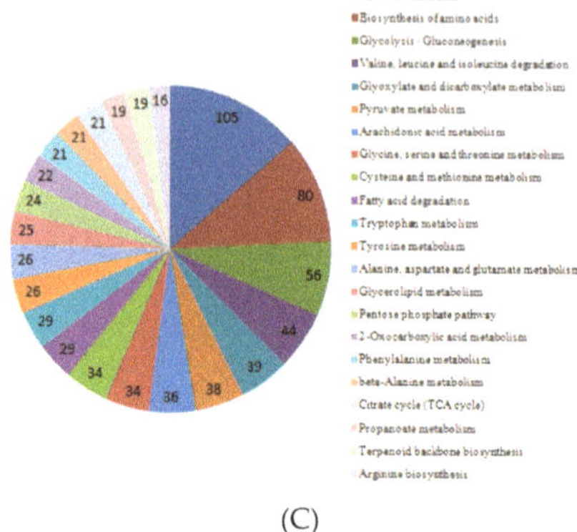

(C)

Figure 7. Significant pathways and numbers based on KEGG analysis of DEGs in cotton roots, stems, and leaves. (**A**) Significant pathways and numbers of DEGs in roots; (**B**) significant pathways and numbers of DEGs in stems; (**C**) significant pathways and numbers of DEGs in leaves.

2.7. Expression of Transcription Factors under Cd Stress of Cotton

It was reported that transcription factors played an important role in response to stresses [34]. To understand the behavior of transcription factors in cotton in response to Cd stress, we analyzed transcription factors of different genes in the roots, stems, and leaves (Figure S2). Many transcriptional factors expressed under Cd stress have been reported in previous studies, such as NAC, bHLH, WRKY, etc [35–37]. In this study, many unreported transcriptional factors were discovered, such as C3H, C2H2, Orphans, and MYB, etc, whereas C2H2 has been previously reported associated with osmotic stress in *Arabidopsis thaliana* [38].

2.8. Verification of Sequence Data

To verify the reliability of sequencing, the RNA samples previously collected for RNA sequencing were used for quantitative real-time PCR. Twenty differential genes were randomly selected for qRT-PCR validation. *GhActin* was selected as the reference gene and the $2^{-\Delta\Delta CT}$ method was applied to calculate the relative gene expression level. Fluorescence quantitative results and transcriptome sequencing data were as listed in Table 2, which was highly correlated by correlation analysis (Figure 8).

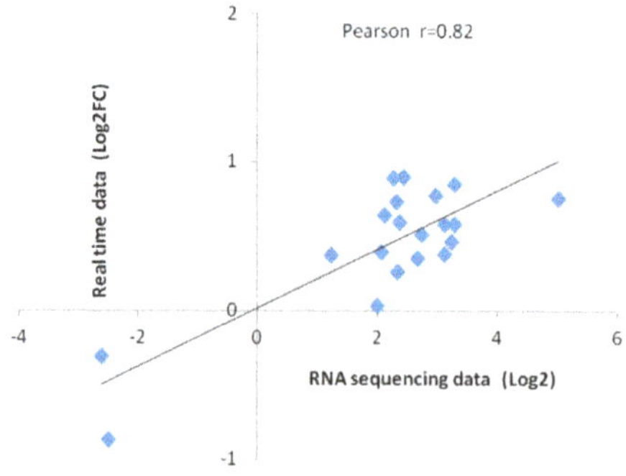

Figure 8. Correlation between sequencing data and quantitative RT-PCR data. Each point represents a value of expression level. Both the X and Y-axes are shown in $\log_2$ scale. And r indicates the correlation coefficient.

Table 2. The DEGs validated by qRT-PCR.

Gene ID	RNA-Seq (Log_2FC)	qRT-PCR (FC)	Description
Gh_A01G1234	3.24	2.89	PLANT CADMIUM RESISTANCE 2
Gh_D11G2939	3.12	2.42	metallothionein 3
Gh_D06G0421	2.97	5.86	NRAMP metal ion transporter 6
Gh_D04G1180	2.68	2.24	Fes1A
Gh_A11G2566	2.37	3.90	metallothionein 3
Gh_A04G0265	2.33	5.37	metallothionein 3
Gh_D08G1950	1.25	2.37	heavy metal atpase 5
Gh_A04G0713	2.12	4.38	Fes1A
Gh_A08G2485	5.03	5.67	AUX/IAA transcriptional regulator family protein
Gh_D07G2124	2.28	7.62	AUX/IAA transcriptional regulator family protein
Gh_D08G2126	2.45	7.96	Heavy metal transport/detoxification superfamily protein
Gh_D04G0262	3.29	3.84	Auxin-responsive GH3 family protein
Gh_D04G0260	3.29	7.16	Auxin-responsive GH3 family protein
Gh_A01G2049	3.13	3.80	Arabidopsis thaliana gibberellin 2-oxidase 1
Gh_A07G1285	−2.49	0.13	Heavy metal transport/detoxification superfamily protein
Gh_D07G1640	2.34	1.80	Heavy metal transport/detoxification superfamily protein
Gh_D10G0078	−2.61	0.49	Heavy metal transport/detoxification superfamily protein
Gh_D13G1609	2.74	3.26	Arabidopsis thaliana gibberellin 2-oxidase 1
Gh_D09G1816	2.01	1.08	Heavy metal transport/detoxification superfamily protein
Gh_A05G0151	2.08	2.43	Heavy metal transport/detoxification superfamily protein

2.9. Silencing of *GhHMAD5*: Phenotype and Expression

The cotyledon flattened cotton seedlings were injected with *Agrobacterium tumefaciens*. After albino symptoms appeared in the leaves of positive plants, the infected plants were treated with 4 mM Cd solution. After being exposed to Cd stress for 9 h, the phenotypic changes were obvious. The seedlings, which were soaked in Cd solution, showed the infection, wilted extensively, and the stem darkened and the veins turned brown. The phenotypic symptoms of non-disseminated plants were not obvious (Figure 9A). The expression level of the *GhHMAD5* gene in the plants decreased significantly on exposure to Cd stress (Figure 9B). Furthermore, results showed that after the gene *GhHMAD5* was silenced, the Cd resistance of cotton seedlings decreased.

A

B

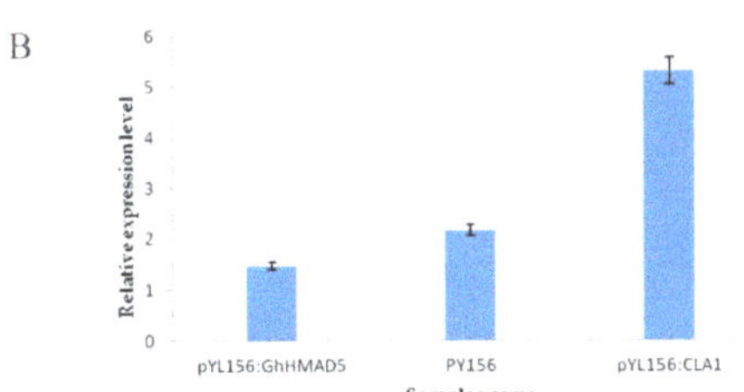

Figure 9. Phenotypic and expression analysis of 4 mM seedlings exposed to Cd stress and the control. (**A**) Phenotypic symptoms of cotton seedlings under 4 mM Cd stress for 9 h. (**B**) The expression analysis of *GhHMAD5* gene of cotton seedlings under 4 mM Cd stress for 9 h.

2.10. Overexpression of *GhHMAD5* Could Enhance the Cd Tolerance of *Arabidopsis Thaliana*

To examine the function of *GhHMAD5*, the seeds of *GhHMAD5* overexpression plants and WT plants were seeded on 1/2 MS solid medium containing 250 and 350 μM Cd solution for 12 days (Figure 10A) and 22 days (Figure 10B). The viability of WT decreased sharply and significantly with increasing Cd concentration, while the germination of *GhHMAD5* overexpression plants was significantly higher than WT plants. In addition, the germination rate of transgenic plants was not significantly affected by the increase in concentration of Cd stress (Figure 10C), which indicates that overexpression of the *GhHMAD5* gene could improve the germination rate of transgenic *Arabidopsis thaliana*.

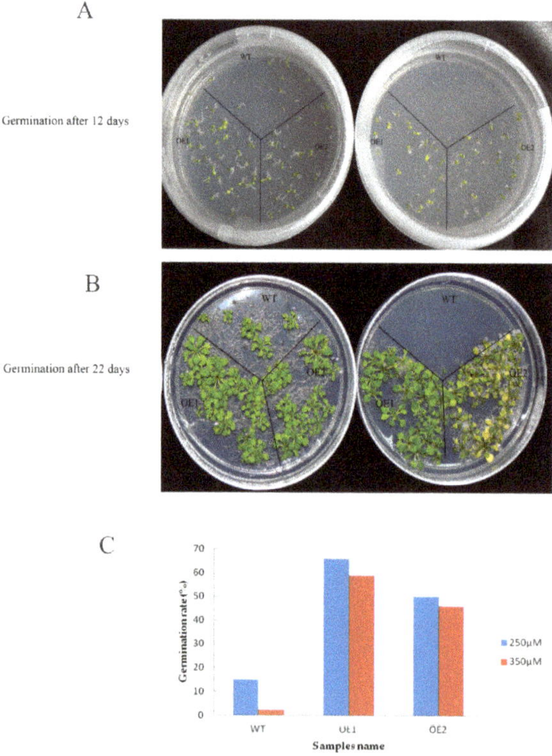

Figure 10. The germination of transgenic *Arabidopsis* and wild type (WT) under Cd stress. (**A**) 12 days of seed germination of transgenic *Arabidopsis thaliana* plants and WT; (**B**) 22 days of seed germination of transgenic *Arabidopsis thaliana* plants and WT. (**C**) The germination rate of transgenic *Arabidopsis thaliana* plants and WT under 250 and 350 μM Cd stress for 22 days. WT represents the wild *Arabidopsis thaliana*. OE1 and OE2 represent the transgenic *Arabidopsis thaliana*.

Transgenic plants and WT seeds grew on 1/2 MS solid medium containing 25, 50 and 100 μM Cd solution for 12 days (Figure 11). There was no difference in root length between transgenic seedlings and WT plants under 25 μM Cd stress, while under 50 μM and 100 μM Cd stress, the root length of transgenic seedlings was significantly longer than WT plants. These results indicate that overexpression of *GhHMAD5* gene can enhance the Cd tolerance of transgenic *Arabidopsis thaliana*.

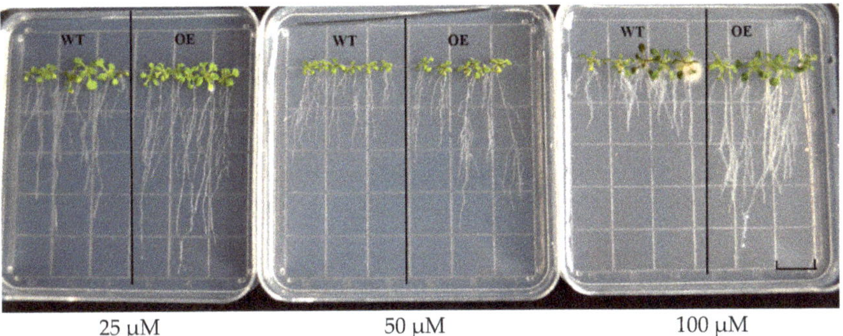

Figure 11. The phenotypic analysis of root length in wild type (WT) and transgenic *Arabidopsis thaliana*. WT represents the wild *Arabidopsis thaliana*, OE represents the transgenic *Arabidopsis thaliana*. Scale bars: 1.4 cm.

3. Discussion

Cadmium, one of the common toxic heavy metals that pollute most farmland when exceeding 3.4 mg/kg in the soil, represents a severe threat to plants, animals, and human beings, and is a major problem concerning our health in daily life [39]. Plants suffering from Cd stress showed toxic symptoms, which subsequently resulted in poor quality and diminishing yield [40]. Root tips of broad bean seedlings contaminated with Cd exhibited necrosis and turned dark brown [41]. The morphological indexes of soybean infected by Cd, such as the root length, number of lateral roots, and total volume of the roots decreased as compared to those of the control. Compared with normal plants, the leaf color turned lighter, and the biological yield decreased, and ultimately, the plant died [42]. In this study, we discovered a series of phenotypic symptoms of the cotton root system under Cd stress (Figure 1A). The basal part of the stem lost its water potential and turned black, the petiole of the cotyledon turned black, and the veins became brown, all of which indicate the poisonous effects of Cd stress. However, no significant change in root color and root hair color were observed in cotton under Cd stress, followed by no obvious change in root length and number of roots, which may be related to the time and concentration of Cd stress. In Figure 1B, the Cd content in roots was very high. This indicated that cotton has a strong cadmium accumulation capacity, and cotton roots possess a different cadmium tolerance transport mechanism compared to other crops.

Transcriptome analysis of cotton roots, stems and leaves under Cd stress was carried out in this study. A large number of DEGs were found in roots, followed by leaves and stems. A series of enzymes related to oxidative stress were expressed in cotton under Cd stress, including many emergency proteins, such as heat shock protein and ubiquitin enzyme. The expression of phytochelatin genes was observed in roots, but not in stems and leaves, which shows its tissue specificity for Cd stress. Heavy metal transport proteins played an important role in plant tolerance to Cd stress. It was very interesting that heavy metal transport/detoxification (HMAD) superfamily proteins were found in cotton under Cd stress, which was unknown to previous studies. These proteins contain a heavy-metal-associated domain (HMAD). Some heavy metal transport/detoxification proteins have been shown to be involved in tolerance to toxic metals, such as Pb and Cd [43].

In this study, 30 DEGs (root: 14 DEGs, stem: 9 DEGs, leaf: 7 DEGs) encoding heavy metal transporters/detoxification superfamily proteins were found (Figure 12). Different tissues contained different HMAD genes (Table S2). These results indicate that the genes belonging to heavy metal transport/detoxification superfamily proteins responded well to Cd stress, and these genes were able to transport and detoxify the heavy metal ions [44]. By silencing the *GhHMAD5* gene in cotton and overexpressing it in *Arabidopsis thaliana*, we found that the *GhHMAD5* gene enhanced Cd resistance, which provided a basis for molecular breeding to remove heavy metal pollution.

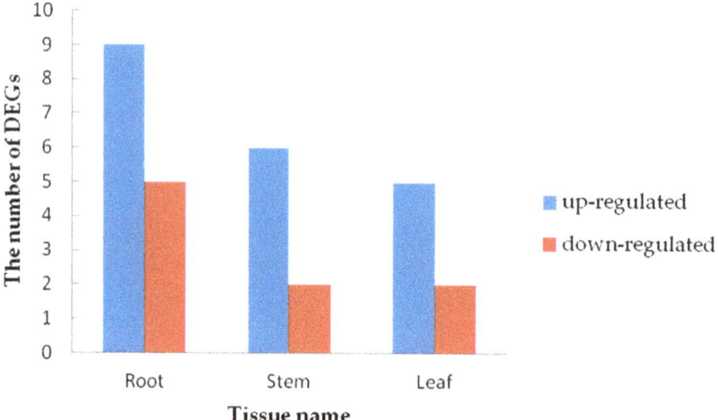

Figure 12. The number of heavy metal transporters/detoxification genes up- and down-regulated in the roots, stems and leaves under Cd Stress.

Aldolase-type TIM barrel family proteins are divided into two categories based on their functional analysis. One is chlorophyllin, which participates in starch synthesis, and the other is cytoplasmic, involved in the biosynthesis of sucrose [45]. Overexpression of *fructose 1,6-bisphosphate aldolase* (*FBA*) increased the proline content in transgenic plants under salt stress, which enhanced the salt tolerance in tobacco [46]. Gh_D12G1971 was highly expressed in the roots, stems, and leaves of cotton under Cd stress, indicating that HSP protein is related to Cd stress. Cotton could induce the production of an emergency protein kinase in the plant under short-term high Cd stress, which is consistent with previous studies. The function of these two genes needs further investigation.

Metabolic pathways of cotton under Cd stress were found to be the same as those reported in other crops, such as carbon metabolism, amino acid biosynthesis, the calcium signaling pathway and the MAPK pathway. Unlike previous reports, 11 DEGs in the roots were enriched in the influenza A pathway (Figure S3), which may indicate a certain correlation between influenza and Cd stress. Cd uptake by plants could lead to DEGs being differentially expressed in the influenza pathway, a necessary process for plants to fight against stress. We found 9 DEGs in cotton which induced pyruvate metabolism (Figure S4), which is the intermediate product playing a key role in the metabolism of sugars, fats and amino acids in plants. Detoxification complexation, thickening of physical barriers and oxidation stress were the main mechanisms in cotton under Cd stress [47]. In this study, we found *GhHMAD5* played an important role in cotton under Cd stress. The *GhHMAD5* gene may be associated with detoxification. According to the DEGs in cotton, we identified the main factors of cotton tolerance to Cd stress (Figure 13). There were complex relationships between PP2C, MAPK and ABA under Cd stress. Ferredoxin 3 and ubiquitin protein also responded to Cd stress in cotton. Above all, we have developed a new prospect for understanding Cd tolerance in cotton (Figure 14), and a new approach for cotton to exploit the Cd tolerance mechanism, which may provide a novel strategy to decode the mechanism of Cd resistance in cotton.

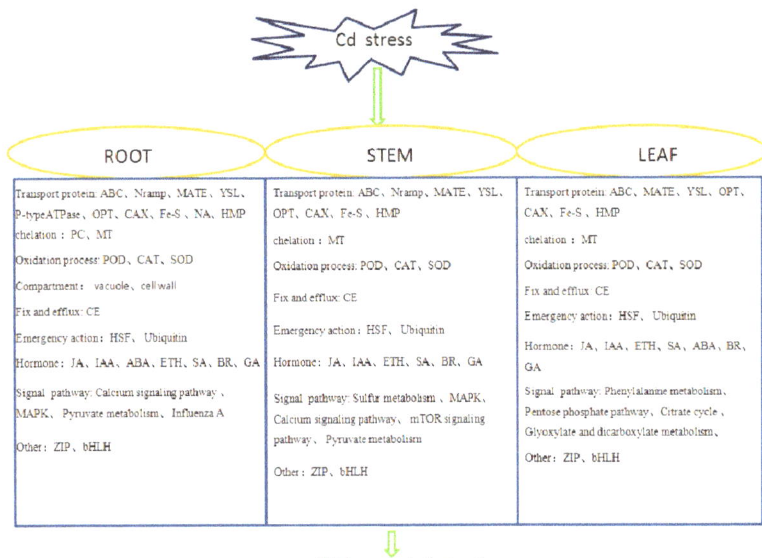

Figure 13. Key factors of Cd tolerance in roots, stems, and leaves of cotton.

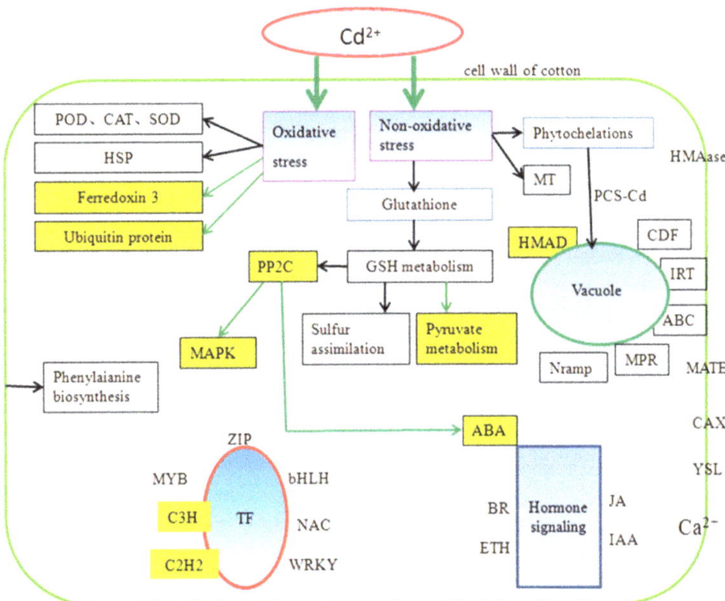

Figure 14. The regulatory network of Cd stress in cotton. The blue fill indicates the main reaction caused by Cd stress. The yellow fill shows a result different from other plants. The yellow background with black characters represents what was previously unknown.

As shown in Figure 14, there were many correlations among the regulatory mechanisms of Cd stress in cotton plants, which coincides with previous studies in other species. The previous study showed that Cd stress caused damage to the antioxidant system of Chinese cabbage, and influenced the expression of genes related to ascorbic acid synthesis to varying degrees [48]. Salicylic acid

(SA) was reported to enhance the Cd-tolerance in grapes [49]. A correlation between oxidative stress and ethylene (ETH) signal transduction was investigated under Cd stress [50]. In this study, many hormone-related genes were differentially expressed (Table S3) including SA, ETH, ABA and so on. PP2Cs played a key role in various transduction pathways, such as ABA, pathogeny, stress, and growth, which was consistent with the results of a previous study [51]. The mechanism of Cd tolerance correlates with carbon metabolism, the MAPK signal pathway, and transduction of hormone signals, which implies that Cd stress signals are transduced by hormone networks instead of by a single hormone.

As shown in Table S4, methyltransferase and methyltransferase-related proteins were expressed differentially in cotton under Cd stress, which coincided with previous studies in rice. It was reported that a great number of methyltransferase genes and DNA-methylation-modified genes were expressed differentially in rice under Cd stress [52]. This study suggests DNA methylation is a vital mechanism among complex pathways under Cd stress in plants.

4. Materials and Methods

4.1. Plant Materials and Cd Stress Treatment

Initially, the experiment was performed on 86 different cotton varieties to check their germination response against several concentrations of Cd stress. Han 242, cotton cultivars (*Gossypium hirsutum* L.), provided by Cotton Research Institute of CAAS, showed the highest germination rate among all varieties under Cd stress. Seeds were washed thoroughly and sown in pots containing well washed, clean, and sterilized sand (121 °C for 8 h). Four seedlings in each pot were cultivated in a 28 °C/14 h light and 25 °C/10 h dark cycle with a light intensity of 150 μmol m^{-2} s^{-1} and 75% relative humidity. After 35 days, cotton plants with three true leaves and one heart-shaped leaf were washed out carefully, and transferred into conical flasks containing 0.5, 1, 2, 4, 6, 8 and 10 mM $CdCl_2$ solution to observe various phenotypic changes in a time series of 1, 3, 6, 9, 12, 24 h. Then 4 mM $CdCl_2$ over 9 h was chosen as the most appropriate treatment for further study. Seedlings transplanted into ddH_2O were used as the controls. After exposure for 9 h, leaf, stem and whole root samples were collected. Samples from whole root hair, young stalks and antepenultimate leaves were collected with three replications for each treatment and control to measure the Cd content and the transcriptome analysis.

The wild type *Arabidopsis thaliana* was used for overexpression tests. All *Arabidopsis thaliana* seeds (treated with 70% alcohol for 15 min, and rinsed with double distilled water 6–8 times, before 0.05% agarose was used to suspend the seeds) were cultured in MS medium with 4 °C/24 h dark for 3 days in refrigerator, then transferred into the growth chamber at 20 °C/16 h light and 18 °C/8 h dark cycle for one week. Seedlings were transplanted into soil pots with the environmental conditions of 20 °C/8 h light and 18 °C/16 h dark for one week. After that, growth conditions were replaced by 20 °C/16 h light and 18 °C/8 h dark until their maturation.

4.2. Measurement of Relative Cd Content

All samples were oven-dried at 80 °C for 3 days followed by the measurement of Cd content under Cd stress and controls. The concentration of Cd in the filtrate was determined by inductively coupled plasma atomic emission spectroscopy (ICP-P6300) following the standard procedures.

4.3. RNA Extraction, cDNA Library Construction, and RNA-Seq

The total RNA was extracted using Trizol reagent (Invitrogen, Carlsbad, CA, USA) following the manufacturer's procedure. Total RNA quantity and purity were measured with Bioanalyzer 2100 and RNA 6000 Nano LabChip Kit (Agilent, Santa Clara, CA, USA) with RIN number >7, followed by gel extraction with 1% agarose gel electrophoresis. Then approximately 10 μg of total RNA was purified using poly-T oligo-attached magnetic beads cleaved into smaller fragments with fragmentation buffer. Then the cleaved RNA fragments were transcribed to first-strand cDNA fragments using reverse transcriptase and a high concentration of random hexamer primer. The cDNA library was developed by the protocol for the RNA-Seq sample preparation kit (Illumina, San Diego, CA, USA). The average insert size for the paired-end libraries was 150 bp (±50 bp). Paired-end sequencing was performed on an Illumina Hiseq4000 (LC Sciences, San Diego, CA, USA) following the protocols.

4.4. Quality Control, and DEG Analysis

Raw data in the fastq format was first processed through in-house perl scripts followed by removal of adaptor and low-quality sequence reads from the data sets. The TopHat package [53] was used to compare the valid dates with the cotton reference genome (*Gossypium hirsutum* L.). These mapped reads were spliced using *Cufflinks* software based on the reference genome sequence. The gene expression levels were calculated using reads per kilobase per million reads (RPKM) [54], which eliminated the influences of gene length and sequencing level during the calculation of gene expression. DEGs were identified using DESeq software [55]. Considering the biological replication, we used the ballgown package in R language to analyze the gene difference after completion of String Tie assembly and quantitation ($p < 0.05$ or $q < 0.05$) [56]. Fold change ≥ 2 and p value < 0.05 were taken as the thresholds to determine whether a gene had differential expression or not.

4.5. Gene Ontology and Gene Pathway Enrichment Analysis

To study the DEGs of cotton in response to Cd stress, several bioinformatics tools were employed for annotation, classification and metabolic pathway analysis. The GO enrichment analysis of the DEGs was performed based on agriGO. We identified markedly enriched metabolic pathways or signal transduction pathways in the DEGs based on KEGG [57] and KOBAS [58]. KOBAS software was used for testing the enrichment statistics in the KEGG pathway.

4.6. Quantitative Real-Time PCR Analysis

To verify the reliability of the DEGs in RNA-Seq data, we used the same sample as the qRT-PCR analysis. Twenty genes from the DEGs were selected randomly and primers were designed by NCBI (Primer-Blast). The primer sequence of the DEGs and the reference genes are listed in a supplementary table (Table S1). qRT-PCR was performed by using the GeneApplied Biosystems@ 7500 Fast and TransStart Top Green qPCR SuperMix (TransGen Biotech, Beijing, China). Reactions were performed with three technological and biological replications: 0.4 μL of each primer (10 μM), 0.4 μL of passive reference dye, and 10 μL of Top Green qPCR SuperMix at a final volume of 20 μL. The PCR setting was configured as follows: 5 min at 95 °C followed by 40 cycles of amplification at 95 °C for 15 s, then 58 °C for 20 s, and 72 °C for 30 s. The relative fold change of the DEGs was calculated by the $2^{-\Delta\Delta Ct}$ [59]. The *GhActin* gene was used as the control, and the correlation analysis between qRT-PCR and RNA-Seq was performed.

4.7. Overexpression and Virus Induced Gene Silence (VIGS) Construction of GhHMAD5

The cloning of the gene, named as GhHMA1-t, was done successfully in the early stage by using pEASY-Blunt Kit. Two vectors, pBI 121 vector and pYL 156 were used for overexpression and gene silencing, respectively. The enzyme sites for the pBI 121 vector were *Sac*I and *Sma*I, whereas for pYL 156, they were *Eco*RI and *Xma*I. The linearized gene was connected with its vectors separately and transformed into *E. coli* bacteria by In-Fusion connecting under the heat shock methodology and the results were checked by gel electrophoresis. For further verification and confirmation, samples were sequenced by Sangon Biotech (Shanghai, China) Co., Ltd. The recombinant expression vector plasmids pBI 121: GhHMAD5 and pYL 156: GhHMAD5 were introduced into the *Agrobacterium tumefaciens* strain GV3101 by the freezing and thawing method. Following that, they were stored −80 °C with 50% glycerin to protect bacteria. Overexpression of *GhHMAD5* was carried out in wild type *Arabidopsis thaliana* with pBI 121 vector, while silencing of the gene *GhHMAD5* was accompanied in the cotton cotyledon.

4.8. Analysis of Cd Tolerance in Transgenic Arabidopsis Thaliana

Wild-type *Arabidopsis thaliana* was cultivated until florescence, the top inflorescence was cut off for the first time. *Agrobacterium tumefaciens* containing the gene was transferred into 120 mL LB solution with overnight shake until its OD_{600} values reached up to 1.2–1.6. The bacterial solution was centrifuged and sediments were transferred in suspension liquid solution (1/2 MS salt, 5% sucrose, pH = 5.8) and the OD_{600} was set to 0.8. Silwet-L-77 (0.02%) was added into the infection solution uniformly just before its application up to 60 s, while in the second infection time, it was prolonged up to 90 s. The infection plants were wrapped with clean plastic film to maintain humidity. After 16–24 h of dark culture, the infected plants were transferred to previous normal environmental conditions. In order to improve the transformation rate, they were re-infected 5–7 days later. After maturation, the transgenic seeds of the T_0 generation were obtained.

Cephalosporin (200 mg/L) and kanamycin (50 mg/L) were used to identify the Cd tolerance seedlings from the T_0 generation seeds. At the 5–7 leaves stage, one leaf was taken as a sample from *Arabidopsis thaliana* plants. DNA extraction was done by the CTAB method for further molecular analysis. Then the T_1 seeds were obtained. Kanamycin was used to identify the T-DNA insertion line from T_1 generation seeds. The kanamycin tolerance seedlings isolation rate was 3:1, which proved that the transgenic line was a single T-DNA insertion line. Finally, the transgenic *Arabidopsis thaliana* seeds were obtained.

The seeds of transgenic and wild *Arabidopsis thaliana* were sown on a 1/2 MS solid medium containing 250, 350 μM $CdCl_2$. After 12 days and 22 days, the germination rate was observed and analyzed.

Wild type and transgenic *Arabidopsis* seeds were seeded on 1/2 MS solid medium. After 12 days, all seedlings, including wild type and transgenic *Arabidopsis*, were transplanted to a 1/2 MS solid culture plate containing 25, 50 and 100 μM of $CdCl_2$ to observe their root development.

4.9. The VIGS Analysis of GhHMAD5

Cotton Seeds were washed thoroughly and sown in sand soil pots and the sands were sterilized (121 °C for 8 h). Pots were transferred to incubators with the controlled conditions of 28 °C/14 h light and 25 °C/10 h dark cycle with a light intensity of 150 μmol m^{-2} s^{-1} and 75% relative humidity. After 5 days, a gas-bacilli began to sprout (in cotton seedlings two cotyledons flattened first and then grew into leaves). The saved *Agrobacterium* solution was shaken in 60 mL LB liquid medium, until its OD_{600} value reached 1.5, which was followed by centrifugation. *Agrobacterium* was dispersed in suspension solution (10 mM $MgCl_2$ + 10 mM MES + 200 μM acetosyingone). Suspension solutions of pYL 156: GhHMAD5, pYL 156: GhCLA1 and pYL 156 were mixed with equal volumes of assistant bacteria auxiliary carrier pYL 192 prior to injection in cotyledons. Cotyledons were injured on their back with a sterilized needle tab and the bacterial suspension solution was injected in this hole for microbial proliferation on the whole cotyledon. Cotton plants were transferred to an incubator in darkness for 24 h at 23 °C. Later on, it was replaced by normal environmental conditions (28 °C/14 h light, 25 °C/10 h dark). After being subjected to blanching, plants were treated with 4 mM Cd stress which showed the symptoms of positive seedlings very clearly as compared to negative ones. Relative expressions were measured at this stage.

5. Conclusions

In this study, 11,503 DEGs were discovered under Cd stress in cotton with RNA-Seq analysis. The *GhHMAD5* gene was cloned and identified as enhancing the resistance to Cd stress. A novel regulation network of Cd stress was constructed, including complex pathways, in cotton. This study suggests a preliminary understanding of Cd tolerance mechanisms in upland cotton, which implies a potential use of cotton is to remediate Cd-polluted soil.

Supplementary Materials: The following are available online at http://www.mdpi.com/1422-0067/20/6/1479/s1.

Author Contributions: Conceived and designed the experiments: M.H., W.Y., J.Y. Performed the experiments: M.H., D.W., C.C., R.C. Analyzed the data: M.H., X.W., X.Y. Wrote the paper: M.H., W.Y., X.L., X.C., J.Y., W.A.M. Contributed materials/analysis tools: J.W., S.W., L.G.

Funding: This project was supported by the National Key Research and Development Program (2016YFD0101401) and USDA-ARS Project (3091-21000-044-01-N).

Acknowledgments: This research was supported by the National Key Research and Development Program (2016YFD0101401), the State Key Laboratory of Cotton Biology, and USDA-ARS Project (3091-21000-044-01-N).

Conflicts of Interest: The authors declare no conflict of interest.

References

1. Kaplan, O.; Ince, M.; Yaman, M. Sequential extraction of cadmium in different soil phases and plant parts from a former industrialized area. *Environ. Chem. Lett.* **2011**, *9*, 397–404. [CrossRef]
2. Groppa, M.D.; Ianuzzo, M.P.; Rosales, E.P.; Vázquez, S.C.; Benavides, M.P. Cadmium modulates NADPH oxidase activity and expression in sunflower leaves. *Biol. Plant.* **2012**, *56*, 167–171. [CrossRef]
3. Satarug, S.; Baker, J.R.; Urbenjapol, S.; Haswell-Elkins, M.; Reilly, P.E.B.; Williams, D.J.; Moore, M.R. A global perspective on cadmium pollution and toxicity in non-occupationally exposed population. *Toxicol. Lett.* **2003**, *137*, 65–83. [CrossRef]
4. Toppi, L.S.D.; Gabbrielli, R. Response to cadmium in higher plants. *Environ. Exp. Bot.* **1999**, *41*, 105–130. [CrossRef]
5. Brooks, R.R.; Lee, J.; Reeves, R.D.; Jaffre, T. Detection of nickeliferous rocks by analysis of herbarium specimens of indicator plants. *J. Geochem. Explor* **1977**, *7*, 49–57. [CrossRef]
6. Zhao, Y.; Hu, H.X.; Chen, Z.F.; Hu, L.J.; Xu, D.Y. The potential for phytoextraction of hexachloro-cyclohexanes contaminated calcareous soils in eastern Suburb of Beijing by cotton. *Fresenius Environ. Bull.* **2012**, *21*, 1948–1955.
7. Ling, L.I.; Chen, J.H.; Qiu-Ling, H.E.; Zhu, S.J. Accumulation, transportation, and bioconcentration of cadmium in three upland cotton plants under cadmium stress. *Cotton Sci.* **2012**, *24*, 535–540.
8. Kolevavalkova, L.; Vasilev, A. Physiological parameters of young cotton plants, grown on heavy metal contaminated soils. *Agrarni Nauki* **2015**, *VII*, 61–66.
9. Daud, M.K.; Sun, Y.; Dawood, M.; Hayat, Y.; Variath, M.T.; Wu, Y.X.; Raziuddin; Mishkat, U.; Salahuddin; Najeeb, U. Cadmium-induced functional and ultrastructural alterations in roots of two transgenic cotton cultivars. *J. Hazard. Mater.* **2008**, *161*, 463–473. [CrossRef]
10. Liu, L.T.; Sun, H.C.; Chen, J.; Zhang, Y.J.; Wang, X.D.; Li, D.X.; Li, C.D. Cotton seedling plants adapted to cadmium stress by enhanced activities of protective enzymes. *Plant Soil Environ.* **2016**, *62*, 80–85.
11. Ibrahim, W.; Ahmed, I.M.; Chen, X.; Cao, F.; Zhu, S.; Wu, F. Genotypic differences in photosynthetic performance, antioxidant capacity, ultrastructure and nutrients in response to combined stress of salinity and Cd in cotton. *Biometals* **2015**, *28*, 1063–1078. [CrossRef] [PubMed]
12. Angelova, V.; Ivanova, R.; Delibaltova, V.; Ivanov, K. Bio-accumulation and distribution of heavy metals in fibre crops (flax, cotton and hemp). *Ind. Crops Prod.* **2004**, *19*, 197–205. [CrossRef]
13. Wu, F.; Wu, H.; Zhang, G.; Dml, B. Differences in growth and yield in response to cadmium toxicity in cotton genotypes. *J. Plant Nutr. Soil Sci.* **2004**, *167*, 85–90. [CrossRef]
14. Hartwell, L.H.; Culotti, J.; Reid, B. Genome-wide expression profiling of maize in response to individual and combined water and nitrogen stresses. *BMC Genomics.* **2013**, *14*, 3. [CrossRef]
15. Yu, S.; Zhang, F.; Yu, Y.; Zhang, D.; Zhao, X.; Wang, W. Transcriptome profiling of dehydration stress in the Chinese Cabbage *(Brassica rapa L.* ssp. pekinensis) by tag sequencing. *Plant Mol. Biol. Rep.* **2012**, *30*, 17–28. [CrossRef]
16. Liu, T.; Zhu, S.; Tang, Q.; Tang, S. Genome-wide transcriptomic profiling of ramie (*Boehmeria nivea L. Gaud*) in response to cadmium stress. *Gene* **2015**, *558*, 131–137. [CrossRef] [PubMed]
17. Yue, R.; Lu, C.; Qi, J.; Han, X.; Yan, S.; Guo, S.; Liu, L.; Fu, X.; Chen, N.; Yin, H. Transcriptome analysis of cadmium-treated roots in maize (*Zea mays L.*). *Front. Plant Sci.* **2016**, *7*, 1298. [CrossRef] [PubMed]

18. Oono, Y.; Yazawa, T.; Kanamori, H.; Sasaki, H.; Mori, S.; Handa, H.; Matsumoto, T. Genome-wide transcriptome analysis of cadmium stress in rice. *BioMed Res. Int.* **2016**, *2016*, 1–9. [CrossRef]
19. Liu, Y.J.; Yu, X.F.; Feng, Y.M.; Zhang, C.; Wang, C.; Zeng, J.; Huang, Z.; Kang, H.Y.; Fan, X.; Sha, L.N.; et al. Physiological and transcriptome response to cadmium in cosmos (Cosmos bipinnatus Cav.) seedlings. *Sci. Rep.* **2017**, *7*, 14691. [CrossRef]
20. Aprile, A.; Sabella, E.; Vergine, M.; Genga, A.; Siciliano, M.; Nutricati, E.; Rampino, P.; De Pascali, M.; Luvisi, A.; Miceli, A.; et al. Activation of a gene network in durum wheat roots exposed to cadmium. *BMC Plant Biol.* **2018**, *18*, 238. [CrossRef] [PubMed]
21. Shukla, D.; Huda, K.M.; Banu, M.S.; Gill, S.S.; Tuteja, R.; Tuteja, N. OsACA6, a P-type 2B Ca 2+ ATPase functions in cadmium stress tolerance in tobacco by reducing the oxidative stress load. *Planta* **2014**, *240*, 809–824. [CrossRef]
22. Paulsen, I.T.; Saier, M.H.A., Jr. Novel family of ubiquitous heavy metal ion transport proteins. *J. Membr. Biol.* **1997**, *156*, 99–103. [CrossRef] [PubMed]
23. Rivetta, A.; Negrini, N.; Cocucci, M. Involvement of Ca2+-calmodulin in Cd2+ toxicity during the early phases of radish (*Raphanus sativus L.*) seed germination. *Plant Cell Environ.* **1997**, *20*, 600–608. [CrossRef]
24. Hegedüs, A.; Erdei, S.; Horváth, G. Comparative studies of $H_{(2)}O_{(2)}$ detoxifying enzymes in green and greening barley seedlings under cadmium stress. *Plant Sci.* **2001**, *160*, 1085–1093. [CrossRef]
25. Chen, B.C.; Wang, P.J.; Ho, P.C.; Juang, K.W. Nonlinear biotic ligand model for assessing alleviation effects of Ca, Mg, and K on Cd toxicity to soybean roots. *Ecotoxicology* **2017**, *26*, 1–14. [CrossRef]
26. Gong, B.; Nie, W.; Yan, Y.; Gao, Z.; Shi, Q. Unravelling cadmium toxicity and nitric oxide induced tolerance in Cucumis sativus: Insight into regulatory mechanisms using proteomics. *J. Hazard. Mater.* **2017**, *336*, 202–213. [CrossRef] [PubMed]
27. Liu, H.; Zhao, H.; Wu, L.; Liu, A.; Zhao, F.J.; Xu, W. Heavy metal ATPase 3 (HMA3) confers cadmium hypertolerance on the cadmium/zinc hyperaccumulator Sedum plumbizincicola. *New Phytol.* **2017**, *215*, 687–698. [CrossRef] [PubMed]
28. Ma, Z.; An, T.; Zhu, X.; Ji, J.; Wang, G.; Guan, C.; Jin, C.; Yi, L. GR1-like gene expression in Lycium chinense was regulated by cadmium-induced endogenous jasmonic acids accumulation. *Plant Cell Rep.* **2017**, *36*, 1–20. [CrossRef] [PubMed]
29. Dai, H.; Shan, C.; Zhao, H.; Jia, G.; Chen, D. Lanthanum improves the cadmium tolerance of Zea mays seedlings by the regulation of ascorbate and glutathione metabolism. *Biol. Plant.* **2016**, *61*, 1–6. [CrossRef]
30. Huilan, W.; Chunlin, C.; Juan, D.; Hongfei, L.; Yan, C.; Yue, Z.; Yujing, H.; Yiqing, W.; Chengcai, C.; Zongyun, F. Co-overexpression FIT with AtbHLH38 or AtbHLH39 in Arabidopsis-enhanced cadmium tolerance via increased cadmium sequestration in roots and improved iron homeostasis of shoots. *Plant Physiol.* **2012**, *158*, 790–800.
31. Tiwari, M.; Sharma, D.; Dwivedi, S.; Singh, M.; Tripathi, R.D.; Trivedi, P.K. Expression in Arabidopsis and cellular localization reveal involvement of rice NRAMP, OsNRAMP1, in arsenic transport and tolerance. *Plant Cell Environ.* **2014**, *37*, 140–152. [CrossRef] [PubMed]
32. Sekhar, K.; Priyanka, B.; Reddy, V.D.; Rao, K.V. Metallothionein 1 (CcMT1) of pigeonpea (*Cajanus cajan, L.*) confers enhanced tolerance to copper and cadmium in Escherichia coli and Arabidopsis thaliana. *Environ. Exp. Bot.* **2011**, *72*, 131–139. [CrossRef]
33. Shukla, D.; Kesari, R.; Tiwari, M.; Dwivedi, S.; Tripathi, R.D.; Nath, P.; Trivedi, P.K. Expression of Ceratophyllum demersum phytochelatin synthase, CdPCS1, in Escherichia coli and Arabidopsis enhances heavy metal(loid)s accumulation. *Protoplasma* **2013**, *250*, 1263–1272. [CrossRef]
34. Fujita, Y.; Fujita, M.; Shinozaki, K.; Yamaguchishinozaki, K. ABA-mediated transcriptional regulation in response to osmotic stress in plants. *J. Plant Res.* **2011**, *124*, 509–525. [CrossRef] [PubMed]
35. Wang, Y.; Gao, C.; Liang, Y.; Wang, C.; Yang, C.; Liu, G. A novel bZIP gene from Tamarix hispida mediates physiological responses to salt stress in tobacco plants. *J. Plant Physiol.* **2010**, *167*, 222–230. [CrossRef] [PubMed]
36. Chmielowska-Bąk, J.; Lefèvre, I.; Lutts, S.; Deckert, J. Short term signaling responses in roots of young soybean seedlings exposed to cadmium stress. *J. Plant Physiol.* **2013**, *170*, 1585–1594. [CrossRef]
37. Wei, W.; Zhang, Y.; Han, L.; Guan, Z.; Chai, T. A novel WRKY transcriptional factor from Thlaspi caerulescens negatively regulates the osmotic stress tolerance of transgenic tobacco. *Plant Cell Rep.* **2008**, *27*, 795–803. [CrossRef]

38. Mao, J.; Li, W.; Mi, B.; Dawuda, M.M.; Calderón-Urrea, A.; Ma, Z.; Zhang, Y.; Chen, B. Different exogenous sugars affect the hormone signal pathway and sugar metabolism in "Red Globe" (*Vitis vinifera L.*) plantlets grown in vitro as shown by transcriptomic analysis. *Planta* **2017**, *246*, 1–16. [CrossRef] [PubMed]
39. Gupta, O.P.; Sharma, P.; Gupta, R.K.; Sharma, I. MicroRNA mediated regulation of metal toxicity in plants: Present status and future perspectives. *Plant Mol. Biol.* **2014**, *84*, 1–18. [CrossRef]
40. Sghayar, S.; Ferri, A.; Lancilli, C.; Lucchini, G.; Abruzzese, A.; Porrini, M.; Ghnaya, T.; Nocito, F.F.; Abdelly, C.; Sacchi, G.A. Analysis of cadmium translocation, partitioning and tolerance in six barley (*Hordeum vulgare L.*) cultivars as a function of thiol metabolism. *Biol. Fertil. Soils* **2015**, *51*, 311–320. [CrossRef]
41. Salim, R.; Al-Subu, M.M.; Douleh, A.; Khalaf, S. Effects on growth and uptake of broad beans (*Vicia fabae L.*) by root and foliar treatments of plant with lead and cadmium. *Environ. Lett.* **2012**, *27*, 1619–1642. [CrossRef]
42. Moussa, H.R. Effect of cadmium on growth and oxidative metabolism of faba bean plants. *Acta Agron. Hungarica* **2004**, *52*, 269–276. [CrossRef]
43. Bull, P.C.; Cox, D.W. Wilson disease and Menkes disease: New handles on heavy-metal transport. *Trends Genet.* **1994**, *10*, 246–252. [CrossRef]
44. Gitschier, J.; Moffat, B.; Reilly, D.; Wood, W.I.; Fairbrother, W.J. Solution structure of the fourth metal-binding domain from the Menkes copper-transporting ATPase. *Nat. Struct. Biol.* **1998**, *5*, 47–54. [CrossRef] [PubMed]
45. Lebherz, H.G.; Leadbetter, M.M.; Bradshaw, R.A. Isolation and characterization of the cytosolic and chloroplast forms of spinach leaf fructose diphosphate aldolase. *J. Biol. Chem.* **1984**, *259*, 1011–1017. [PubMed]
46. Yamada, S.; Komori, T.; Hashimoto, A.; Kuwata, S.; Imaseki, H.; Kubo, T. Differential expression of plastidic aldolase genes in Nicotiana plants under salt stress. *Plant Sci.* **2000**, *154*, 61–69. [CrossRef]
47. Chen, H.; Li, Y.; Ma, X.; Guo, L.; He, Y.; Ren, Z.; Kuang, Z.; Zhang, X.; Zhang, Z. Analysis of potential strategies for cadmium stress tolerance revealed by transcriptome analysis of upland cotton. *Sci. Rep.* **2019**, *9*. [CrossRef]
48. Lin, Z.; Cui, H.M.; Wang, J.J.; Hou, X.L.; Ying, L.I. Effects of Cadmium stress on genes expression in L-Galactose pathway involved in VC biosynthesis and antioxidant system of Brassica campestris ssp. chinensis. *Plant Physiol. J.* **2015**, *51*, 1099–1108.
49. Shao, X.J.; Yang, H.Q.; Ran, K.; Jiang, Q.Q.; Sun, X.L. Effects of Salicylic acid on plasma membrane ATPase and free radical of grape root under cadmium stress. *Sci. Agric. Sin.* **2010**, *43*, 1441–1447.
50. Schellingen, K.; Straeten, D.V.D.; Remans, T.; Vangronsveld, J.; Keunen, E.; Cuypers, A. Ethylene signalling is mediating the early cadmium-induced oxidative challenge in Arabidopsis thaliana. *Plant Sci.* **2015**, *239*, 137–146. [CrossRef]
51. Tougane, K.; Komatsu, K.; Bhyan, S.B.; Sakata, Y.; Ishizaki, K.; Yamato, K.T.; Kohchi, T.; Takezawa, D. Evolutionarily Conserved regulatory mechanisms of abscisic acid signaling in land plants: characterization of -like type 2C protein phosphatase in the liverwort. *Plant Physiol.* **2010**, *152*, 1529–1543. [CrossRef] [PubMed]
52. Feng, S.J.; Liu, X.S.; Tao, H.; Tan, S.K.; Chu, S.S.; Oono, Y.; Zhang, X.D.; Chen, J.; Yang, Z.M. Variation of DNA methylation patterns associated with gene expression in rice (*Oryza sativa*) exposed to cadmium. *Plant Cell Environ.* **2016**, *39*, 2629–2649. [CrossRef]
53. Trapnell, C.; Williams, B.A.; Pertea, G.; Mortazavi, A.; Kwan, G.; Van Baren, M.J.; Salzberg, S.L.; Wold, B.J.; Pachter, L. Transcript assembly and quantification by RNA-Seq reveals unannotated transcripts and isoform switching during cell differentiation. *Nat. Biotechnol.* **2010**, *28*, 511–515. [CrossRef] [PubMed]
54. Cheadle, C.; Vawter, M.P.; Freed, W.J.; Becker, K.G. Analysis of microarray data using Z score transformation. *J. Mol. Diagn. JMD* **2003**, *5*, 73–81. [CrossRef]
55. Anders, S.; Huber, W. Differential expression analysis for sequence count data. *Genome Biol.* **2010**, *11*, R106. [CrossRef] [PubMed]
56. Bi, R.; Liu, P. Sample size calculation while controlling false discovery rate for differential expression analysis with RNA-sequencing experiments. *BMC Bioinform.* **2016**, *17*, 1–13. [CrossRef] [PubMed]
57. Kanehisa, M.; Goto, S. KEGG: Kyoto Encyclopedia of Genes and Genomes. *Nucl. Acids Res.* **1999**, *27*, 29–34. [CrossRef]

58. Xie, C.; Mao, X.; Huang, J.; Ding, Y.; Wu, J.; Dong, S.; Kong, L.; Gao, G.; Li, C.Y.; Wei, L. KOBAS 2.0: A web server for annotation and identification of enriched pathways and diseases. *Nucl. Acids Res.* **2011**, *39*, 316–322. [CrossRef]
59. Qian, W.; Tao, T.; Zhang, Y.; Wu, W.; Li, D.; Yu, J.; Han, C. Rice black-streaked dwarf virus P6 self-interacts to form punctate, viroplasm-like structures in the cytoplasm and recruits viroplasm-associated protein P9-1. *Virol. J.* **2011**, *8*, 1–15.

© 2019 by the authors. Licensee MDPI, Basel, Switzerland. This article is an open access article distributed under the terms and conditions of the Creative Commons Attribution (CC BY) license (http://creativecommons.org/licenses/by/4.0/).

MDPI
St. Alban-Anlage 66
4052 Basel
Switzerland
Tel. +41 61 683 77 34
Fax +41 61 302 89 18
www.mdpi.com

International Journal of Molecular Sciences Editorial Office
E-mail: ijms@mdpi.com
www.mdpi.com/journal/ijms

www.ingramcontent.com/pod-product-compliance
Lightning Source LLC
LaVergne TN
LVHW071554170726
843515LV00008B/2286

* 9 7 8 3 0 3 9 3 6 6 3 3 0 *